AF614990

Series Editor: Rhodes W. Fairbridge
Columbia University

Volume

1 ENVIRONMENTAL GEOMORPHOLOGY AND LANDSCAPE CONSERVATION, Volume 1: Prior to 1900 / *Donald R. Coates*
2 RIVER MORPHOLOGY / *Stanley A. Schumm*
3 SPITS AND BARS / *Maurice L. Schwartz*
4 TEKTITES / *Virgil E. Barnes and Mildred A. Barnes*
5 GEOCHRONOLOGY: Radiometric Dating of Rocks and Minerals / *C. T. Harper*
6 SLOPE MORPHOLOGY / *Stanley A. Schumm and M. Paul Mosley*
7 MARINE EVAPORITES: Origin, Diagenesis, and Geochemistry / *Douglas W. Kirkland and Robert Evans*
8 ENVIRONMENTAL GEOMORPHOLOGY AND LANDSCAPE CONSERVATION, Volume III: Non-Urban / *Donald R. Coates*
9 BARRIER ISLANDS / *Maurice L. Schwartz*
10 GLACIAL ISOSTASY / *John T. Andrews*
11 GEOCHEMISTRY OF GERMANIUM / *Jon N. Weber*
12 ENVIRONMENTAL GEOMORPHOLOGY AND LANDSCAPE CONSERVATION, Volume II: Urban Areas / *Donald R. Coates*
13 PHILOSOPHY OF GEOHISTORY: 1785–1970 / *Claude C. Albritton, Jr.*
14 GEOCHEMISTRY AND THE ORIGIN OF LIFE / *Keith A. Kvenvolden*
15 SEDIMENTARY ROCKS: Concepts and History / *Albert V. Carozzi*
16 GEOCHEMISTRY OF WATER / *Yasushi Kitano*
17 METAMORPHISM AND PLATE TECTONIC REGIMES / *W. G. Ernst*
18 GEOCHEMISTRY OF IRON / *Henry Lepp*
19 SUBDUCTION ZONE METAMORPHISM / *W. G. Ernst*
20 PLAYAS AND DRIED LAKES: Occurrence and Development / *James T. Neal*
21 GLACIAL DEPOSITS / *Richard P. Goldthwait*
22 PLANATION SURFACES: Peneplains, Pediplains, and Etchplains / *George F. Adams*
23 GEOCHEMISTRY OF BORON / *C. T. Walker*
24 SUBMARINE CANYONS AND DEEP-SEA FANS: Modern and Ancient / *J. H. McD. Whitaker*
25 ENVIRONMENTAL GEOLOGY / *Frederick Betz, Jr.*
26 LOESS: Lithology and Genesis / *Ian J. Smalley*

27 PERIGLACIAL PROCESSES / *Cuchlaine A. M. King*
28 LANDFORMS AND GEOMORPHOLOGY: Concepts and History / *Cuchlaine A. M. King*
29 METALLOGENY AND GLOBAL TECTONICS / *Wilfred Walker*
30 HOLOCENE TIDAL SEDIMENTATION / *George deVries Klein*
31 PALEOBIOGEOGRAPHY / *Charles A. Ross*
32 MECHANICS OF THRUST FAULTS AND DÉCOLLEMENT / *Barry Voight*
33 WEST INDIES ISLAND ARCS / *Peter H. Mattson*
34 CRYSTAL FORM AND STRUCTURE / *Cecil J. Schneer*
35 OCEANOGRAPHY: Concepts and History / *Margaret B. Deacon*
36 METEORITE CRATERS / *G. J. H. McCall*
37 STATISTICAL ANALYSIS IN GEOLOGY / *John M. Cubitt and Stephen Henley*
38 AIR PHOTOGRAPHY AND COASTAL PROBLEMS / *Mohamed T. El-Ashry*
39 BEACH PROCESSES AND COASTAL HYDRODYNAMICS / *John S. Fisher and Robert Dolan*
40 DIAGENESIS OF DEEP-SEA BIOGENIC SEDIMENTS / *Gerrit J. van der Lingen*
41 DRAINAGE BASIN MORPHOLOGY / *Stanley A. Schumm*
42 COASTAL SEDIMENTATION / *Donald J. P. Swift and Harold D. Palmer*
43 ANCIENT CONTINENTAL DEPOSITS / *Franklyn B. Van Houten*
44 MINERAL DEPOSITS, CONTINENTAL DRIFT AND PLATE TECTONICS / *J. B. Wright*
45 SEA WATER: Cycles of the Major Elements / *James I. Drever*
46 PALYNOLOGY, PART I: Spores and Pollen / *Marjorie D. Muir and William A. S. Sarjeant*
47 PALYNOLOGY, PART II: Dinoflagellates, Acritarchs, and Other Microfossils / *Marjorie D. Muir and William A. S. Sarjeant*
48 GEOLOGY OF THE PLANET MARS / *Vivien Gornitz*
49 GEOCHEMISTRY OF BISMUTH / *Ernest E. Angino and David T. Long*
50 ASTROBLEMES—CRYPTOEXPLOSION STRUCTURES / *G. J. H. McCall*
51 NORTH AMERICAN GEOLOGY: Early Writings / *Robert Hazen*
52 GEOCHEMISTRY OF ORGANIC MOLECULES / *Keith A. Kvenvolden*
53 TETHYS: The Ancestral Mediterranean / *Peter Sonnenfeld*
54 MAGNETIC STRATIGRAPHY OF SEDIMENTS / *James P. Kennett*
55 CATASTROPHIC FLOODING: The Origin of the Channeled Scabland / *Victor R. Baker*
56 SEAFLOOR SPREADING CENTERS: Hydrothermal Systems / *Peter A. Rona and Robert P. Lowell*
57 MEGACYCLES: Long-Term Episodicity in Earth and Planetary History / *G. E. Williams*

A BENCHMARK® Books Series

MEGACYCLES
Long-Term Episodicity in Earth and Planetary History

Edited by
G. E. WILLIAMS
The Broken Hill Proprietary Co. Ltd.

Hutchinson Ross Publishing Company
Stroudsburg, Pennsylvania   Woods Hole, Massachusetts

Benchmark Papers in Geology, Volume 57
Library of Congress Catalog Card Number: 79-19908
ISBN: 0-89733-366-9

83 82 81 1 2 3 4 5
Manufactured in the United States of America.

LIBRARY OF CONGRESS CATALOGING IN PUBLICATION DATA
Main entry under title:
Megacycles.
(Benchmark papers in geology; 57)
Bibliography: p.
Includes indexes.
1. Geology—Periodicity—Addresses, essays, lectures.
2. Solar system—Addresses, essays, lectures.
I. Williams, George Ellis.
QE33.M38 550 79-19908
ISBN 0-87933-366-9

Distributed world wide by Academic Press, a subsidiary of Harcourt Brace Jovanovich, Publishers.

For having, in the natural history of this earth, seen a succession of worlds, we may from this conclude that there is a system in nature; in like manner as, from seeing revolutions of the planets, it is concluded, that there is a system by which they are intended to continue those revolutions. But if the succession of worlds is established in the system of nature, it is in vain to look for any thing higher in the origin of the earth. The result, therefore, of our present enquiry is, that we find no vestige of a beginning,—no prospect of an end.

["Theory of the Earth," James Hutton, 1788]

Nature vibrates with rhythms, climatic and diastrophic, those finding stratigraphic expression ranging in period from the rapid oscillation of surface waters, recorded in ripple-mark, to those long-deferred stirrings of the deep imprisoned titans which have divided earth history into periods and eras. The flight of time is measured by the weaving of composite rhythms—day and night, calm and storm, summer and winter, birth and death—such as these are sensed in the brief life of man. But the career of the earth recedes into a remoteness against which these lesser cycles are as unavailing for the measurement of that abyss of time as would be for human history the beating of an insect's wing. We must seek out, then, the nature of those longer rhythms whose very existence was unknown until man by the light of science sought to understand the earth.

["Rhythms and the Measurement of
Geologic Time," J. Barrell, 1917]

CONTENTS

SERIES EDITOR'S FOREWORD

The philosophy behind the Benchmark Papers in Geology is one of collection, sifting, and rediffusion. Scientific literature today is so vast, so dispersed, and, in the case of old papers, so inaccessible for readers not in the immediate neighborhood of major libraries that much valuable information has been ignored by default. It has become just so difficult, or so time consuming, to search out the key papers in any basic area of research that one can hardly blame a busy person for skimping on some of his or her "homework."

This series of volumes has been devised, therefore, as a practical solution to this critical problem. The geologist, perhaps even more than any other scientist, often suffers from twin difficulties—isolation from central library resources and immensely diffused sources of material. New colleges and industrial libraries simply cannot afford to purchase complete runs on all the world's earth science literature. Specialists simply cannot locate reprints or copies of all their principal reference materials. So it is that we are now making a concerted effort to gather into single volumes the critical materials needed to reconstruct the background of any and every major topic of our discipline.

We are interpreting "geology" in its broadest sense: the fundamental science of the planet earth, its materials, its history, and its dynamics. Because of training in "earthy" materials, we also take in astrogeology, the corresponding aspect of the planetary sciences. Besides the classical core disciplines such as mineralogy, petrology, structure, geomorphology, paleontology, and stratigraphy, we embrace the newer fields of geophysics and geochemistry, applied also to oceanography, geochronology, and paleoecology. We recognize the work of the mining geologists, the petroleum geologists, the hydrologists, and the engineering and enviromental geologists. Each specialist needs a working library. We are endeavoring to make the task of compiling such a library a little easier.

Each volume in the series contains an introduction prepared by a specialist (the volume editor)—a "state of the art" opening or a summary of the object and content of the volume. The articles, usually some twenty to fifty reproduced either in their entirety or in significant extracts, are selected in an attempt to cover the field, from the key papers of the last century to fairly recent work. Where the original works are in foreign languages, we have endeavored to locate or commission translations. Geologists, because of their global subject, are often acutely aware of the oneness of our world. The selections cannot therefore be restricted to any one country, and whenever possible an attempt is made to scan the world literature.

To each article, or group of kindred articles, some sort of "highlight commentary" is usually supplied by the volume editor. This commentary should serve to bring that article into historical perspective and to emphasize its particular role in the growth of the field. References, or citations, wherever possible, will be reproduced in their entirety—for by this means the observant reader can assess the background material available to that particular author, or, if desired, he or she too can double check the earlier sources.

A "benchmark," in surveyor's terminology, is an established point on the ground that is recorded on our maps. It is usually anything that is a vantage point, from a modest hill to a mountain peak. From the historical viewpoint, these benchmarks are the bricks of our scientific edifice.

RHODES W. FAIRBRIDGE

PREFACE

This Benchmark volume considers geological and planetary episodicity and rhythm that are long-term—some 10^7–10^9 years in period. Events of more rapid pulse and unique events are excluded. The key word *megacycle* in the title serves to distinguish this volume from works on short-term geocyclicity (e.g., Duff et al., 1967; Trofumik and Karozobin, 1976) but does not imply that the events considered here necessarily are strictly cyclic and periodic.

The classic work of the Dutch geologist J. H. F. Umbgrove, *The Pulse of the Earth* (1942, 1947), focused attention on long-term rhythm and episodicity in several areas of earth history. Since then a rapid expansion of studies on this subject, the development of new disciplines in the earth sciences, and improvement of the geochronological time-scale have provided much additional and more reliable evidence of long-term episodicity in earth history. Recent studies have revealed episodicity also in the evolution of the terrestrial planets. Although several books (e.g., Balukhovskiy, 1966; Logvinenko, 1976) and symposia proceedings (e.g., Nalivkin and Tupitsyn, 1963; Trofumik et al., 1977) dealing with long-term geocyclicity have appeared in the USSR since 1960, Umbgrove's book is the only major work on this subject to have been published in the West. We hope that the present volume will help fill this gap.

The extensive literature soon convinced me that I could not include early works and also do justice to developments over the past two decades. As Umbgrove's book is a landmark in this field and reviews earlier works, I have gathered for this volume only papers published after 1959. The *theme* of the volume therefore could be "a generation after Umbgrove;" the *aim* is to provide access to this important subject, to offer some synthesis of the collected papers, and to point out areas requiring further endeavor. Earlier works are discussed in the Introduction and in several of the Editor's Comments and included papers.

The forty papers gathered here are grouped by topic into seven parts dealing, in order, with geotectonism, geomagnetism, glaciation, stratigraphy, geochemistry, biological evolution, and planetary history. Epochs of economic mineralization are not considered, since they are discussed by W. Walker (ed., *Metallogeny and Global Tectonics,* Benchmark Papers in Geology, vol. 29, 1976). The Introduction provides perspective to the subject as a whole, whereas the Editor's Comments review each topic and draw attention to apparent relationships among different events. These comments contain discussions of nine of the included papers by their respective authors (Papers 1, 5, 6, 10, 13, 15, 26, 32 and 38). The references at the end of the volume include important works in Russian, German, French, and Chinese.

In selecting papers of suitable length, consideration was given to providing a balance among the different topics and viewpoints and a blend of original and later, updated or more embracing works. It is regretted that several papers are abridged, but only thus could the subject be adequately covered. Authors hail from the United States, Britain, Canada, the USSR, Brazil, Australia and Portugal.

I thank K. A. W. Crook, V. A. Gostin, V. Ye. Khain and J. Steiner for assistance with literature; Drs. Gostin and Steiner and my wife Olena for translations; R. W. Fairbridge, P. A. Jell, M. Jones, R. W. R. Rutland and A. D. Stewart for commenting on my discussions; and my family for their forbearance while I completed this task.

G. E. WILLIAMS

CONTENTS BY AUTHOR

MEGACYCLES

INTRODUCTION

HISTORY OF CONCEPTS

The concept of the geological cycle, which forms the foundation of the modern science of geology, has its roots in James Hutton's "Theory of the Earth" (1788). Hutton perceived that the cycle of erosion, deposition, lithification and uplift produces a succession of worlds that has "no vestige of a beginning, no prospect of an end." During the nineteenth century the concept of geological revolutions and cyclicity, as distinct from the catastrophism of paleontology, was advanced by M. L. Elie de Beaumont, James D. Dana, E. Hébert, Sir William Dawson, Eduard Suess and T. C. Chamberlin, among others. Elie de Beaumont (1829, 1830) proposed the existence in Europe of twelve distinct mountain systems of different ages; he considered that rapid elevation resulted from the continuous application of horizontal forces in the earth's crust. Dana (1856, pp. 342–344) wrote of continental evolution, growth, and stabilization through "long secular vibrations" of the crust that were "again and again repeated." Dana and also Hébert (1857) and Dawson (1868) perceived that sedimentary cycles could result from such crustal oscillations. As discussed by Suess (1904, p. 10), by the late 1850s most workers attributed the differences displayed by successive sedimentary formations and by their faunas to the "slow and extensive oscillations of the continents, and to repeated changes of climate, perhaps connected with these oscillations." (For further discussion of early ideas on catastrophism and cyclicity in geology, see C. C. Albritton, ed., *Philosophy of Geohistory: 1785–1970*, Benchmark Papers in Geology, vol. 14, 1975.)

In 1870 Dana (pp. 722–723) proposed that mountain chains have formed through the ages as the result of successive epochs of "plication, uplift, and metamorphism" separated by long intervals during which crustal tension accumulated. Toward the end of the nineteenth century Suess, in his monumental *Das Antlitz der Erde* (*The Face of the Earth*, English translation by H. B. C. Sollas, 1904–1924), first formulated as a general working hypothesis the idea of relative changes of sea level on a global scale. Suess (1904, p. 10) also perceived "a great and as yet unknown rhythm in the evolution of living things—a rhythm dependent on periodic changes in the inorganic environment." In 1898 T. C. Chamberlin proposed that the three main grounds for the division of geological time are the postulated periodicity of: (1) global diastrophism; (2) fluctuations of relative sea level and of evolutionary response; and (3) variations of atmospheric composition and consequent climatic change. By the close of the nineteenth century, therefore, ideas of rhythm and episodicity in orogeny, epeirogeny and crustal oscillation, sea level fluctuation, sedimentation, biological evolution, and climatic change already were well formulated.

During the first half of the twentieth century such concepts were advanced repeatedly and at times vigorously debated. The list of protagonists contains many well-known names including T. C. Chamberlin, Eliot Blackwelder, Joseph Barrell, R. S. Lull, Hans Stille, Charles Schuchert, John Joly, Arthur Holmes, Walter Bucher, A. W. Grabau, David Griggs, J. H. F. Umbgrove, Serge von Bubnoff, Ph. H. Kuenen, Sir Harold Jeffreys, Francis Shepard, James Gilluly, H. Shapley, V. Ye. Khain, V. V. Beloussov, and N. M. Strakhov. During this interval long-term rhythm or episodicity was proposed for tectonic movements (e.g., T. C. Chamberlin, 1909; R. T. Chamberlin, 1914; Blackwelder, 1914; Schuchert, 1914; Joly, 1923, 1925, 1928; Stille, 1924, 1936; Holmes, 1925, 1926, 1937; Bucher, 1933, 1939; Griggs, 1939; Umbgrove, 1939a; Beloussov, 1940; Bubnoff, 1941; Kuenen, 1941), epeirogeny and crustal oscillations (e.g., Haarmann, 1930; Born 1936; Khain 1939), changes of sea level (e.g., Schuchert, 1910, 1914, 1916, 1932; Grabau, 1936a and b, 1940; Bubnoff, 1949), climatic change, particularly glaciation (e.g., Huntington, 1914; Schuchert, 1914; Coleman, 1926; Forbes, 1931; Shapley, 1949), sedimentation (e.g., Schuchert, 1914; Barrell, 1917; Umbgrove, 1945; Strakhov, 1949; Willard, 1950), and biological evolution (Lull, 1918, 1929). Certain authors discussed cyclicity in several of the mentioned fields (e.g., Schuchert, 1914; Snider, 1932; Masarovich, 1940; Umbgrove, 1939b, 1942, 1947; Bubnoff, 1948a and b).

Space does not permit detailed discussion of the numerous concepts of geological episodicity and cyclicity proposed during these fifty or so years. Useful summaries are provided by Dunbar and Rodgers (1957, pp. 302–304), Fairbridge (1961), Beloussov (1962, pp. 739–750), and Goguel (1962, pp. 272–277); further discussion appears in Papers 1, 9, 22, and 32 and in the Editor's Comments for Parts III, IV, and VI. Diagrammatic synopses dating from near the start and the close of this interval are shown in Figures 1 and 2, respectively. Debate centered around ideas on diastrophic cycles: Shepard (1923) and Berry (1929) criticized the concept of periodic global diastrophism as proposed by T. C. Chamberlin (1909), R. T.

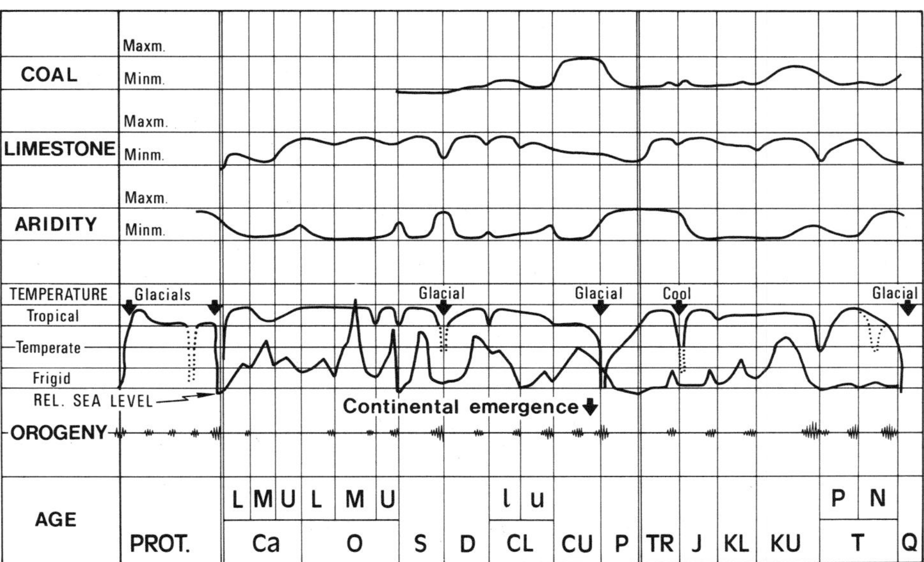

Figure 1 Chart of episodic geological events according to Schuchert (1914, Figure 90).

Chamberlin (1914), Joly (1925), and others, considering instead that diastrophism is regional and episodic; Jeffreys (1926, 1928) raised geophysical objections to Joly's (1925) hypothesis of global magmatic and diastrophic cycles resulting from radioactive heating of the earth's interior; and in one of geological science's best-known debates, Gilluly (1949, 1950) and Rutten (1949) objected to Stille's (1924, 1936, 1950) scheme of episodic tectonic phases of global extent, concluding that orogeny is of regional significance only.

Since 1950, and particularly since 1960, the number of works on long-term episodicity and cyclicity has increased greatly in the four fields of geotectonics, paleoclimatology, stratigraphy, and paleontology. At the same time, similar studies have been undertaken in the three new disciplines of paleomagnetism, stable isotope geochemistry, and planetary science. Such studies in these seven fields (represented by the seven Parts of this volume) have

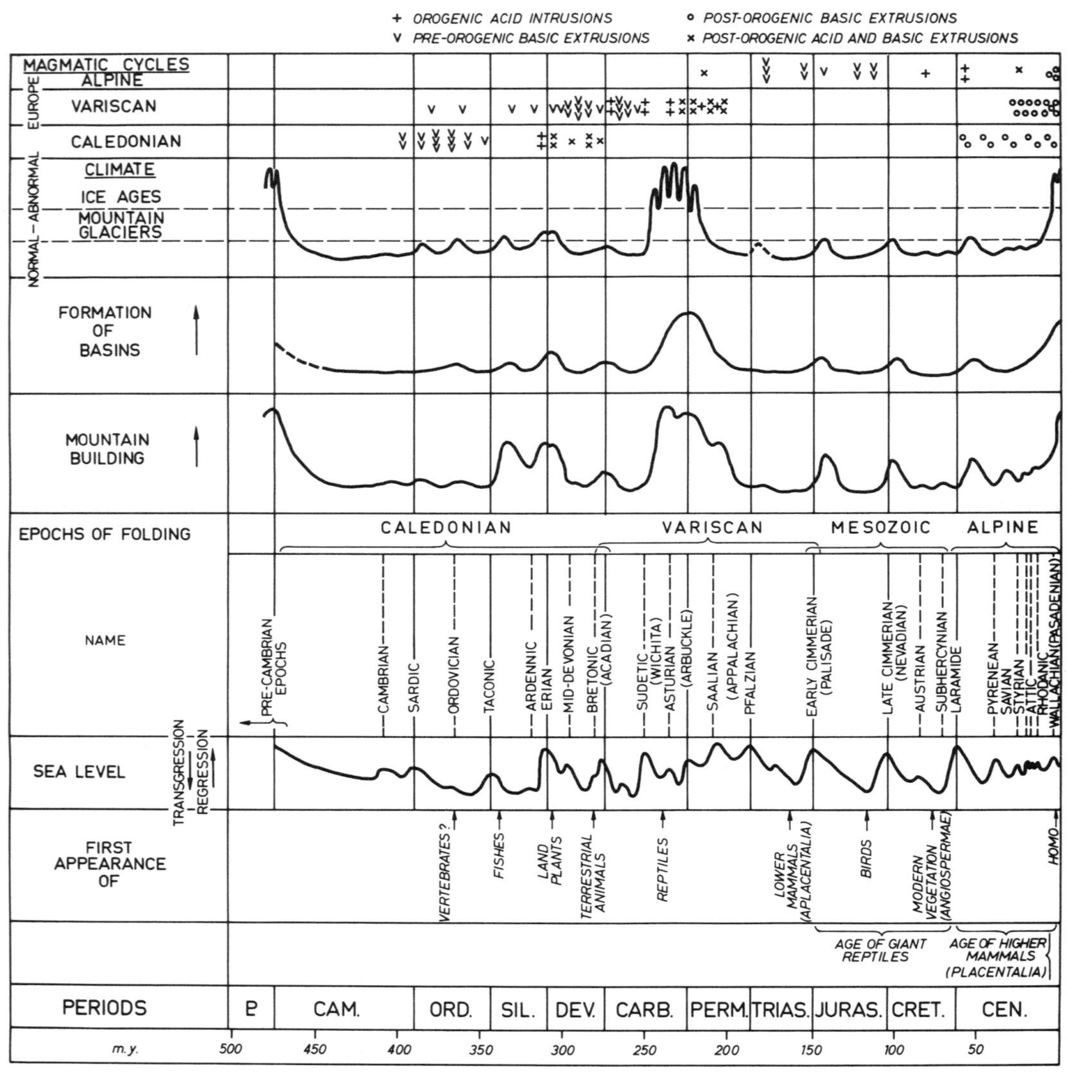

Figure 2 Diagrammatic synopsis of "the pulse of the earth," adapted from Umbgrove (1947, Table II). The absolute time scale shown is now obsolete.

provided, with the aid of an improved geochronological time scale, much additional and more reliable evidence of long-term episodicity (and perhaps in some cases, periodicity) in earth history and revealed episodicity also in the evolution of the terrestrial planets. Many old ideas have found support in the modern, more quantitative approach: the concept of Stille and Joly, among others, of global cycles of alternating orogenesis and crustal tension has again been advanced (Paper 7); Holmes' prescience of geotectonic megacycles in the Precambrian has been confirmed (Papers 1–7), as has Schuchert's proposal (Figure 1) of episodic glaciation (Papers 16 and 17); Grabau's principle of cyclic "pulsation" of sea level is supported by seismic stratigraphy (Paper 22); Barrell's concept of stratigraphic gaps and nonsequence applies to the ultimate repository of sediments, the ocean deeps (Papers 26, 27); and Lull's "pulse of life" seems established (Paper 32 and Figure 9). Moreover, evidence of long-term episodicity has emerged in many areas unknown to Umbgrove and his predecessors—namely, ocean basin tectonism and sedimentation, geomagnetic polarity bias, directions and rates of apparent polar wander, stable isotope geochemistry, Precambrian biological evolution, growth rhythms and geochronometry, and the early history of the moon and other terrestrial planets.

This emphasis on new findings is evident when Figures 1 and 2, dating from 1914 and 1947, respectively, are compared with Figure 3, the diagrammatic synopsis of data presented in this volume. Little new evidence of long-term episodicity accrued during the thirty-three years between 1914 and 1947, whereas twelve of the sixteen diagrams in Figure 3 represent disciplines established during an equal time interval since 1947. The Editor's Comments draw attention to apparent relationships among events illustrated in Figure 3 and discussed in the forty included papers.

EPISODICITY AND UNIFORMITARIANISM

How is the considerable evidence for long-term episodicity in earth and planetary history to be reconciled with the doctrine of uniformitarianism? Three of the four concepts contained in Lyell's doctrine—the temporal uniformity of the rates of geological processes and conditions and the temporal continuity of geological processes—have long troubled those geologists who recognized the importance of episodicity in earth history. Thus, Le Conte (1895, p. 315), realizing the polarization of views that had occurred with regard to catastrophism and uniformitarianism, favored "a substantial reconciliation of these two extremes, viz: that of gradual evolution both of the earth and of organic forms, *but not at uniform rate*." Barrell (1917, pp. 746–747) considered that "the doctrine of uniformitarianism has ignored the presence of age-long rhythms, and where they were obtrusive has sought to smooth them out;" he regarded this doctrine only as "a datum plane against which may be sought out and measured the amplitudes of

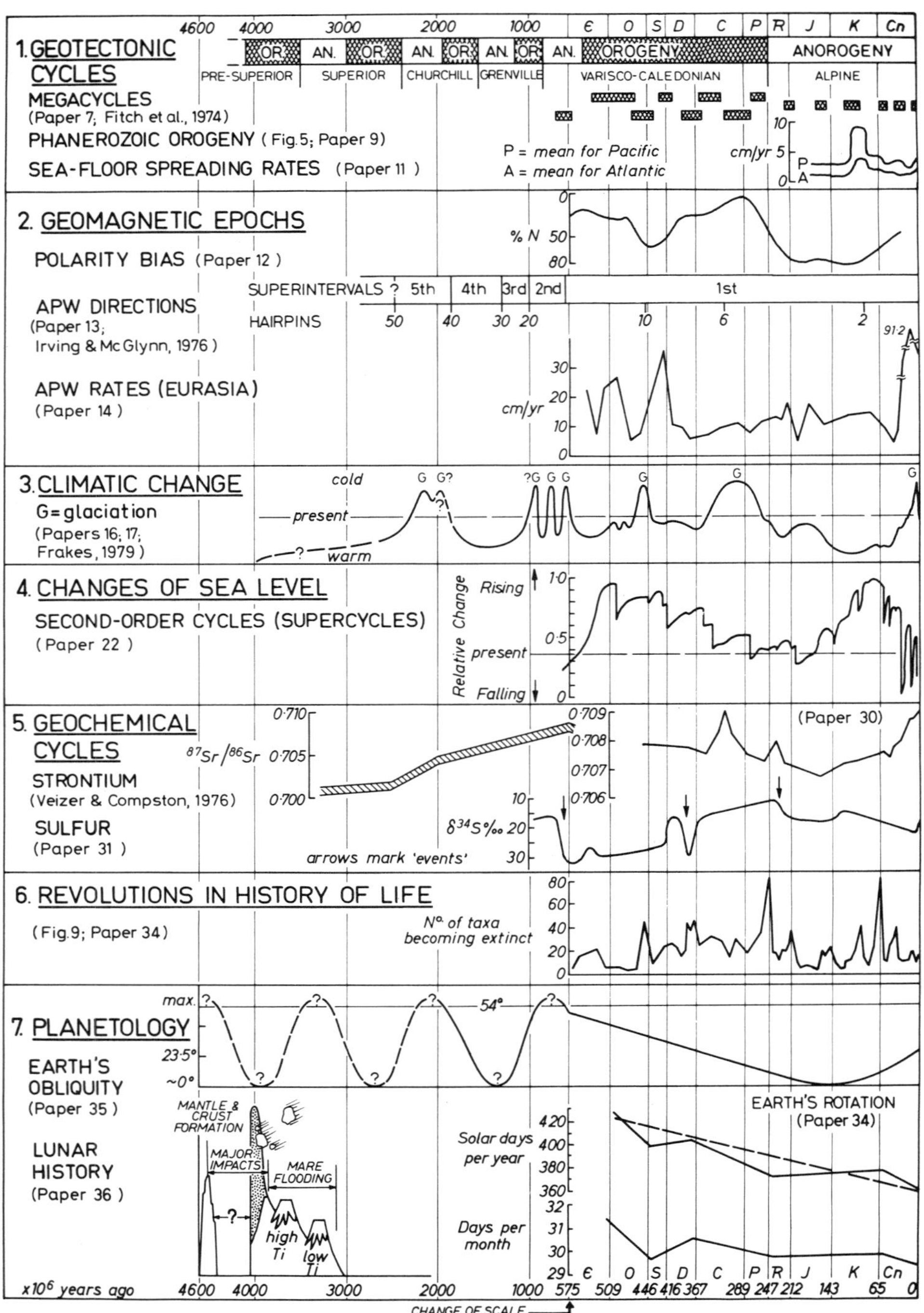

Figure 3 Diagrammatic synopsis of some of the long-term episodic events discussed in this Benchmark volume. The main sources of data are the included papers and figures; other references are updated or complementary works not included herein. Note the change of scale at the base of the Cambrian. Phanerozoic time scale after Armstrong and McDowall (1975).

oscillations." Similar difficulties arise today and can prompt semantic gems. Moorbath (1977a), for example, in acknowledging that major rock-forming events appear to have occurred episodically, coined "episodic uniformitarianism," and Ager (1973) advocated a "catastrophic uniformitarianism" in stratigraphy. Dissatisfaction with the doctrine of uniformitarianism has extended to study of the moon and terrestrial planets. Masursky (1978), for example, stated that the recognition of episodic events in their histories "must be based, at least partly, on nonuniformitarian thinking," since the events recorded cannot be related to those processes operating now.

The dissatisfaction with uniformitarianism prompted Gould (1965) and Austin (1978) to reevaluate Lyell's doctrine. Gould and Austin both concluded that the only unassailable concept inherent in the doctrine is the remaining principle of "methodological uniformitarianism," which asserts the invariance of natural laws in time and space. Newell (Paper 32) observed that many modern geologists unconciously embrace this principle together with the concept of nonuniformity of rates of geological processes. As pointed out by Gould (1965, p. 223), however, the methodological principle "belongs to the definition of science and is not unique to geology." Many will be inclined to agree with Austin that the term *uniformitarianism* "should be abandoned when describing formal assumptions used in modern geologic inquiry."

CAUSES OF LONG-TERM EPISODICITY

The many proposed causes of the episodicity of geological events are discussed in the Editor's Comments. Such proposals all too commonly reflect a rather ad hoc approach, with evidence from fields beyond the worker's immediate area of interest being overlooked. Comprehensive studies are required as to how the different successions of events relate one to another in cause and effect, and as to the importance of feedback mechanisms.

Perhaps "catastrophe theory"—a mathematical technique developed principally by Thom (1975) for modeling natural phenomena that contain discontinuity—will have future application to the study of long-term episodicity. Thom himself pointed out that his theory could throw light on the origin of geological phenomena such as faults, folds, and bed forms, and the theory's possible application to geology was further discussed by Cubitt and Shaw (1976) and Henley (1976). Catastrophe theory has been applied also to the modeling of events more akin to those discussed in the present volume, such as mass extinction (Lantzy et al., 1977), geomagnetic reversals (Chillingworth and Furness, 1975), and plate tectonics (Thom, 1978), while Moorbath (1977b) has suggested that eventually it may have application to the study of long-term diastrophic episodicity. The theory is not without its critics among mathematicians (e.g., Zahler and Sussmann, 1977) and geologists (Borchardt and Platt, 1978), however, and as concluded by Lantzy et al. (1978), its value to the physical sciences "can only be determined when sufficient data are available for a more rigorous mathematical application of the theory to physical processes."

THE SEARCH FOR CYCLICITY

The search for periodicity or cyclicity (sensu stricto) in the successions of episodic events characterizing earth history has occupied a long line of geologists that includes T. C. Chamberlin, Barrell, Lull, Joly, Holmes, Grabau, Griggs, Umbgrove, and Bubnoff, among others. Until the last two decades, however, reliable geochronological data were insufficient to allow more than subjective interpretations. With the improvement of the absolute geochronological time scale and the advances in other analytical techniques and in computation since the 1950s, more quantitative if not as yet conclusive evidence in favor of long-term periodicity has emerged. As exemplified by the papers included in this volume, approximate periodicity is proposed for or is apparent in geotectonism (Papers 1, 3–7, 9, and accompanying Editor's Comments), geomagnetic polarity bias (Paper 12), plots of APW rates (Paper 14), times of major glaciation (Papers 16, 17, and Part III), changes of sea level (Paper 22), temporal distribution of sedimentary facies (Paper 24), geochemical data (Papers 28, 29), and changes in planetary dynamics (Papers 34, 35) and in the gravitational "constant" (Paper 40). Moreover, the apparent close association of times of net extinction with Phanerozoic Period boundaries (Paper 32, and Part VI) would indeed indicate a "pulse of life" if the assertions of Hudson (1964) and Steiner (1979) of temporal regularities in the sequence of Periods are correct.

It is of course possible that some of these rhythms or near rhythms are no more than artifacts of incomplete data, or arise from unwarranted interpretation. On this latter point, Zeller (1964), writing of "Cycles and psychology," argued that the need of a scientist to explain and organize, and which provides him with the basic drive to become a scientist, may cause him in some cases to read too much between the (geological) lines. On the other hand, scientists tend to perceive discontinuity as the "poor relation" of continuity (Chadwick, 1977). Provided the researcher maintains "his objectivity about his work, and perhaps even more rigorously, about himself" (Zeller, 1964, p. 631), the search for long-term regularities in earth and planetary history could break important new ground.

Such research could, for example, explore the possiblity of commensurability among different proposed periods. A commensurability of 2 is indeed suggested by proposed rhythms of events discussed in this volume (Table 1). This scheme finds similarities in Khain's (1964) conclusion that globally important cycles 35–40, 150–200 and about 500 m.y. long are due to a "resonance effect" of interior and cosmic processes; in the long-term tectonic and sedimentary cycles of 10–20, 30–60, 150–200, and 500–600 m.y. duration proposed by Longvinenko (1976, p. 10), which have a commensurability of about 3; and in the observation of Stashkov (1977) that the two main periods of 160 ± 20 and 640 m.y. that he perceived for tectonic processes are commensurable. Furthermore, as discussed in Part VII, confirmation of the long-term cyclicity (s.s.) of certain planetary events could have important implications for cosmology, since it could mean that the gravitational parameter *G* is essentially constant over long intervals of time.

ENDOGENETIC VERSUS EXOGENETIC PROCESSES

The eighteenth-century Russian poet and scientist Mikhail Vasilevich Lomonosov was, according to Khain (1963), the first to recognize that geological processes are divisible into two principal groups, internal and external (or as now termed, *endogenetic* and *exogenetic*). Today the distinction between endogenetic processes (those arising within or upon an earth regarded essentially as a closed system) and exogenetic processes (those of external or astronomical origin affecting an earth regarded as an open system) persists, and disagreement exists as to their relative importance in shaping earth history.

As discussed in Part VII, an understanding of the histories of the other terrestrial planets would help resolve whether the successions of events that characterize earth history are reflections principally of endogenetic or exogenetic processes. Consonance of the rhythms of similar events determined for the earth and its

Table 1 Proposed periods of long-term rhythms discussed in this volume, and their possible commensurability

Approximate period (m.y.)	*Factor*	*Events*	*Reference in this volume*
30–50	X 1	Phanerozoic diastrophism	Part I; Paper 9
70–80	X 2	Changes in APW rates	Paper 14
74		Changes in earth's rotation	Paper 34
140–170	X 4	Major glaciations	Part III; Papers, 16, 17
300	X 8	Dominant period of geomagnetic polarity bias	Part II; Paper 12
200–300		First-order changes of sea level	Part IV; Paper 22
600–800	X 16	Geotectonic megacycles	Part I; Papers 1, 6, 7
600–800		Major lunar events	Part VII; Paper 36
1200–1300	X 32	Very long-term climatic changes	Part VII; Paper 35
2500	X 64	Proposed obliquity change	Part VII; Paper 35

planetary neighbors would provide strong evidence for the importance of exogenetic influences, whereas dissimilarity of meters would argue in favor of endogenetic controls. Clear answers in this area would greatly illuminate our understanding of the earth. Paper 38 by W. K. Hartmann exemplifies the interdisciplinary approach required.

CYCLES AND THE GALAXY

As reviewed in Part III, since Forbes (1931) and Umbgrove (1942, 1947) suggested that a causal relationship might exist between the timing of earth glaciation and the period of the sun's orbit about the galactic center (termed the cosmic year), numerous persons have sought empirical connections between the apparent rhythms of terrestrial events and the calculated durations of elements of the sun's galactic orbit. The cosmic year having been regarded as 200–300 m.y. long, such hypotheses have been concerned with apparent earth rhythms of about this duration, or fractions thereof. Such comparisons have generally lacked conviction because of poor geochronological control and uncertainty as to the duration of the cosmic year. Using the recent geological data assembled in the present volume and a duration of about 300 m.y. for the cosmic year as recalculated in Paper 17, it is tempting, however, to again seek geogalactic coincidences.

Thus, taking at face value the proposed earth rhythms listed in Table 1, the periods of ~300 m.y., ~155 (±15) m.y., ~70–80 m.y., and ~40 (±10) m.y. may be equated respectively with the period, half-period, quarter-period, and eighth-period of the cosmic year. Moreover, the ~70–80 m.y. and ~40 m.y. rhythms are in close agreement with the period and half-period of oscillation of the solar system normal to the galactic plane (see Papers 16 and 17).

Scientists have widely assumed that the cosmic year is the longest orbital parameter of our galactic system. Recent astronomical research indicates, however, that a much longer galactic rhythm exists—namely, the orbital period of the Large Magellanic Cloud around the Galaxy. (The Large and Small Magellanic Clouds [LMC and SMC] are satellite galaxies of our own galaxy. Numerous astronomers believe that the flexure exhibited by the plane of the Galaxy is a tidal distortion caused by the LMC [see Paper 17].) Fujimoto and Sofue (1977), in a computer simulation of the tidal distortion of the Galaxy by the Magellanic Clouds, made sixty-five determinations of the orbital period of the LMC using different

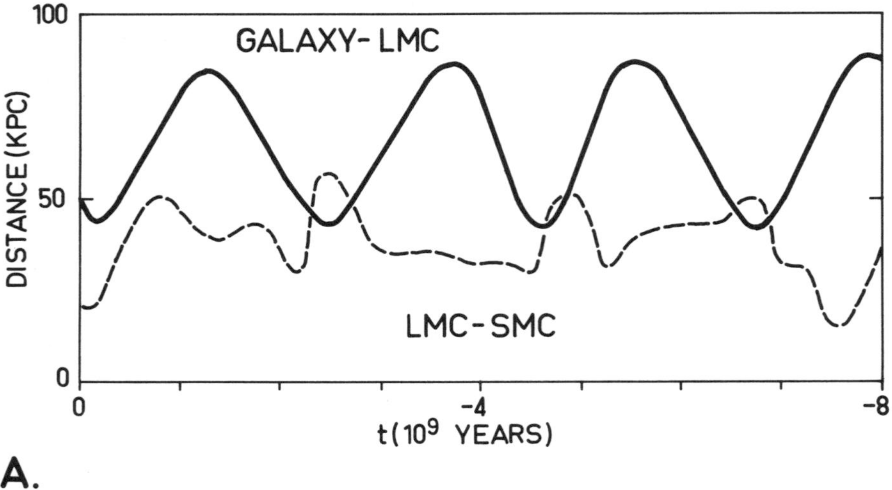

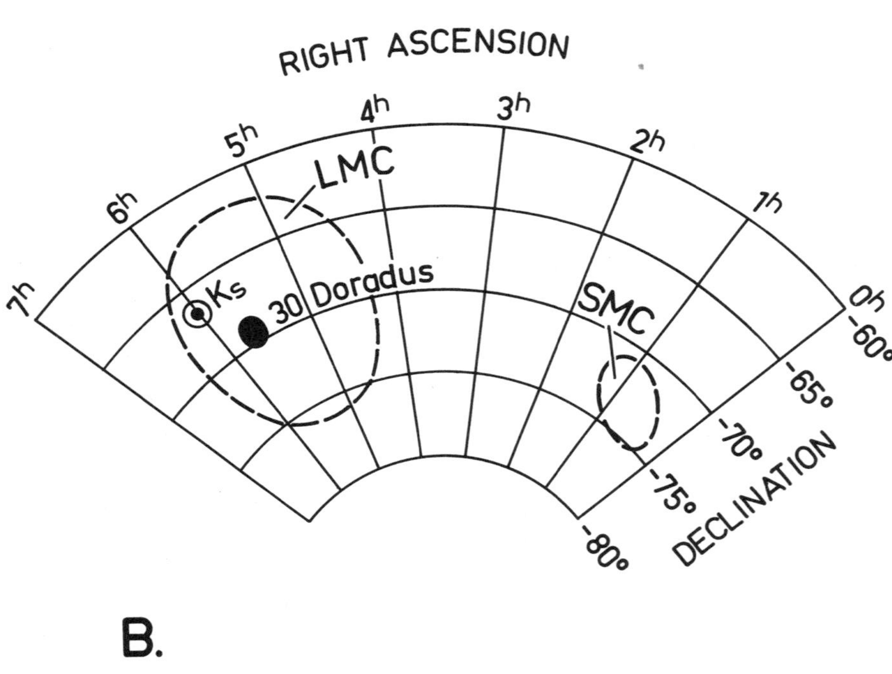

Figure 4 A: Distances in kiloparsecs between the galactic center and the Large Magellanic Cloud (LMC) and between the LMC and the Small Magellanic Cloud (SMC) as a function of time (after Fujimoto and Sofue, 1977, Figure 4a). B: South pole of the ecliptic (K_S) relative to the approximate limits of the LMC and SMC and to the position of the massive 30 Doradus complex (data for Clouds from Mathewson and Healey, 1964).

values for its orbital geometry and dynamics and the galactic mass. The mean of the sixty-three finite periods obtained is 2.49 b.y., with a standard deviation of 0.81 b.y. For their Cases III and IV, which Fujimoto and Sofue considered best simulate the observed galactic flexure, the LMC's mean orbital period is 2.59 ($\pm$0.37) b.y. Since the LMC appears to be in a highly eccentric orbit, it makes a close approach to the Galaxy about every 2.5 b.y. (see Figure 4A). It is intriguing that the terrestrial geotectonic megarhythm of ~600–800 m.y., the meter of proposed very long-term climatic change (Paper 35), and the full period of postulated secular change in the earth's obliquity (Paper 35) are, respectively, about one quarter, one half, and about equal to the LMC's proposed orbital period. A geometrical link between the LMC and the solar system is provided by another intriguing coincidence—that is, the south pole of the ecliptic (the projection to the galactic south of the pole of the earth's orbital plane) passes through the LMC near 30 Doradus, the most important mass component in our galactic system (Figure 4B). One can imagine the pole of the ecliptic (and of the plane of the solar system) tracking the LMC in its orbit around the Galaxy, thereby causing secular changes in the obliquity of the ecliptic and consequent very long-term changes in the earth's paleoclimates (Paper 35) and tectonism (Williams, 1973). The torque due to the LMC appears too small, however, to so affect the solar system.

Evidence exists, therefore, of broad agreement between proposed periods of important events in earth and planetary history and calculated galactic periods or their whole fractions. In view of the evidence for commensurability among certain proposed terrestrial rhythms, one may speculate that this complex "symphony of the earth" is, in part at least, a response in some way to resonant solar-galactic motions, or to resonances between interior and cosmic processes (Khain, 1964). Doubtless the data can be interpreted in other ways, however, and these ideas are presented in the same spirit as that of Umbgrove's (1942, 1947) early comments on the same topic in the hope that they might likewise stimulate further studies.

CONCLUSIONS

The evidence assembled and discussed in this Benchmark volume in my opinion establishes on firm ground the principle of long-term episodicity in many areas of earth and planetary history.

Numerous problems remain unresolved, however, and future work should be directed toward providing answers to the following questions:

1. How do the differing successions of terrestrial events relate one to another in cause and effect, and how important are feedback mechanisms?
2. Are certain patterns of events common to the earth, the moon, and other planets?
3. Is the repetition of certain events truly periodic and cyclic? If so, are commensurabilities present among the different rhythms?
4. Can certain apparent rhythms be related to galactic periods?

Answers to such questions may greatly illuminate our understanding not only of the earth and planets but also of our galactic system and in cosmology. Those who seek answers should recognize the many patterns of events displayed in earth and planetary evolution, for as Fairbridge (1972) observed, "nature abhors the *ad hoc.*"

Part I

GEOTECTONIC CYCLES

Editor's Comments on Papers 1 Through 4

EVIDENCE FOR GEOTECTONIC MEGACYCLES

During the 1950s evidence arose that the isotopic ages obtained from rocks were not uniformly distributed in time but seemed to cluster into definite groups that were very similar from continent to continent. York and Farquhar (1972, p. 116) stated that such age groups or "Magic Numbers" had been "evident to many people in geochronology and it is not possible to assign their recognition to any one individual." However, studies by Holmes (1951), who drew attention to geochronological similarities among the histories of African, Fennoscandian, North American, and Indian Precambrian provinces, as well as the studies of Voitkevich (1958) and Tilton and Davis (1959), who recognized mineral age groupings at ~2800–2500, ~2000–1800, ~1150–1000, and ~600–500 m.y. ago, prepared the ground for the major syntheses of the following decade.

A major step forward was made in 1960 with the appearance of Gastil's compilation and analysis of 413 isotopic age determinations (Paper 1). From a plot of mineral date abundance Gastil proposed that crustal adjustments are episodic, with major cycles of ~350 to ~500 m.y. duration. Gastil's results have, in principle, been confirmed by later studies, and in the opinion of Sutton (1977, p. 162) "constitute one of the major advances of this century in the understanding of crustal evolution." Two other important geochronological compilations are included here. In 1966 Dearnley (Paper 2) presented a histogram based on just over 3400 age determinations, which shows three particularly well-defined changes of slope at 2750±50, 1950±50, and 1075±50 m.y., and peak abundances at ~2600, ~1800, ~950, and ~500 m.y. ago. Paper 3 by Stockwell (1968) was the culmination of studies on the K-Ar geochronology of stratified rocks of the Canadian Shield (Stockwell, 1961, 1963a and b, 1964). Four orogenies divide the succession into five natural time-stratigraphic units: the Kenoran orogeny with a mean at 2490 m.y., the Hudsonian at 1735 m.y., the Elsonian with a very weak peak at 1370 m.y., and the Grenvillian at 945 m.y. ago. Papers 2 and 3 therefore supported Gastil's conclusion of the long-term episodicity of crustal processes, but their analyses of much larger numbers of dates failed to confirm the presence of significant peaks of activity at ~2100 and ~1360 m.y. Other studies of isotopic age abundance (e.g., Sabine and Watson, 1965; Choubert, 1967; Stockwell et al., 1970; Garrels and Mackenzie, 1971, p. 79) likewise identified broad statistical peaks at ~2700–2500, ~1900–1700, ~1100–900, and ~600–400 m.y. Overall, the average of the estimates of significant dates in earth history made by numerous workers between 1958 and 1965 gives 2690, 1840, 1090, and 570 m.y. (York and Farquhar, 1972, p. 116).

The above groupings of isotopic ages commonly have been interpreted as reflecting global orogenic-plutonic megacycles with a mean duration of ~600–800 m.y. Moorbath (1967, p. 123), however, has cautioned that "the popular exercise of plotting analytical dates on histograms is quite unjustified unless (a) the analytical dates refer to the true age of a geological event, and (*b*) the nature and significance of that event is fully understood." He pointed out that the majority of such histograms contain "a high proportion of cooling dates or overprinted dates," whereas the more significant intrusive and metamorphic dates from a basement complex (from particularly retentive minerals, from low-grade metamorphic zones, and from whole rocks [Rb-Sr]) are poorly represented. York and Farquhar (1972, pp. 125–126) likewise drew attention to the risks in interpreting such histograms, but they concluded that "the persistence of the Magic Numbers, defined earlier, as many more data have been added, and the similarity of the geochronological pictures of the different continents indicate that a genuine structure is to be seen in the age distribution."

In other studies since 1960 such geotectonic megacycles have been recognized, or utilized in stratigraphic subdivision, in North

America (Goldich et al., 1961; Goldich, 1968; James, 1972; Stockwell, 1973; King, 1976), the USSR (Vinogradov and Tugarinov, 1961; Salop, 1968; Semenenko et al., 1968, 1972), Australia (Rutland, 1973a), Africa (Martin, 1965), and India (Sarkar, 1972). Their common occurrence in several continents, or their global nature, has been proposed by Sutton (Paper 6, 1967, 1973, 1977), Salop (1964, Paper 4, 1977), Semenenko (1964, 1972), Fitch and Miller (Paper 7), Burwash (1969), Kölbel (1971), Allègre (1972), Rutland (1973b), Williams (1973), and Fitch et al. (1974). The Russians A.P. Vinogradov, A. I. Tugarinov, L. J. Salop, and N. P. Semenenko have made major contributions in this field, and Paper 4 by Salop (of the All-Union Geological Research Institute, Leningrad) provides a fairly recent (1972) Russian viewpoint on the existence of diastrophic megacycles of global significance and their employment in the subdivision of the Precambrian.

Significant new age and isotope data bearing on the existence and nature of diastrophic megacycles are provided by Moorbath (1975, 1976, 1977a) and McCulloch and Wasserburg (1978). Moorbath (1977a, p. 172) considered that relatively short intervals of accelerated growth of the sialic crust "may have occurred episodically, rather than randomly or semi-continuously, during the earth's history. There seems to be a grouping of major terrestrial rock-forming events at *approximately* 3800–3500, 2900–2600, 1900–1600, 1200–900, and 600–0 m.y. ago." Moreover, McCulloch and Wasserburg (1978, p. 1007) concluded that "some truly major periods of continental crustal growth are sharply episodic." These findings contrast with those of Hurley and Rand (1969), who considered that the continental crust has developed at an accelerating rate with no evidence of discrete pulses at times of widespread orogeny. Isotope data also indicate a multistage history of mantle evolution (e.g., Cooper et al., 1967; Ulrych, 1969; Oversby and Gast, 1970; Brooks et al., 1976). Future synthesis may identify a pattern of mantle fractionation events that can be related to the pattern of other important events in earth history.

1

Reprinted from pages 1–16 and 30–35 of *Am. Jour. Sci.* **258**:1–35 (1960)

THE DISTRIBUTION OF MINERAL DATES IN TIME AND SPACE

G. Gastil

ABSTRACT. Most igneous and metamorphic mineral dates indicate times of rock cooling, the terminal events of crustal adjustment. Accordingly, the distribution of mineral dates in time indicates the periodicity of such events, and the distribution in space indicates the geometry of their occurrence. By this we may test the validity of such concepts as "orogenic periodicity", "cyclic orogeny", and "continental accretion".

A plot of mineral date abundance against age shows that crustal adjustments are periodic and roughly cyclic. Intervals for which abundant mineral dates have been preserved are about 175 to 250 million years in length, with cycles of about 350 to 500 million years. Intervals of date abundance fall in the ranges: –2710 to –2490 m.y., –2220 to 2060 m.y., –1860 to –1650 m.y., –1480 to –1300 m.y., –1100 to –930 m.y., –620 to –280 m.y., and –120 to the present.

Mineralogenic events of very different ages are commonly recorded in the same area, not infrequently in the same rock. Large areas, however, can be characterized by the mineral dates of the last interval of mineralogenic activity by which they were seriously effected. In North America these mineralogenic provinces show a crudely concentric pattern, younger dates outward. Similar arrangements have not been found in other continents.

INTRODUCTION

The temporal distribution of crustal adjustment and the geometry of unstable areas have long been topics of geologic discussion and hypothesis. To utilize mineral dates in resolving these questions it is first necessary to decide what events they record, which mineral dates are applicable to the study of large-scale crustal adjustment, and to what extent we may rely on existing mineral dates. After choosing criteria for the selection and evaluation of dates it is possible to tabulate and plot them according to their time and space distribution. Naturally we must satisfy ourselves that the sample is sufficiently large and areally distributed to be representative of a region. Finally, it will be possible to examine the degree of temopral and spatial order, check the statistical significance of this order, and use it as a test of geologic concepts. If our conclusions are valid it should be possible to predict the time and space distribution of mineral dates as yet undetermined.

ACKNOWLEDGMENTS

The fund of mineral age data upon which this paper is based is the contribution of many researchers in laboratories all over the world. The compilation of this material has been significantly aided by the work of Holmes and Cahen (1957) and Gheith (1958a, 1958b). Correspondence and discussion with H. W. Fairbairn, Massachusetts Institute of Technology, A. H. Lang, Geological Survey of Canada, Leon Silver, California Institute of Technology, and Henry Faul, U. S. Geological Survey, are gratefully acknowledged. Especial thanks is due Henry Faul and John Crowell, University of California at Los Angeles, for reading the paper, and Mrs. Opal Kurtz for drafting the illustrations.

THE EVENTS RECORDED BY MINERAL DATES

Most geologic events involve the growth or regrowth of crystalline minerals. Uplift and erosion, sinking and sedimentation, faulting and folding, all trace their origin to physical-chemical adjustments which, through magmatization or metamorphism, ultimately require mineral crystallization.

Where a crystal contains atoms which disintegrate spontaneously, the chemical composition of the mineral changes as a function of time, almost independently of environment. The altered atoms, thus formed, may or may not be accommodated by the crystal. Some are able to escape as soon as formed (radon), some escape only after a threshold concentration has been achieved (argon in feldspar?), and others can accumulate almost indefinitely. As long as they do not escape from the crystal in which they were formed their abundance is a measure of the time elapsed since crystallization.

If, after crystallization, a change in physical environment requires recrystallization to the same or a different mineral, most of the daughter atoms not required by the new form will be excluded. The abundance of daughter atoms, then, tells us how much time has elapsed since the crystal in question was capable of expelling them. Unfortunately we do not know the limiting physical-chemical conditions which control the retention or expulsion of daughter atoms. Comparative studies of different age-dating methods (references below) show us that the same environment which brings about the removal of daughter atoms from one element ratio, or from one mineral, is not necessarily effective in removing a different daughter element, or in removing the same element from an adjacent mineral of a different species (M.I.T., 1958a; Grunenfelder, 1958; Tilton, et al., 1958a, b; Aldrich, 1958). It is this overlap in retention stability that makes it possible to obtain the date of more than one geologic event from the same rock.

When the environment of a crystal requires the expulsion of daughter atoms, does the crystal necessarily undergo rapid and total cleansing so that its "age" is neatly reestablished as that of the time of cleansing? Or is it possible that only a portion of the unaccommodated atoms are expelled, leaving the crystal with an intermediate "age value" that does not date any event? Obviously a composite crystal, portions of which have grown at measurably different times, will not yield significant age data unless the different portions can be analyzed individually. If the parent-daughter isotope ratios in a rock have been affected by gradual daughter-escape, partial daughter-cleansing, or composite crystallization, different mineral species, and different radiogenic isotope ratios within those minerals would not be expected to yield the same "age". However, many cases of date concordance have been found, and these must record physical events in the history of the earth's crust.

To utilize mineral dates geologists must make assumptions as to the nature and circumstances of the causal event. This event may have been crystallization from a melt, metamorphism, or deposition from an aqueous fluid. Since these processes may require millions of years for completion, and may more or less continuously accompany tectonic instability over still longer intervals of time, the mineral dates which are preserved represent only the final achievement of crystal stability, the terminal cooling. This final stability is achieved individually by each crystal, and due to the heterogeneity of orogenic belts, and the continual structural dislocation within them, mineral dates representing the entire history of a mineralogenic province may be preserved. The spread of mineral dates ascribed to a mineralogenic epoch may or may not represent the entire interval of crustal instability.

THE SELECTION OF MINERAL DATES SIGNIFICANT TO THE STUDY OF CRUSTAL STABILITY

In dating intervals of crustal adjustment we are most concerned with the crystallization of deep-seated igneous and metamorphic rocks. Micas, potassium feldspar, and accessory minerals such as zircon and monazite are primary constituents of such rocks and contain one or more naturally radioactive elements. Pegmatites and associated hydrothermal veins which are satellitic to centers of magmatization or metamorphism should yield dates no older or younger than the active life of their source. These contain micas, potassium feldspar, and a variety of primary uranium and thorium minerals. Medium- to low-temperature vein and replacement hydrothermal deposits, as well as supergene deposits, may occur in areas marked by crustal stability, and therefore do not appear to be necessarily related to fundamental crustal adjustment. Some mineralogenesis in stable regions, however, appears temporally related to distant crustal adjustment (Cummings, 1955). Uranium mineralization on the Colorado Plateau during the deformation of the Rocky Mountains is one example.

A table of mineral dates is included as an appendix. This table is designed to include every locality for which a mineral date, attributed to fundamental crustal adjustment, has been determined by the use of methods and materials currrently regarded as reliable. Where several similar dates have been obtained for materials from one locality, more than one representative date is presented.

The dates have been classified as either *confirmed* or *tentative*. *Confirmed* dates are based upon corroborative determinations by different methods, materials, or samples. *Tentative* dates are those based upon a single determination, upon several highly discordant ones, or upon unsatisfactory or unknown methods or materials.

Concordant uranium-thorium-lead isotope dates and lead 206/207 dates for medium- to high-temperature minerals, rubidium 87/strontium 87 dates for low-calcium minerals, and potassium 40/argon 40 dates for micas and young ($<$100m.y.) potassium feldspar are considered to be reliable and significant. Dates determined by non-isotopic chemical analysis, common lead, helium, crystal damage, and lead-alpha methods, while frequently correct and useful as confirmations, are not considered reliable by themselves. Dates determined for older feldspars, secondary minerals, mineraloids, total rock, or sedimentary materials are likewise, by themselves, either unsatisfactory or unapplicable. Dates for which the mineral has not been given have generally been omitted. To make the dates directly comparable they have been adjusted to the decay constants currently used by North American laboratories (Wetherill, 1957; Aldrich, et al., 1956a).

A spread of dates for a given locality may represent a sequence of closely spaced mineralogenic events, a difference in response to the same event by different minerals or elements, a difference in response to post-crystallization environment by different minerals, elements, or different crystals of the same mineral, or a lack of analytic precision. Thus, when we wish to consider individual localities in terms of representative dates there is no uniquely satisfactory or entirely objective procedure for generalizing the data. Where an analyst or reviewer has published a "preferred date" the preference has generally been followed and acknowledged. For a pair of determinations by different methods the average is used. For a pair of determinations on different materials by the same method the higher value is used. Where applicable, a superior method or a more recent determination is given preference. Where more than two determinations must be considered, those dates forming the most significant cluster are averaged. The method of generalization employed is indicated by a suffix attached to each composite date given in the table.

THE VALIDITY OF MINERAL DATES

Attempts to measure mineral age began soon after the discovery of natural fission (Marble, 1954, p. 3). During each decade since then analysts have discarded most of the age determinations made in the preceding one. Certainly many of the dates used in this paper are destined for revision, but generalizations based upon groups of dates have suffered a lower mortality rate than have the individual age values. In 1927 Holmes and Lawson presented a time scale of which the post-Cambrian portion differs little from that in use today. In 1932 Ellsworth showed that uraninites of the Grenville orogenic province were about one billion years old and much younger than those of the adjoining Keewatin province. At that time his conclusions differed from common geologic thought, but many subsequent determinations by improved methods have only confirmed and complemented his discovery.

THE STUDY OF MINERAL DATES ON GRAPHS AND MAPS

Plots of mineral date abundance against age.—In figure 1 mineral ages are plotted along the abscissa. Time is subdivided into intervals of ten million years, and the number of mineral date localities tabulated for each interval is plotted in histogram form parallel to the ordinate. Both *confirmed* dates (represented on the graph by solid squares), and *tentative* dates (represented by open squares) are plotted for the individual continents, but only the *confirmed* dates are plotted for the earth as a whole. The histograms are supplemented by a curve which draws attention to the date clusters. This curve is constructed by adding 4 vertical scale units for each date falling in the time interval in question, two vertical units for each of those falling in the two adjacent intervals, one unit for each of those in the four next closest intervals, and one-half for each in the next closest eight intervals. The area beneath the *confirmed* curve is shaded, whereas the curve representing *confirmed* plus *tentative* dates is

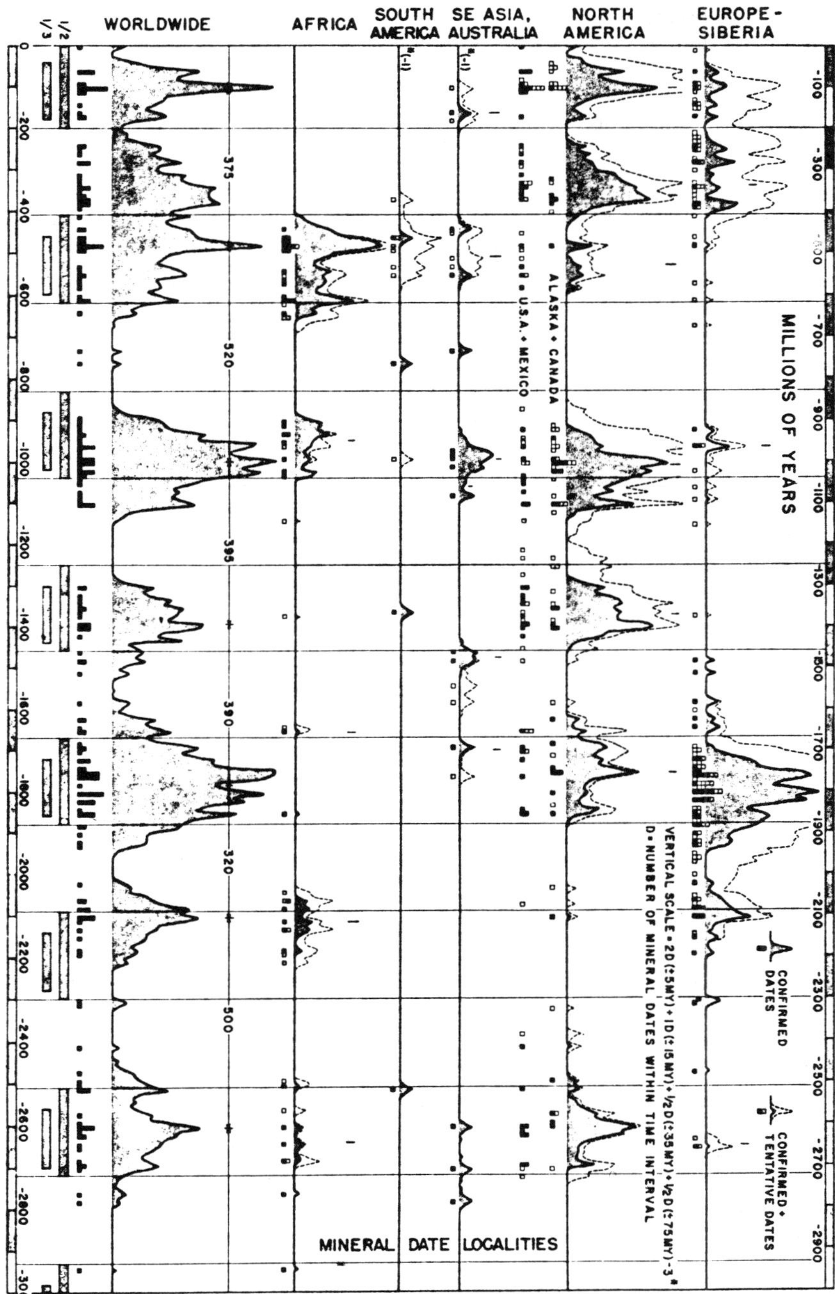

Fig. 1. Mineral date abundance plotted against age.

dashed. The three lowest vertical scale units are subtracted from all of the graphs except those for Southeast Asia—Australia, and South America, for which only one scale unit has been subtracted. The single date from Antarctica was included with Southeast Asia and Australia.

Sample representativeness.—If a coordinate grid were superimposed upon a map of North America and a mineral age determination were made for the closest appropriate material to each grid intersection the representativeness of the data would depend only upon the coordinate interval. The selection of material for mineral dating, however, has depended upon proximity to laboratories, the location of economic mineral deposits, roadcuts, and other man-dependent variables. As a result the number of determinations reported for some small areas and individual localities is out of proportion to their areal significance. In attempting to minimize this effect we have treated each mineral district or township as a single locality. The *area-compensated* graph for North America (fig. 2) is a further attempt to minimize the effect of disproportionate mineral distribution. To construct this graph North America was divided according to coordinates of longitude and latitude (fig. 3). Each grid area was then treated as a locality and one or more representative dates were selected according to the same criteria described previously for individual localities. Another check of sample representativeness was made by dividing North America along the geologically arbitrary United States-Canadian border (fig. 1). The marked agreement of both portions with the *area-compensated* graph for the entire continent shows that the date sample is representative.

It could be argued that the crystallization of a particular mineral species, or of minerals of a particular element, has not necessarily been distributed representatively through geologic time, and that therefore the respective mineral dates represent only those adjustments involving particular minerals or elements. To investigate this possibility the age values for each locality were tabulated according to method and mineral used. Where several determinations by the same method and material were available for a single locality they were compounded by the previously explained method. This treatment involves the use of many individual determinations and therefore produces greater dispersion on the distribution graph. In figure 2 there are separate plots for potassium-argon, rubidium-strontium, and a composite of uranium-thorium-lead method dates. The latter graph is divided into five histograms: (A) concordant uraninite determinations, (B) lead 206/207 determinations for uraninite, (C) concordant uranium-thorium-lead determinations for zircon, monazite, xenotime, allanite and other accessory minerals, (D) determinations by lead 206/207 for the accessory minerals, (E) determinations by the common lead method, lead-alpha method, non-isotopic chemical analysis of uranium and thorium minerals, and the isotopic analysis of pitchblende. Although there is a marked absence of uranium dates at about 350 m.y. and an absence of mica dates at 2100 m.y., the graphs for the different methods and materials show great similarity in date distribution. This coincidence indicates that radioactive mineral dates are representative of deep-seated mineral genesis as a whole.

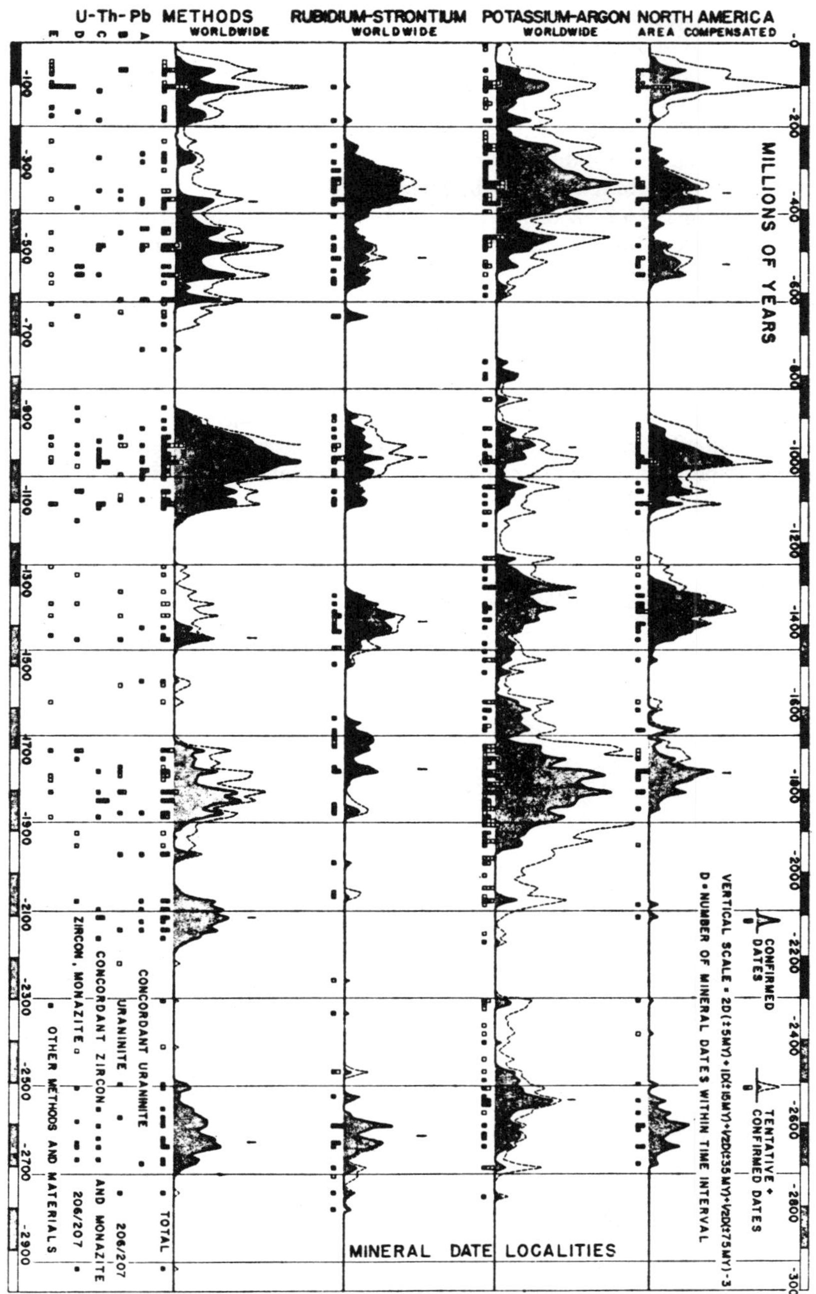

Fig. 2. Mineral date abundance plotted against age.

Finally consider the distribution of mineral ages with depth. In a crustal belt of recurrent mineral genesis (an orogenic belt) crystals of zero age are continuously or spasmodically being formed at all depths. Some crystals which cooled at great depth are carried toward the surface by structural dislocation, others by erosion and uplift. Processes of fusion, solution, and metamorphism that continuously re-establish mineral ages during the active life of the belt are increasingly important with depth. The surface rocks may contain mineral ages representing many or all of the past mineralogenic events of the area, while at great depth all minerals may bear the age of the most recent mineralogenic event. Thus, deeply eroded terranes like the Grenville orogenic province and the Pacific batholithic belt are relatively uniform in mineral age, whereas at Sudbury and in New England, where variably metamorphosed rocks are cut by many small intrusions, a wide spread of mineral ages has been preserved. The younger the terrane, the smaller the chance that it has been eroded to its mineralogenic roots; thus, the dispersion of mineral ages in recently unstable provinces should, all other factors neglected, be greater than in more ancient terranes. The Precambrian mineralogenic intervals can be expected to correspond more closely to the closing phases of orogeny, whereas post-Cambrian dates tend to represent the entire span of crustal instability.

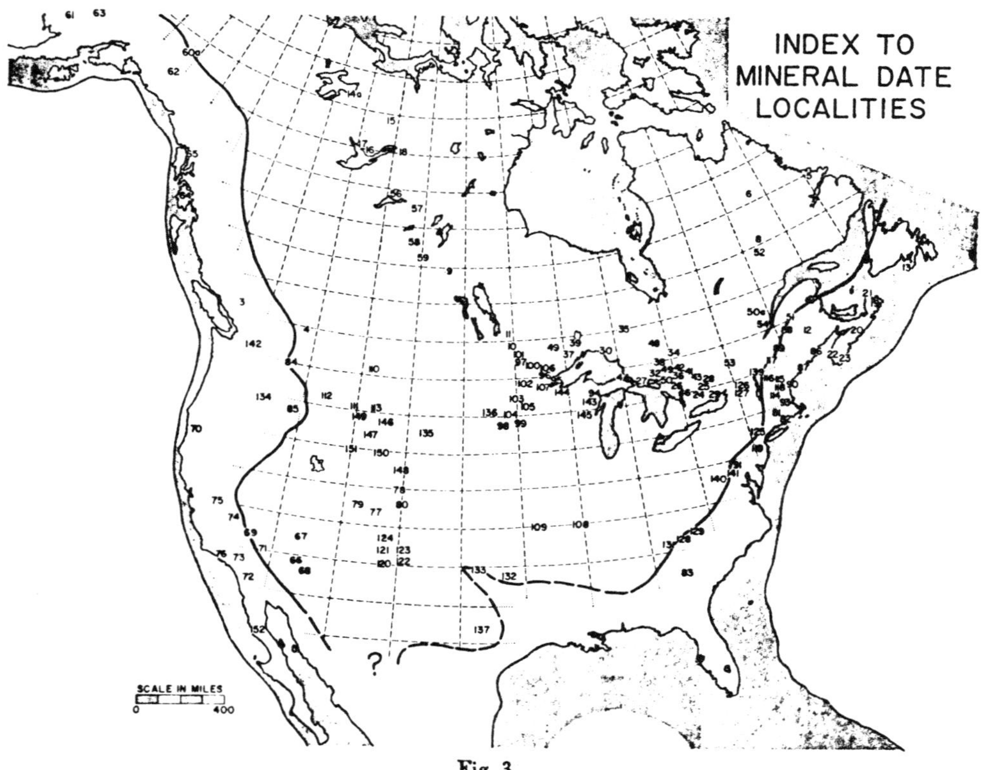

Fig. 3.

Mineral date provinces in North America.—Figure 3 locates the tabulated mineral dates for North America. Figure 4 shows the regional lineaments and structural grain of North American Precambrian rocks, as abstracted from the literature. The dotted lines indicate the apparent boundaries of Precambrian structural provinces. In constructing figure 5 we have sought a coincidence between these apparent structural boundaries and those of the mineral date provinces. The patterns by which the different provinces are distinguished are diagrammatic extrapolations of the structural patterns indicated by figure 4.

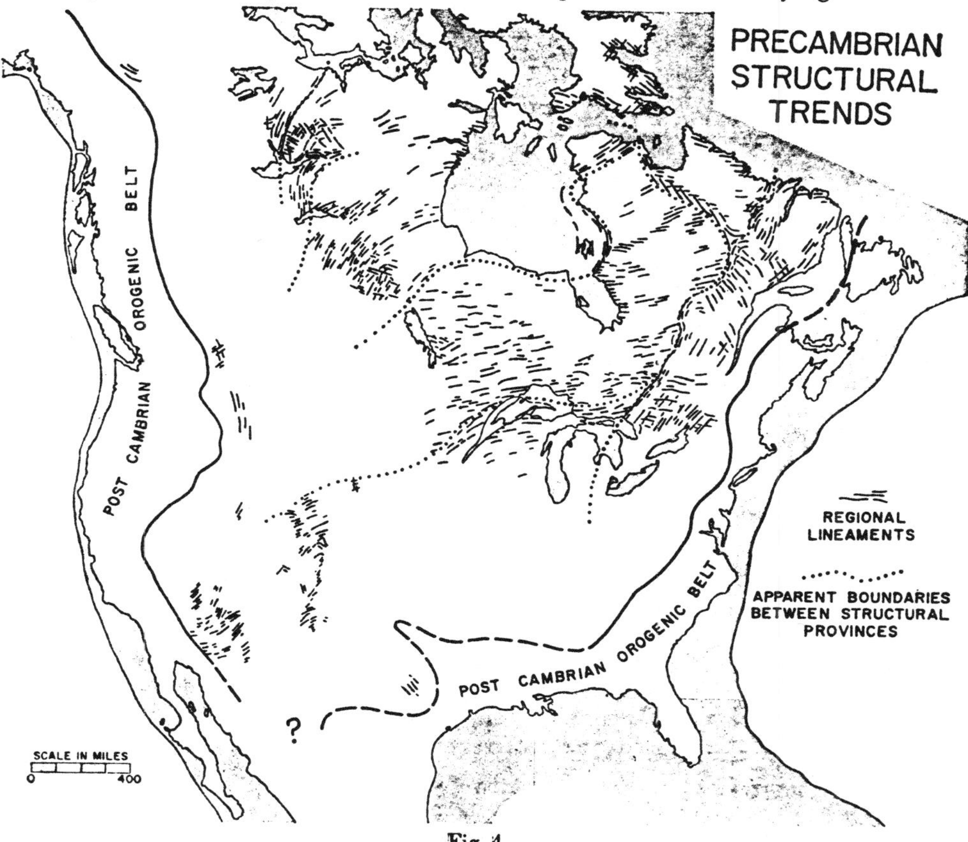

Fig. 4.

In figure 5 the mineral date localities of North America have been divided according to the abundance intervals which appear in figure 1. The intervals are designated by the letters A to G, younger to older. The distribution of mineral dates combined with considerations of regional geologic structure serves to define the mineral age provinces. Each province is characterized by the spread of dates representing the last widespread interval of orogenic mineralogenesis by which it was severely effected. Thus, the Pacific margin province is labeled 0-150, the Atlantic margin province 250-550, and the Grenville orogenic province 950-1100. Other provinces are characterized by the intervals 1250-

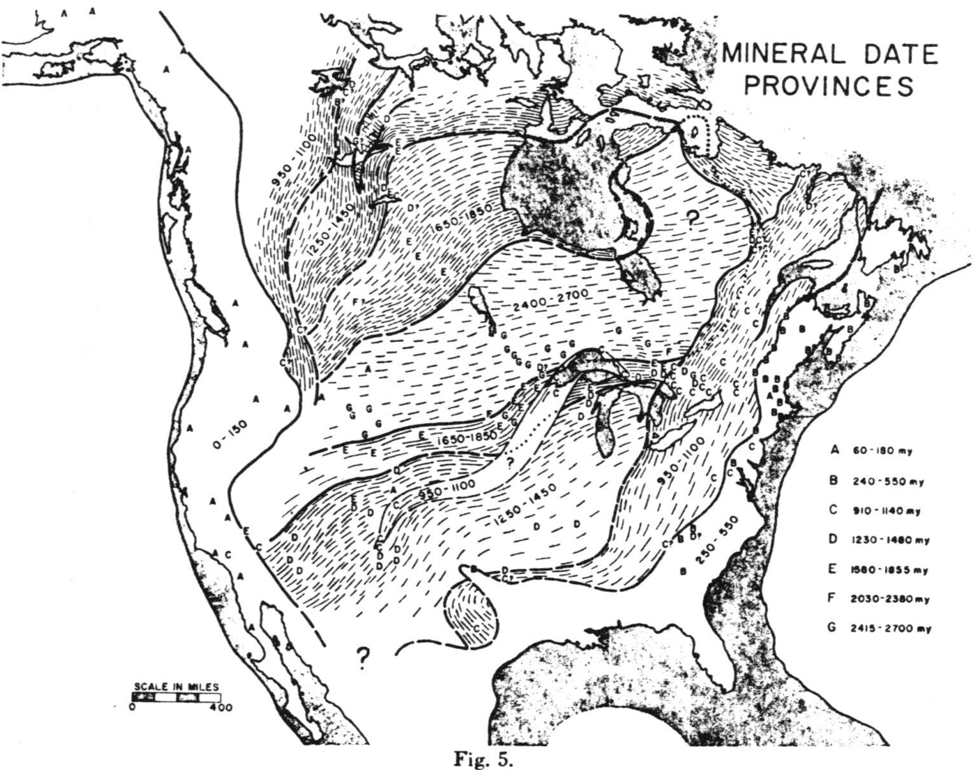

Fig. 5.

1450, 1650-1850, and 2400-2700. The complete spread of dates assigned to the respective intervals is given in the legend.

The absence of mineral dates from the Hudson Bay region, most of the Ungava Peninsula, the arctic islands, and much of south-central United States makes it impossible to present anything but a speculative working hypothesis.

Distance from continental escarpment plotted against age.—Figure 6 is a graphical demonstration of the observation made by J. Tuzo Wilson (1949) and others that the mineral age provinces of North America are progressively younger as they approach the present continental margin. In the graph the assigned age for each locality is plotted along the abscissa and its minimum distance inland from the continental escarpment is plotted parallel to the ordinate. The points define a diffused but definite trend toward greater age with greater inland distance. Part of the point scatter can be attributed to the tendency to seek out exceptionally young intrusions in older terranes and exceptionally old gneisses in younger provinces.

GEOLOGIC CONCEPTS IN THE LIGHT OF MINERAL DATE DISTRIBUTION

Episodes.—In 1958 Hurley (p. 1) wrote:

> The remarkable appearance of certain age values that occur in almost every continent has raised considerable speculation on the possible reasons behind these

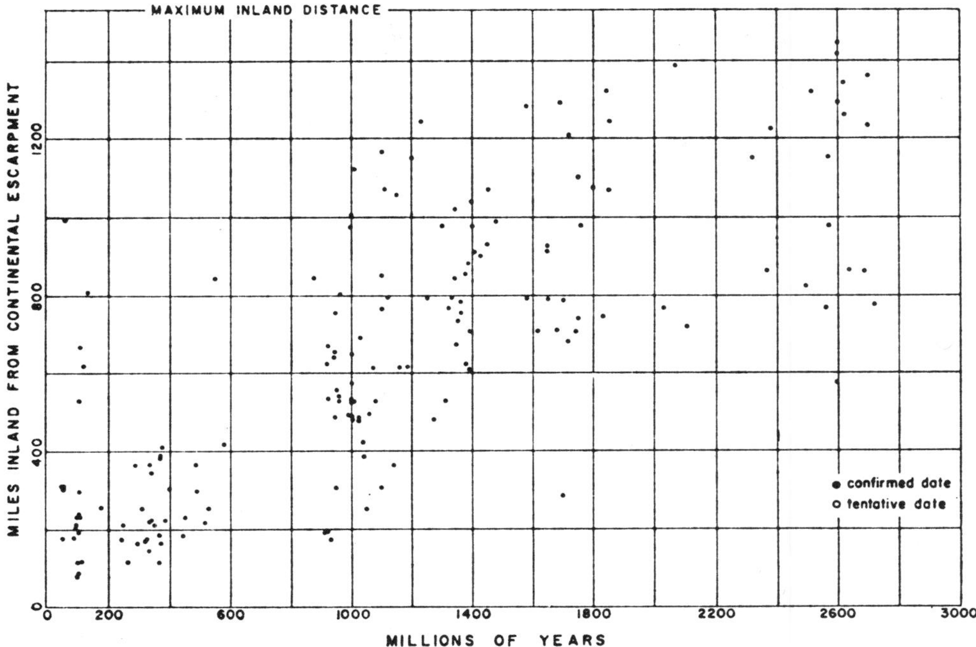

Fig. 6. Distance from continental escarpment plotted against age.

> world-wide orogenic events. Although not definitely proven to be the rule yet the evidence is becoming quite strong that these values are truly prominent points in time on a global scale.

And again (p. 2):

> The 2600 m.y. age has now appeared in Ontario, Quebec, the Northwest Territories (Yellowknife belt), Karelia, south and central Africa, India, Australia, and finally Brazil. The 1700 (± 100) m.y. is appearing in northern Manitoba, south Greenland, central Fennoscandia, the metamorphic belt involving the Huronian section of the Great Lakes region extending from the Sudbury district into Minnesota, at least parts of Labrador, India, central Africa, and the northern territories of Australia. Similar statements may be made regarding the typical Grenville age of about 1000 m.y., and the typical latest Precambrian age in the vicinity of 500 m.y.

Episodes of mineral date abundance, indicative of terminal crustal adjustment, comprise intervals of 160 to 340 m.y. (figs. 1, 2). Such age spreads are considered to be geologic rather than analytical (Fairbairn, 1958). The interval during which a given crustal belt has been unstable can, and probably does, far exceed the interval for which abundant dates have been preserved. For example, the Pacific margin batholithic belt, yielding Mesozoic ages, was already active in the Paleozoic, and the Appalachian orogenic belt, yielding predominately mid-Paleozoic ages, has its origin in the Precambrian.

Figure 1 shows the following intervals of mineral date abundance: 0 to –120 m.y., –280 to –620 m.y., –930 to –1100 m.y., –1300 to –1480 m.y., –1650 to –1860 m.y., –2060 to –2220 m.y., –2490 to –2710 m.y., and –2940 + m.y. These intervals are 120+, 340, 170, 180, 210, 160, 220, and 60+ million years in length and together comprise 45 percent of the past 3 billion years. Table 1 shows the percentages of mineral dates which fall into these intervals. The average duration of the complete intervals is 240 m.y. with a standard deviation of 60 m.y.

TABLE 1

The Occurrence of Mineral Dates in Intervals of Abundance

	dated localities	confirmed + tentative	dated localities	confirmed
Europe-Siberia	146	70%	62	69%
North America	177	86%	98	94%
N. A. (area compensated)	109	88%	81	94%
SE Asia-Australia	23	70%	15	73%
South America	11	91%	3	—
Africa	56	82%	39	90%
Worldwide	413	78%	218	84%
Potassium/argon	231	67%	130	76%
Rubidium/strontium	86	86%	71	90%
U-Th-Pb methods	160	81%	107	86%

Table 2 lists the peaks of mineral date abundance indicated in figures 1 and 2. Mineral date interval B (see fig. 5) is indicated by two peaks, B and B′.

TABLE 2

Peaks of Mineral Date Abundance

	A	B	B	C	D	E	F	G
Europe-Siberia	—	345	—	965	—	1800	2105	(2650)*
North America	100	365	530	1010	1370	1755	(2105)*	2605
N.A. (area compensated)	105	360	530	1005	1360	1760	(2090)	2605
SE Asia-Australia	(165)	—	510	995	(1475)	(1700)	—	(2695)
South America	—	—	475	—	(1365)	—	—	(2515)
Africa	—	—	485	950	—	(1655)	2115	(2640)
Worldwide	105	365	485	1000	1395	1785	2105	2605
Potassium/argon	105	335	470	970	1330	1790	—	2545
Rubidium/strontium	(105)	350	515	995	1390	1750	—	2630
U-Th-Pb methods	105	360	490	1005	1430	1815	2105	2645

* parentheses indicate a very small number of dates.

The average peak values (excluding those in parentheses) for the five continental areas are 105 m.y., 352 m.y. (standard deviation 12 m.y.), 500 m.y. (s.d. 20 m.y.), 980 m.y. (s.d. 25 m.y.), 1360 m.y., 1780 m.y. (s.d. 20 m.y.), 2110 m.y. (s.d. 5 m.y.), and 2605 m.y. Omitting B′, the intervals between these peaks are 395 m.y., 480 m.y., 390 m.y., 410 m.y., 332 m.y., and 493 m.y., giving an average interval of 417 m.y. (s.d. 47 m.y.). Very similar values can be obtained for each of the dating methods.

The existence of date abundance intervals does not necessarily mean that mineral growth has been episodic. These could result from an episodic redistribution in the extent of earth-wide mineralogenesis. By analogy, shoreline terraces are dated by the terminal activity of waves prior to a change in their space of activity. The waves have worked continuously, but the terrace levels that they left behind are evidences of an episodic earth history.

Cycles.—Since the intervals between peaks of date abundance average 417 m.y. the distribution of mineral dates has been compared with a hypothetical 420 m.y. cycle in which 210 m.y. intervals of mineral date abundance alternate with like intervals of mineral date scarcity. Except for the mid-Paleozoic mineralogenic interval (B′) found in the northern hemisphere, the distribution of dates agrees well with the cycles. The 210 m.y. half-cycles are indicated on figures 1 and 2 by vertical coordinates, and the intervals of greater mineral date abundance are shown as solid bars at the bottom of figure 1. Table 3 shows the percentages of dates which correspond to this cyclic pattern.

TABLE 3

The Cyclic Occurrence of Mineral Dates in Intervals of Abundance

	dated localities	confirmed tentative	dated localities	confirmed
Europe-Siberia	146	58%	62	60%
North America	177	71%	98	73%
N.A. (area compensated)	109	72%	81	73%
SE Asia-Australia	23	70%	15	67%
South America	11	82%	3	—
Africa	56	75%	39	85%
Worldwide	413	67%	218	71%
Potassium/argon	231	57%	130	59%
Rubidium/strontium	86	63%	71	61%
U-Th-Pb methods	160	74%	107	76%

By comparing the distribution of mineral dates with fixed cyclic intervals it is possible to investigate its statistical significance in the same manner as the flipping of a coin. Where p is the probability of occurrence of an event on a single trial, the Probability (P) that the event will occur at least r times in the course of n independent trials is:

$$P \geqslant r = \sum_{i\,0}^{i\,n-r} = \frac{n'}{i'(n-i)'} p^{n-i} q^{i} \qquad (\text{where } q = 1\text{-}p)$$

Considering each mineral date as a single trial, the chance of its falling into a set of alternate intervals representing half of geologic time is one in two ($p = \frac{1}{2}$). The probability (P) that 71 percent of the 218 worldwide confirmed mineral date localities should give dates falling into the same set of intervals is less than 1×10^{-12}. Similarly, the probability for confirmed North American dates is about 1×10^{-8}, for European-Siberian 1×10^{-5}, and for African 1×10^{-6}. Similar calculations can be made setting p equal to 1/3, as suggested by the short bars in the lower part of figure 1.

If intervals of mineral date abundance are cyclic, they indicate a cyclic withdrawal of large portions of the earth's crust from active igneous and metamorphic mineralogenesis, a cyclic phenomenon quite different from that treated by Stille (1936) and Umbgrove (1942). The orogenic cycles which Stille, Umbgrove, and others recognized in the post-Cambrian record are events of a much smaller magnitude than those considered in this paper. The possible existence of long cycles in the earth's orogenic history has been suggested by many authors. Veining Meinez (1934, p. 531) writes:

> It is rather probable that in this whole phenomena we have to distinguish between currents of two sizes, a smaller one responsible for the different periods of tectonic activity, and a larger one taking place over the entire depths of the mantle which, at greater intervals than the first, would bring about the periods of especially strong tectonic activity (occurring at intervals of some 150-250 m.y.).

Gill (1955), when he spoke of 20 to 30 Precambrian orogenies, must have had in mind intervals similar to those of Veining Meinez. Cummings (1955) concluded that the active life of an orogenic belt is about 500 m.y. Holmes (1948) appears to have had in mind cycles of about 300 m.y. Griggs (1939) suggested a mechanism for producing orogenic cycles of about 560 m.y. in a given area, and M.I.T. (1958b) observes that orogenic belts develop over a period of 200 or more million years.

The concepts of periodic and cyclic orogeny to which Gilluly (1949) and Rutten (1949) objected concerned relatively short intervals of orogenic activity alternating with relatively long periods of quiescence. Their objections were not in conflict with the views of periodists such as Holmes, who wrote in 1932 (p. 186): "The testimony of geology is that compression seems always to have been more or less active somewhere or other in the world, periodicity being a matter of climaxes in particular regions heralded by crescendoes and followed by diminuendoes." Gilluly (1949, p. 588) says: "Of course I do not contend that intense crustal deformation has been either continuous or ubiquitous. Doubtless there have been variations in the rates from the statistical average, and perhaps present conditions deviate from the average in one direction or the other."

The concept of mineral date provinces and their meaning.—Following Ellsworth's (1932) pioneer use of mineral dates to distinguish the Grenville and Keewatin provinces, a succession of writers has attached characterizing mineral ages to "orogenic belts." In 1945 Holmes compared supposed intervals of North American orogeny with dates to the orogenic belts of south and central Africa and compared this succession of dates with supposed successions of orogeny in North America, Europe, and India. In 1949 J. Tuzo Wilson began using the age subdivision of the Canadian shield as evidence of "continental accretion." During these same years the existence of a Pacific batholithic belt with mineral ages of –100 ± 30 m.y. and an Appalachian belt of –240 to –500 m.y., long suspected by field geologists, were confirmed by mineral dating.

In 1955 and again in 1957 Holmes and Cahen proposed a sequence of orogenic intervals for Africa. In 1955 Holmes expanded on his earlier time and space subdivision of India and Ceylon according to belts of Precambrian orogeny. In the Symposium on Precambrian correlation and dating held by the Geological Association of Canada in 1955 Derry, Gill, Cummings, and others amplified the mineral age subdivision of the Canadian shield.

In 1956 Tugarinov used mineral dates to subdivide the various Precambrian shields of the world according to successive belts of mountain building. In 1958 Gerling characterized the Precambrian orogenic belts of the Soviet Union by mineral dates, Goldich reported mineralogenic epochs for the Lake Superior region, Davis recognized an areal distribution of age "belts" in North America, and Hurley and Fairbairn reported the wordwide occurrence of "suspiciously similar dates."

Thus, in the past three decades, in each of the continents, clusters of mineral ages of areal significance and perhaps worldwide coincidence have been found. In 1956 J. Tuzo Wilson wrote (p. 550): "From the results now attainable it is quite apparent that the ages and lead isotope ratios found for any one area tend to be uniform but that they differ markedly from one geologic province to another. This regularity is much greater than had been anticipated . . . "

The picture, however, is not a simple one, for just as Tertiary intrusions are found well within the continent, rocks a billion years old intrude still older Precambrian terranes of Colorado, Minnesota, and Ontario. In the opposite regard each of the mineralogenic provinces preserves within it at least a few crystals which retain the ages of still more ancient mineralogenic events. Such examples of igneous rejuvenation and relicts from past epochs do not invalidate the concept of dated mineralogenic provinces, but they do indicate that old sialic crust is not immutably stable and that many orogenic belts are founded upon still older continental rocks.

Recently mineral ages of a billion years and more have been discovered on the coast of Labrador (Lang, 1958), in eastern Newfoundland (M.I.T., 1958a), in New Jersey (Davis, 1958b), near Baltimore (Tilton et al., 1958b), and in southern California (Larsen, see Faul, 1954; Davis, 1958a, Wasserburg, 1959) each within a relatively short distance of the present continental margin. This evidence is in direct conflict with the hypothesis of *continental accretion* (Wilson, 1949, 1950, 1956). It is not in conflict, however, if we attribute the concentric arrangement of mineralogenic provinces to a gradual withdrawal of crustal instability from the continent's interior. By this explanation the continent has not grown, but the portion of it involved in mineralogenesis has gradually decreased.

The distribution of mineral dates in other continents not only fails to support the hypothesis of accretion, but shows little evidence of an orderly withdrawal of instability from the continental centers. In Eurasia Precambrian terranes border both the Arctic and Indian oceans while mineral provinces of Paleozoic to Tertiary age extend through the heart of the continent. The mineral date map of Australia (M.I.T., 1958b, p. 94) locates the oldest dates near the ocean and younger dates in the continent's interior. The mineral date map of Africa (Holmes, see M.I.T. 1958b, p. 118) shows little relation between age and interior distance. And in South America Precambrian terranes extend to the coast in several places.

Orogenic rejuvenation.—Have continental areas, once stable for hundreds of millions of years, remobilized to participate in widespread magmatization and metamorphism? Areas with complex histories, such as Yellowknife (Henderson, 1955), Lake Superior (Goldich, 1958), Sudbury (M.I.T., 1958a), the Appalachian Piedmont (Tilton, et al., 1958a, b; Grunenfelder, 1958), Kouvo, Finland (Davis, 1958), the southern Ukraine (Vinogradov, 1956), and Karelia (Gerling, 1957) retain mineral dates from two or more widely separated times. Either these areas underwent periodic rejuvenation or significant portions of their mineralogenic record have yet to be discovered. For example, the Grenville orogenic province became stable about 1000 m.y. ago, and the adjacent Appalachian orogenic province 250 to 300 m.y. ago. Since the Appalachian province also participated in the Grenville orogeny, did it also stabilize 1000 m.y. ago and later remobilize, or was it continuously unstable until late Paleozoic time? If the area was continuously active it is difficult to explain the preservation of –900 to –1100 m.y. dates, considering the almost total absence of –550 m.y. to –900 m.y. dates.

If continents have neither grown nor become progressively stable toward their margins, either the picture is confused by the fragmentation of continents, or belts of mobility can trend across all portions of the globe, irrespective of the boundaries between continents and oceans.

CONCLUSIONS

Igneous and metamorphic mineral dates from all parts of the world tend to occur in the same intervals of cyclic abundance. This suggests that the earth's history is marked by episodes during which the spatial distribution of intense mineralogenic activity shifts, resulting in the selective preservation of "withdrawal dates" for the areas becoming stable.

The spatial distribution of mineral dates in North America defines a sequence of age provinces which are younger nearer the continental margin but does not support the hypothesis of continental accretion. An hypothesis envisioning the gradual outward solidification of an originally mobile continent is more satisfactory but is not supported by the spatial distribution of mineral dates on other continents and does not explain repeated, widespread rejuvenation of stable areas.

[*Editor's Note:* The appendix of mineral date localities on pages 17 to 29 has been omitted.]

References

1 Ahrens, L. H., 1955a, The convergent Pb ages of the oldest monazites and uraninites: Geochim. et Cosmochim. Acta, v. 7, p. 294-300.

2 ———, 1955b, Implications of the Rhodesia age pattern: Geochim. et Cosmochim. Acta, v. 8, p. 1-15.

3 ———, 1956, Radioactive methods for determining geological age: Reports on Progress in Physics, v. 19, p. 80-106.

4 Aldrich, L. T., 1956, Measurement of radioactive ages of rocks: Science, v. 123, p. 871-875.

5 ———, 1958, Mineral age measurements in metamorphic zones of Iron and Dickinson counties, Michigan: Geol. Soc. America Bull., v. 69, p. 1528.

6 Aldrich, L. T., Davis, G. L., Graham, J. W., Nicolaysen, L. O., and Tilton, G. R., 1954, after Faul, H., Nuclear Geology: New York, John Wiley, p. 262-281.

7 Aldrich, L. T., Tilton, G. R., Davis, G. L., Nicolaysen, L. O. and Patterson, C. C., 1955, Comparison of U-Pb, Pb-Pb and Rb-Sr ages of Precambrian minerals: Geol. Assoc. Canada Proc., v. 7, p. 7-13.

8 Aldrich, L. T., Davis, G. L., Tilton, G. R., and Wetherill, G. W., 1956a, Radioactive ages of minerals from the Brown Derby Mine and the Quartz Creek Granite near Gunnison, Colorado: Geophys. Res. Jour., v. 61, p. 215-232.

9 Aldrich, L. T., Davis, G. L., Tilton, G. R., Wetherill, G. W. and Jeffrey, P. M., 1956b. Evaluation of mineral age measurements I: Nat. Acad. Sci., Nat. Res. Council, Pub. 400, p. 147-150 (Part II, p. 151-156).

10 Aldrich, L. T., Wetherill, G. W., and Davis, G. L., 1956c, Determination of radiogenic Sr^{87} and Rb^{87} of an interlaboratory series of lepidolites: Geochim. et Cosmochim. Acta, v. 10, p. 238-240.

11 Aldrich, L. T., Wetherill, G. W., Tilton, G. R., and Davis, G. L., 1956d, Halflife of Rb^{87}: Phys. Rev., v. 103, p. 1045-1047.

12 Aldrich, L. T., Wetherill, G. W., and Tilton, G. R., 1958, Radioactive ages of micas from granitic rocks by Rb-Sr and K-A methods: Am. Geophys. Union Trans., v. 39, no. 6, p. 1124-1134.

13 Aswathanarayana, U., 1956, Absolute ages of the Archaean orogenic cycles of India: Am. Jour. Sci., v. 254, p. 19-31.

14 Bugge, J. A. W., 1943, Geological and petrological investigations in the Kongsberg-Bamble formation: Norges Geol. Undersok., 160, 150 p.

14a Bullwinkle, H. J., Fairbairn, H. W., Pinson, W. H., Jr., and Hurley, P. M., Age investigation of syenites from Coldwell, Ontario: Geol. Soc. America Bull., v. 69, p. 1543.

15 Baadsgaard, H., Nier, A. O. C., and Goldich, S. S., 1957, Investigations in K-A dating: Am. Geophys. Union Trans., v. 38, p. 385.

16 Baadsgaard, H., Goldich, S. S., Nier, A. O. C. and Hoffman, J. H., 1957, The reproducibility of A^{40}/K^{40} age determinations: Am. Geophys. Union Trans., v. 38, p. 539-542.

17 Carr, D. R. and Kulp, J. L., 1957, The potassium-argon method of geochronometry: Geol. Soc. America Bull., v. 68, p. 763-784.

18 Chapman, R. W., Gottfried, D. and Waring, C. L., 1954, after Faul, H., Nuclear Geology: New York, John Wiley, p. 262-281.

19 ———, 1955, Age determinations on some rocks from the Boulder Batholith and other batholiths of western Montana: Geol. Soc. America Bull., v. 66, p. 607-610.

20 Collins, C. B., Lang, A. H., Robinson, S. C. and Farquhar, R. M., 1952, Age determinations for some uranium deposits in the Canadian Shield: Geol. Assoc. Canada Proc., v. 5, p. 15-41.

21 Collins, C. B., Farquhar, R. M. and Russell, R. D., 1954, Isotopic constitution of radiogenic leads and the measurement of geological time: Geol. Soc. America Bull., v. 65, p. 1-22.

22 Cumming, G. L., Wilson, J. Tuzo, Farquhar, R. M. and Russell, R. D., 1955, Some dates and subdivisions of the Canadian Shield: Geol. Assoc. Canada Proc., v. 7, pt. 2, p. 27-79.

23 Curtis, G. H., Lipson, J., and Evernden, J. F., 1956, Potassium-argon dating of plio-pleistocene intrusive rocks: Nature, v. 178, p. 1360.

24 Damon, P. E. and Kulp, J. L., 1957, Argon in mica and the age of the Beryl Mountain, N. H. pegmatite: Am. Jour. Sci., v. 255, p. 697-704. (Abstract in Geol. Soc. America Bull., v. 68, p. 1713).

25 Davis, G. L., and Aldrich, L. T., 1953, Determination of the age of lepidolites by the method of isotope dilution: Geol. Soc. America Bull., v. 64, p. 379-380.

25a Davis, G. L., Aldrich, L. T.. Tilton, G. R., Wetherill, G. W., and Jeffery, P. M., 1956, The age of rocks and minerals: Ann. Rept. of the Director of the Geophys. Lab., 1955-56, p. 161-168.

26 Davis, G. L., Aldrich, L. T., Tilton, G. R., Wetherill, G. W., and Faul, H., 1957, The age of rocks and minerals: Ann. Rept. of the Dir. of the Geophys. Lab., 1956-57, p. 164-171.

27 Davis, G. L., Aldrich, L. T., Tilton, G. R., and Wetherill, G. W., 1958a, The age of rocks and minerals: Ann. Rept. of the Director of the Geophys. Lab., 1957-58, p. 176-181.

28 Davis, G. L., Wetherill, G. W., Tilton, G. R., and Hopson, C. A., 1958b, Age of the Baltimore gneiss: Geol. Soc. America Bull., v. 69, p. 1550-1551.

29 De Azcona, J. M. L., 1957, Age of the uraninites of the Sierra Albarrana: Notas y Comuns, Inst. Geol. y Minero España, no. 45, p. 3-14 (Chem. Abst., v. 51, p. 14498).

30 Demay, Andre, 1953, Determination de l'age absolu d'une pechblende du gisement filonien de la Crouzille dans le massif granulitique de St. Sylvestre, au nord de Limoges: Compts Rendus, v. 237, p. 48-50.

31 Derry, D. R., 1950, Tectonic map of Canada: Geol. Assoc. Canada.

32 ————, 1955, Preface to symposium on Precambrian correlation and dating: Geol. Assoc. Canada Proc., v. 7, pt. II, p. 5-6.

33 Eberhardt, P., Geiss, J., and Houtermans, F. G., 1955, Uber Blei und Schwefelisotopenverhaltnisse in Bleiglanzen (Lead and sulfur isotope ratios in Galenas): Helv. Phys. Acta, v. 28, p. 339-341.

34 Eckelmann, F. D. and Gast, P. W., 1957, Plutonic history and absolute age of the Huron Claim-Johnston Lake area, SE Manitoba: Geol. Soc. America Bull., v. 68, p. 1720.

35 Eckelmann, W. R. and Kulp, J. L., 1956, Uranium-lead method of age determination, Part I; The Lake Athabasca problem: Geol. Soc. America Bull., v. 67, p. 35-54.

36 ————, 1957, Uranium-lead method of age determination. Part II, North American localities: Geol. Soc. America Bull., v. 68, p. 1117-1140.

37 Ehrenberg, H. Fr., 1953, Isotopenanalysen au Blei aus Mineralen: Z. Physik, v. 134, p. 317-333.

38 Ellsworth, H. V., 1932, Rare-element minerals of Canada: Canada Geol. Survey Econ. Geol. ser., no. 11, 272 p.

39 Everanden, J. F., Curtis, G. H., and Lipson, J., 1957, Potassium-argon dating of igneous rocks: Am. Assoc. Petroleum Geologists Bull., v. 41, no. 9, p. 2120-2127.

40 Fairbairn, H. W. and Hurley, P. M., 1957, Radiation damage in zircon and its relation to ages of Paleozoic igneous rocks in northern New England and adjacent Canada: Am. Geophys. Union Trans., v. 38, p. 99-107.

41 Fairbairn, H. W., 1958, Nova Scotia age program: NYO-3938, 5th Ann. Progress Report, Mass. Inst. Tech., p. 4-16.

42 Faul, H., 1954, Nuclear geology: New York, John Wiley, p. 262-281.

43 Folinsbee, R. E., Lipson, J. and Reynolds, J. H., 1956, Potassium-argon dating: Geochim. et Cosmochim. Acta, v. 10, p. 60-68.

44 Folinsbee, R. E., Baadsgaard, H., and Lipson, J. I., 1958, Potassium-argon age dating of sediments: Am. Assoc. Petroleum Geologists, Ann. Meeting, Los Angeles, California, March 10.

45 Gast, P. W. and Long, L. E., 1957, Absolute age determinations from the basement rocks of the Beartooth Mountains and Bighorn Mountains: Geol. Soc. America Bull., v. 68, p. 1732-1733.

46 Gast, P. W., Kulp, J. L., Long, L. E., 1958, Absolute age of early Precambrian rocks in the Bighorn Basin of Wyoming and Montana, and southeastern Manitoba: Am. Geophys. Union Trans., v. 39, no. 2, p. 322-334.

47 Gentner, W. and Klev, W., 1957, Argonbestimmungen an Kaliummineralien, IV, Die Frage der Argonverluste in Kalifeldspaten und Glimmermineralian: Geochim. et Cosmochim., Acta, v. 12, p. 323-329.

48 Gerling, E. K., 1956, Argon method of age determination and its use for differentiation of Precambrian formations of the Baltic and Ukrainian shields: Geochimica, no. 5, p. 30-42.

48a Gerling, E. K., Yashchenko, M. L., and Ermoulin, G. M., 1957, The argon method of age determination and its use: Byull. Komissi Opredelen, Absolyut, Vozrasta Geol. Formatsii, Akad. Nauk S.S.S.R., Otdel. Geol.-Geograf., Nauk, 1957, no. 2, p. 8-27 (Chem. Abst., v. 51, p. 17661).

49 Giletti, B. J. and Kulp, J. L., 1955, Radon leakage from radioactive minerals: Am. Mineralogist, v. 40, p. 481-496.

50 Gill, J. E., 1955, Precambrian history of the Canadian shield with notes on correlation and nomenclature: Geol. Assoc. Canada Proc., v. 7, pt. 2, p. 117-124.

51 Gilluly, J., 1949, Distribution of mountain building in geologic time: Geol. Soc. America Bull., v. 60, p. 561-590.

52 Gheith, M. A., 1958a, Tabulated index and bibliography of published age measurements of North America: NYO-3938, 5th Ann. Progress Rept., Mass. Inst. Tech., p. 167-207.

53 ————, 1958b, Tabulated index and bibliography of world age measurements exclusive of North American localities: NYO-3939, 6th Ann. Progress Rept., Mass. Inst. Tech., p. 132-191.

54 Goldich, S. S., Baadsgaard, H. and Nier, A. O., 1957a, Investigation in A^{40}/K^{40} dating: Am. Geophys. Union Trans., v. 38, p. 547-551.

55 ————, 1957b, A^{40}/K^{40} Dating of rocks of the L. Superior region: Am. Geophys. Union Trans., v. 38, p. 392.

56 Goldich, S. S., Nier, A. O., Krueger, H. W., and Hoffman, J. H., 1958, K-A Dating of Precambrian iron formations: Am. Geophys. Union Trans., v. 39, no. 3, p. 516.

56a Goldich, S. S., Nier, A. O., and Washburn, A. L., 1958, A^{40}/K^{40} Age of gneiss from McMurdo Sound, Antarctica: Am. Geophys. Union Trans., v. 39, no. 5, p. 506-517.

57 Grace, J. N. A., 1940, Occurrence of xenotime in Western Australia: Roy. Soc. W. Australia Jour., v. 15, p. 95-98.

57a Griggs, D., 1939, A theory of mountain building: AM. JOUR. SCI., v. 237, p. 609-660.

57b Grunenfelder, M., and Silver, L. T., 1958, Radioactive age dating and its petrologic implications for some Georgia granites: Geol. Soc. America Bull., v. 69, p. 1574.

58 Hamilton, W. B., 1956, Precambrian rocks of the Wichita and the Arbuckle Mountains, Oklahoma: Geol. Soc. America Bull., v. 67, p. 1319-1330.

59 Hee, A., Coche, A., Jarovoy, M., and Kraemer, R., 1957, The age of two granites of the Vosges Mtns.: Acad. Sci. Nat., Comptes Rendus, v. 244, no. 7, p. 922-923.

60 Henderson, J. F., 1955, Discussion, Symposium on Precambrian correlation and dating: Geol. Assoc. Canada Proc., v. 7, pt. 2, p. 132-133.

61 Herzog, L. F., 1956, Rb-Sr and K-Ca analyses and ages: Nat. Acad. Sci., Nat. Res. Council, pub. 400, p. 114-131.

61a Holland, H. D., 1956, Radiation damage and age measurements in zircons: Nat. Acad. Sci., Nat. Res. Council, pub. 400, p. 85-89.

62 Holmes, A., and Lawson, R. W., 1927, Factors involved in the calculation of the ages of radioactive minerals: AM. JOUR. SCI., v. 13, no. 76, p. 327-344.

62a Holmes, A., 1932, The thermal history of the earth: Wash. Acad. Sci. Jour., v. 23, p. 169-195.

63 ————, 1945, Principles of physical geology: New York, Ronald Press Co.

64 ————, 1948, The sequence of Precambrian orogenic belts in South and Central Africa: 18th Internat. Geol. Cong., Great Britain, Rept. 14, p. 254-269.

65 Holmes, A., Leland, W. T. and Nier, A. O., 1950, Age of uraninite from a pegmatite near Singar, Gaya district, India: Am. Mineralogist, v. 35, p. 19-28.

66 Holmes, A. and Besairie, H., 1954, Sur quelques measures de geochronologie a Madagascar: Acad. Sci. Nat., Comptes Rendus, v. 238, p. 758-760.

67 Holmes, A., 1955a, African geochronology: Colonial Geol. and Mineral Resources, v. 5, no. 1, p. 3-38.

68 ————, 1955b, Dating the Precambrian of Peninsular India and Ceylon: Geol. Assoc. Canada Proc., v. 7, part II, p. 81-106.

69 Holmes, A., Shillibeer, H. A., and Wilson, J. Tuzo, 1955, Potassium-argon ages of some Lewisian and Fennoscandian pegmatites: Nature, v. 176, no. 4478, p. 390-392.

70 Holmes, A. and Cahen, L., 1957, Academie Royale des Sciences Coloniales, classe des sciences naturelles et medicales, Memoires in-8, Nouvelle serie, tome 5, fasc. 1, 169 p. (Contains in table form all the ages reported in 67).

71 Hurley, P. M., Larsen, E. E., Jr., and Gottfried, D., 1956, Comparison of Radiogenic Helium and Lead in zircon: Geochim. et Cosmochim., Acta, v. 9, p. 98-102.

72 Hurley, P. M., Pinson, W. H., Jr., Fairbairn, H. W. and Cormier, R. F., 1957, Comparison of A^{40}/K^{40} and Sr^{87}/Rb^{87} ages on biotite: Am. Geophys. Union Trans., v. 38, p. 396.

73 Hurley, P. M., 1958, Variations in isotopic abundances of strontium, calcium, and argon and related topics: NYO-3938 5th Ann. Prog. Rept., Mass. Inst. Tech., p. 1-3.

74 Jaffe, H. W., 1955, Precambrian monazite and zircon from the Mountain Pass Rare Earth district, San Bernardino County, California: Geol. Soc. America Bull., v. 66, p. 1247-1256.

75 Jamieson, R. T., and Schreiner, G. D. L., 1957, Ages of some African lepidolites determined from the rubidium 87-strontium 87 decay: Roy. Soc. (London) Proc., B. 146, p. 257-269.

76 Jeffrey, P. M., 1956, The radioactive age of four Western Australian pegmatites by the potassium and rubidium methods: Geochim. et Cosmochim. Acta, v. 10, p. 191-195.

77 Kantor, J. and Kupco, G., 1956, The absolute age of lepidolites from Rozna Township in Moravia determined by the radiogenic Sr method: Geol. Prace Zpravy, v. 7, p. 3-12.

78 Kerr, P. F. and Kulp, J. L., 1952, Precambrian uraninite, Sunshine Mine, Idaho: Science, v. 115, p. 86-88.

79 Knopf, Adolph, 1956, Argon-potassium determination of the age of the Boulder Bathylith, Montana: AM. JOUR. SCI., v. 254, p. 744-745.

80 Komlev, L. V., Danllevich, S. I., Ivanova, K. S., Mikhalevskaya, A. D., Savonenkov, V. G., Filippov, M. S., 1957, Age of geological formations of southwest Ukranian Precambrian: Geochimiya no. 7, p. 566-572.

81 Kulp, J. L. and Eckelmann, W. R., 1955, The U-Pb age studies in the Lake Athabasca region: Am. Geophys. Union Trans., v. 36, p. 517.

82 Kulp, J. L., Bate, G. L. and Bruno, J. G., 1955, New age determinations by the lead method: Geol. Assoc. Canada Proc., v. 7, part II, p. 15-24.

83 Kulp, J. L., Long, L. E. and Eckelmann, F. D., 1957a, Age of the Piedmont and southern Appalachians: Geol. Soc. America Bull., v. 63, p. 1758-1759.

84 ————, 1957b, Discordant U-Pb ages and mineral types: Am. Mineralogist, v. 42, p. 154-164.

85 Lang, A. H., 1958, Personal communication of values obtained by the Age Determination Laboratories of the Geological Survey of Canada.

86 Larsen, E. S., Jr., Waring, C. L. and Berman, J., 1953, Zoned zircon from Oklahoma: Am. Mineralogist, v. 38, p. 1118-1125.

87 Larsen, E. S., Jr., Gottfried, D., Jaffe, H. W., and Waring, C. L., 1958, Lead-alpha ages of the Mesozoic batholiths of western North America, U. S. Geol. Survey Bull., 1070-B (Also see 42).

88 Lipson, J., 1958, Potassium-argon dating of sedimentary rocks: Geol. Soc. America Bull., v. 69, p. 137-150.

89 Lyons, J. B., Jaffe, H. W., Gottfried, D., and Waring, C. L., 1957, Lead-alpha ages of some New Hampshire granites: AM. JOUR. SCI., v. 255, p. 527-546.

90 Mahadevan, C. and Aswathanarayana, U., 1955, Age levels of Archaean structural provinces: Current Science, v. 24, p. 73-74.

91 Marble, J. P., 1954, Recent analyses of Brazilian radioactive minerals: Rept. Com. Measurement of Geol. Time 1952-53, Nat. Res. Council, Washington, Pub. 319, p. 143-153.

92 Mass. Inst. Tech., Dept. of Geology and Geophysics, 1958a, Variations in isotopic abundances of strontium, calcium, and argon, and related topics: NYO-3938 5th Ann. Progress Rept.

93 ————, 1958b, Variations in isotopic abundances of strontium, calcium, and argon, and related topics: NYO-3939 6th Ann. Progress Rept.

94 Matzko, J. J., Jaffe, H. W., and Waring, C. L., 1958, Lead-alpha age determinations of granitic rocks from Alaska: AM. JOUR. SCI., v. 256, no. 8, p. 529-539.

95 Nicolaysen, L. O., Aldrich, L. T., and Dosk, J. B., 1953, Age measurements on African micas by Sr/Rb method: Am. Geophys. Union Trans., v. 34, p. 342-343.

96 Nicolaysen, L. O., 1957, Solid diffusion in radioactive minerals and the measurement of absolute age: Geochim. et Cosmochim. Acta, v. 11, p. 41-59.

97 Nier, A. O., 1939, The isotopic constitution of radiogenic leads and the measurement of geological time, II: Phys. Rev., v. 55, p. 153-163.

98 Nier, A. O., Thompson, R. W., and Murphey, B. F., 1941, The isotopic constitution of lead and the measurement of geological time: III: Phys. Rev., v. 60, p. 112-116.

99 Parwel, A., and Wickman, F. E., 1954, Några preliminära resultat av åldersbestämningar på svenska pegmatitmineral: Geol. Fören. Stockholm, Förh., v. 76, p. 353-354.

100 Phair, G., After Faul, H., 1954, Nuclear geology: New York, John Wiley, p. 262-281.

101 Pinson, W. H., Jr., Fairbairn, H. W., and Cormier, R. F., 1958, Sr/Rb age measurements on hornblende and feldspar, and the age of syenite at Chicoutimi, Quebec, Canada: Geol. Soc. America Bull., v. 69, p. 590-602.

102 Polavaya, N. I., 1956, Absolute age of some magmatic rocks of the USSR according to data of the argon method: Geochimiya no. 5, p. 43-53.

103 Polkanov, A. A., and Gerling, E. K., 1958, K-A and Sr-Rb methods and age of Precambrian of USSR: Am. Geophys. Union Trans., v. 39, no. 4, p. 713-715.

104 Reynolds, J. H., 1956, Comparative study of argon content and argon diffusion in mica and feldspar: Geochim. et Cosmochim., Acta, v. 12, p. 177-184.

105 Riley, George C., Jr., 1956, The geology of the Cumberland Sound region, Baffin Is.: McGill Univ., unpubl. doctoral thesis.

106 Rodgers, John, 1952, Absolute ages of radioactive minerals from the Appalachian region: AM. JOUR. SCI., v. 250, p. 411-427 (Abstract in Geol. Soc. America Bull., v. 62, p. 1561).

107 Russell, R. D., Shillibeer, H. A., Farquhar, R. M. and Mousuf, A. K., 1953, Branching ratio of K^{40}: Phys. Rev., v. 91, p. 1223-1224.

108 Rutten, L. M. R., 1949, Frequency and periodicity of orogenic movements: Geol. Soc. America Bull., v. 60, p. 1755-1770.

109 Schürmann, H. M., Bot, A. C., Steenama, J. J., Suringa, R., Eberhardt, P., Geiss, J., von Gunter, H. R., Houtermans, F. G. and Signer, P., 1956, Age determinations of magmatic rocks by means of radioactivity (in English): Geol. en Mijnbouw, v. 18, p. 312-330.

110 Silver, L. T., Stehli, F. G., and Allen, C. R., 1956, Lower Cretaceous pre-batholithic rocks of northern Baja California, Mexico: Resumenes de los Trabajos Presentados XX Congresso Geologico Internacional, p. 30.

111 Starik, I. E., 1957, Trudy IV, Komissiipo opredelenlyu absolutnoyo vosrasta geol. formazii: Akad. Nauk. SSSR, Otdel. Geol.-Geograt. Nauk., no. 2, p. 6-7.

112 Stieff, L. R. and Stern, T. W., 1952, Identification and lead-uranium ages of massive uraninities from the Shinarump conglomerate, Utah: Science, v. 115, p. 706-708.

113 Stieff, L. R., Stern, T. W. and Milkey, R. G., 1953, A preliminary determination of the age of some uranium ores of the Colorado plateau by the Pb-U method: U. S. Geol. Surv. Circ. 271, 19 p. and some data atfer Faul, H., 1954, Nuclear Geology, p. 262-281.

114 Stieff, L. R., Stern, T. W., Cialella, C. M., and Warr, J. J., 1956, Preliminary age determinations of some uranium ores from the Blind River area, Algoma district, Ontario, Canada: Geol. Soc. America Bull., v. 67, p. 1736.

115 Stevens, J. R., and Shillibeer, H. A., 1956, Loss of argon from minerals and rocks due to crushing: Geol. Assoc. Canada Proc., v. 8, p. 71-76.

116 Stille, H., 1936, The present tectonic state of the earth: Am. Assoc. Petroleum Geologists Bull., v. 20, p. 849-880.

117 Studienkova, A. V., and Knorre, K. G., 1957, Age of granites of the northern Caucasus: Geokhimiya, no. 7, p. 573-579.

118 Tilton, G. R., 1956, The interpretation of lead age discrepancies by acid washing experiments: Am. Geophys. Union Trans., v. 37, p. 224-230. (Same data in Nat. Acad. Sci., Nat. Res. Council, Pub. 400, p. 79-84).

119 Tilton, G. R. and Aldrich, L. T., 1955, The reliability of zircons as age indicators: Am. Geophys. Union Trans., v. 36, p. 531.

120 Tilton, G. R., Davis, G. L., Wetherill, G. W. and Aldrich, L. T., 1957a, Isotopic ages of zircon from granites and pegmatites, Am. Geophys. Union Trans., v. 38, p. 360-371.

121 Tilton, G. R. and Nicolaysen, L. O., 1957b, The use of monazites for age determination: Geochim. et Cosmochim., Acta, v. 11, p. 28-40.

122 Tilton, G. R., Davis, G. L., and Wetherill, G. W., 1958a, The age of the Baltimore gneiss: Am. Geophys. Union Trans., v. 39, no. 3, p. 534.

122a Tilton, G. R., Wetherill, G. W., Davis, G. L., and Hopson, C. A., 1958b, Ages of minerals from the Baltimore gneiss near Baltimore, Maryland: Geol. Soc. America Bull., v. 69, p. 1469-1474.

123 Tomlinson, R. H. and Das Gupta, A. K., 1953, The use of isotope dilution in determination of geologic age of minerals: Can. Jour. Chemistry, v. 31, p. 909-914.

124 Tugarinov, A. I., 1956, Epochs of mineral formation in the Precambrian: Izvest. Akad. Nauk. SSSR. Ser. Geol. no. 9, p. 3-26.

125 Umbgrove, J. H. F., 1942, Rhythm and synchronism of tectonic movement: Am. Jour. Sci., v. 248, p. 521-526.

126 Veining-Meinez, F. A., 1951, Convection currents in the mantle: Am. Geophys. Union Trans., v. 32, no. 4, p. 527-528.

127 Vinogradov, A. P., 1956, Comparison of data on rock ages obtained by different methods, and geological conclusions: Geokhimiya 1956, no. 5, p. 3-17.

128 Vinogradov, A. P., Tugarinov, A. I., Silov, S. I., and Fedorova, V. A., 1957, Age determinations of granitoides of the Ukraine: Geokhimiya, no. 7, p. 559-565.

129 Wasserburg, G. J. and Hayden, R. J., 1954, The branching ratio of K^{40}: Phys. Rev., v. 93, p. 645. (Also some data after Faul. H., 1954, Nuclear Geology, Wiley, p. 262-281.)

130 Wasserburg, G. J., Hayden, R. J. and Jensen, K. J., 1956, A^{40}-K^{40} dating of igneous rocks and sediments: Geochim. et Cosmochim., Acta, v. 10, p. 153-165. (Part of same data also found in Nat. Acad. Sci., Nat. Res. Council Pub. 400, p. 131-134.)

130a Wasserburg, G. J., Wetherill, G. W., and Wright, L. A., 1959, Ages in the Precambrian of Death Valley, California: Program 40th Ann. Meeting, Am. Geophys. Union, p. 40-41.

131 Webber, G. R., Hurley, P. M. and Fairbairn, H. W., 1956, Relative ages of eastern Massachusetts granites, by total lead ratios in zircon: Am. Jour. Sci., v. 254, p. 574-583.

132 Webster, R. K., Morgan, J. W. and Smales, A. A., 1957, Some recent Harwell analytical work on geochronology: Am. Geophys. Union Trans., v. 38, p. 543-546.

133 Wetherill, G. W., 1957, Radioactivity of potassium and geologic time: Science, v. 126, p. 545-549.

134 Wetherill, G. W., Aldrich, L. T. and Tilton, G. R., 1956, Comparisons of K-A ages with concordant U-Pb ages of pegmatites: Am. Geophys. Union Trans., v. 37, p. 362.

135 Wetherill, G. W., Wasserburg, G. J., Aldrich, L. T., Tilton, G. R. and Hayden, R. J., 1956, Decay constants of K^{40} as determined by radiogenic argon content of potassium minerals: Phys. Rev., v. 103, p. 987-989.

136 Wetherill, G. W., Tilton, G. R., Davis, G. L. and Aldrich, L. T., 1956, Evaluation of mineral age measurements II: Nat. Acad. Sci., Nat. Res. Council, Pub. 400, p. 151-156.

137 Wilson, J. Tuzo, 1949, The origin of continents and Precambrian history: Royal Soc. Canada Trans., v. 43, ser. 3, p. 157-182.

138 ————, 1951, On the growth of continents: Roy. Soc. Tasmania Pap. and Proc., p. 85-111.

139 Wilson, J. Tuzo, Farquhar, R. M., Gretener, P., Russell, R. D. and Shillibeer, H. A., 1954, Ages of some African minerals: Nature, v. 174, p. 1006-1007.

140 Wilson, J. Tuzo, Russell, R. D., Farquhar, R. M., 1956, Economic significance of basement subdivision and structures in Canada: Canadian Inst. Min., Met. Bull., v. 49, p. 550-558.

142 Work Presented at the Annual Meeting of the Geol. Soc. America, Nov. 1957, Atlantic City, N. J. by Miller, D. S., and Gast, P. W. (see Geol. Soc. America Bull., v. 68, p. 1767-1768) and Kulp, J. L., et al. (1957a), but not reported in published abstracts. Data reported by Gheith, M. A. (1958b).

143 Zhirov, K. K., Zikov, S. I., Zhivova, V. V., and Stupnikova, N. I., 1957, Effect of processes of hydrothermal alteration on the determination of age of radioactive minerals: Geokhimiya, no. 8, p. 657-665.

144 Zykov, S. I., and Stupnikova, N. I., 1956, Determination of the age of the pegmatite vein of Koit-Tundra from cyrtolite, allanite and uraninite: Geokhimiya 1956, no. 8, p. 35-38.

2

Reprinted from pages 3–7 of *Phys. and Chem. Earth* **7**:1–114 (1966)

OROGENIC FOLD-BELTS AND A HYPOTHESIS OF EARTH EVOLUTION

By R. DEARNLEY

Institute of Geological Sciences, South Kensington, London, S.W.7

INTRODUCTION

Adequate information has only recently become available on which to base a preliminary consideration of dated Precambrian fold-belts in relation to the theory of continental drift. The objects of this paper are to examine some features of the distribution and history of these fold-belts and secondly, since these major tectonic features may be regarded as a direct crustal response to the underlying activity in the mantle, to attempt to deduce from them the possible patterns of this activity.

Prior to the availability of comparatively widespread and relatively abundant age determinations, the almost complete impossibility of correlation within the Precambrian precluded any consideration of this portion of the geological record in the study of geotectonics, and consequently little or no account has been taken of the Precambrian in the continental drift reconstructions which have been made by WEGENER (1924), TAYLOR (1928), DU TOIT (1937), CAREY (1958), KING (1958) and others. These reconstructions, therefore, although based upon a large body of data from independent fields, rest almost entirely on the evidence of the last 600 m.y. out of a total "readable" geological time span in places greater than 3000 m.y. Thus, the application of this highly critical data, in so far as it has not been considered in previous continental drift theories or reconstructions, may be regarded as a completely independent check on the general concept of continental drift as presently developed (see DEARNLEY, 1965a, b).

Various different lines of evidence, some independent, some complementary, bear upon the problems of continental drift. At the present state of knowledge, the evidence derived from some of these different approaches clearly carries more weight than others when used as a basis for precise continental reconstructions.

Palaeomagnetic evidence has supplied a general confirmation of the original continental reconstructions of Wegener and du Toit, and although in future this method promises to supply the most detailed information for precise testing of the reconstructions derived by other methods, at the present time some of the continental relationships suggested by the palaeomagnetic results cannot be reconciled with the geological and tectonic data. Palaeoclimatic evidence supplies a general check on other methods, but could not be used in isolation to give more than a very approximate continental reconstruction, and this applies also to palaeontological methods involving faunal provinces. In order of reliability and degree of precision the different methods may be used as follows: (1) structural evidence mainly based on orogenic geosynclinal fold-belts; (2) detailed matching of the morphology of the boundaries of the continental crust as constituted by the outer limits of the continental shelves; (3) palaeomagnetic evidence; and (4) palaeoclimatological and palaeontological evidence.

Where palaeomagnetic determinations presently conflict with the structural evidence of dated fold-belts there is no doubt that the latter evidence should be given the most weight. Similarly, palaeomagnetic evidence is generally more reliable than that derived from palaeoclimatic or palaeontological data.

Investigation of the distribution of dated geosynclinal fold-belts is the only method of geotectonic analysis that can be used appreciably far back into the Precambrian, and since this evidence must be directly related to the former deep-seated activity in the mantle, the age and distribution patterns of the fold-belts may be used to examine the possible fundamental mechanisms of Earth evolution.

PRECAMBRIAN CHRONOLOGY AND CLASSIFICATION

The age of the Earth has been calculated as 4550 m.y. by MASUDA (1958), and as 4530 ± 30 m.y. and 4540 ± 20 m.y. by OSTIC *et al.* (1963). Corresponding meteorite ages of 4500–4600 m.y. have been determined by the lead and rubidium–strontium methods by PATTERSON (1955, 1956), HERZOG and PINSON (1956), and by STARIK and SOBOTOVICH (1961).

It now seems that a figure of 4550 m.y. is as reliable as can be obtained for the "age of the Earth", and the oldest known crustal rocks extend to about 3500 m.y. (POLKANOV and GERLING, 1960; CATANZARO, 1963; SLAWSON *et al.*, 1963), which represents the lower limit of the known Precambrian.

The broad terms "Archaean" and "Proterozoic" have lost their original meanings, and the previously termed Proterozoic of one area may be the chronological equivalent of the Archaean of another. Hence these terms, if they are to have any usefulness, should be redefined on a world-wide basis. Such a classification has been proposed by VINOGRADOV and TUGARINOV (1961a, b) and, as pointed out by SUTTON (1963), this scheme is supported by the range of dates within different Precambrian provinces throughout the world (Table 1). Broadly these major geochronological divisions were given as: Katarchaean 3600 ± 200 to 2700 ± 150; Archaean 2700 ± 150 to 1900 ± 100; Lower Proterozoic 1900 ± 100 to 1100 ± 100; Upper Proterozoic 1100 ± 100 to 600 ± 50; and Phanerozoic 600 ± 50 to the present.

Two recent studies which contribute much to the understanding of the Precambrian have been made by RUNCORN (1962b, c) and SUTTON (1963). Runcorn has outlined a theory of continental evolution in which the gradual growth of the Earth's core, in a manner previously suggested by UREY (1952), has caused periodic changes in the convection cell pattern excited at marginal stability in the mantle. These changes to systems characterized by harmonics of progressively higher degree were calculated to have occurred at about 2600 m.y., 1800 m.y., 1100 m.y. and 200 m.y., dates which coincide with some of the world-wide peaks of igneous and metamorphic age determinations recognized by GASTIL

Table 1. *Geochronological Classifications*

Age, m.y.	Vinogradov and Tugarinov (1961)	Sutton (1963)	Runcorn (1963)	Proposed classification: Regime		Proposed classification: Convection Mode
0			$n = 5$		Late	n=4 (4 cell)
	Phanerozoic		200			
500		Grenville	$n = 4$	Grenville		$n = 4$
	600 ± 50			regime	Early	(8 cell)
	Upper Proterozoic	900 ± 100				
1000	1100 ± 100		*c.* 1100	1075 ± 50		
		Svecofennid				
1500	Lower Proterozoic		$n = 3$	Hudsonian regime		$n = 3$?
		1800 ± 100	*c.* 1800			
	1900 ± 100			1950 ± 50		
2000			$n = 2$			
		Shamvaiian				
2500	Archaean			Superior regime		
	2700 ± 150		*c.* 2600			$n = 2$
3000		2800 ± 100	$n = 1$			
	Katarchaean					
		Kola		— — —?— — —		— —?— —
3500						$n = 1$?
4000						
4500						

(1960). Sutton (1963) has suggested that the periods 2900–2700 m.y., 1900–1700 m.y. and 1000–800 m.y. mark the beginnings of periods of evolution of three major structural provinces ("chelogenic" provinces), each one comprising a number of essentially parallel or sub-parallel fold-belts which truncate the older provinces and which in turn are cut by the succeeding major cluster of fold-belts.

The variation of metamorphic and plutonic activity throughout geological time was studied by Gastil (1960), and a number of broad peaks of activity were noted at about 2600 m.y., 2100 m.y., 1780 m.y., 1360 m.y., 980 m.y., 500 m.y., 352 m.y. and 105 m.y. on the basis ofabout 400 age determinations. Since then, however, many more age determinations have become available and the graph of Fig. 1 shows the results of plotting just over 3400 determinations taken from

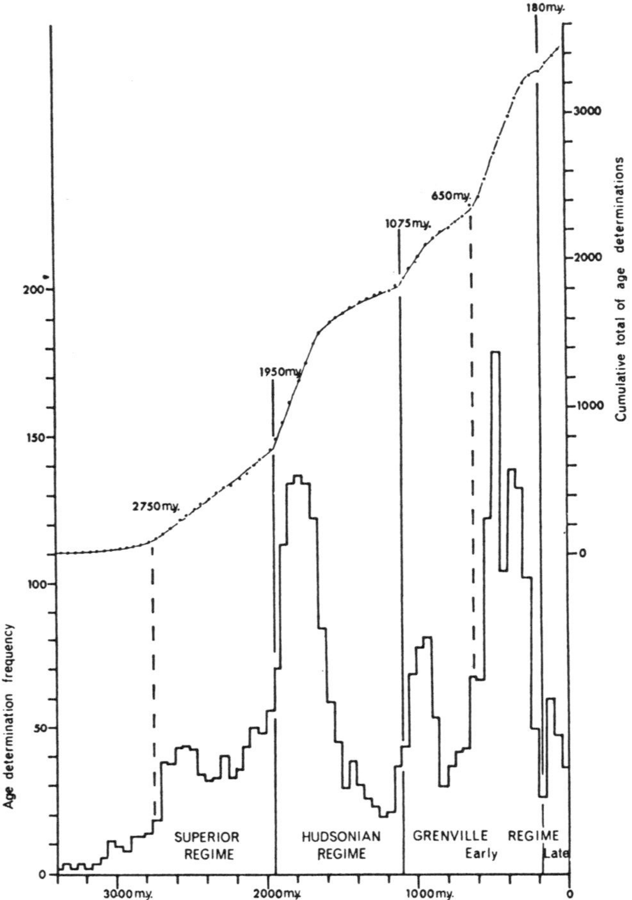

FIG. 1. Frequency histogram and cumulative curve of igneous and metamorphic age determinations plotted against geological time, showing division into Superior, Hudsonian and Grenville regimes (compare with Table 1).

the following publications: *North America:* LEECH *et al.* (1963), HURLEY *et al.* (1962), DAMON and GILETTI (1961), GILETTI and GAST (1961); *Antarctica:* PICCIOTTO and COPPEZ (1963, 1964); *Africa:* HOLMES and CAHEN (1957), NICOLAYSEN (1962), ALLSOPP *et al.* (1962), BONHOMME (1962), SCHÜRMANN (1964), NOAKES (1963), CAHEN *et al.* (1964); *Europe and Asia:* FAUL (1962), GOLDENFELD (1962), JÄGER (1962), POLKANOV and GERLING (1961), NEUMANN (1960), KOUVO (1958), MAGNUSSON (1960), LONG and LAMBERT (1963), VOYTKEVICH and ANOKHINA (1961), KOMLEV *et al.* (1962), GERLING and POLKANOV

(1958), GERLING (1956), VINOGRADOV *et al.* (1960), POLEVAYA (1956), VINOGRADOV (1956), TARASOV *et al.* (1963), KRYLOV *et al.* (1963), LI *et al.* (1960), KULP and NEUMANN (1961), KRYLOV (1961), KOUVO and KULP (1961); *India:* ASWATHANARAYANA (1964a, b); *South America*: NOAKES (1963), CHOUBERT (1964), HERZ (1961); *Australia:* NOAKES (1963). This collection is by no means exhaustive, but is considered to be a large enough sample to avoid any undue overall place—or age—bias.

The major peaks on Fig. 1, derived from a 50 m.y. interval frequency distribution histogram, are only slightly different from those of GASTIL (1960) indicating the representative nature of this earlier analysis which was based upon far fewer age determinations. The cumulative curve of Fig. 1 shows three particularly well-defined changes of slope at 2750±50 m.y., 1950±50 m.y., and 1075±50 m.y. and the divergence of the cumulative curve between 1950 and 1075 m.y. from a mean curve for the remainder of the determinations is also notable. These major changes of slope correspond closely to the Katarchaean–Archaean (2700±150 m.y.), Archaean–Lower Proterozoic (1900±100 m.y.) and Lower Proterozoic–Upper Proterozoic (1100±100 m.y.) boundaries of VINOGRADOV and TUGARINOV (1961a, b) and to the periods suggested by SUTTON (1963) as marking the beginnings of evolution of the three major structural provinces. It may be suggested that if the age determinations plotted in Fig. 1 are representative of the incidence of metamorphic and igneous activity throughout geological time, then these well-marked changes of slope mark the boundaries between the major structural provinces, and thus probably also the change-over periods between convection systems characterized by harmonics of progressively higher degree. The major peaks of activity fall within these provinces and the change-over periods between provinces are marked by a falling off of the activity.

The terminology of the classification proposed here (see Table 1) is based on the structural provinces of North America where the cross-cutting relations of the fold-belts of the different age groups are particularly well displayed. The terms "Superior regime", "Hudsonian regime" and "Grenville regime" are used to include all the crustal fold-belts, wherever situated, which were produced during the periods >2750 to 1950±50 m.y., 1950±50 to 1075±50 m.y., and 1075±50 m.y. to 0 m.y. respectively. The terms "Superior province" or "Grenville province" have not been used because they would imply restriction to relatively limited regions on the North American continent only, and the term "chelogenic" is not used because the *cyclic* nature inherent in its original use by Sutton is not substantiated by the evidence outlined here.

In the following sections the world-wide orogenic fold-belt distribution are outlined from the earliest recognizable Precambrian structural regime, and their significance is examined in relation to the theory of continental drift and to the construction of an evolutionary Earth model.

[*Editor's Note:* In the remainder of this paper Dearnley develops an hypothesis of earth evolution that brings together the convection current systems postulated by Runcorn (Paper 5), the geotectonic groupings proposed by Sutton (Paper 6), and the concept of earth expansion.]

REFERENCES

Allsopp, H. L., H. R. Roberts, G. D. L. Schreiner, and D. L. Hunter. 1962. Rb-Sr age measurements on various Swaziland granites. *Jour. Geophys. Research* **67**:5307.

Aswathanarayana, U. 1964a. Age determination of rocks and geochronology of India. *Internat. Geol. Congr., 22nd,* 1.

Aswathanarayana, U. 1964b. Isotopic ages from the Eastern Ghats and Cuddapahs of India. *Jour. Geophys. Research* **69**:3479.

Bonhomme, M. 1962. Contribution à l'étude géochronologique de la Plateforme de l'ouest Africain. *Clermont Univ. Fac. Sci. Annales, Géologie et Mineralogie* **5**:1.

Cahen, L., G. Choubert, and D. Ledent. 1964. Premiers résultats de Géochronologie sur le Précambrien de l'Anti-Atlas (Sud Marocain) par la méthode strontium-rubidium. *Acad. Sci. (Paris) Comptes Rendus* **258**:635.

Carey, S. W. 1958. The tectonic approach to continental drift. In *Continental Drift, a Symposium,* S. W. Carey, ed. Hobart: Univ. of Tasmania, p. 177.

Catanzaro, E. J. 1963. Zircon ages in southwestern Minnesota. *Jour. Geophys. Research* **68**:2045.

Choubert, B. 1964. Âges absolus du Précambrien guyanais. *Acad. Sci., (Paris) Comptes Rendus* **258**:631.

Damon, P. E., and B. J. Giletti. 1961. The age of the basement rocks of the Colorado Plateau and adjacent areas. *New York Acad. Sci. Annals* **91**:443.

Dearnley, R. 1965a. Orogenic fold-belts and continental drift. *Nature* **206**:1083.

Dearnley, R. 1965b. Orogenic fold-belts, convection and expansion of the earth. *Nature* **206**:1284.

Du Toit, A. L. 1937. *Our Wandering Continents.* Edinburgh: Oliver & Boyd.

Faul, H. 1962. Age and extent of the Hercynian complex. *Geol. Rundschau* **52**:767.

Gastil, G. 1960. The distribution of mineral dates in time and space. *Am. Jour. Sci.* **258**:1.

Gerling, E. K. 1956. The argon method of age determination and its application to the subdivision of the Precambrian formations of the Baltic and Ukrainian shields. *Geochemistry* **5**:458.

Gerling, E. K., and A. A. Polkanov. 1958. The absolute age determinations of the Precambrian of the Baltic Shield. *Geochemistry* **8**:867.

Giletti, B. J., and P. W. Gast. 1961. Absolute age of Precambrian rocks in Wyoming and Montana. *New York Acad. Sci. Annals* **91**:454.

Goldenfeld, I. V. 1962. Division of the Archaean of the southwestern Ukraine into two age groups. *Geochemistry* **6**:553.

Herz, N., P. M. Hurley, W. H. Pinson, and H. W. Fairbairn. 1961. Age measurements from a part of the Brazilian Shield. *Geol. Soc. America Bull.* **72**:1111.

Herzog, L. F., and W. H. Pinson, 1956. Rb-Sr age, elemental and isotopic abundance studies of stony meteorites. *Am. Jour. Sci.* **254**:555.

Holmes, A., and L. Cahen. 1957. Géochronologie Africaine, 1956—Résultats acquis au ler juillet 1956. *Acad. Royale Sci. Coloniales (Brussels) Cl. Sci. nat. Méd.* t. V, fasc. **1**, 1.

Hurley, P. M., H. Hughes, G. Faure, H. W. Fairbairn, and W. H. Pinson. 1962. Radiogenic strontium-87 model of continental formations. *Jour Geophys. Research* **67**:5315.

Jäger, E. 1962. Rb-Sr age determinations on micas and total rocks from the Alps. *Jour. Geophys. Research* **67**:5293.

King, L. C. 1958. Basic palaeogeography of Gondwanaland during the late Palaeozoic and Mesozoic eras. *Geol. Soc. Lond. Quart. Jour.* **114**:47.

Komlev, L. V., V. G. Savonenkov, S. L. Danilevich, K. S. Ivanova, G. N. Kuchina, and A. D. Mikhalevskaya. 1962. Geological significance of regional rejuvenation of ancient formations in the southwestern part of the Ukrainian crystalline shield. *Geochemistry* **3**:219.

Kouvo, O. 1958. Radioactive ages of some Finnish Pre-Cambrian minerals. *Finlande Comm. Géol. Bull.* **182**.

Kouvo, O., and J. L. Kulp. 1961. Isotopic composition of Finnish galenas. *New York Acad. Sci. Annals* **91**:476.

Krylov, A. Y. 1961. The possibility of utilizing the absolute age of metamorphic and fragmental rocks in paleogeography and paleotectonics. *New York Acad. Sci. Annals* **91**:325.

Krylov, A. Y., A. N. Vishnevskii, Y. I. Silin, L. Y. Atrashenok, and G. V. Avdzeiko. 1963. Absolute ages of rocks of the Anabar Shield. *Geochemistry* **12**:1193.

Kulp, J. L., and H. Neumann. 1961. Some potassium-argon ages on rocks from the Norwegian basement. *New York Acad. Sci. Annals* **91**:469.

Leech, G. B., J. A. Lowdon, C.H. Stockwell, and R. K. Wanless. 1963. Age determinations and geological studies (including Isotopic Ages—Report 4). *Canada Geol. Survey Paper 63,* p. 17.

Li P'u, Chen, Tu Gon Chzhi Yu-Ni, A. I. Tugarinov, S. I. Zykov, N. I. Stupnikova, K. G. Knorne, N. I. Polyevaya, and S. B. Brandt. 1960. Rock ages in the Chinese Peoples Republic. *Geochemistry* **7**:682.

Long, L. E., and R. St. J. Lambert. 1973. Rb-Sr isotopic ages from the Moine Series. In *The British Caledonides,* M. R. W. Johnson and F. H. Stewart, eds. Edinburgh: Oliver & Boyd, p. 217.

Magnusson, N. H. 1960. Age determinations of Swedish Precambrian rocks. *Geol. Fören. Stockholm Förh.,* No. 503, Bd. **82**:407.

Masuda, A. 1958. Isotopic composition of primitive lead of the earth. *Geochim. et Cosmochim. Acta* **13**:143.

Neumann, H. 1960. Apparent ages of Norwegian minerals and rocks. *Norsk Geol. Tidsskr.,* **40**:173.

Nicolaysen, L. O. 1962. Stratigraphic interpretation of age measurements in southern Africa. In *Petrologic Studies: A Volume in Honor of A.F. Buddington,* Geol. Soc. America, p. 569.

Noakes, L. C. 1963. Age determinations in the British Commonwealth, 1962. *B.C.S.O. (Lond.) Liaison Symposium.*

Ostic, R. G., R. D. Russell, and P. H. Reynolds. 1963. A new calculation of the age of the earth from abundances of lead isotopes. *Nature* **199**: 1150.

Patterson, C. C. 1955. The Pb^{207}/Pb^{206} ages of some stone meteorites. *Geochim. et Cosmochim. Acta* **7**:151.

Patterson, C. C. 1956. [Not listed.]

Picciotto, E., and A. Coppez. 1963. Bibliographie des mesures d'ages absolus en Antarctique. *Soc. Géol. Belgique Annales* **85**:263.

Picciotto, E., and A. Coppez. 1964. Bibliographie des mesures d'ages absolus en Antarctique. *Soc. Géol. Belgique Annales* **87**:115.

Polevaya, N. I. 1956. The absolute age of some igneous complexes of the U.S.S.R. according to the Argon Method data. *Geochemistry* **5**:473.

Polkanov, A. A. and E. K. Gerling. 1960. The Precambrian geochronology of the Baltic Shield. *Internat. Geol. Congr., 21st,* Part IX, 183.

Polkanov, A. A., and E. K. Gerling. 1961. The Precambrian geochronology of the Baltic Shield. *New York Acad. Sci. Annals* **91**:492.

Runcorn, S. K. 1962b. Convection currents in the Earth's mantle. *Nature* **195**:1248.

Runcorn, S. K. 1962c. Palaeomagnetic evidence for continental drift and its geophysical cause. In *Continental Drift,* S. K. Runcorn, ed. New York: Academic Press, p. 1.

Runcorn, S. K. 1963. [Not listed.]

Schürmann, H. M. E. 1964. Rejuvenation of Pre-Cambrian rocks under epeirogenetical conditions during Old Palaeozoic times in Africa. *Geologie en Mijnbouw* **43**:196.

Slawson, W. F., E. R. Kanasewich, R. G. Ostic, and R. M. Farquhar. 1963. Age of the North American crust. *Nature* **200**:413.

Starik, I. Y., and Y. V. Sobotovich. 1961. The age of meteroic bodies and of the Earth according to radioactive data. *Acad. Sci. U.S.S.R. Izv.* (trans.) **10**:53.

Sutton, J. 1963. Long-term cycles in the evolution of the continents. *Nature* **198**:731.

Tarosov, L. S., E. Y. Gavrilov, and V. I. Lebedev. 1963. Absolute ages of the Precambrian rocks on the Anabar Shield. *Geochemistry* **12**:1199.

Taylor, F. B. 1928. Sliding continents and tidal and rotational forces. In *Theory of Continental Drift,* Tulsa, Ok.: Am. Assoc. Petroleum Geologists, p. 158.

Urey, H. C. 1952. *The Planets, Their Origin and Development*. New Haven, Conn.: Yale Univ. Press.

Vinogradov, A. P. 1956. Comparison of data on the age of rocks obtained by different methods and geological conclusions. *Geochemistry* **5**:427.

Vinogradov, A. P., L. V. Komlev, I. G. Danilevich, A. V. Savonenkov, A. I. Tugarinov, and M. S. Filippov. 1960. Absolute geochronology of the Ukrainian Pre-Cambrian. *Internat. Geol. Congr., 21st,* 116.

Vinogradov, A. P., and A. I. Tugarinov. 1961a. The geologic age of Pre-Cambrian rocks of the Ukrainian and Baltic shields. *New York Acad. Sci. Annals* **91**:500.

Vinogradov, A. P., and A. I. Tugarinov. 1961b. Geochronology of the Pre-Cambrian. *Geochemistry* **9**:787.

Voytkevich, G. V., and L. K. Anokhina. 1961. Ages of some rock complexes in the Krivoi Rog iron region. *Geochemistry* **2**:212.

Wegener, A. 1924. *The Origin of Continents and Oceans* (trans. of 3d ed.). London:Methuen.

3

Reprinted from *Canadian Jour. Earth Sci.* **5**:693–698 (1968)

Geochronology of stratified rocks of the Canadian Shield[1]

C. H. STOCKWELL
Geological Survey of Canada, Ottawa, Canada

Introduction

The Canadian Shield covers an area of 1 864 000 square miles (4 828 000 sq. km) and is one of the large areas of Precambrian rocks of the world. It is dotted with innumerable lakes, making almost all parts readily accessible by float-equipped aircraft, and much of the area is suitable for helicopter landings. Outcrops are glaciated with the removal of weathered material and it is easy to collect fresh samples for isotopic dating.

More than 750 samples have been dated so far, mainly in laboratories of the Geological Survey of Canada and of several universities. The samples are widely distributed to give a good coverage over almost the whole area, but special emphasis has been placed on dating rocks of particular stratigraphic significance. Most dating has been done by the potassium-argon method, mainly on biotite and muscovite, but with a few on hornblende and a considerable number on whole-rock samples, mainly of basic dikes. Relatively few determinations have been made by the rubidium-strontium whole-rock isochron method and these, for the most part, confirm the potassium-argon dates, but some give ages significantly older.

Principles

Almost the whole of the Canadian Shield has been mapped geologically and the rocks record an orderly succession of depositional and orogenic events dating from the base of the Cambrian back to more than 2500 million years ago. Four profound unconformities are present and these divide the succession into five, natural, mappable, time-stratigraphic units. Following the principles proposed by the American Commission on Stratigraphic Nomenclature (1961), the time classification is based fundamentally on type localities of these concrete units and not on abstract time as given by isotopic dates. It is an application of the law of superposition of strata. The isotopic dates enable the geological divisions to be placed into a scale of years and become extremely useful, as are fossils in the Phanerozoic, for purposes of correlation. The type localities serve as standards for reference and any improvements in their dating or any mistakes made in correlation do not invalidate the units.

Two ways of dating stratified rocks have been used in the shield, the one direct and the other indirect. However, direct dating of age of sedimentation or volcanism has been largely unsuccessful. Glauconite, for example, is found only rarely and has been of little use. Some whole-rock rubidium-strontium ages on slightly metamorphosed sedimentary and volcanic rocks give about the same age as the regional metamorphism and can be interpreted as either the age of deposition or the age of metamorphism. Virtually unmetamorphosed basic volcanic rocks have generally proven to be chemically unsuitable for the rubidium-strontium whole-rock isochron method and their whole-rock ages by the potassium-argon method are too erratic to be useful. Because of general lack of success in direct dating of time of deposition of stratified rocks, it has not been possible to date the time from beginning to end of deposition and, consequently, it has not been practical to define series, epochs, or periods. Lack of success in direct dating may be due to the predominance, in the Canadian Shield, of metamorphic and plutonic rocks.

[1]Presented at the International Conference "Geochronology of Precambrian stratified rocks," held in Edmonton, Alberta, June 12–14, 1967.

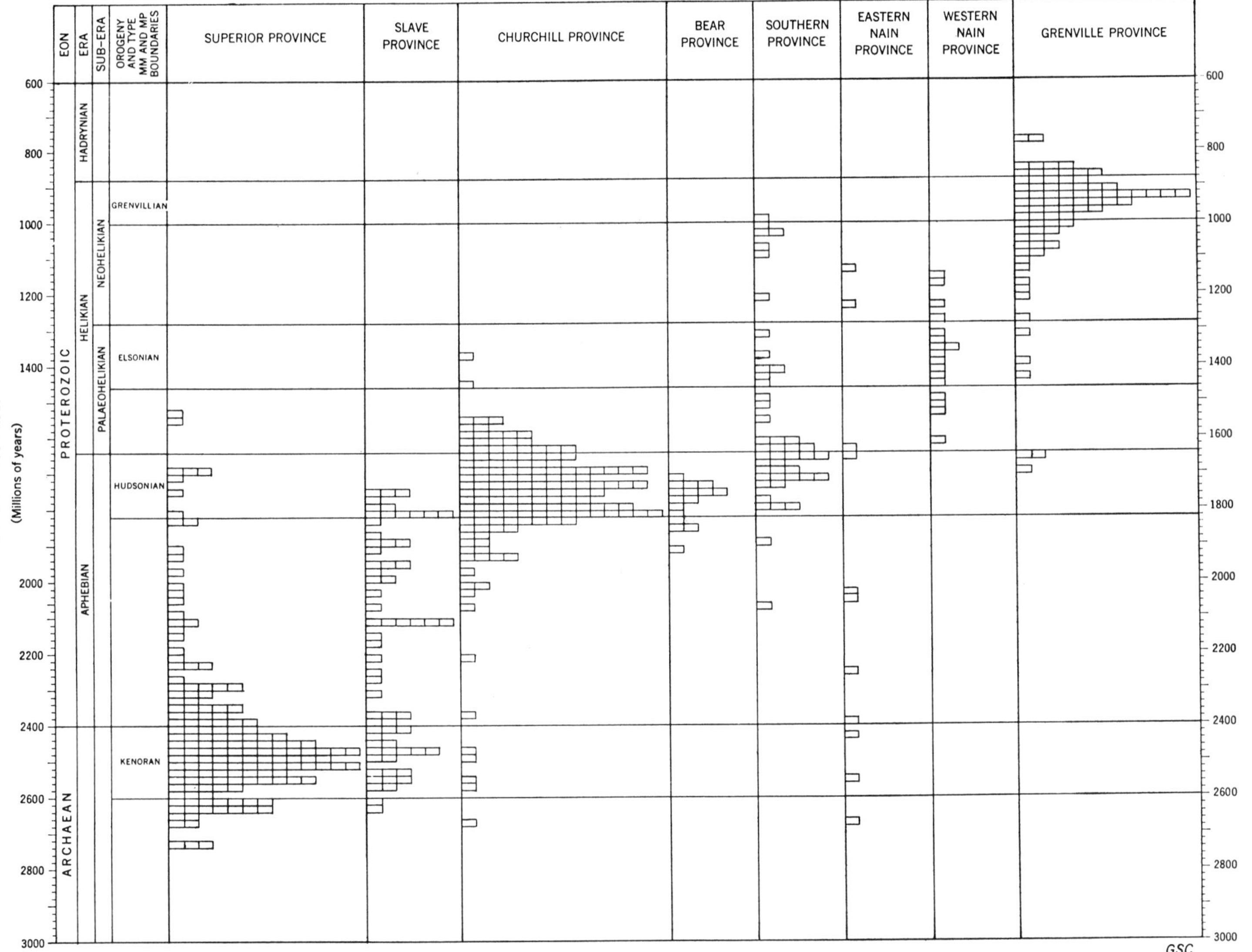

FIG. 1. Potassium-argon ages on orogenic minerals of the Canadian Shield.

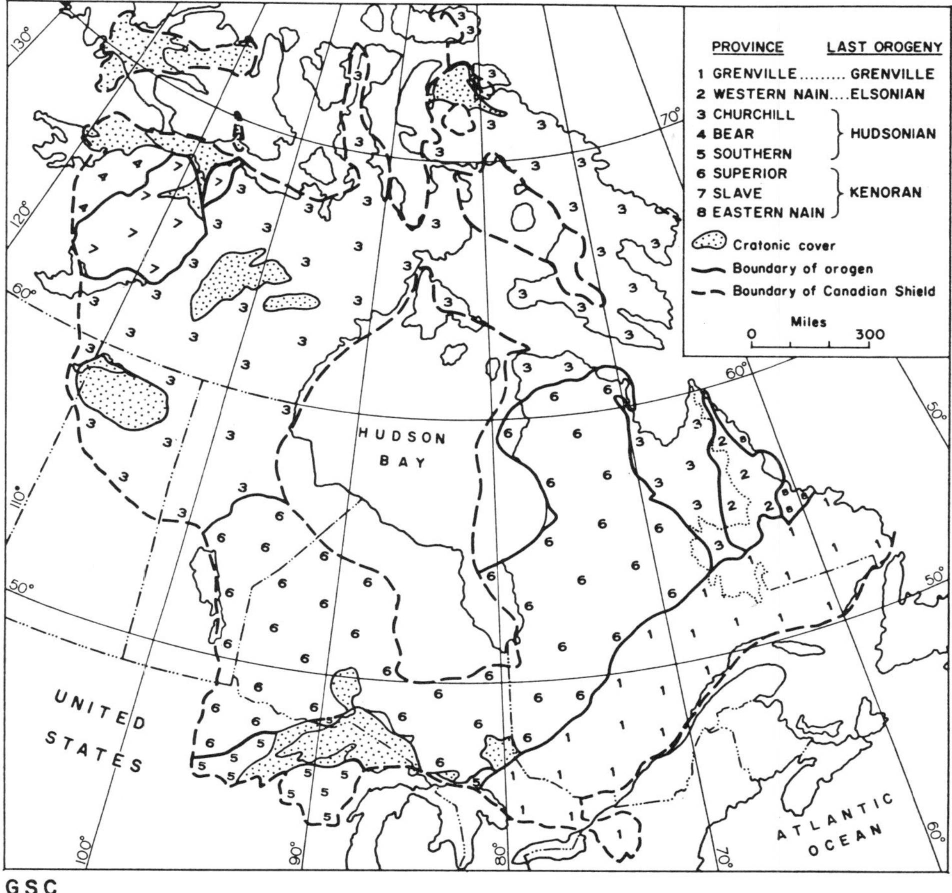

FIG. 2. Orogenies and structural provinces of the Canadian Shield.

Although the direct method has been of little use, the indirect method has been remarkably successful. By the indirect method the age of deposition of a stratified sequence is bracketed between maximum and minimum limits. For example, the maximum age of an unconformity-bounded sequence is obtained by dating the metamorphism and intrusions of its basement, and the minimum is given by dating the metamorphism of the stratigraphic sequence itself and by obtaining ages on intrusions that cut it. In this way all of the five main stratigraphic units have been placed in a scale of years and long-range correlations have become possible. Difficulty arises where an orogeny that deformed a sequence overrides the basement as well, but already some progress has been made in penetrating this superimposed metamorphism and so obtaining the older age.

Orogenies

The potassium-argon dates on orogenic minerals, mainly micas from metamorphic and granitic rocks, have been plotted on a histogram (Fig. 1), where the ages are arranged, objectively, according to structural provinces. The provinces are separated from one another by unconformities or by orogenic fronts. The isotopic ages within them reflect mainly the last important orogeny producing the structural features that characterize the province. In each province there is a considerable spread in isotopic dates due in part to analytical inaccuracies, in part to the time span of the orogeny itself, in part to the effects of earlier and later events, and in part to anomalous ages near orogenic fronts. Accordingly, the time span of each orogeny is uncertain because the spread in ages could be due very largely to analytical inaccuracies alone, which may commonly be from ±40 to ±80 million years. However, the peak values for the several orogenies are apparent and they are distinct from one another. The orogenies have been named: the Kenoran with a mean at 2490 m.y.; the Hudsonian at 1735 m.y.; the Elsonian (less well defined) at 1370 m.y.; and the Grenvillian at 945 m.y. These are potassium-argon-orogenic-mica dates. Ages by some other method, or on some other mineral might be different but this would not change the natural classification nor the terminology. The structural provinces and orogenies are illustrated in Fig. 2.

Time-stratigraphic Classification

Table I depicts the time-stratigraphic classification in relation to the orogenies, each of which is followed by a profound unconformity. For convenience in description, mapping, and discussion, each of the units is named. The large first-order units, called the Archean and Proterozoic (Alcock 1934), are useful for comparing or contrasting the geological features of each and they also permit a gross time-classification of those rocks for which information is too meager for more detailed assignment. The Archean includes rocks involved in the Kenoran orogeny and all older rocks. Profound unconformities are known within the Archean rocks but, because of the strong overprinting of the Kenoran orogeny, it has not yet been possible to date them nor to subdivide the Archean on a time-stratigraphic basis.

TABLE I
Precambrian time classification

EON	ERA	SUB-ERA	OROGENY	K-Ar AGE on MICA m.y.
				570
PROTEROZOIC	HADRYNIAN			
	HELIKIAN	NEOHELIKIAN	GRENVILLIAN	945
		PALEOHELIKIAN	ELSONIAN	1370
	APHEBIAN		HUDSONIAN	1735
ARCHEAN			KENORAN	2490

GSC

The Proterozoic is subdivided into second order stratigraphic units that in general span the time from the close of one orogeny to the close of the next. These have been named: the Aphebian, the Helikian (subdivided where possible into the Paleohelikian and Neohelikian), and the Hadrynian (Stockwell 1964, 1965). The Hadrynian ends, generally conformably and without orogeny, at the base of the Cambrian. The stratigraphic units so defined have the advantage that no part of the geological column or any gaps in it remain unnamed. Although the mean age of an orogeny is obtained quite readily, it is difficult to define, in terms of years, the precise age of its closing. Consequently, the age of the beginning of the next overlying stratigraphic units is not known with precision, a difficulty that is unavoidable at present, mainly because of the large analytical errors.

The distribution of the several time-stratigraphic units and their lithology are shown on the geological map (Fig. 3). The Archean sedimentary and volcanic sequences and their metamorphic equivalents are cut by widespread intrusions of mainly granitic rocks.

This was followed by a long period of deep erosion and by the unconformable deposition of Aphebian sediments and volcanics. The Aphebian is widespread and the unconformity at its base is firmly established at many localities. In several of these areas the Aphebian is unmetamorphosed close to the basement but, farther out, becomes folded, metamorphosed, and cut by granite. At some places Aphebian rocks form a flat cover on the Archean. Elsewhere much of the Aphebian strata are converted to gneisses and in such places evidence of the basal unconformity is destroyed. However, these gneisses are classed with some confidence as Aphebian because they can be traced into firmly established Aphebian rocks and retain some of their lithological character.

The areas of folded Aphebian rocks are, in places, overlain unconformably by the Paleohelikian and undivided Helikian. These rocks are much less abundant than the Aphebian and occur for the most part as virtually unfolded cratonic cover. Bodies of Paleohelikian anorthosite and

FIG. 3. Generalized geological map of the Canadian Shield.

related intrusions are widespread in eastern Canada and are overlain unconformably by the Neohelikian. In the western and southern parts of the Canadian Shield, conformable successions span the Neohelikian-Hadrynian boundary and, at present, cannot be separated. It is to be hoped that, in the future, direct methods of dating horizons within the conformable succession may fix the position of the boundary.

In the eastern part of the shield, some sediments lie unconformably on the Grenvillian orogen, but their minimum age has not been determined. They may be either Hadrynian or Cambrian. The best development of the Hadrynian lies beyond the limits of the Canadian Shield, in the Appalachians of eastern Canada and in the Cordilleran of the west. Discussion of these localities is beyond the scope of the present paper.

The time-stratigraphic classification as applied to the Canadian Shield has proven to be very useful for purposes of mapping, description, and discussion of the geological history. Further progress may be expected in the future.

ALCOCK, F. J. 1934. Report of the National Committee on Stratigraphical Nomenclature. Trans. Roy. Soc. Can., Ser. 3, **28**, Sec. 4, pp. 113–121.

AMERICAN COMMISSION ON STRATIGRAPHIC NOMENCLATURE. 1961. Code on stratigraphic nomenclature. Bull. Am. Assoc. Petrol. Geol., **45**, pp. 645–665.

STOCKWELL, C. H. 1964. Fourth report on structural provinces, orogenies, and time-classification of rocks of the Canadian Precambrian Shield. *In* Age determinations and geological studies. Geol. Surv. Can., Paper 64–17, Part II, pp. 1–21.

——— 1965. Tectonic map of the Canadian Shield. Geol. Surv. Can., Map 4–1965.

4

Reprinted from *Internat. Geol. Congr., 24th Session* **sect. 1**:253–259 (1972)

A Unified Stratigraphic Scale of the Precambrian

L. J. SALOP,
U.S.S.R.

ABSTRACT

The records of isotope dating of plutonic and metamorphic rocks are indicative of the existence of five diastrophic cycles of global significance in Precambrian time that took place about 3.5, 2.6-2.8, 1.9-2.0, 1.0-1.1 and 0.65-0.68 x 10^9 yeatrs ago.

Supracrustal formations formed between two diastrophic cycles, together with plutonic rocks of the final diastrophism, form natural complexes that can be conventionally compared with the largest subdivisions of the stratigraphic groups. The Precambrian groups of the same age in various regions are characterized by similar formations, tectonic structures and organic remains.

The greatest difference is between the most ancient group (> 3.5 × 10^9 years) and younger groups. In accordance with this, two eons (supergroups) are distinguished in the Precambrian: Archean and Protozoic. The Archean eon is composed of one Archean group devoid of definable organic remains; disconformities within it are lacking, although there are evidences of the existence of an earlier gneiss-granite basement that has not been detected anywhere for certain as yet. The Protozoic eon is composed of four groups for which the following names have been proposed: Paleoprotozoic, Mesoprotozoic, Neoprotozoic and Epiprotozoic. Neoprotozoic is divided into three subgroups: the two lower subgroups characterized by global platform magmatism (1.6-1.7 x 10^9 years), the two upper by characteristic complexes of stromatoliths. The worldwide display of two-phase glaciation is characteristic of Epiprotozoic. Above the Epiprotozoic, Eocambrian deposits occur, as a rule, unconformably; in most regions there is a gradual transition to the Lower Cambrian, which is characterized paleontologically. A short description of the subdivisions is given.

THE ABSENCE of an accepted scheme of Precambrian subdivision makes the systematization and generalization of data related to the early stages of the geological development of the Earth extremely difficult. Meanwhile, in recent years, much geological and geochronological material has been accumulated that enables us to consider as quite practicable the task of the creation of a unified stratigraphic scale for the Precambrian, if only as a first approximation. In this paper I shall formulate the main principles which in my opinion constitute the basis for such a scale, and then propose a draft.

First of all, it seems indisputable that the Precambrian divisions should be based upon the recognition of natural stages of geological history; such stages are characterized by specific features of the tectonic environment and geochemical media which together determine definite types of lithogenesis and to some degree define organic evolution. Precambrian divisions as defined on the basis of geochronological (radiometric) or biostratigraphic data only cannot be considered satisfactory, as they do not conform to this main principle; also, radiometric dates are difficult to interpret, and biostratigraphic data are almost lacking for Early Precambrian time and are insufficient for Late Precambrian time. It is evident that the establishment of natural stages is possible only by means of the application of data obtained from all methods of study, primarily the historic-geologic ones.

As is known, the stratigraphic subdivision of ancient complexes is based on tectonics; in accordance with tectonic principles, diastrophic (tectono-plutonic) cycles, which terminate relatively calm and long stages of geological development, are accepted as the boundaries of large subdivisions of the Precambrian. It is assumed that the end of the diastrophic cycle is simultaneously conformable to the beginning of a new evolutionary stage.

Radiometric dating of plutonic and metamorphic rocks indicates that in Precambrian time, in contrast to the Phanerozoic, diastrophic cycles were comparatively long-lived and approximately contemporaneous in different continents. This enables us to contend that the tectonic principle subdivision is methodically sufficiently well grounded.

The analysis of well-known geochronological data indicates that in Precambrian time five world-wide diastrophic cycles took place: about 3500, 2800-2600, 2000-1900, 1100-1000 and 650-680 m.y. ago (Salop, 1970). These are the Saamian, Kenoran, Karelian, Grenville and Katanga diastrophic cycles respectively. It should be noted that the determination of a time of diastrophism is a complicated task because isotope dating of abyssal rocks in most cases shows the "rejuvenated" values (especially with K-Ar dating). To solve this problem it is necessary to correlate the data of different isotopic methods and use, in a number of cases, the so-called "relict dating" which, with some approximation, shows the time of initial metamorphism or crystallization of the rock. Statistical methods (histograms) are utterly useless for this purpose.

"Globality" of diastrophic cycles of the Precambrian should be understood only in the sense of their simultaneity, not of their world-wide distribution. In those regions where folding and accompanying intrusive magmatism have not taken place, complete sections of sedimentary series may be observed, though even there traces of world-wide tectonic events would be imprinted on them in some form (stratigraphic hiatuses, alternation of types of lithogenesis, etc.).

A comparative study of Precambrian formations originating in the intervals between the diastrophic cycles has enabled us to establish that complexes of the same age are essentially similar, whereas complexes of different ages are correspondingly different. These natural complexes might be ranked as groups — the largest subdivisions of the stratigraphic scale—but these Precambrian groups represent significantly longer periods of time than the Phanerozoic groups. However, to avoid the introduction of a new taxonomic unit, they can be provisionally called groups.*

Thus, in the Precambrian, five groups may be distinguished that are characterized by many features of gross composition, tectonic structure, metamorphism and plutonism, and of organic remains as well. These groups were distinguished (Salop, 1964) at first under the names of: Archean, Lower Proterozoic, Middle Proterozoic, Upper Proterozoic and Epiproterozoic. In addition, a relatively small Eocambrian complex was distinguished, the formation of which took place before the Cambrian. Later, Proterozoic groups were renamed by us as Paleoprotozoic, Mesoprotozoic, Neoprotozoic and Epiprotozoic (Salop, 1968, 1970).

---

*Besides supracrustal rocks, plutonic formations of the final cycle of diatrophism are included in the composition of the group.

The names have been changed for the following reasons. Firstly, the former names gave the impression that the large subdivisions are parts of a single group (subgroups). Secondly, in connection with the division of Upper Proterozoic into two groups — Upper Proterozoic proper and Epiproterozoic — ambiguity often appeared in the understanding of this term. Thirdly, the names of the groups should be short and be compiled according to the same principle as the names of Phanerozoic groups. In this respect, new terms seem to be preferable to those proposed by Stockwell (1964), because they give immediately a clear impression of the succession of deposits. The term "Riphean" is also unsatisfactory, not only because of its different construction from the other names of the groups but also, and mainly, because the volume of Riphean (in particular its Uralian stratotype) is excessively great, and it includes deposits that are related to three groups of Precambrian and Eocambrian. Finally, the application of new names eliminates many linguistic inconveniences in the application of old terms.

Let us consider the main features of each of the distinguished Precambrian groups in a short and general form.

ARCHEAN GROUP (older than 3500 m.y.)

The gneiss-granulite complex of the Aldan shield, the Aldan series, might be its stratotype. The following series and complexes belong to it: Anabar of North Siberia; Belomorian and Kola of the Baltic shield; Pridneprovian of the Ukraine; Lewisian of Scotland; Peninsular gneisses of Hindustan; Taishan of China; Ungava (Pre-Keewatin) complex of Canada; Pre-Swaziland complex of South Africa; Kasila of Western Africa; the Pre-Pilbara (Pre-Yilgara) complex of Western Australia, etc.* In the composition of all these complexes, gneisses of different types, amphibolites and basic crystalline schists are widely developed, and in some bands — marbles, calcareous crystalline shales, quartzites, graphitic gneisses, etc. There are also banded magnetite rocks, closely associated with basic metavolcanics (orthoamphibolites and orthoschists); magnetite ores are often observed directly in metavolcanics. For the Archean of North Eurasia, a succession of similar composition has been established in the sequences of different widely separated regions, the most ancient series containing thick (to 1000 m) layers of quartzites (Salop, 1968_2). The presence of the latter enables us to suspect the existence of a Pre-Archean sialic crust, although this has not been established yet. High grade metamorphism of rocks of the granulite and amphibolite facies, as well as the widespread occurrence of ultrametamorphism (granitization and anatexis) are characteristic features of Archean complexes. Structure-formational and metamorphic zoning is not observed, or is extremely poorly defined. Stratigraphic breaks and disconformities within complexes are lacking. Conglomerates have not been established anywhere for certain, although psephitic texture is well preserved, even under conditions of intensive recrystallization.

Grouping of compressed linear folds in the form of large closed oval or amoeboid systems up to 600 km in diameter is typical mainly of the Archean tectonic structures. Large areas of gneiss-granites are found in the interior parts of these systems. Between folded ovals, sections are located that are characterized by the development of gneiss domes and of large diversely oriented brachyfolds (Salop, 1971).

*Here and in the following, well-known subdivisions are given as examples.

According to known data, thermal and tectonic regimes in the Archean were very specific and unlike those that are known for the Post-Archean. Therefore, it is possible to agree with the opinion of those investigators who consider the distinguishing of geosynclines and platforms to be unjustified at this time. We propose to call this stage of the Earth's development permobile and to distinguish the Archean group as a special Archean or Cryptozoic supergroup (eon).

PALEOPROTOZOIC GROUP

The formations that originated between the Saamian (3500 m.y.) and Kenoran (2800-2600 m.y.) cycles of diastrophism are related to this group. The Keewatin-Timiskaming of the Canadian shield might be taken as a stratotype of the group with which the following series and complexes are parallel: Bulawayan and Shamvaian, and Swaziland of South Africa; Gimola and Tundra of the Baltic shield; Bazuvluk (metabasite) of the Ukraine; Muya and Olonda of Eastern Siberia; Anshan of China; Dharwar of Hindustan; Pilbara and Yilgarn of Western Australia, etc.

Deposits of various types are known but eugeosynclinal deposits are especially widely developed, represented by greenstone igneous series and slate-greywacke series overlying them, oftn with a stratigraphic break. Banded iron ores (Algoman type) are closely related to volcanics; similar ores are found from time to time in sedimentary series that are located in the margins of zones of volcanic activity. The rocks are metamorphosed irregularly under the conditions of the greenschist and amphibolite facies; regionally altered rock of the granulite facies are lacking. Granitization phenomena occur locally.

In Paleoprotozoic rocks, the first determinable organic remains, microscopic spherical bodies containing amino acid, are observed and isolated stromatoliths, poorly studied as yet, are found as well.

Well-defined structure-formational and metamorphic zoning of a linear type is clearly observed and grouping of folds in the form of extensive belts or arched systems within which linear folds are usually combined with domes is characteristic; the latter often form elongated swarms. Abyssal fractures play an exceptionally great part in tectonic structure.

It follows from the aforesaid that in the Paleoprotozoic a sharp change of tectonic regime occurred that was expressed in the change of the permobile stage of the Earth's development to the platform-geosyncline stage in which evolution has continued up to the present. This circumstance and the appearance of organic remains in the deposits enables us to relate Paleoprotozoic and all the younger groups of the Precambrian to the Protozoic supergroup (eon).

MESOPROTOZOIC GROUP

Its lower and upper age boundaries are 2800-2600 m.y. and 2000-1900 m.y. (Karelian diastrophism). The Karelian complex of the Baltic shield may be proposed as its stratotype. The following series are synchronous: the Huronian, Kaniapiskau, Goulburn and Animikie of Canada; Minas of Brazil; Dominion Reef + Witwatersrand + Ventersdorp + Transvaal of South Africa; Krivorogian series of the Ukraine; Burzyan series of the Urals; Teya of Enisei ridge; Udokan of Eastern Siberia; Liacho and Huto of China; Aravalli and Bijawar of Hindustan; Nullaginian of North Australia, etc. As in the Paleoprotozoic, all the main types of deposits are present, but miogeosyncline and platform deposits are much more widely represented. Among both of these types of deposits are quartzites or quartzite-sandstones, often obliquely laminated; intercalations of conglomerates are very widely developed and contain syngenetic gold-uranium mineralization (Blind River, Witwatersrand, Jacobina, Kaltimo, etc.). Gold-uranium-bearing conglomerates and quartzites are usually confined to the lower part of sedimentary complexes; in their middle part, dolomite horizons, often with stromatoliths, are found, and above them red sandstones are widely developed. As lack of oxygen is an indispensable condition for the formation of uranium mineralization in conglomerates, and as red beds could originate only under oxidizing conditions, it is conceivable that in Mesoprotozoic time an important evolution of the composition of the atmosphere occurred, possibly in connection with an increase in photosynthesis, marked by the development of algae.

Iron ores are also very characteristic of the Mesoprotozoic, stratified hematite and siderite ores being widely distributed; banded jaspilites also occur (Lake Superior type), and oölitic varieties are observed in both of them. Attention is drawn to the fact that iron ore formations are found only among sedimentary (terrigenous and carbonate) rocks exclusively in miogeosyncline complexes. Thus, the third iron-ore epoch was characterized by new geochemical and tectonic conditions.

In some sedimentary series of Canada and South Africa, conglomerates are present that are sometimes considered as tillites, but their glacial genesis cannot be considered strictly proved.

In Mesoprotozoic formations stromatoliths are widely developed which, in some places, formed large bioherms; for the first time microphytoliths appear that are represented only by oncolites. The mentioned organic remains make up the first or Lower Riphean paleontological complex (*Kussiella kussiensis, Collenia frequens, Osagia libidinosa,* etc.) (Krylov, 1963; Zhuravleva, 1964).

The character of metamorphism and tectonic style of the Mesoprotozoic are, on the whole, similar to those mentioned for the Paleoprotozoic. However, some decrease in the metamorphism of rocks in geosynclinal belts and the presence of almost unaltered deposits in platforms are observed; the role of gneiss domes in the structure of folded zones has decreased noticeably.

NEOPROTOZOIC GROUP

The deposits of this group were formed during the time interval between the Karelian (2000-1900 m.y.) and Grenville (1100-1000 m.y.) diastrophisms. In many regions of the world within this interval, mainly about 1600-1700 m.y., the occurrence of peculiar plutonism of a platform or "undeveloped" orogenic type is related to the formation of differentiated metal-bearing intrusions of gabbro-norites, labradorites, granophyre granites, alkaline syenites and, especially, intrusions of rapakivi-granites. This global magmatic stage enables us to divide the Neoprotozoic into two parts. The upper part, in its turn, is divided into two parts according to characteristic complex organic remains. Thus, three subgroups might be distinguished in the composition of the Neoprotozoic.

The Subjotnian of the Baltic shield is a stratotype of the lower subgroup. The following series are of the same age and partly analogous: the Mashak of the Urals; Akitkan and Teptorgo of Eastern Siberia; Lower Dabawnt, Et-Then and Letitia of Canada; Davenport and Cliffdale of North Australia; Waterberg of South Africa, etc. Many of the subdivisions are made up of subaerial volcanics of acid, less often of basic compositions, that alternate with red sandstones and conglomerates. Later comagmatic intrusions of rapakivi-granites and granophyres as well as sills and dykes of diabases are commonly associated with volcanics.

The Middle and Upper Riphean of the Urals, composed of the Yurmatin and Karatau series, characterized by stromatoliths and microphytoliths of the second (*Baicalia, Anabaria, Osagia undosa,* etc.) and third (*Gymnosolen, Minjaria, Osagia grandis,* etc.) complexes, accordingly should be taken apparently as a stratotype of the Upper and Middle subgroups of the Neoprotozoic. The following series are parallel to them: the Jotnian + Lower Sparagmite of the Baltic shield; Sukhopit + Tungusik + Oslyan of the Enisei ridge; Billyakh of North Siberia; Ballaganakh + Kadalikan of Eastern Siberia; Sanvon of Korea; Lower and Middle Sinian of China; Raialo and Delhian series of India; Kibara, Mine and Damara of Africa; Lower and Middle Keweenawan, Seal, Athabasca (*s. str.)* and Belt of North America; Ytacolumi of South America; Carpentaria of North Australia; Kalana + Burra of South Australia, etc.

The formations composing the two upper subgroups are, on the whole, similar to the formations of the Mesoprotozoic, but still, a definite evolution is observed which is expressed in the decrease of the role of ferruginous and dolomite deposits and in the almost complete disappearance of jaspilites. Parallel with it, the significance of rudaceous rocks is increased; for the first time typical molasse appears, and flysch is widely developed; the role of limestones increases. In the platforms, red clastic rocks are widely distributed.

At the boundary of the middle and upper subgroups, stratigraphic unconformities are sometimes observed; but in most regions the section is continuous. In Siberia, in boundary deposits, stromatoliths and microphytoliths of the "mixed" second and third complexes are found. Noticeable complication of structure-formational zoning, increase of the amplitude of movements, increase of contrast of morphostructures, and a further decrease in the role of gneiss domes are all characteristic of tectonic evolution. In many geosynclinal systems foredeeps appear or deepen which were scarcely outlined before.

EPIPROTOZOIC GROUP

Its lower boundary is determined by the completion of Grenville diastrophism (1000 m.y.) and the upper boundary by Katanga diastrophism (650-680 m.y.). The Kundelungu series of Equatorial Africa may be taken as a stratotype of this group. In other regions the following series are related to it: Upper Sparagmite of Scandinavia; Vilchan and Pachelma of the Russian platform; Churochin of the North Urals; Taseyersk of the Enisei ridge; Nantou + Doushanto + Danin of South China; Semri and Kaimur of India; Damara (Chuos, Otavi), Tidimen + Uarzazat + Gbelia of Africa; Karanday of South America; Windermere and Upper Keweenawan of North America; Upper Kimberlay (Duerdin) of North Australia; Umberatana (Sturtian) of South Australia, etc. The presence of glacial formations in the composition of Epiprotozoic deposits on all the continents, and usually two horizons of tillites, which are good datum levels for the intraregional correlation, is their most characteristic feature. In other respects, the Epiprotozoic formations inherit the features of the preceding era, but those tendencies of the evolution of lithogenesis and tectonesis that were already outlined before are reflected in them more sharply.

The organic remains of the Epiprotozoic are stromatoliths and microphytoliths of the fourth complex (*Sacculia, Linella, Osagia minuta, Vesicularites concretus,* etc.) and different microphytofossils. Besides that, for the first time the remains of skeleton-less animals, as represented by medusoid imprints, are found.

EOCAMBRIAN COMPLEX

The deposits that were formed after Katanga diastrophism (650 m.y.) but before the beginning of the Cambrian period (570 m.y.) are related to this complex; thus, the duration of the Eocambrian (80 m.y.) is quite comparable with that of the Cambrian period. The Yudoma formation of Siberia might be the best stratotype of this complex, which is characterized by stromatoliths and microphytoliths; the Wilpena Group (Upper Marinoan) of South Australia might be proposed as a parastratotype in the upper part of which, in Pound quartzites, numerous imprints of skeleton-less animals are found. The Valdayan of the Russian platform, Asha of the Urals, Ringsak quartzites of Scandinavia, Upper Sinian of China, Abudu of Anti-Atlas, Nama of South Africa, Conception + Hodgewoter of North America, etc. are stratigraphic analogues of these subdivisions. In most regions of the world the lower boundary of the complex is marked by a significant unconformity; the upper boundary is gradual and is determined only on the basis of the appearance of the remains of skeletal fauna in the lowermost of Lower Cambrian (hyoliths, gastropods, brachiopods, archaeocyaths — upwards in the section, trilobites and others appear). Thus, in geohistorical respects, the Eocambrian is closer to the Paleozoic group, but in biostratigraphic respects it is distinctly a part of the Protozoic group. Stromatoliths and microphytoliths of the Eocambrian belong to the same fourth complex which is characteristic of the Epiprotozoic, but in some deposits algae (*Epiphyton* and *Renalcis*) are found for the first time. Besides medusoids, rangeids, vermiculars and others are known among remains of skeleton-less animals.

The proposed stratigraphic scale of the Precambrian is not notable for great detail, because only the largest subdivisions — groups and subgroups (for Neoprotozoic) — are distinguishable in regional correlation. The creation of a more detailed scale is impossible at present. However, it appears that even such a rough division will be an aid in advancing our knowledge of the most ancient formations of the Earth.

REFERENCES

Krylov, I. N., 1963. Columnar branching stromatoliths of the Riphean deposits of the South Urals and their significance for Upper Precambrian stratigraphy. Isd. A. N. SSSR., Moscow.

Salop, L. I., 1964. Geochronology of Precambrian and some features of the early stage of the Earth's geological development. XXI sec. MGK. Geologia dokembria. Isd. Nedra, Moscow.

————, 1968. The ways of the creation of the unified stratigraphic scale of Precambrian. V kn. "Geol. stroenie SSSR", t. 5, isd. Nedra, Moscow.

————, 1968_2. Precambrian of the USSR. XXIII Int. Geol. Congr. Rep., Geology of Precambrian. Prague.

————, 1970. Revision of the geochronological scale of Precambrian. Bull. MOIP, Otd. geol. N 4 and 5.

————, 1971. Two types of Precambrian structures: gneiss folded ovals and gneiss domes. Bull. MOIP, otd. geol., N 4.

Stockwell, C. H., 1964. Age determination and geological studies. Geol. Surv. Can. Pap.. 64-17.

Zhuravleva, Z. A., 1964. Oncolites and katagraphies of Riphean and Lower Cambrian of Siberia and their stratigraphic significance. Isd. Nauka, Moscow.

Editor's Comments on Papers 5 Through 8

5 **RUNCORN**
Convection Currents in the Earth's Mantle

6 **SUTTON**
Long-Term Cycles in the Evolution of the Continents

7 **FITCH and MILLER**
Major Cycles in the History of the Earth

8 **WILSON**
Did the Atlantic Close and Then Re-Open?

MEGACYCLES AND THE MOBILE EARTH

S. K. Runcorn was first to propose, in 1962, a general earth model in explanation of geotectonic megacycles. He postulated that the radius of the earth's core has been increasing since the formation of the earth and ascribed post-Paleozoic continental drift to a change in the pattern of whole-mantle convection resulting from core growth (Runcorn, 1962a). Runcorn (Paper 5) then associated the transitions between postulated successive whole-mantle convection patterns due to core growth with peaks in isotopic age abundance at ~2600, ~1800, ~1000, and ~200 m.y. ago, which implied earlier epochs of continental displacements. This hypothesis is further discussed by Runcorn (1962c, 1966). Despite a recent renewal of interest in whole-mantle convection (see Smith, 1977), Runcorn's hypothesis meets with several objections. Sutton (Paper 6) commented that Runcorn's hypothesis "does not appear to account for the distribution of orogenic belts, especially for the distribution in space and time of those formed over the past thousand million years." Moreover, paleomagnetic measurements from very old (~3800–2700 m.y.) rocks suggest that core formation was well advanced early in the earth's history (see Part II). The possibility of early formation in the earth's core is recognized by Professor Runcorn (personal communication,

13 June 1978) in the following comment on Paper 5:

> The hypothesis of the growing core has not over the years attracted support, and the interpretation of lead ages might indicate that the core formed in the first 10^8 years of earth history (although there are alternative explanations). However, the idea that convection patterns could be relatively stopped over long periods and then suddenly changed still seems to be a plausible explanation of why Wegenerian drift started so recently in geological time. However, the slowly changing geometry of the convecting mantle which was postulated is not by any means the only or most plausible way of explaining sudden changes in the convection pattern: it simply is a useful illustration of the general principle that drastic changes in the number of convection cells can occur with cell changes with a number of physical parameters as is shown in many laboratory convection experiments.

Walzer (1974) and Busse (1978) presented alternative models of discontinuous mantle convection as a cause of long-term diastrophic episodicity.

An important series of papers published in *Nature* between 1963 and 1966 (Papers 6–8) laid the foundations for the blending of concepts of the geotectonic megacycle and the orogenic cycle, although more than a decade later the topic remains controversial.

In 1963 Sutton (Paper 6) advanced the concept of the "chelogenic cycle," which was the first attempt to explain long-term geochronological episodicity principally in terms of then emerging ideas of crustal mobility and the disruption, dispersal, and reassembly of continents. This concept, which has been further advanced by Sutton (1967, 1970, 1973, 1977), is reviewed as follows by Professor Sutton (personal communication, 14 November 1978):

> Chelogenic cycle was a term devised to describe a sequence of events which periodically leads to a peak of igneous activity. The peak at about 1.7 b.y. is an example. At such a peak a greater part of the continental crust was affected by igneous activity than immediately before. In other words such a peak marks an interruption in the process of stabilisation that has continued throughout geological time. The subsequent analysis of apparent polar wander curves indicates that major shifts in the sense of movement of large blocks of lithosphere have taken place [see Paper 13 and Part II]. If the two phenomena are linked the chelogenic cycle could turn out to describe a periodic disturbance of the circulation of the mantle though this remains to be proved.

Sutton's concept of the chelogenic cycle was critically appraised in 1965 by Fitch and Miller (Paper 7). Through the separation of dates obtained for orogeny, regional metamorphism, and orogenic magmatism from dates determined for anorogenic events, they presented a revised division of geological time into individual cycles. Their revised cyclic scheme of non-orogenic followed by orogenic events, while strongly supporting Sutton's general thesis of long-term cyclical processes, better accords with the long-standing concept of the geosynclinal/orogenic cycle and with the new ideas of crustal tension, sea-floor spreading, and continental drift. Such separation of dates obtained on orogenic rocks from those characteristic of anorogenic environments was commended by York and Farquhar (1972) and has been applied to the study of geotectonic megacycles also by Burwash (1969), Williams (1973) and Fitch et al. (1974). These studies have confirmed the view of Sabine and Watson (1965) that the low points or troughs in isotopic-age histograms are as important to the interpretation of earth history as are the peaks associated with orogenic activity. Further evidence of long-term episodicity of anorogenic events is provided by the grouping of ages of Archean greenstone belts (e.g., Rutland, 1973a; Sutton, 1976a; Wilson et al., 1978) and of Proterozoic basic dikes in the same region (e.g., Baadsgaard and Mueller, 1973; Gates and Hurley, 1973; Stueber et al., 1976).

In 1966 Wilson (Paper 8) presented comprehensive evidence that: (1) in early Paleozoic time a "proto-Atlantic Ocean" existed in approximately the same location as the present North Atlantic; (2) during the middle and late Paleozoic this proto-Atlantic Ocean gradually closed; and (3) since Early Cretaceous time the present Atlantic Ocean has been opening, but not along the precise suture formed by the closing of the former ocean. This hypothesis of the opening and closing of ocean basins led Wilson (1968, p. 312) to formalize "the life-cycle of ocean basins." He wrote then:

> If continental drift has been going on for an appreciable part of geological time, at such rapid rates as recent work suggests, it means that a succession of ocean basins may have been born, grown, diminished, and closed again. Since ocean basins are the largest features of the earth's surface and would dominate other features it seems useful to outline the stages in their life cycle in terms of present examples. This makes it apparent that each stage has its own characteristic rock types and structures [as outlined in Table 2].

Wilson (1970, 1974) further discussed and enlarged upon this concept. He emphasized that this cycle has *two* phases: a first phase during ocean-basin birth and growth, marked by crustal tension, sedimentation, and basaltic magmatism, and a second phase during

Table 2 Stages in the life-cycle of ocean basins and their properties (after Wilson, 1968)

Stage	*Example*	*Motions*	*Sediments*	*Igneous rocks*
Embryonic	East African rift valleys	Uplift	Negligible	Tholeiitic flood basalts, alkalic basalt centers
Young	Red Sea and Gulf of Aden	Uplift and spreading	Small shelves, evaporites	Tholeiitic sea-floor, basaltic islands
Mature	Atlantic Ocean	Spreading	Great shelves (miogeosynclinal type)	Tholeiitic ocean-floor, alkalic basalt islands
Declining	Western Pacific Ocean	Compression	Island arcs (eugeosynclinal type)	Andesitic volcanics, granodiorite-gneiss plutonics
Terminal	Mediterranean Sea	Compression and uplift	Evaporites, red beds, clastic wedges	Andesitic volcanics, granodiorite-gneiss plutonics
Relic scar (geosuture)	Indus line, Himalayas	Compression and uplift	Red beds	Negligible

ocean basin closure, characterized by compression, uplift, and andesitic magmatism. Dewey and Burke (1974, p. 57) subsequently proposed the term *Wilson cycle* for Wilson's "perceptive conception that oceans open and close," which "has provided a remarkably successful framework upon which to analyze geosynclinal and orogenic evolution." The concept of the Wilson cycle was further discussed and applied particularly to Precambrian events by Dewey and Spall (1975), Burke et al. (1976, 1977), Baragar (1977), Burke (1977), and Hoffman et al. (1978). Dietz (1977) considered the concept of the Wilson cycle to be Professor Wilson's most important contribution to the theory of plate tectonics.

How then does the concept of the Wilson cycle relate to that of the globally episodic geotectonic megacycle? Proponents of such megacycles consider that they are explicable in terms of plate tectonic theory. Rutland (1973a, p. 1021), for example, suggested that the Proterozoic chelogenic cycle in Australia "may have been related to ocean floor spreading and subduction in a similar way to Cordilleran belts in the Phanerozoic." Moreover, Professor Gordon Gastil (personal communication, 19 April 1978) provides the following addendum to his conclusions in Paper 1:

> In 1960 the distribution of mineral dates in time indicated an episodicity (and possible cyclicity) on the order of hundreds of millions of years. Left unresolved was the question of whether these mineral dates meant that magmatic (and metamorphic) mineral generation was episodic or that the preservation of mineral dates was episodic. In either case, however, the episodicity called for some fundamental reorganization of crustal tectonics every few hundred million years. The theory of plate tectonics allows a model which may explain the alternate intervals of preserved and unpreserved mineral dates. In this model we assume that rises are active at a steady rate balanced by steady subduction. However, during intervals of low date preservation, all of the subduction is along oceanic island arcs; highly differentiated magma is minor, and the igneous rocks produced are largely shallow-level volcano-plutonic. Such arc material is eventually driven against the edge of a continent, metamorphosed, intruded, and effectively removed from the mineral-date record. However, when these arcs collide with the continental edges, or when the continental areas themselves collide, and subduction takes place beneath the edge of established continental crust, the resultant mineral genesis often becomes part of the permanent mineral-date record. Continent-continent collisions force a rearrangement of the plate-motion pattern initiating a new epoch (or cycle).

The ages of proposed Wilson cycles (e.g., Wilson, Paper 8;

Baragar, 1977; Hoffman et al., 1978) fit in with Fitch and Miller's (Paper 7) scheme of chelogenic cycles. The chelogenic cycles as perceived by Fitch and Miller therefore may reflect overall consonance of important Wilson cycles throughout the world. Paleomagnetism of pre-Mesozoic rocks will have a decisive role in determining the magnitude of past continental separations and the former widths of now-deformed sedimentary basins, but at present such data are open to conflicting interpretations (e.g., Burke et al., 1976; McElhinny and McWilliams, 1977).

5

Reprinted from *Nature* **195**:1248–1249 (1962)

CONVECTION CURRENTS IN THE EARTH'S MANTLE

By Prof. S. K. RUNCORN

King's College, Newcastle upon Tyne, and Jet Propulsion Laboratory, California Institute of Technology, Pasadena, California

IN a recent attempt to provide an explanation of continental drift[1], I postulated that convection currents were occurring in the Earth's mantle. This supposes the mantle to have zero strength under stresses persisting over millions of years and to have a Newtonian viscosity. I then applied Chandrasekhar's[2] theory of convection in spherical shells under a radial gravitational field, in which he showed that instability in the marginal state occurs through a convection pattern described by a spherical harmonic, the degree of which depends on the ratio (η) of the radii of the inner to the outer spherical boundaries. Vening Meinesz[3] had argued that spherical harmonic analysi[s] of the present topography of the globe reflects th[e] convection pattern in the Earth's mantle, as th[e] continents would tend to lie above the descendin[g] currents. He showed that the present positions o[f] continents and oceans implied that the convectio[n] currents in the mantle had strong harmonics of thir[d], fourth, and fifth degree. Chandrasekhar's calculatio[n] showed that, as η for the Earth's mantle is at presen[t] 0·55, these harmonics are those through which th[e] thermally induced instability would be expected t[o] occur.

Therefore, I argued that if the accretion theory of the origin of the Earth was accepted and it was supposed that the Earth's iron core had only separated gradually through the Earth's life, as originally suggested by Urey[4], and its radius was still increasing, a change in the number of convection cells in the mantle might recently have occurred. This increase in the degree of the convection was taken to be the mechanism for the dispersal of the continents since the Palæozoic. Thus the surprisingly recent time—compared with the life of the Earth—at which continental drift, as understood by geologists and palæomagnetic workers, occurred is given a rational explanation.

It has recently been possible to obtain new evidence for this process. The ratio η of core radius to the Earth's radius, at which the Rayleigh number of a given harmonic becomes less than that of the harmonic of next lower degree and therefore may be expected to develop in place of the latter, can be calculated from Chandrasekhar's theory; but the values depend on the boundary conditions used. It is clear that for the Earth these are a free inner boundary and a rigid outer one, for shearing stresses at the core-mantle boundary must vanish and there can be no doubt of the strength of the rigid crust of the Earth. If these boundary conditions are used one can derive[5], from Chandrasekhar's work, that the ratio of the radii of core to Earth at which the second harmonic replaces the first harmonic as the mode through which instability occurs is 0·06; for the transition from second to third the value of η is 0·36; and for third to fourth 0·49. During these transitions the continents will be under considerable stress as the convection-pattern changes to its new form. The last such transition from fourth harmonic to the fifth is observed in the geological column as continental drift.

The question now arises whether there is any record in the Earth's crust of the earlier transitions; these mode changes would be expected to cause world-wide orogeny, involving extensive recrystallization of crustal material. Radioactive age determinations have been made on igneous and metamorphic rocks which have been formed in the deeper parts of the crust as the result of orogenic forces. It is therefore of great interest that the ages obtained for such rocks have been found to be grouped in peaks of about ± 100 m.y. around the dates 200 my., 1,000 m.y., 1,800 m.y., and 2,600 m.y. ago. The most recent peak covers much of the geological record since the middle Palæozoic, and we have seen that this is associated with continental drift and the transition from a fourth to fifth degree convection pattern in the mantle. It is natural to identify the 1,000-m.y. peak with the transition of the third to fourth degree convection pattern, the 1,800-m.y. peak with the transition from the second to third degree, and the 2,600-m.y. peak with the transition from the first to second. Gastil[6] has reviewed the radioactive age dates and has shown that the peaks are present whether the ages are determined by the rubidium-strontium method, the lead method, or the potassium-argon method. American rocks give a peak at 1,300 m.y., which is thought to result from later metamorphism. He also shows that in spite of the poor coverage in some continents, the peaks appear to represent world-wide events. Davis and Tilton[7] had also come to a similar conclusion concerning the reality of these events: they are not simply due to deficiencies of sampling. It must, however, be emphasized that the peaks are broad (of the order of 200 m.y.) and further proofs that they are truly world-wide are much needed[8].

In this way the dates at which the core attained these critical values is determined as shown in Fig. 1. Before this new information was available, I had assumed that the core started its growth 4,600 m.y. ago—at the time of the Earth's origin[1]. Fig. 1 shows that it must have started its growth a little over 3,000 m.y. ago. But this, of course, is in fact entirely reasonable: one must allow time for the Earth to heat up to temperatures at which the creep processes necessary for convection will become important. As these depend on a factor exp (E/kT), where E is an activation energy, k Boltzmann's constant, and T the temperature, it may be supposed that the creep process rather suddenly became important and then convection commenced. The source of the heat is the uniform distribution of radioactivity. Thereafter, it may be supposed that the convection just carries off the heat generated by radioactivity within the Earth and by the release of potential energy due to the settling of iron into the core.

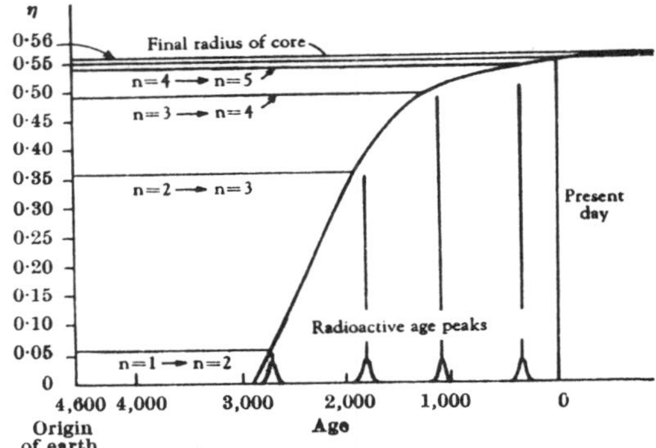

Fig. 1. Growth of radius of Earth's core

The law of growth of the core found by the above argument fits well with the simple model suggested previously by me[1], that is, that the rate of movement of iron into the core is proportional to the surface area of the core and to the mass of iron remaining in the mantle. Initially, the radius will grow linearly with time as equal increases in radius represent increments of mass proportional to the surface area of the core. Eventually, the curve will become asymptotic to the final radius (g) of the core when all the iron has left the mantle. It is therefore worth fitting by least squares the theoretical equation previously derived on this basis[1] :

$$\log_e \frac{(r/g-1)}{(r/g)^2 + (r/g)+1} + 2\sqrt{3}\tan^{-1}(-2\,r/g-1)/\sqrt{3} = Kt + c$$

where K and c are constants and r the radius of the core at time t. The fitted curve is shown in Fig. 1. The value of g is found to be 0·56: thus about 6 per cent of the free iron still remains in the mantle.

[1] Runcorn, S. K., *Nature*, **196**, 311 (1962).
[2] Chandrasekhar, S., *Hydrodynamic and Hydromagnetic Stability* (Oxford, 1961).
[3] Vening Meinesz, F. A., *Proc. Kon. Ned. Akad. Weten.*, **55**, 527 (1952).
[4] Urey, H., *The Planets* (Oxford Univ. Press, 1952).
[5] Runcorn, S. K., *Continental Drift*, Chap. 1 (Academic Press, 1962).
[6] Gastil, G., *Amer. J. Sci.*, **258**, 1 (1960).
[7] Davis, G. L., and Tilton, G. R., *Researches in Geochemistry*, edit. by Abelson, P. H., 190 (John Wiley, New York, 1959).
[8] *Carnegie Inst. Washington Year Book*, **59**, 208 (1960).

6

Reprinted from *Nature* **198**:731–735 (1963)

LONG-TERM CYCLES IN THE EVOLUTION OF THE CONTINENTS

By Prof. J. SUTTON

Department of Geology, Imperial College of Science and Technology, London, S.W.7

SEVERAL of the long-term geological processes which occur in the Earth's crust are known to have fluctuated over periods of time longer than the periods of 200 million years or so often accepted as the intervals between successive orogenies. Gastil[1], for example, has shown that over the past 3,000 m.y. igneous and metamorphic minerals were formed at unusually high rates during seven episodes, each of 200–400 m.y. duration, as compared with the intervening periods. Radiometric determinations show that certain parts of the crust have acted as mobile belts (that is, they have been intermittently the site of thick sedimentation, folding, regional metamorphism, migmatization or widespread granitization) over periods of as long as 750 m.y. before finally reverting to a more stable condition; such regions are exemplified by the Urals[2] or the Appalachians[3]. Structural analyses, coupled with radiometric age-determinations, have now made it possible to break down most of the large terrains of crystalline rocks into their constituent structural provinces and it has become apparent that some of these provinces took as long as 800 m.y.[44] to evolve. The recent increase in radiometric data has brought out the striking fact that in widely separate parts of the world major structural provinces developed in the last 3,000 m.y. began to form during one or other of three periods, none of which appears to have greatly exceeded 200 m.y. in duration. As each province is made up of a cluster of fold-belts which truncate older chains and are in turn cut by younger structures, this relationship indicates that in these three periods the pattern of mobile and stable regions throughout the crust underwent more drastic rearrangement than in the intervening longer stretches of time. Since the number of orogenic cycles which have occurred in the past 3,000 m.y. may be in the region of fifteen or more[4], this suggests that we are dealing with some structural rhythm of longer duration than the orogenic cycle.

The probable existence of such a long-term rhythm first emerged when investigators familiar with the structural and magmatic history of Africa, Western Australia, Canada, India and the U.S.S.R. began to make intercontinental comparisons. Holmes[5] was the first to direct attention to possible geochronological similarities between African, Indian and Australian provinces. Wilson *et al.*[6] have pointed out that the major structural divisions in Western Australia were evolved over periods of time comparable with those postulated for the formation of structural provinces in the Canadian Shield, a comparison confirmed by more recent work in Canada[7]. In discussing this recent work, Stockwell[8] has pointed out that the dating of the Superior, Churchill and Grenville Provinces of the Canadian Shield indicates chronological divisions similar to those proposed by Voitkevich[9] on the basis of reliable age determinations from the Baltic, Ukrainian, African, Indian and Australian Precambrian. Voitkevich distinguished four main geochronological divisions, starting at 2,650 ± 150 m.y.; 1,800 ± 90 m.y.; 1,030 ± 50 m.y.; 550 ± 10 m.y., which he named the Shamvaian, Svecofennid, Grenville and Indian Ocean orogenic cycles. The 'obvious similarity' between the sequence of events in the Ukraine and Baltic Shields and that in other Precambrian continental masses led Vinogradov and Tugarinov[10] to propose a sub-division of the Precambrian, rather similar to that of Voitkevich.

SCHEME OF SUB-DIVISIONS OF PRECAMBRIAN (Vinogradov and Tugarinov, 1961)

Boundary	Division
600 ± 50	
	Upper Proterozoic
1,100 ± 100	
	Lower Proterozoic
1,900 ± 100	
	Archaean
2,700 ± 150	
	Katarchaean
3,600 ± 200	

Vinogradov and Tugarinov's suggestion that these divisions might prove to be valid throughout the world receives support from the information summarized in Table 1, which gives the range of dates (youngest and oldest dated minerals from each structural province) reported from the major Precambrian terrains; with the exception of South America (for which adequate data are not available) the figures cover every continent. It appears from Table 1 that the main structural units established by geologists working in widely separated areas began their development in one of the following three periods:

2,900 — 2,700 m.y.
1,900 — 1,700 m.y.
1,200 — 1,000 m.y.

Table 1. RANGES OF DATES IN TECTONIC PROVINCES (DATES ONLY FROM ANTARCTICA)

	m.y.	m.y.	m.y.	Ref.
Scotland	2,800—2,200	1,700—1,300	?1,000—750 —400	38, 39
Baltic	2,900—1,900	1,900—1,250	1,100—400	28
Central Africa	2,700—2,000	1,850—1,500	1,100—475	4, 18, 40
Canada	2,750—2,300	1,850—1,550	1,200—800	7
Western U.S.A.	2,700—2,350	1,750—1,200	1,000—18	13, 41, 42, 43
East Asia (N. China; Aldan)	2,700—2,250	1,900—1,100	1,100—650	30
India	2,450—2,300	1,650—1,300	1,150—735	35, 45
Australia	2,950—2,300	1,700—?1,000	1,100—100	6, 7
Antarctica		1,800—	—375	22
Central N. America	2,800—2,000	2,000—1,500 —1,200	1,200—500	44

Each period marks the start of a chain of connected events continuing for between 750 and 1,250 m.y.; the youngest cycle is still continuing, and there was no break at the beginning of the Cambrian.

A different type of periodic change in the tectonic history of the Earth is marked by the incidence of continental drift, or the movement of individual continental masses as opposed to polar wandering or shift of the crust as a whole relative to the mantle. In a paper[11] which introduced the suggestion that continental drift might have occurred in Precambrian times, Runcorn developed a theory of continental drift in which periodic changes in the convection system of the mantle were related to expansion of the Earth's core. Runcorn[12] afterwards pointed out that the times at which these changes in the mantle should, on his calculations, be expected, in fact coincided with three marked peaks of igneous and metamorphic activity recognized by Gastil[1]. These peaks occur between 2,750 and 2,450 m.y. between 1,900 and 1,600 m.y. and between 1,150 and 900 m.y. Each covers the earliest 200–300 m.y. of the chronological divisions of Vinogradov and Tugarinov[10] and marks an early stage in the evolution of one of the structural divisions listed in Table 1. Within the crust, we thus find fluctuations in the rate at which igneous and metamorphic minerals were created and in the incidence of continental drift, together with periodic displacements of the mobile or orogenic regions on an unusually large scale. Runcorn's theory provides an explanation of the first two phenomena, but in its present form does not appear to account for the distribution of orogenic belts, especially for the distribution in space and time of those formed over the past thousand million years.

The purpose of this article is to suggest that the three phenomena may be related to a cycle of events with a periodicity of 750–1,250 m.y. which has been repeated at least four times. The existence of the earliest cycle is indicated only by a small number of age-determinations giving values for minerals between 3,600 and 2,900 m.y. It will not be considered further. Structural maps of provinces formed during each of the three younger cycles show that although the effects of several orogenies may be detected within a province, the resulting fold-belts rarely truncate each other and never abruptly cut off an earlier belt of the same cycle. The various fold-belts of a single cycle are usually found in close proximity in several of the continents, and it is this spatial grouping of fold-belts formed over a certain time-span which gives rise to the great structural provinces such as those of the Canadian Shield[7]. If the products of the fifteen or more orogenic periods all followed unrelated courses through the crust, we should expect a much more complex system of intersecting belts. As the long-term cycles lead ultimately to the production of shields, the name 'chelogenic cycle' is proposed (Gr., literally a tortoise, and thence the carapace of shields held over besieging soldiers while sapping fortifications).

The four cycles at present recognized are named from a shield area where rocks of the appropriate age are exposed, just as Tertiary orogenic belts are said to belong to the Alpine orogeny. The four cycles are named, from oldest to youngest, the 'Kola', 'Shamvaian', 'Svecofennid' and 'Grenville' cycles. The four names can be used for describing four major geochronological divisions and, therefore, so far as possible, Voitkevich's[9] nomenclature is retained; his Grenville cycle is re-defined and the Kola peninsula used to provide a name for the earliest cycle.

The Grenville Chelogenic Cycle

This cycle began between 1,200 and 1,000 m.y. ago and has continued to the present day. The relationship between successive belts of a single chelogenic cycle and the long-term structural changes throughout the cycle are best illustrated by the arrangement of the Grenville, Caledonian, Hercynian, Nevadan and Alpine mountain belts formed during the past thousand million years.

The type Grenville orogenic belt occupies the coastal region of eastern Canada and radiometric dating shows that it extends as far as the borders of New Mexico and Texas[13,44]. Scattered dates of about 1,000 m.y. indicate that some thermal event of Grenville age (metamorphism or igneous intrusion) affected also at least part of the western coast of North America[13]. Thick sequences of late Precambrian sediments occur in eastern and northern Greenland[14]. In late Precambrian time it would appear that the continental mass of North America was flanked to east and west by mobile belts and this condition continued into the Lower Palæozoic[15]. Geosynclines were extended or initiated in Cambrian and Ordovician times along the western, south-eastern, north-eastern and northern margins of the North American craton, truncating rocks formed in the Shamvaian and Svecofennid cycles[14,44].

Between 1,000 m.y. and 500 m.y., a network of fold-belts was formed on every continent, isolating a number of stable blocks of unmodified older rocks. At least nine such blocks can be identified and more may have existed in oceanic areas.

In south-west Europe, a stable block is flanked on the north-west by the British Caledonides and on the east by the Caledonian belt which runs from Scandinavia to central and southern Europe. The Baltic Shield and Russian platform make a third block which is separated from the Siberian platform by a zone of late Precambrian geosynclinal sediments extending southward from the Arctic coast of the U.S.S.R. into central Kazakhstan[15]. Eastward continuations of these sediments folded in the late Precambrian and Lower Palæozoic divide the Siberian platform from the North China–Korean Shield[15] and this in turn from the South China–Vietnam platform[16].

Australia[17] displays fragments of two stable blocks one of which may have been united with that now exposed in south-east Africa. The latter is flanked to the south, west and north by late Precambrian or Lower Palæozoic orogenic belts[18] which extend from the south-west coast through Rhodesia into East Africa, where they make the Mozambique Belt. North of Kenya this belt divides, one branch continuing into the Sudan and Egypt[19,20] and the other running north-eastward into Somaliland and the Horn of Africa. Yet another extends from south-west Africa along the west coast of the Congo and thence into Nigeria[21] and the western Sahara. This complex of African mobile belts, only recently shown to be of late Precambrian or Lower Palæozoic age, divides the older rocks into four blocks—that of south-east Africa, partly covered by Karroo, the central African block, a fragment seen in Mauretania and the extreme west of northern Africa which presumably continued into South America, and an Arabian block, not yet established on radiometric evidence, which must lie north of the Red Sea and the Gulf of Aden. The point to be emphasized is that by Lower Palæozoic times at least nine stable blocks had been isolated by the development of mobile belts in what are now the continental regions.

In Lower Palæozoic times, orogeny came to an end throughout Africa, except in the extreme north, in Antarctica except in Graham Land[22] and in Australia except in the east[23]. By the end of the Mesozoic, activity had ceased in the Urals and in the zone between the Siberian and North China[15] platforms. In north-west Europe and Greenland orogeny ended in the Palæozoic and in the south-eastern United States in Mesozoic times.

The ending of orogenic activity in these regions led to the union of stable blocks and as a consequence the nine isolated continental blocks of early Palæozoic times were reduced by the Jurassic to two continental blocks, each surrounded by zones of continuing activity on or near the sites of surviving Palæozoic mobile belts. One block underlay Eurasia and North America and was flanked by the mountain belts of Eurasia and the northern Pacific; a second underlay the great mass of Gondwanaland, and a third presumably lay in the Pacific, surrounded by the circumpacific Mesozoic mobile belts. Because so little is known of the Palæozoic history of the Pacific it is not clear whether this block evolved from smaller stable blocks. Fragmentation of the two stable continental masses suggests further enlargement of the convection cells below the North American–Eurasian and Gondwanaland blocks. A later stage of the chelogenic cycle is seen at the present day when, as Wilson[25] has pointed out, the mid-Oceanic ridges of the Atlantic, the Indian Ocean, the Southern Ocean south of Australasia, and those of the south and east Pacific appear to delineate a single convection cell of world-wide extent. A further development which might be anticipated would be the junction of the east Pacific and Atlantic Oceanic ridges to form a continuous structure around the Earth. (See Dietz[46] and Menard[26] for discussion of structure below North America and in Caribbean.) If this should occur the continents might, in time, be expected to form two clusters in the northern and southern hemispheres, recalling the way in which Gondwanaland and Laurasia are believed to have been arranged at the start of the present cycle.

If the assumption that fold-belts and regions of regional metamorphism and granite formation mark the zones where downward-moving convection currents have thickened the crust is correct, a simple conclusion emerges. During the past 1,000 m.y. the size of the largest convection cells below the continents has increased intermittently, initially by the fusion of neighbouring cells, latterly by fusion coupled with the outward migration of cell margins as indicated by continental drift. As this change occurred the pattern of orogenic belts became simpler, the network established early in the cycle gradually collapsed from the central parts of the continents outwards, and the three stable areas of Laurasia and Gondwanaland and the Pacific developed. In the later stages of the chelogenic cycle disruption of the areas where orogeny had ceased began to occur. The fact that the fold-belts on the peripheries of the two expanding continental masses were succeeded by younger fold-belts or themselves remained active throughout the cycle while those in the interiors of the masses became inactive in the latter part of the cycle indicates a close link between continental drift and the changing pattern of orogenic activity. Some smaller cells appear to be present throughout the cycle, and are indicated, for example, as H. Cloos[26] has suggested, by the relief of the African continent. The comparative regularity of the successive mountain belts in North America, with which has been contrasted the complexity of Europe, Africa and Australia[44], appears to result from the fact that the present-day North American continent evolved from a single stable block of Palæozoic times. Each of the three other continents contain fragments of several Palæozoic stable blocks.

Over much of the crust the network of mobile belts established early in a chelogenic cycle appears to follow a pattern independent of any older structure. This is particularly true in regions where a period of stability occurred between the ending of orogeny connected with one cycle and the return of orogeny at the start of the next chelogenic cycle. However, in regions where folding continued until the end of a chelogenic cycle, the new orogenic belts of the succeeding cycle might have followed the lines of the older structures. More information is needed on this question. What does seem clear is that the start of a cycle is marked by the development of many new convection cells and that in consequence a network of mountain chains forms and orogeny returns to previously stable areas. The new fold-belts truncate those of earlier cycles in the way seen at the north-west margin of the British Caledonian belt or the western margin of the Mozambique belt in East Africa[5]. This relationship is in strong contrast to the arrangement of successive fold-belts of one cycle. Over much of the early part of the Grenville cycle (and possibly also in other cycles) there appears to have been little migration of fold-belts, and the orogenic pattern established about 1,000 m.y. ago was maintained until Lower Palæozoic times, in many areas for periods of as much as 600 m.y. Much of the Palæozoic Appalachian[12,27] chain lies on a basement of rock metamorphosed at 1,000 m.y., though it is not precisely superimposed on the Grenville belt[44].

In north-west Europe[26,28,29], the Urals[29], the Baikal region of Siberia[30] and in parts of Africa[18] and the Antarctic[22], thermal events occurred sporadically between 1,000 m.y. and 400 m.y. and their effects are often superimposed on one another in the same rocks. In some areas such as northern Scotland, where the stratigraphy is well known, it seems clear that there was no general period of uplift leading to sedimentation of a molasse type until the deposition of the Old Red Sandstone which preceded the final return to stable conditions. The fold-belts formed in the early part of the cycle differ from those formed in the final stages in making a more intricate pattern, in showing less lateral migration (a feature perhaps associated with the absence of continental fragmentation) and possibly also in their history of sedimentation. The material filling the troughs was in some instances (for example, the Torridonian of Scotland) derived from uplifted areas of much older rocks on the flanks, which were apparently not regions of active folding. Although the Torridonian closely resembles the Old Red Sandstone in facies, the latter was derived from erosion of a recently active fold-belt.

The forms and positions of the continental masses undergo considerable modifications during the chelogenic cycle. The initial arrangement inherited from the earlier cycle is first modified by the development of the network of orogenic belts and then more drastically changed when the convection cells become sufficiently large to promote continental drift. In time, the continental masses are strung out along the downward flowing margin of a single cell of world-wide extent. The sites of the upward moving convection currents of this cell are marked by the oceanic ridges with their abnormally high heat-flow[36,46]. The cell at the present day, the existence of which Wilson[25] has recently pointed out, has developed from parts of the smaller cells underlying the Pacific and the two continental masses in very much the same way as the latter cells in turn had developed during the Mesozoic from the still smaller cells which underlay the continents in late Precambrian and Early Palæozoic times.

The Shamvaian and Svecofennid Cycles

So far as is known, the relationships of the orogenic belts formed during those cycles resemble those of the Grenville cycle. The oldest fold-belts of each cycle cut a variety of still older belts indicating that the Shamvaian and Svecofennid chelogenic cycles started with a return of orogenic conditions to regions which had previously been stable. In both cycles one can find belts which have migrated outwards from regions of older rocks. It seems reasonable to conclude that each cycle was marked, like the Grenville cycle, by a period when a network of mountain chains became established throughout the continents followed by a long period when orogenic activity withdrew from large areas in the centres of the continents. Late in the Grenville cycle (Jurassic–Tertiary) widespread basic vulcanicity took place in the disrupted fragments of Gondwanaland and in the North Atlantic region. Widespread systems of basic dykes are known to cut rocks of the Shamvaian cycle in south-western Greenland[31,32], Scotland[33], Karelia[34] and southern India[35]. Some are deformed in the succeeding Svecofennid cycle and must, therefore, be late Shamvaian or early Svecofennid in age.

It is suggested that these rocks were intruded during a phase of continental disruption towards the end of the Shamvaian cycle and more accurate knowledge of their distribution would indicate the positions from which dispersal of continental fragments took place at that time. There appears to be nothing petrologically distinctive about these basic rocks and the diagnostic feature of the vulcanicity at this stage of the cycle is the scale on which it occurs. Where this reaches the magnitude of the Arcto-British Tertiary province or of the Mesozoic basalts and dolerites of the southern hemisphere we may well have evidence of past disruption of continents. Such conditions may have been reached in what is now Greenland, north-west Europe and Finland about 2,000 m.y. ago in late Shamvaian times.

Nature of the Chelogenic Cycle

The hypothesis attempts to account for three groups of related phenomena: changes in the distribution of mountain chains over continents, changes in the number and shape of continental masses and changes in the nature of the convection system believed to exist in the mantle.

The cycle postulated here consists of a sequence of events (Table 2) which leads to the displacement and

Table 2. TYPICAL CHELOGENIC CYCLE

Stage	State of continents	State of mountain chains	Possible state of convection system in mantle
4	Two clusters develop once more	Further restriction of orogeny	Development of new world-wide cell from major cells of stage 3
3	Maximum dispersion of continental masses.	Further restriction of orogeny	Major cell under each fragmenting continent. No single cell of world-wide extent.
2	Fragmentation of the clusters starts	Restricted to outer parts of continent	Major cell under each continent. World-wide cell weak or non-existent.
1	In two clusters	Network over continent. Orogeny returns to previously non-orogenic regions.	Many cells of sub-continental size below each continent + world-wide cell originated in previous cycle.
Start			

(Vertical arrow beside the stages: 750–1,250 million years)

Typical cycle. Oldest stage at bottom of table. At the stage of maximum dispersal (stage 3) individual continents may contain fragments derived from both clusters as they existed at the start of the cycle (for example, India now part of Asia[37]).

disruption of the continents as they existed at the start of the cycle and later to the re-grouping of these disrupted masses of continental crust. At the end of the Svecofennid cycle the continental masses appear to have been assembled in two clusters and as there are indications that the present cycle could lead to a comparable arrangement it is tentatively concluded that it is an inherent feature of the cycle (one cluster may greatly exceed the other in size). If this conclusion is correct, the continents have throughout geological time been periodically dispersed from two clusters, one in each hemisphere, and then reassembled once more in two groups. It is possible that the relative position of the clusters is changed early in the cycle before fragmentation starts. The validity of this supposition can be tested by considering the distribution of mountain chains during a chelogenic cycle.

The process is cyclic, but a convenient break is recorded in regions where orogenic activity returned to areas previously free of mountain building for at least several hundred million years. The start of the sedimentation which preceded this return of mountain building may be taken as marking the beginning of the cycle. At an early stage, mountain chains form a network over the continents and an essential feature is the regularity of their distribution. This condition was maintained in the Grenville cycle for some 600 m.y. and during this time it seems unlikely that any major break-up of the continents can have occurred for such an event would have interrupted the evolution of this network of mountain chains (though the clusters might have moved *en bloc* relative to each other). As it is known that in Lower Palæozoic times towards the end of this phase the present-day continents were still closely clustered, it is assumed that this phase of mountain building characterizes periods in the Earth's history when the continents were grouped in large masses.

It is suggested that as the cycle develops, orogeny apparently ends in large parts of the interiors of the continents and in contrast to the previous arrangement we can distinguish a large non-orogenic region in each continent surrounded by a peripheral region where orogeny continues in parts of the original network of mountain chains or in their close proximity. In the Grenville cycle this development led to the formation of two stable (that is, non-orogenic) continental areas ringed by mountain chains by Jurassic times in place of the nine or more smaller non-orogenic regions encircled by the network of mountain chains which covered the continents in Lower Palæozoic times.

It is argued that this change comes about by the fusion of the sub-continental convection cells, established early in the cycle, to form the very much larger cells such as developed in the Mesozoic, when two major cells formed under Gondwanaland and Laurasia respectively. As a result of the development of the two cells, outward movement of continental masses occurred towards the peripheries of the cells, where folding continued. The onset of continental fragmentation is thus on this hypothesis directly related to the appearance of large convection cells in place of the more numerous smaller cells which previously underlay the continents. At this stage, individual continental masses may move outwards from one fragmented cluster to join with parts of another disrupted continent. India[37] and probably Arabia moved from Gondwanaland to Asia in this way during the Mesozoic.

To-day, as Wilson has pointed out, a single world-wide convection cell has been established, and if in the future a link between the east Pacific and mid-Atlantic ridges should form, the Earth would have returned to a condition where an uninterrupted upward moving convection current (indicated by the abnormally high heat-flow marking such oceanic ridges) had been set up once more. Given such a situation, movement of continents through convection currents flowing outwards from the line of this oceanic ridge would eventually lead to the re-grouping of the sialic crust in two clusters around the poles, the southern group on this occasion consisting of South America and Antarctica. It is suggested that throughout the geological history of the Earth some such cycle has operated. Essentially the hypothesis postulates the periodic disturbance of a world-wide single cell convection system in the mantle by the growth of smaller convection cells which increase in size, fuse and ultimately give rise to a single cell system once more, after a cycle lasting 750–1,250 m.y.

The evidence for the changes is recorded in the continental crust, and in Table 2 an attempt is made to sketch the history of a typical cycle in terms of changes in continental structure. Since the cycle involves major changes in the convection system of the mantle, comparable changes presumably occurred below the oceans, but no attempt has been made to indicate these in Table 2. As the size of the cells below the continents increases it is suggested that orogeny ends in the central parts of continents and fragmentation of the sialic crust occurs. Ultimately, as at the present day, a single cell system is established once more and the continental fragments return, though in a different arrangement, to form new clusters and the cycle begins again. It is suggested that this cycle has been repeated at least four times during the geological history of the Earth.

[1] Gastil, G., *Proc. Twenty-first Sess. Intern. Geol. Congr.*, Pt. 9, 162 (1960).
[2] Ovchinnikov, L. N., and Harris, M. A., *Proc. Twenty-first Session Intern. Geol. Congr.*, Pt. 3, 33 (1960).
[3] Faul, H., Stern, T. W., Thomas, H. H., and Elmore, P. L. D., *Amer. J. Sci.*, **261**, 1 (1963).
[4] Holmes, A., and Cahen, L., *Colonial Geol. Min. Resour.*, **5**, 3 (1955).
[5] Holmes, A., *Proc. Eighteenth Session Intern. Geol. Congr.*, Pt 14, 254 (1952)
[6] Wilson, A. F., *et al.*, *J. Geol. Soc. Austral.*, **6**, 179 (1960).
[7] Lowdon, J. A., *Geol. Surv. Canad. Paper* 61–17, 1 (1961).
[8] Stockwell, C. H., *Geol. Surv. Canad. Paper* 61–17, 108 (1961).
[9] Voltkevich, G. H., *Priroda*, 77 (1958).
[10] Vinogradov, A. P., and Tugarinov, A. I., *Ann. N.Y. Acad. Sci.*, **91**, 50 (1961).
[11] Runcorn, S. K., *Nature*, **193**, 311 (1962).
[12] Runcorn, S. K., *Nature*, **195**, 1248 (1962).
[13] Wasserburg, G. J., Wetherill, G. W., Silver, L. T., and Flawn, P. T., *J. Geophys. Res.*, **67**, 4021 (1962).
[14] Kay, M., *Geol. Soc. Amer. Mem.* 48, 1 (1951).
[15] Shatsky, N. S., and Bogdanoff, A. A., *Intern. Geol. Rev.*, **1**, 1 (1959).
[16] Schüller, A., and Ying Szu-Huai, *Geologie*, **8**, 699 (1960).
[17] Wilson, A. F., Compston, W., and Jeffery, P. M., *Ann. N.Y. Acad. Sci.* **91**, 514 (1961).
[18] Cahen, L., *Ann. N.Y. Acad Sci.*, **91**, 535 (1961).
[19] Gheith, M. A., *Ann. N.Y. Acad. Sci.*, **91**, 530 (1961).
[20] Schürmann, H. M. E., *Geol. Fören Stockh.*, **83**, 109 (1961).
[21] Bonhomme, A., *Ann. Fac. Sci. l'Univ. de Clermont*, **5**, 1 (1962).
[22] Picciotto, E., and Coppez, A., *Ann. Soc. Géol. Belg.*, **85**, 263 (1962).
[23] Evernden, J. F., and Richards, J. R., *J. Geol. Soc. Austral.*, **9**, 1 (1962).
[24] Haller, J., *Medd. Om Grønland*, **154** (1956).
[25] Wilson, J. T., *Nature*, **197**, 236 (1963).
[26] Cloos, H., *Gesprach mit der Erde* (Munchen, 1947).
[27] Kulp, J. L., and Eckelmann, F. D., *Ann. N.Y. Acad. Sci.*, **91**, 408 (1961)
[28] Polkanov, A. A., and Gerling, E. K., *Trav. Lab. Pre-Cam. Geol. Acad Sci. U.S.S.R.*, **12**, 7 (1961).
[29] Harris, M. A.. *Mat. Geol. Ural*, **8**, 127 (1961).
[30] Vinogradov, A. P., and Tugarinov, A. I., *Geochim. Cosmoch Acta*, **26** 1283 (1962).
[31] Noe Nygaard, A., *Rep. Eighteenth Session Intern. Geol. Congr.* (1948).
[32] Berthelsen, A., *Geol. Rund.*, **52** (1962).
[33] Sutton, J. and Watson, J., *Tran. Roy. Soc. Edin.*, **65**, 89 (1962).
[34] Eskola, P., *Quart. J. Geol. Soc. Lond.*, **104**, 461 (1949).
[35] Pichamuthu, C. S., *J. Geol. Soc. India*, **3**, 106 (1962).
[36] Menard, H. W., *Science*, **132**, 1737 (1960).
[37] Clegg, J. A., Radakrishnamurty, C., and Sahasrabudhe, P. W., *Nature* **181**, 830 (1958).
[38] Giletti, B. J., Lambert, R. St. J., and Moorbath, S., *Ann. N.Y. Acad Sci.*, **91**, 464 (1961).
[39] Giletti, B. J., Moorbath, S., and Lambert, R. St. J., *Quart. J. Geol. Soc Lond.*, **117**, 233 (1961).
[40] Cahen, L., *Ann. Soc. Géol. Belge*, **85**, 185 (1963).
[41] Damon, P. E., and Giletti, B. J., *Ann. N.Y. Acad. Sci.*, **91**, 443 (1961).
[42] Giletti, B. J., and Gast, P. W., *Ann. N.Y. Acad. Sci.*, **91**, 454 (1961).
[43] Lipson, J., Folinsbee, R. E., and Baadsgaard, H., *Ann. N.Y. Acad. Sci.* **91**, 459 (1961).
[44] Tilton, G. R., Davis, G. L., Hart, S. R., and Aldrich, L. T., *Ann. Rep. Geophys. Lab. Washington*, 1961–62, 173 (1962).
[45] Sarkar, S. N., and Saha, A. K., *Geol. Mag.*, **100**, 69 (1963).
[46] Dietz, R. S., *Nature*, **190**, 854 (1961).

7

Reprinted from *Nature* **206**:1023–1027 (1965)

MAJOR CYCLES IN THE HISTORY OF THE EARTH

By F. J. FITCH

Department of Geology, Birkbeck College, University of London

AND

DR. J. A. MILLER

Department of Geodesy and Geophysics, University of Cambridge

SUTTON[1] has developed a hypothesis of long-term cycles in the evolutionary history of the continents, which he calls 'chelogenic' (or shield forming) cycles. In each cycle he recognizes a characteristic sequence of geological events involving continental disruption, drift and reconstruction, leading, in certain more or less restricted belts, through a stage of mobility and mountain-building to the eventual renewal of continental shield stability. Any evaluation of Sutton's analysis must depend largely on the latest findings in four broad fields of research in the Earth sciences—geochronology, geophysics, petrology and classical geology.

For such an evaluation, the most important of these fields of knowledge is the comparatively new and rapidly expanding science of geochronology. Before the advent of radiometric methods of rock and mineral age determination, age studies within individual continental shields could produce, even under optimum conditions, little more than a relative sequence of events, and comparison between separated shield areas was a form of scientific speculation, difficult to prove or disprove. To a large extent, isotopic age determination provides the missing key to our understanding of the Pre-Cambrian past of the Earth. With this new tool it becomes possible to make comparative studies of the geological histories of the continental shields based on the probable identity of dated events in each separate shield area. Sutton quotes the pioneer work of numerous workers[2-10] in this field.

Research in petrology and classical geology over the past 50 years has produced a considerable advance in our knowledge of the Pre-Cambrian. Major structural provinces, such as those mapped in the Canadian Shield[11], and in Africa[12], have been investigated and delimited in many areas of the globe, and it is suggested by both Gastil and Sutton that the culminations of these large-scale periods of structural development may be broadly synchronous on a world-wide scale. The concept of a 'geological cycle' is not new: it has been discussed in various forms by many geologists over the past 150 years. Examples of recent developments of the classical petrological and geological points of view are those of Tyrrell[13] and Harpum[14]. The important addition to this theory made by Sutton is his assertion that the major long-term cycles identifiable in each of the structural units of the crust can be correlated and brought together in a single world synthesis.

Another relevant field of enquiry is that of geophysics, for any concept which postulates a unitary cyclical process governing the visible geology of the outer crust must look for an explanation of this phenomenon on an Earth scale. Runcorn[15,16] has suggested one possible mechanism by a further development of the well-known hypothesis of convection currents in the mantle, and his variant was the basis of Sutton's original conception of chelogenic cycles.

In a critical appreciation of Sutton's hypotheses we have reviewed the evidence quoted by him and have repeated and extended the type of analysis of isotopic age data originally made by Gastil. On the geological side, we consider that evidence in favour of continental drift during Phanerozoic time is now strong (see our contribution[17] to the Royal Society symposium on "Continental Drift", 1964). Evidence for numerous periods of Pre-Cambrian continental drift, as advocated by Sutton, is, however, far less conclusive. Ancient basic dyke swarms, intruded during phases of tensional stress in the crust, may sometimes be taken as indicative of a period of continental drift during Pre-Cambrian times, but in most instances it is not possible to relate the distribution of the exposed dyke swarms to any convincing history of complete disruption of the continental shields (see, for example, the work of Fahrig and Wanless[18] on the Canadian Shield). In addition, the more fundamental problem of how and when the discontinuous continental sialic skin of the Earth evolved has not been solved. The work of Patterson and Tatsumoto[19] on lead isotopes in detrital feldspar, for example, supports the hypothesis of an early evolution of the bulk of the continental substance, while the work of Hurley *et al.*[20] on the distribution of rubidium and strontium isotopes in rocks suggests a continuous process of sialic evolution that would be compatible with a hypothesis of continental accretionary growth similar to that proposed by Wilson[21]. We favour a compromise view, in which it is postulated that continuous differentiation of continental material has gone on throughout the Earth's history, but that the rate of production of new sialic substance has fallen off rapidly from an initially high value. In an alternative view, a large increase in the surface diameter of the Earth over geological time has been suggested as a possible explanation of the discontinuous nature of the sialic skin[22-24]. The fact that all oceanic rocks are considerably younger than those of the continental basement would appear at first sight to support this hypothesis, but Lyttleton[25] and Runcorn[26] have advanced arguments which apparently disprove the possibility of Earth expansion during geological time. It would appear that agreement regarding the fundamental mechanism which controls Earth history, and may be the cause and explanation of geological cycles, continental drift and igneous activity, cannot be achieved in our present knowledge. Although it is to-day the most favoured explanation, there are as strong geophysical arguments against the hypothesis of convection currents in the mantle as there are for it (see, for example, the discussions of gravity and other data from artificial satellites in the Royal Society symposium on "Continental Drift", 1964).

Recent work, especially that on the mid-oceanic ridges, appears to indicate, however, that any really satisfactory explanation of oceanic geology and volcanism, of continental drift and of the many more ancient episodes of crustal tension known from the geological record, must involve some form of intermittent ocean floor spreading, as proposed by Heezen[26], Wilson[27] and others. If continuous convection in the mantle is not possible, then perhaps convective overturn (half-cycle convection) may still be possible as an intermittent event. If such a process occurs on occasion in the outer layers of the Earth, it need not be necessarily incompatible with the overall Earth contraction suggested by classical structural geology and now so strongly supported by the work of Lyttleton on the core and by the inertia studies of Runcorn.

Within this present-day confusion of conflicting hypotheses, it is apparent that the rapidly expanding field of geochronology holds promise of important new evidence, and, as an extension of our critical review of Sutton's work, we have commenced a long-term investigation of the isotopic age evidence from rocks and minerals in its relation to the findings of field geology. There has been a tremendous increase in the volume of isotopic ages from rocks since Gastil made his pioneer analysis in 1962, and we believe that it is now possible to separate from the data at present available indications of greater geological significance than was then feasible. The results of a preliminary re-investigation of the isotopic age data are summarized diagrammatically in column 5 of Fig. 1. For the purpose of this preliminary analysis the world data have been divided into two categories. The first category consists of all those ages which we interpret as indicating past episodes of orogenic earth movements, regional metamorphism and orogenic magmatism (such as ages from regional metamorphic rocks and minerals, synorogenic granites, pegmatites and from basic magmatic rocks closely associated with orogenesis): the second category includes the ages which we interpret as being the initial age of extrusion or intrusion of some non-orogenic magmatic body (such as ages from tensional dyke swarms, continental flood basalts and ignimbrite provinces, mid-oceanic ridges, rift complexes, post-orogenic granites and from alkaline plutonic complexes). As a result of analysing the mineral and rock ages in this manner a striking cyclical array of age concentrations becomes apparent, and it is considered likely that, even allowing for major deficiencies in the available data (for large areas of the globe remained unsampled), these results are meaningful. They definitely suggest that the dominating factor in the long-term geological history of the Earth has been a major alternating cycle of predominantly non-orogenic intervals followed by predominantly orogenic episodes. In addition, there are indications, which may be verified by future work on as yet unsampled areas, of minor alternations within the major cycles (for example, the episode of late Scourian metamorphism which occurred during the dominantly non-orogenic first half of the Churchill cycle—as defined in this article—has analogues in the Grenville, Varisco-Caledonian and Alpine cycles).

The success of this preliminary analysis has encouraged us to undertake a more rigorous investigation of the isotopic age evidence, in which it is intended that the evidence from each continental shield area will be analysed separately before the world pattern is re-examined. A programme of rock dating from some of the unsampled and poorly sampled areas has been instigated as part of this project.

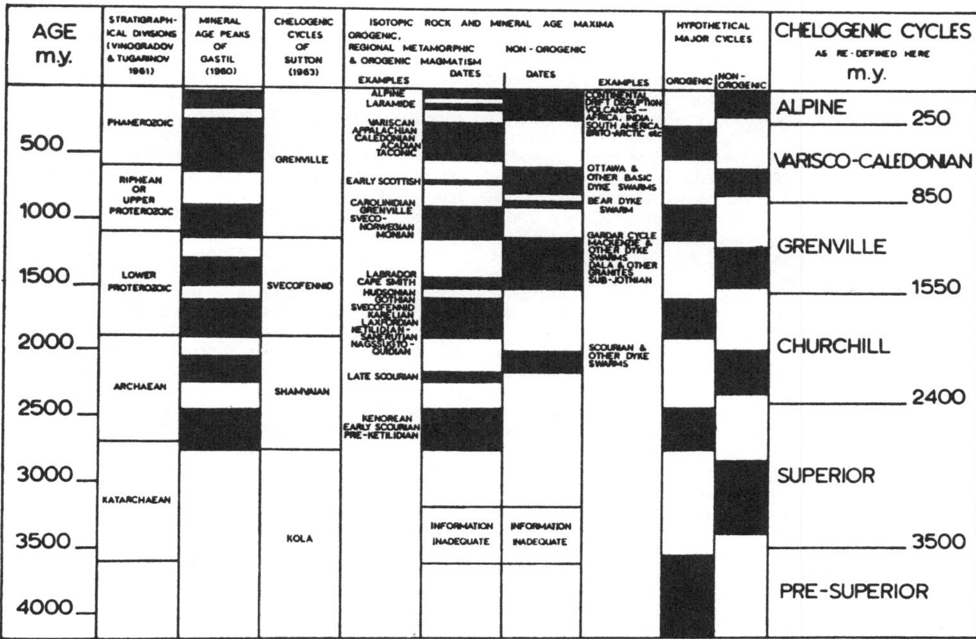

Fig. 1. Chelogenic cycles (time-units of Earth history)

Certain important conclusions are already clear from the preliminary work. While Sutton's general thesis of long-term cyclical processes controlling Earth history is strongly supported, it is seen that his division of geological time into individual cycles requires revision. On the basis of the present preliminary analysis it is possible tentatively to re-define the major divisions of Pre-Cambrian time into a series of chelogenic or geological cycles, each of which comprises a 'non-orogenic' period followed by an 'orogenic' period (see column 7 of Fig. 1). Pre-Cambrian units, as defined in the past, begin with an orogenic episode. It is thought preferable to follow the more usual stratigraphical practice (for example, as seen in the geosyncline/orogen concept), and to define each unit so that it develops through an early, dominantly sedimentary, non-orogenic, phase towards an orogenic culmination close to its end. Each chelogenic cycle and half-cycle so defined is a time-unit of Earth history. During the first, predominantly anorogenic half of the cycle, the thermal activity of the globe appears to be on the increase, leading to cratogenic magmatism of increasing extent—in the recently consolidated mobile belts of the previous cycle, post-orogenic granites and andesite–rhyolite volcanism; elsewhere, the appearance of syenitic and carbonatite alkaline intrusive complexes, extensive continental ignimbrite volcanism, flood basalt and oceanic basic volcanism. Rifting (and continental drift in appropriate circumstances) is typical of this half-cycle, which is characterized by the dominance of a tensional régime in the outermost layers of the Earth. Cratogenic magmatism appears to reach its maximum immediately before the inception of the succeeding orogenic half-cycle, for example, in North America, during the Grenville cycle (as defined in this article), we have the intrusion of the Duluth gabbro, the Muskox complex and the Lackner alkaline complex all just before the beginning of the Grenville orogeny. The dominantly orogenic half-cycle sees the full development of mobile belts which begin to extend across the Earth's crust during the latter part of the preceding anorogenic half-cycle. Within the mobile belts separate orogenic systems can be distinguished: each orogenic system is a space–time unit. Within each of these units a history of individual phases or events—geosynclinal, parorogenic, magmatic, evorogenic—can be recognized. In each mobile belt the crustal basement becomes metamorphosed and re-activated to a greater or lesser extent during the culminating orogenesis, and the final volume of sialic material in any given continental sector may be augmented by new differentiate from sub-crustal magmatic sources. A dramatic change from a world-tensional régime to one dominated by compression occurs during the orogenic half-cycle, and at its close the Earth as a whole appears to have expended the energy which has accumulated beneath its outer crust in the earlier parts of the chelogenic cycle. The end of each cycle is therefore characterized by a return to stability over the greater part of the Earth and by an extension of existing shield areas.

The concept of chelogenic cycles as outlined here does not represent an unqualified return to the classical geological hypothesis of alternating periods of world-wide tension and compression as developed by Stille[28]. The alternate halves of the chelogenic cycle which are predominantly anorogenic and orogenic, respectively, are of a different order of magnitude from the periods of Stille, and only become recognizable on a long-term world-scale. Furthermore, they must represent the effects of a cycle of events in the outer mantle to which the crust reacts passively. Continental drift disruption of sialic shield areas as a result of ocean floor spreading is a manifestation of world tension in the outermost layers of the Earth, but in certain regions of the crust, especially where two continental shields are being driven against each other, the results may be violently orogenic. At any single continental locality it may be possible to recognize a unique sequence of compressive and tensional events within either half of the chelogenic cycle. In some areas, and for long periods, it may seem, as it did to Gilluly[29] in his review of the tectonic history of the western United States, that there is an essential uniformity in tectonic and volcanic activity throughout geological time. Even in this area, however, Gilluly notes the occurrence of catastrophic plutonism that is apparently not related to either of the other processes. The essential character of the chelogenic cycle only becomes apparent when geological history is viewed on a world-scale in both time and space. The world concentration of major orogenic events towards the end of each cycle, such as that between 1,600 and 1,900 m.y. during the Churchill cycle, does not necessarily imply world-wide synchroneity of orogeny on a smaller time-scale. It is most probable that paroxysmal failure of mobile belts occurs only along relatively short lengths of a belt at any one time. As the independent sequences, of slow, then suddenly more violent, orogenic activity, seen at individual localities along mobile belts, become unified within a single orogenic system when seen on a larger scale in space and time, so the independent orogenic systems, when viewed together on a world-scale, can be seen to fit into the pattern of the chelogenic cycle.

Differences of opinion regarding the detailed interpretation and application of the research reported in this article are probable, but the main conclusion, that the geological history of the Earth as a whole reveals a large-scale cyclical repetition of predominantly anorogenic periods followed by predominantly orogenic periods, appears certain. Again, although views regarding the fundamental mechanisms which may control this cyclical history must, owing to the very nature of the problem, remain largely speculative at present, one major inference appears to be valid: this is that each chelogenic cycle represents an energy-release cycle affecting at least the outermost layers of the Earth. The intensity of energy release through the continental crust appears to rise over the cycle towards a maximum during the phase of widespread evorogenesis in the mobile belts (the phase of intense folding and regional metamorphism: isostatic uplift and the formation of topographically extreme mountains occurs later). This phase in each cycle may not represent the period of maximum energy release by the Earth as a whole, however, for it may well be that this occurs towards the end of the first half-cycle, when cratogenic magmatism, ocean floor spreading and continental drift are at their greatest intensity. At this time considerable quantities of heat energy could be dissipated directly into the atmosphere and through the ocean floors and the sea. The nadir of each cycle would occur at its beginning and end, following the final burst of rapid energy release through the crust during evorogenesis in the various mobile belts. In the first and last phases of a chelogenic cycle topographic contrast between oceans and continents would be at a maximum and heat-flow through the crust at a minimum. These conditions would favour the inception of ice-ages, and it is not surprising that ancient tillites are known from the Permo-Carboniferous (between the Varisco-Caledonian and Alpine cycles) and from rock sequences that may be early Varisco-Caledonian and early Grenville in age respectively. The current (Pleistocene-Recent) 'ice-age' occurred following a subsidiary compressive maximum (Alpine: Mid- and Late-Tertiary orogenesis), and there have been similar intra-cyclical ice-ages in the past. Further isotopic age data on these rocks are required before any firm conclusions concerning the cyclical incidence of ice-ages can be made, but the present indications suggest a confirmation of the chelogenic cycle hypothesis.

A further factor to be considered is the apparent reduction in time-span of successive chelogenic cycles. Taking the origin of the Earth to have been some 4,750–5,000 m.y. ago, the time spans of the cycles recognized in this article are approximately as follows:

Pre-Superior: 1,250–1,500 m.y. Superior: 1,100 m.y. Churchill: 850 m.y. Grenville: 750 m.y. Varisco-Caledonian: 600 m.y. Alpine: ? (say, 525 m.y.).

This progressive shortening in time-span of chelogenic cycles can be interpreted either in terms of an intensifying or of a decaying system. Initially we favoured the latter hypothesis for two reasons: first, because the structural evidence from the earliest continental rocks suggests that orogenesis in early times was a more widespread and diffuse event (for example, in the Dodoman Province of East Africa, where the relatively small steep folds have a parallelism of strike and uniformity over hundreds of miles), suggesting a higher overall energy-level than obtained in more recent cycles, when orogenic failure was restricted to more localized belts; and secondly, because an ultimate decline in the Earth's total energy output from radioactivity must be expected with time. It now appears that the solution to this problem may not be quite so direct.

The recent study by Runcorn[26] of past changes in the Earth's moment of inertia appears to confirm the general hypothesis of Lyttleton[25], in which Earth contraction consequent on core formation is suggested as the principal factor controlling geological history. Lyttleton remarks on the extremely poor thermal conductivity of surface rock material, which should effectively prevent release of thermal energy accumulating deep in the globe as a result of radioactive decay. He envisages an initially cool Earth composed of mantle substance in which internal heat accumulation leads to the growth of a liquid core. From a consideration of the available physical data on the core and mantle, Lyttleton suggests that, at the pressures and temperatures involved, conversion of mantle substance to core substance results in an actual reduction in volume (possibly due to the Ramsey effect; he also suggests that only the inner core, 7 per cent of the entire core mass, is specifically enriched in nickel–iron as a result of gravitative differentiation). Lyttleton shows that the surface area contraction of the Earth consequent on the growth of a liquid core to its present size might be as much as 60×10^6 km^2. This, he suggests, is the principal cause of surface mountain-building phenomena.

If Lyttleton's Earth model were correct, chelogenic cycles would not occur, for in this model the crust is in a perpetual state of compressive stress. The fact of chelogenic cycles in the Earth's history suggests, at the very least, that a subsidiary cyclical mechanism operates in the outer layers of the Earth, most probably involving convective overturn in the mantle, and that its effects are superimposed on any overall contraction caused by core growth or for other reasons. If Lyttleton's views on core growth are confirmed by future high-pressure experiments, it will be necessary to examine a more complex Earth model which can explain chelogenic cycles as well as overall Earth contraction. The frequent release of magma into the crust and the voluminous outpourings of lava at its surface demonstrate the reality of a potentially more effective mechanism of release of heat energy from the mantle than by conduction through rock layers, which, as Lyttleton rightly supposes, would inhibit release of heat energy from more than a few hundred kilometres of the Earth's surface layers. Little is known of the temperature gradient within the Earth, but it is reasonably certain from volcanic and other evidence that it is not far from the melting-point of outer mantle substance at the pressure involved. We know that, well above the critical level below which Lyttleton suggests that melting results in volume reduction rather than expansion, local accumulations of heat energy do produce partial and complete melting of rock with consequent decrease in density and upward migration of magma. From these facts it could be suggested that accumulation of heat energy deep in the mantle may at times cause very widespread partial melting or near melting, with consequent volume increase and reduction in density, especially in any Earth shell in which the temperature gradient remains close to the melting-point over a considerable depth range. The effect of widespread increases in volume within certain annular zones of the mantle would be to induce a state of tension in the outermost layers of the Earth, followed by tensional failure of this relatively brittle surface layer, leading to cratogenic magmatism and, very probably, to the evacuation of much of the partially or very nearly molten, and therefore, potentially mobile, segments of the mantle along 'world-cracks' such as the mid-oceanic ridges. This would result in ocean floor spreading, and, occasionally, when the cracks in the outer layer of the mantle happened to be beneath a continent, to continental disruption and drift. Evacuation of the mobile segments can be regarded as being, in effect, a half-cycle convective overturn, and would result in a great dissipation of heat from the mantle as a whole and a return to more rigid conditions of Earth contraction until renewed heat accumulation at the appropriate depth was sufficient to cause a further period of expansion and mobility. Thus the cyclical sequence of surface geological events outlined in the chelogenic cycles concept of the earlier part of this article would find a satisfactory explanation. It is interesting to note that owing to the much smaller gravitational forces involved, release of similar mobile material generated within the Moon would most probably be by diapiric rise and evacuation from discrete centres, only occasionally aligned closely together along global cracks.

In the modified Lyttleton Earth model each chelogenic cycle would begin at a time of maximum crustal compression and minimum crustal heat-flow. As suggested by Lyttleton, the cold and relatively rigid surface layers would tend to ride over one another (in the same way as Asia is at present being thrust over the Pacific along the Japanese arc), causing extreme continental topography and the possibility of extensive ice-ages. As the temperature in the mantle recovered from the lowering effects of the previous evacuation cycle, volume increases within this shell would eventually reverse the effect of global contraction in the nearer surface layers. Cratogenic magmatism and ocean floor spreading, the characteristic features of the anorogenic first half of a chelogenic cycle, would then begin, and rise, through minor regressive phases, to a climax. The heat loss from the level of evacuation could be extremely rapid at this climax. After evacuation was completed, the temperature gradient in this shell would fall significantly below the melting-point curve, and the outermost layers would no longer be protected from the effects of Earth contraction resulting from core growth. Mobile belts, formed in the relatively rigid continental crust, where it was most weakened by heat release during the period of evacuation, would, on the reappearance of contraction stresses in the surface layers, act as the principal continental loci of compression failure of increasing intensity, culminating in the final evorogenesis of the chelogenic cycle. Compression failure of the oceanic crust would most probably be distributed and much less dramatic. With the gradual growth of the core and the consequent reduction in mantle thickness and surface diameter, the level of evacuation in each successive cycle must be closer to the core boundary. This process would explain the observed reduction in time-span of successive chelogenic cycles. In this Earth model, it can be expected that the tensional first half of a cycle will reach its climax quicker, and that the succeeding compressive sub-cycle will be more and more localized in its effects as time goes on. An eventual steady state of the Earth would occur when the core boundary extends to the critical level outside which expansion occurs on liquefaction. Accumulation of heat in the core could then be dissipated directly to the surface by magma derived from the mantle/core boundary. The surface of the Earth would then experience a state of almost perpetual volcanism unaccompanied by any significant Earth movements. Further evolution could only occur as the Earth's source of radioactive heat begins to wane. Failure in the Earth's energy supply would lead to the gradual extinction of surface volcanism and to the development of a permanent tensional régime on the surface as the core finally cooled and solidified.

[1] Sutton, J., *Nature*, **198**, 731 (1963).
[2] Holmes, A., *Proc. Eighteenth Session Intern. Geol. Congr.*, Pt. 14, 254 (1952).
[3] Gastil, G., *Proc. Twenty-first Session Intern. Geol. Congr.*, Pt. 9, 162 (1960).
[4] Wilson, A. F., *et al.*, *J. Geol. Soc. Austral.*, **6**, 179 (1960).
[5] Lowdon, J. A., *Geol. Surv. Canada Paper*, **61**, 17, 1 (1961).
[6] Stockwell, C. H., *Geol. Surv. Canada Paper*, **61**, 17, 108 (1961).
[7] Voitkevich, G. H., *Priroda*, **77** (1958).
[8] Vinogradov, A. P., and Tugarinov, A. I., *Ann. N.Y. Acad. Sci.*, **91**, 500 (1961).
[9] Ovchinnikov, L. N., and Harris, M. A., *Proc. Twenty-first Session Intern. Geol. Congr.*, Pt. 3, 33 (1960).
[10] Faul, H., Stern, T. W., Thomas, H. H., and Elmore, P. L. D., *Amer. J. Sci.*, **21**, 1 (1963).
[11] Stevenson, J. S., *The Tectonics of the Canadian Shield* (University of Toronto Press, Toronto, 1962).
[12] Quennell, A. M., and Haldemann, E. G., *Twenty-first Session Intern. Geol. Congr.*, Pt. 9, 170 (1960).
[13] Tyrrell, G. W., *Bull. Geol. Soc. Amer.*, **66**, 405 (1955).
[14] Harpum, J. R., *Twenty-first Session Intern. Geol. Congr.*, Pt. 9, 201 (1960).
[15] Runcorn, S. K., *Nature*, **193**, 311 (1962).
[16] Runcorn, S. K., *Nature*, **195**, 1248 (1962).
[17] Miller, J. A. (with an appendix by Fitch, F. J.), *Trans. Roy. Soc. Lond.* (in the press).
[18] Fahrig, W. H., and Wanless, R. K., *Nature*, **200**, 934 (1963).
[19] Patterson, C., and Tatsumoto, M., *Geochim. Cosmochim. Acta*, **28**, 1 (1964).
[20] Hurley, P. M., *et al.*, *J. Geophys. Res.*, **67**, 5315 (1962).
[21] Wilson, J. T., *The Earth as a Planet*, 204 (Univ. of Chicago Press, Chicago 1954).
[22] Egyed, L., *Nature*, **178**, 534 (1956).
[23] Creer, K. M., *Nature*, **205**, 539 (1965).
[24] Heezen, B. C., *Proc. Intern. Oceanographic Conf. New York* (1959).
[25] Lyttleton, R. A., *Proc. Roy. Soc. Lond.*, A, **275**, 1 (1963).
[26] Runcorn, S. K., *Nature*, **204**, 823 (1964).
[27] Wilson, J. T., *Nature*, **197**, 536 (1963).
[28] Stille, H., *Grundfragen der vergleichenden Tektonik* (Berlin, 1924).
[29] Gilluly, J., *Quart. J. Geol. Soc., Lond.*, **474**, 133 (1963).
[30] Wilson, J. T., *Nature*, **198**, 929 (1963).

8

Reprinted from *Nature* **211**:676–681 (1966)

DID THE ATLANTIC CLOSE AND THEN RE-OPEN?

By PROF. J. TUZO WILSON

Institute of Earth Sciences, University of Toronto

FOR more than a century it has been recognized that an unusual feature of the shallow water marine faunas of Lower Palaeozoic time is their division into two clearly marked geographic regions, which are commonly referred to as faunal realms. "The faunal assemblages are amazingly uniform throughout each realm so that correlation of any Cambrian section with another in the same realm is usually easy; on the other hand, the difference between the faunas in the two separate realms is so great as to make correlation between them very difficult"[1].

Two aspects of the distribution of these realms are remarkable. For one thing, some regions of similar faunas are separated by the whole width of the Atlantic Ocean; then, on the other hand, some regions of dissimilar faunas lie adjacent to one another. This is illustrated by Fig. 1, which is based on work by Cowie[2], Grabau[3] and Hutchinson[4].

Grabau showed that, if Europe and North America had become separated by continental drift, a simple reconstruction could explain the first anomaly in the distribution of the faunal realms in that, before the opening of the Atlantic Ocean, each realm would have been continuous, with no large gaps between outcrops of similar facies (Fig. 2).

It is the object of this article to show that drift can also explain the second anomaly. It is proposed that, in Lower Palaeozoic time, a proto-Atlantic Ocean existed so as to form the boundary between the two realms, and that during Middle and Upper Palaeozoic time the ocean closed by stages, so bringing dissimilar facies together (Fig. 3). The supposed closing of the Tethys Sea by northward movement of India into contact with the rest of Asia, and the partial closing of the Mediterranean by northward movement of Africa, can be regarded as a similar but more recent event. The figures are based on a reconstruction by Bullard, Everett and Smith[5], but because those authors pointed out that no allowance had been made for the construction of post-Jurassic shelves, the continents have been brought more closely together.

Four lines of evidence suggest that this proposal is reasonable. (Unfortunately, so far as I can ascertain, palaeomagnetic evidence which might bear on this problem does not exist.)

First, this reconstruction of geological history is held to provide a unified explanation of the changes in rock types, fossils, mountain building episodes and palaeoclimates represented by the rocks of the Atlantic region.

Second, wherever the junction between contiguous parts of different realms is exposed, it is marked by extensive faulting, thrusting and crushing.

Third, there is evidence that the junction is everywhere along the eastern side of a series of ancient island arcs (Fig. 3).

Fourth, the fit appears to meet the geometric requirement that during a single cycle of closing and

Fig. 1. The North Atlantic region showing the present distributions of the 'Atlantic' faunal realm (horizontal shading) and the 'Pacific' faunal realm (vertical shading). (After J. W. Cowie, A. W. Grabau and R. D. Hutchinson)

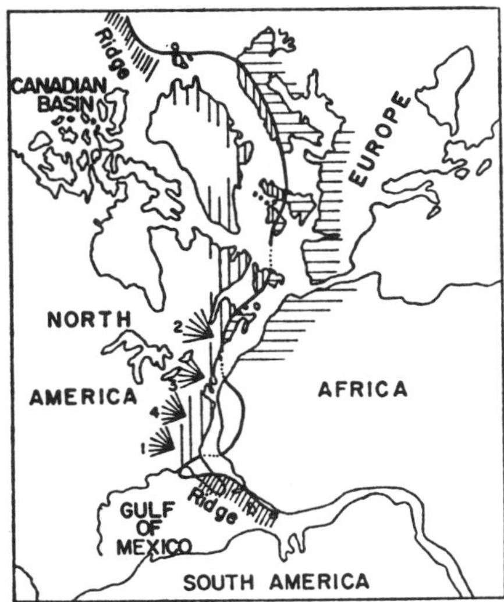

Fig. 2. The North Atlantic region in Upper Palaeozoic and Lower Mesozoic time showing that of the present Atlantic Ocean only the Canadian Basin and the Gulf of Mexico then existed. Four fans are shown which were formed: (1) during Middle Ordovician; (2) during Upper Ordovician; (3) during Upper Devonian; (4) during Pennsylvanian. The heavy line separates 'Pacific' and 'Atlantic' faunal realms. The two ridges are considered to have formed when the modern Atlantic started to open

reopening of an ocean, and in any latitudinal belt of the ocean, only one of the pair of opposing coasts can change sides (Fig. 4).

Recurrent Drift in the North Atlantic

The history proposed for the North Atlantic region can be stated very briefly as follows: (*a*) From the Late Pre-Cambrian to the close of Middle Ordovician time an open ocean existed in approximately, but not precisely, the same location as the present North Atlantic (Fig. 3). (*b*) From the Upper Ordovician to Carboniferous time, this ocean closed by stages. (*c*) From Permian to Jurassic time there was no deep ocean in the North Atlantic region. The only marine deposits of that time are those connected with the Tethys Sea, with a shallow Jurassic invasion of Europe and with deeper Jurassic seas in the Gulf of Mexico and in the western Arctic Basin (Fig. 2). (*d*) Since the beginning of the Cretaceous period the present Atlantic Ocean has been opening, but this reopening did not follow the precise line of junction formed by the closing of the early Palaeozoic Atlantic Ocean; the result is that some coastal regions have been transposed (Fig. 1).

The Lower Palaeozoic continents may have first touched each other at the end of Middle Ordovician time, for thereafter the distinction between 'Atlantic' and 'Pacific' faunal realms ceases to be marked, but the complete closing of the ancient Atlantic may have required several periods.

For each continent, union meant replacing the open ocean by the other continent. This is offered as an explanation of the borderlands of J. Barrell and C. Schuchert for which there is no clear evidence until Upper Ordovician time. As Kay has suggested[6] concerning Eastern North America: "There has been little discussion of the evidence for borderlands in earlier Paleozoic time, though some have expressed scepticism". Kay's own support for island arcs is muted after Lower Palaeozoic time and he accepts the view that the sediments of the "Late Devonian and Early Mississippian came from the land of Appalachia" —a borderland.

This view that extensive upland source areas lay to the east of the Appalachian geosyncline in the sites of the present coastal plain or ocean has been fully supported by recent work[7–9]. Tens of thousands of cubic miles of quartz-rich sediments, derived from the east, were deposited in shallow marine to sub-aerial deltas.

When the continents were pushed together, they would have touched first at one promontory and then at another. It can be expected that high mountains would have been formed locally and that they would have produced alluvial fans on both continents. As the ocean diminished the climate would have become increasingly arid. Such drastic alterations in the physiography would explain the change from predominantly marine and island arc deposition in the Lower Palaeozoic to conspicuous fans of Queenston, Catskill, Old Red Sandstone, and other deltas of Middle and Upper Palaeozoic time (see refs. 10 and 11, and Fig. 2). It can also be expected that the collision of continents would have produced great local uplifts which, if one continent overrode the other, would have migrated inland, perhaps pushing the Taconic and northern Newfoundland klippen[12] before them.

It would seem that by Permian and Triassic times the Atlantic Ocean was completely closed, because only continental beds, such as the Dunkard and Newark series, are found in North America. In Great Britain the New Red Sandstone is also continental as is the Permian of the Oslo district.

No Jurassic beds are known in eastern North America except in the Gulf of Mexico. Those of Europe were formed by a shallow marine invasion of the continent and are said by Hallam[13] to have fossils that "include many neritic forms that could not have crossed a deep ocean. The paleogeography for the Scottish Jurassic gives no hint of increasingly marine conditions to the west" (personal communication). Most of the available geological evidence suggests that the present Atlantic Ocean started to open at the beginning of Cretaceous time[14,15]. Although objections to this view are still being raised, they seem to be minor in comparison with the other evidence, and it is possible that they can be explained in other ways.

Faulted Contact between Faunal Realms

Starting our considerations in the north, the island of West Spitsbergen is underlain by a thick eugeosynclinal section of Lower Palaeozoic rocks named the Hecla Hoek succession. These strata rest on no known basement and were deformed, metamorphosed and intruded by granites during the Caledonian orogeny[16,17]. They contain fossils of the 'Pacific' fauna similar to those of Scotland and North America[18].

In Nordaustland, the adjacent, eastern island of the Spitsbergen group, Kulling[19] and Sandford[20] have mapped a thin section of unmetamorphosed and gently folded strata which a few fossils indicate to be of about the same age. These beds lie uncomformably on a basement which is regarded as part of the Baltic Shield. These strata do not thicken to the west as the much thicker section of West Spitsbergen is approached, nor do the few fossils necessarily belong to the 'Pacific' faunal realm.

Despite the considerable number of attempts to compare the sections in the adjacent islands, the correlation is not good. Changes in thickness, facies, and degree of metamorphism, basement and type of intrusives are all abrupt and striking. Orvin[21] and others have mapped faults in Hinlopen Strait between the islands. Klitin[22] summarizes the situation thus: "Of particular interest is the junction zone of the alleged Caledonian platform and Caledonian fold system. The transition to typical caledonids takes place in a zone not over 15 to 20 km wide, in the Hinlopen

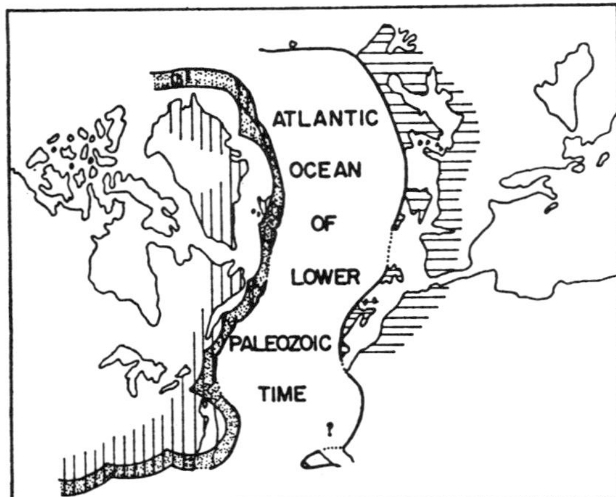

Fig. 3. The North Atlantic region in Lower Palaeozoic time. The proto-Atlantic Ocean would have formed a complete barrier between two faunal realms (shaded). Island arcs (dotted) probably lay along the North American coast. The floor of this ocean could have been absorbed in the trenches associated with these arcs as the ocean closed

Strait area, where the Hecla Hoek section abruptly increases in thickness, by a factor of four, with the appearance of extrusives in its metamorphosed and linearly folded beds. Such radical changes in thickness suggest a fault junction between an ancient platform and the caledonids." Following this interpretation I suggest that the Lower Palaeozoic ocean once separated the two islands and that, whereas Nordaustland formed part of the Baltic shield and shelf, West Spitsbergen was the site of a North American island arc.

In Scandinavia it is well established that the peninsula is divided longitudinally into two different provinces separated by a great zone of nappes and faults overthrust towards the east. According to the descriptions of Holtedahl[23], south-eastern Norway and most of Sweden are underlain by an extension of the Baltic Platform on which lie nearly flat, unfossiliferous rocks and "the eastern facies of the Cambro-Silurian (which) can be classed as miogeosynclinal in the terminology of Stille. The thickness is rather large and there is much terrigeneous material. Caledonian volcanic and intrusive igneous rocks are not found in the deposits of this type. The rocks are unmetamorphosed in the east and are of low metamorphic grade farther to the north-west. The deposits occur in an autochthonous or parautochthonous position above the original Archean basement."

The fossils are repeatedly referred to as being similar to those of the Baltic region, England and Wales. On the other hand, in the Trondheim area of western Norway, a thick succession of pillow lavas, shale and limestone with serpentinites in the lower part of the sequence contains a fauna "of American affinities . . . the limestone of Smøla is similar to the Durness in Scotland and to more or less contemporaneous limestone in Newfoundland, Bear Island and Spitsbergen. The limestone in Smøla thus seems to represent an American-Arctic facies of the Ordovician." In these "mainly pelitic sediments . . . we thus have a eugeosynclinal facies characteristic of the central parts of a geosyncline . . . probably all rocks of the present facies occur in allochthonous positions." An important event in central southern Norway was the close of marine deposition "in Late Ordovician or Early Silurian time brought about by the thrusting of nappes and deposition of the Valdres sparagmite (arkose), which was considered as a deposit of flysch type by Goldschmidt". The zone of nappes is, therefore, held to be the boundary between faunal provinces and the line of closure of the Lower Palaeozoic ocean.

The geological relationship between Spitsbergen, Scandinavia and the British Isles has been discussed by Bailey and Holtedahl[24]. Of the Caledonian structures they state: " . . . in the present land area of Scandinavia we find the eastern part of the orogenetic belt only, while in Great Britain the whole of the orogenetic zone is represented. The Spitsbergen Group seems to lie rather centrally in the zone of deformation." Following Wegener, they consider that Greenland may have been formerly connected with Europe and suggest that, if so, it would have completed the western side of the mountain belt opposite Norway. This relationship has found support in recent years from several authors[14] including Umbgrove[25].

That the boundary between 'Atlantic' and 'Pacific' faunal realms crosses northern England between Scotland and Wales is supported by Walton, who writes of the Scottish occurrences: "The close affinity of the Durness and North America rocks was recognised long ago . . . . The long-recognised affinity of the Girvan Caradocian and Appalachian Mohawkian faunas was re-emphasized by Williams . . . . By contrast the faunas are only remotely connected with the Welsh Ordovician rocks"[26]. George[27] describes the relations thus: "The Salopian geosyncline may thus have extended unbrokenly from conjectural north-western limits in the Highlands to a south-eastern margin or marginal shelf in the English Midlands". It does not require much change to regard this geosyncline as a former ocean.

In Newfoundland the geology of the north-eastern coast has recently been described by Williams[28]. He suggests that in Cambrian time a deep basin or ocean, not underlain by continental crust, crossed the central part of the island and separated two shelves. The north-western shelf underlying the Long Peninsula has a basement of Grenville age overlain by strata with 'Pacific' faunas like those of Scotland, while the south-eastern shelf, which forms the Avalon peninsula, has a younger Pre-Cambrian basement overlain by strata with 'Atlantic' faunas like those of Wales. During Ordovician and subsequent time the intervening sea became filled with eugeosynclinal and volcanic sedimentary rocks, probably representing a former island arc and mountain belt.

Anderson[29] believes that faulting in Hermitage Bay on the south coast of Newfoundland not only separates the rocks of the south-eastern shelf from the central geosyncline, but may also completely divide the two faunas. On the north coast the corresponding fault zone may be in Freshwater Bay, but this point has not been settled as yet[30].

Among those who have recently correlated the Newfoundland and British sections are Dewey and Church[31]. Although Church does not favour continental drift, both he and Dewey make the same correlation as that already given here and extend the Caledonian mobile belt from northern England and central Ireland to central Newfoundland.

South of Newfoundland, the Gulf of St. Lawrence and younger rocks cover the key areas of Lower Palaeozoic formations almost as far as the Maine border. Little can be said except that the faunas of Cape Breton Island and St. John, New Brunswick, have European affinities while those of Gaspé and the Eastern Township of Quebec are typical of Scotland and most of North America[4].

In northern and eastern Maine the older literature is sparse and generalized. Recently a combined group of government and university geologists, including W. B. N. Berry, A. J. Boucot, E. Mencher, R. S. Naylor and L. Pavlides, have discovered new fossil localities there with both European and North American affinities, important Caledonian uplifts and large pre-Silurian faults[32]. Much of the state has been remapped, but for the reason that little of this work has yet been published (R. G. Doyle, personal communication), and because the structure is clearly much more complex than shown on early maps, this is not an opportune time to consider the area in detail.

From the southern part of Maine south across New Hampshire, Massachusetts and Connecticut a major zone can be traced from published accounts. Novotny[33] has described this "major fault zone" where it crosses the New Hampshire–Massachusetts boundary (Fig. 5). To the north it connects with several faults and silicified zones shown on the map of New Hampshire[34,35]. These lie along the line which separates those formations which underlie the greater part of New Hampshire from a suite of completely different formations, underlying south-eastern New Hampshire. This line may be continued northward into south-eastern Maine by the faults which Katz[36] has suggested bound the Berwich gneiss, itself crumpled, "closely folded and overturned". In Massachusetts, Novotny's fault may be exposed in the abandoned Worcester "coal" mine. The description suggests that much of the rock may be carbonaceous mylonite[37]. South of Worcester this fault connects with a major change in formations evident on the geological map of Massachusetts[38].

In Connecticut this boundary appears to join the Honey Hill fault and its northern continuation which bisect the state and separate two major rock sequences. Where best described in the south it "has been mapped for 25 miles eastward from Chester nearly to Preston without apparent repetition of the stratigraphy on either side of the fault. The fault plane dips 10°–35° N parallel to the underlying metasedimentary rocks .... The fault is marked by a zone a mile wide of mylonitized and crushed rocks .... Displacement must have been many miles"[39]. As in Norway, the orogenic belt appears to have been thrust eastwards over the eastern platform. To the south this major fault seems to strike into the Atlantic Ocean and the Appalachian belt, then narrows conspicuously. I have recently learned from L. R. Page and J. W. Peoples that mapping (for the most unpublished) by the federal and state surveys has defined this fault zone more satisfactorily, and that published aeromagnetic maps show a change in the strike of anomalies across it. This evidence suggests that New England is divided by a major fault zone into two provinces underlain by quite different rock formations. Of the few occurrences of the Lower Palaeozoic faunas, all those with European affinities lie to the east of the fault zone; all those typical of North America to the west[40,41].

In the light of this evidence it seems reasonable to suggest that this fault zone marks the line of closure of the Lower Palaeozoic Atlantic Ocean. It may seem strange to propose that a former position of the Atlantic Ocean lies through New England and that its full significance has not been realized, but it must be remembered that throughout the area outcrops are poor and that most of the mapping is old. Surface mapping reveals few faults, but new tunnels have shown that faults abound under the drift-filled valleys and that some of these are major (J. W. Skehan and A. Quinn, personal communications).

Most North American geologists have not accepted the idea of continental drift. Instead they have sought to explain the changes in faunas in terms of different environmental conditions. This has certainly been a factor and no doubt was responsible for the differences between the Durness, Girvan and Moffatt facies in Scotland, although those facies belong to the same faunal realm. In another example, G. Theokritoff (personal communication) has directed my attention to a possible mixing of Atlantic, Pacific and endemic faunas in the Taconic sequence of New York[42]. Christina Lochman[43] has emphasized the difficulties incurred in accepting such an interpretation. She states that the areas of mixed faunas

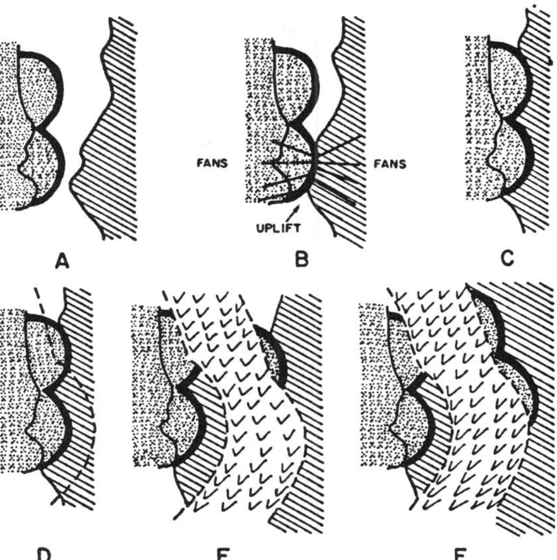

Fig. 4. *A*, A closing ocean, with island arcs on one coast, separating two different faunal realms. *B*, First contact between two opposite sides of a closing ocean. *C*, The ocean closed by overlap of the opposite coasts. *D*, A possible line (dashed) along which a younger ocean could reopen. *E*, A new ocean (checked) opening in an old continent. *F*, A geometrically impossible way for a younger ocean to open. (Note how the arcs overlap.)

lie in deeply down-warped basins between two shallow shelf deposits containing respectively Atlantic and Pacific faunas. These basins, she remarks, had connexions with the Atlantic Ocean and had a benthonic environment "similar to that of the open ocean .... Few normal benthonic species of the coastal shelf could establish themselves in such an alien environment, although, because of the geographic proximity of the two areas, sporadically drifted individuals might be found." She also refers to deep basins separating different faunas and to the evidence for a "biofacies regime ordinarily found on the floors of the continental shelf beyond the inner islands of the volcanic archipelago". These views would seem to admit the possibility of interpreting the palaeogeography as has been done in this article. It is suggested that one of the deep basins, instead of lying on a shelf, might have been an open ocean in Lower Palaeozoic time and that the mixed fauna may have come from the extreme edge of one continent.

The reconstruction (Fig. 2) then leads to Africa. Sougy[44] has described a large plate of metamorphic rocks east of Dakar thrust eastwards over Upper Devonian strata. Farther north in Spanish Morocco there is a folded belt. A large Cretaceous overlap along the coast and lack of diagnostic fossils in the overthrust block make correlation difficult, but Sougy concludes that: "In the future, when geologists study the relationships between America and Africa for evidence bearing on permanence of continents and oceans *versus* continental drift, they will have to consider that the western rim of Africa is made, from Guinea to Morocco, not of a Pre-Cambrian basement but of a mainly Hercynian orogenic belt, in some respects symmetrical to the Appalachian belt." His mention of drift suggests, as do most reconstructions[5], that this West African belt was formerly part of the Appalachians. This view is supported by P. A. Mohr, who writes: "the Cambrian manganese ores of Newfoundland, Wales and Morocco . . . I think were formed in a common geosynclinal trough" (personal communication). The view that these regions were

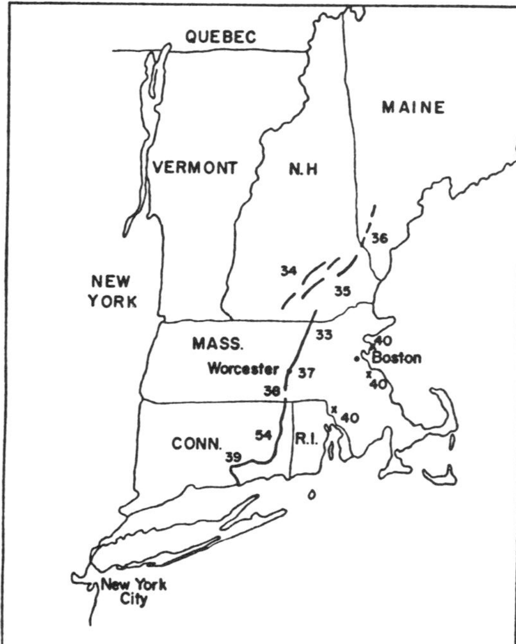

Fig. 5. Sketch map of New England showing location of some major faults and three fossil localities of the 'Atlantic' faunal realm (X). The numbers refer to papers in the list of references from which information was obtained

together until Cretaceous time is supported by the close similarity of Cretaceous fossils from these two regions now on opposite sides of the ocean[45]. Only further investigation, aided by extensive seismic investigations or drilling through the Cretaceous of both continental coasts, can show whether it is possible or likely that West Africa ever fitted into the central east coast of the United States. If so, some rotation of Africa clockwise and a closer fit than that shown by Bullard, Everett and Smith[5] may be indicated.

Although the northern termination of the West African fold belt is covered, its termination to the south is definite; the Pre-Cambrian of the African Shield extends to the Atlantic near Conakry[46]. The reconstruction suggests that this contact might reach Florida. In Florida and Georgia a cross-section of the Appalachians resembles that of Newfoundland in that these are the only parts of the Appalachians with a platform on both sides of the mobile belt. The platform on the south-eastern side of the mobile belt which lies beneath northern Florida and adjacent states is only known from well cores. These consist of Early Ordovician to Middle Devonian sandstones and shales said to have been deposited in shallow water[47]. Published accounts do not correlate the strata or fossils with formations elsewhere. It would seem that the basement has not been penetrated. Thus it is only speculative to suggest that this region was formerly part of Africa.

Island Arcs of Lower Palaeozoic Atlantic

According to a widely adopted hypothesis, island arcs and mountains represent places where the lithosphere is being compressed, while mid-ocean ridges represent places where it is being pulled apart and where new crust is being created. Thus the present Atlantic Ocean is expanding from the Mid-Atlantic Ridge.

For a Lower Palaeozoic Atlantic Ocean to have closed, it must, according to this hypothesis, have been marked not by a central ridge, but by a continuous system of island arcs and mountains. Observation shows that such systems commonly lie at the sides and not in the centre of oceans. Kay's view, that a system of island arcs existed off the eastern and southern coasts of North America in Lower Palaeozoic time, is entirely compatible with this hypothesis. This would have allowed the former Atlantic Ocean to close, and thus to convert the offshore island arcs of Lower Palaeozoic time into the intercontinental Appalachian mountains of Middle and Upper Palaeozoic time. Following Keith, Kay and King, I have sketched the location of seven former arcs in the Appalachians and their south-western continuation past the Ouachita and Marathon Mountains[48]. I see no reason to change this view.

The situation in Scotland has been described by Walton: "It seems likely that both the north-west Highlands and the Appalachians formed part of a very wide stable shelf which also included Greenland and other 'Boreal' regions having very similar Cambrian rock types. . . . it is probable that sedimentation in the area of the southern Highlands during the Cambrian period was mainly of a greywacke, geosynclinal type and contrasted strongly with that in the north-west Highlands . . . it is probable that Cambrian sedimentation stretched unbroken to an unknown distance south of the Highland Boundary fault"[26]. Ordovician volcanism followed and these correlations and descriptions suggest that, during Lower Palaeozoic time, arcs associated with a western land extended across Scotland. That there may have been contemporaneous volcanism and islands along the eastern coast in what is now Wales, merely adds complexity without affecting the history in Scotland. In the Silurian period conditions changed and the marine conditions "gave way to mixed environments . . . at least partly in fluvio-deltaic environments with periods of emergence"[26].

Enough has been described of the conditions in Norway and West Spitsbergen and of their correlations with Scotland to show that they too were the sites of Lower Palaeozoic island arcs associated with a western continent. Thus the Caledonian–Appalachian arcs seem to have formed a continuous system along the western side of the former ocean. It is suggested that it was in the trenches of these arcs that the floor of the Lower Palaeozoic Ocean was swallowed up as that ocean closed.

Geometrical Control of Transposition

If two bodies are brought together so that they unite, and if they are later pulled apart so that they break on a different line from the line of union, then one important geometrical consideration holds. It is that, along any one stretch of the junction, only one fragment can change sides: one cannot transpose pieces from each margin at the same place. This is illustrated in Fig. 4 and a comparison with the other figures shows that our reconstruction obeys this principle. This does not prove that the reconstruction is correct, but the neat fashion in which fragments from either side are alternately transposed meets the geometrical requirements.

Some Possible Extensions

When, as is believed, the present Atlantic Ocean started to open at the beginning of Cretaceous time, it did so by breaking open a continent which was then continuous from West Spitsbergen to Florida (Fig. 2). North of Spitsbergen the coasts of North America and Siberia at that time diverged and the opening ceased to lie wholly within a continent and to have continental blocks on both sides. Following the descriptions of B. C. Heezen and M. Ewing, and Ya. Ya. Hakkel and N. A. Ostenso, I have suggested that the fracture followed the coast of Siberia forming the Lomonosov Ridge on the other side of the opening. Thus the Lomonosov Ridge separates an older Canadian ocean basin from a younger Siberian ocean basin in the Arctic Sea[14].

In the south the situation appears to be similar, but to understand it one must clearly separate the platform of northern Florida from the different geology of southern Florida. South of central Florida a southern Florida–Bahamas Ridge separates a main Atlantic, apparently of Cretaceous age, from a Gulf of Mexico which seems to have been a deep ocean and evaporite basin during the Jurassic period.

Drake, Heirtzler and Hirschman[49] emphasize the importance of the Florida-Bahamas Ridge for sealing off the Gulf of Mexico from the open ocean so that the Jurassic salt deposits could form. They suggest that this ridge is an "extension of the Ouachita system" and that it forms the "foundation for the entire chain of islands and banks". This ignores the earlier interpretations of magnetic and gravity anomaly maps made by Miller and Ewing[50] and by Lee[51]. Both papers suggested that southern Florida and the Bahamas are coral banks built on deeply submerged volcanoes. The Palaeozoic platform-type of sedimentary rocks, described from drill cores in northern Florida, are not like the Ouachita folds. Drilling has revealed no Palaeozoic rocks beneath southern Florida and the Bahamas. The magnetic anomaly map shows a marked change in central Florida. Drake *et al.*, in their Fig. 5, show three trend lines connecting southern Florida with the Bahamas and only one connecting it with northern Florida. It is suggested that the Florida-Bahamas Ridge, like the Lomonosov Ridge, formed when the main Atlantic Ocean started to open at the beginning of Cretaceous time and separated an ocean basin of Cretaceous age from an older ocean basin.

Drake *et al.* hold that the Florida-Bahamas Ridge extends to Navidad Bank, north of Hispaniola. It thus ends at the major zone of faulting which, according to Hess and Maxwell[52], and others, extends from Central America to the northern end of the West Indies arc. The Caribbean Sea and West Indies arc have often been regarded as associated with the Pacific Ocean, and I have discussed elsewhere their possible origin as a tongue, thrust from the Pacific and bounded by transform faults[53].

A complete discussion would require consideration of the Hercynian orogeny and faulting and post-Triassic faulting. This seems feasible but will not be attempted here.

It has been suggested in this article that during Lower Palaeozoic time North America and Europe were approaching each other, that this motion stopped and that it later reversed. If this is true, the onset of the reverse motion and the start of reopening of the Atlantic Ocean must have been an event of very major significance in world geology. The evidence suggests that it occurred at about the close of the Jurassic and the beginning of the Cretaceous periods. It seems reasonable to link it with other major events of that age in the Americas.

McLearn[55] has pointed out that at that time the drainage of much of western North America reversed its direction so that rivers which had been flowing west and building a great shelf along the Pacific coast were interrupted by the rise of the Cordillera and began to follow their present directions. Gilluly[56] has pointed out that "In Cretaceous time plutons probably a thousand times larger than those of all the rest of the Phanerozoic were emplaced". The onset of a relative advance of the Americas over the Pacific Ocean floor might well have caused a crumpling of the shelves along that coast with the creation and rise of extensive batholiths. Gilluly has pointed this out and concluded that the likely cause was that "it is probable that the continent as a whole is moving away from a widening Atlantic".

I am happy to acknowledge the benefits of discussions with W. B. Harland, A. Hallam, J. Dewey, W. R. Church, F. D. Anderson, H. Williams, E. W. R. Neale, H. H. Hess, F. J. Vine, E. Irving, A. Quinn, T. Mutch, J. Sougy and others, and of correspondence with many. I also thank F. W. Beales, C. Harper and D. York for help with the final version of this article.

This work was supported by the National Research Council of Canada, Vela-Uniform Program and Unesco.

1 Hutchinson, R. D., *Geol. Surv. Canada, Mem.*, **263**, 52 (1952).
2 Cowie, J. W., *Intern. Geol. Cong., Sess.* 21, *Copenhagen*, Part 8, 57 (1960).
3 Grabau, A. W., *Palaeozoic Formations in the Light of the Pulsation Theory*. 1 (University Press, National University of Peking, 1936).
4 Hutchinson, R. D., *Intern. Geol. Cong., Sess.* 20, *Mexico, The Cambrian System Symposium*, **2**, 290 (1956).
5 Bullard, E. C., Everett, J. E., and Smith, A. G., *Phil. Trans. Roy. Soc.*, A, **258**, 41 (1965).
6 Kay, M., *Geol. Soc. Amer. Mem.*, **48**, 31, 56 (1951).
7 Pettijohn, F. J., *Bull. Amer. Assoc. Petrol. Geol.*, **46**, 1468 (1962).
8 Yeakel, jun., L. S., *Geol. Soc. Amer. Bull.*, **73**, 1515 (1962).
9 Naylor, R. S., and Boucot, A. J., *Amer. J. Sci.*, **263**, 153 (1965).
10 King, P. B., *The Evolution of North America*, **61** (Princeton University Press, 1959).
11 Clark, T. H., and Stearn, C. W., *The Geological Evolution of North America*, 104, 114 (Ronald Press, New York, 1960).
12 Rodgers, J., and Neale, E. R. W., *Amer. J. Sci.*, **261**, 713 (1963).
13 Hallam, A., in *The Geology of Scotland*, edit. by Craig, G. Y. (Oliver and Boyd, Edinburgh, 1955).
14 Blackett, P. M. S., Bullard, E. C., and Runcorn, S. K., *Phil. Trans. Roy. Soc.*, A, 258 (1965).
15 Furon, R., *The Geology of Africa*, 49 (Oliver and Boyd, Edinburgh, 1963).
16 Odell, N. E., *Quart. J. Geol. Soc. London*, **83**, 147 (1927).
17 Harland, W. B., *Quart. J. Geol. Soc., London*, **114**, 307 (1958).
18 Gobbett, D. J., and Wilson, C. B., *Geol. Mag.*, **97**, 441 (1960).
19 Kulling, O., *Geogr. Annaler.*, *1934*, 161 (1934).
20 Sandford, K. S., *Quart. J. Geol. Soc. London*, **112**, 339 (1956).
21 Orvin, A. K., *Skr. Svalb. og Ishavet*, **78**, 1 (1940).
22 Klitin, K. A., *Izvestiya Acad. Sci., U.S.S.R., Geol. Ser.* (Engl. Trans., Amer. Geol. Inst.), *1960*, 50 (1960).
23 Holtedahl, O. (ed.), *Norges Geol. Undersökelse, Nr.* 208, 128, 153, 157, 165 (1960).
24 Bailey, E. B., and Holtedahl, O., *Regionale Geologie der Erde*, **2** (Abschn. 2), 1 (Akad. Verlags. m. b. H., Leipzig, 1938).
25 Umbgrove, J. H. F., *The Pulse of the Earth*, second ed., 232 (M. Nijhoff, The Hague, 1947).
26 Walton, E. K., in *The Geology of Scotland*, edit. by Craig, G. Y., 167, 177, 201 (Oliver and Boyd, Edinburgh, 1965).
27 George, T. N., in *The British Caledonides*, edit. by Johnson, M. R. W., and Stewart, F. H., 12 (Oliver and Boyd, Edinburgh, 1963).
28 Williams, H., *Amer. J. Sci.*, **262**, 1137 (1964).
29 Anderson, F. D., *Geol. Surv. Canada, Map* 8-1965, (1965).
30 Jenness, S. E., *Geol. Surv. Canada, Mem.* 327 (1963).
31 Church, W. R., *Can. Min. Metal. Bull.*, **58**, 219 (1944).
32 *U.S. Geol. Survey Prof. Papers*, 424-*B*, 65 (1961); 475-*B*, 117 (1963); 501-*C*, 28 (1964); 525-*A*, 74 (1965).
33 Novotny, R. F., *U.S. Geol. Surv. Prof. Paper*, 424-*D*, 48 (1961).
34 Billings, M. P., *Geological Map of New Hampshire* (U.S. Geol. Surv., Washington, 1955).
35 Freedman, J., *Geol. Soc. Amer. Bull.*, **61**, 449 (1950).
36 Katz, F. J., *U.S. Geol. Surv. Prof. Paper*, **108**, 165 (1917).
37 Zartman, R., Snyder, G., Stern, T. W., Marvin, R. F., and Buckman, R. C., *U.S. Geol. Surv. Prof. Paper*, 575-*D*, 1 (1965).
38 Emerson, B. K., *U.S. Geol. Surv. Bull.*, 597 (1917).
39 Lundgren, jun., L., Goldsmitt, R., and Snyder, G. L., *Geol. Soc. Amer. Bull.*, **69**, 1606 (1958).
40 Billings, M. P., *The Geology of New Hampshire, Pt. II*, 105 (New Hampshire State Planning and Devel. Comm., Concord, 1956).
41 Howell, B. F., *Intern. Geol. Cong. Sess.* 20, *Mexico, The Cambrian System Symposium*, **2**, 315 (1956).
42 Bird, J. M., and Theokritoff, G., *Geol. Soc. Amer. Bull.*, **77**, 13 (1966).
43 Lochman, C., *Geol. Soc. Amer. Bull.*, **67**, 1331 (1956).
44 Sougy, J., *Geol. Soc. Amer. Bull.*, **73**, 871 (1962).
45 Reyment, R. A., *Nature*, **207**, 1384 (1965).
46 Bureau Res. Geol. Min., *Carte Géol. Afrique Occid.*, Feuille No. 1 (1960).
47 Carroll, D., *U.S. Geol. Surv. Prof. Paper*, 454-*A*, 1 (1963).
48 Wilson, J. T., in *The Earth as a Planet*, edit. by Kuiper, G. P. (Univ. of Chicago Press, 1954).
49 Drake, C. L., Heirtzler, J., and Hirschman, J., *J. Geophys. Res.*, **68**, 5289 (1963).
50 Miller, E. T., and Ewing, M., *Geophysics*, **21**, 406 (1956).
51 Lee, C. S., *Inst. Petroleum J.*, **37**, 633 (1951).
52 Hess, H. H., and Maxwell, J. C., *Bull. Geol. Soc. Amer.*, **64**, 1 (1953).
53 Wilson, J. T., *Earth and Planetary Science Letters* (in the press).
54 Rodgers, J., Gates, R. M., and Rosenfeld, J. L., *Connecticut Geol. Nat. Hist. Surv. Bull.*, **84**, (1959).
55 McLearn, F. H., *Geol. Surv. Canada, Mem.* (to be published).
56 Gilluly, J., *Quart. J. Geol. Soc. London*, **119**, 133 (1963).

Editor's Comments on Papers 9, 10, and 11

EPISODIC DIASTROPHISM DURING THE PHANEROZOIC

On Continents

Geochronological and stratigraphic studies during the past thirty years have confirmed the episodicity of Phanerozoic orogeny in many regions and provided evidence of broad synchronism of orogeny throughout the world.

Histogram plots of isotopic ages have demonstrated the regional episodicity of orogenesis, metamorphism, and granitic intrusion in North America (e.g., Muller, 1961; Faul et al., 1963; Damon and Mauger, 1966; Damon, 1968 and Paper 9; Gilluly, 1973; Lanphere and Reed, 1973), Europe (Fitch et al., 1969), the USSR (Stashkov, 1977), Asia (Bingquan and Xinhua, 1976; Jahn et al., 1976), Africa (Briden and Gass, 1974), and Australasia (Landis and Coombs, 1967). Regional episodicity of granitic magmatism has been proposed also by Budanov et al. (1962), Kawano and Ueda (1967), Evernden and Kistler (1970), and Kistler et al. (1971). All of these studies indicate that metamorphic or magmatic activity recurs at intervals ranging from several hundred million years (Briden and Gass, 1974) to ~40 m.y. or less (e.g., Damon, Paper 9; Evernden and Kistler, 1970). Geochronological studies have revealed episodicity also in Cenozoic circum-Pacific volcanism (Kennett et al., 1977; Hein et al., 1978), in carbonatite emplacement during the Mesozoic and Cenozoic (Macintyre, 1977), and in Mesozoic basic volcanism (Siedner and Mitchell, 1976).

Stratigraphic studies, principally through the dating of sedimentary sequences and major unconformities, have confirmed the regional episodicity of Phanerozoic tectonic movements (e.g., Sloss, 1964; Gilluly, 1967; Rodgers, 1967). Such episodicity is evident in the plot of times of important Paleozoic orogeny in four continents (Figure 5). The importance of stratigraphy to the determination of times of orogeny is emphasized in Part IV (see Papers 21–23), for the pattern of oscillating sea level during the Phanerozoic may yet prove to be the best indicator of orogenic episodicity.

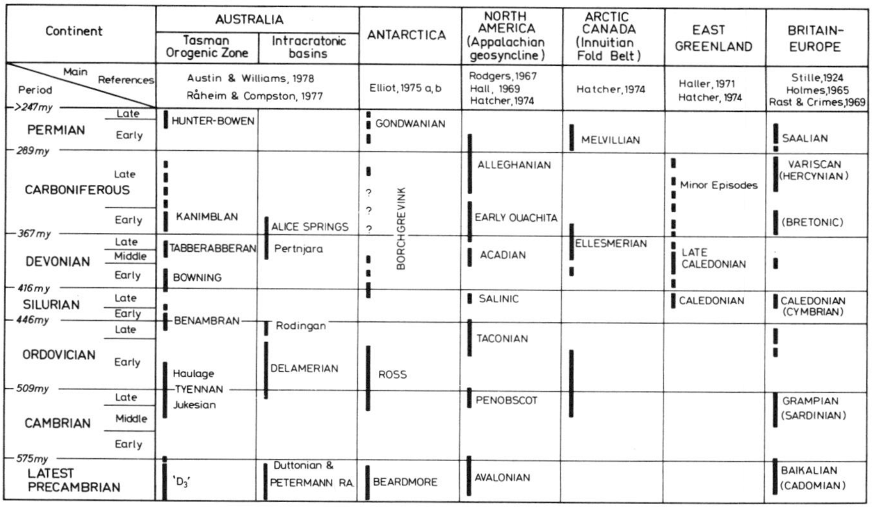

Continent	AUSTRALIA: Tasman Orogenic Zone	AUSTRALIA: Intracratonic basins	ANTARCTICA	NORTH AMERICA (Appalachian geosyncline)	ARCTIC CANADA (Innuitian Fold Belt)	EAST GREENLAND	BRITAIN-EUROPE
Period / Main References	Austin & Williams, 1978; Råheim & Compston, 1977		Elliot, 1975 a, b	Rodgers, 1967; Hall, 1969; Hatcher, 1974	Hatcher, 1974	Haller, 1971; Hatcher, 1974	Stille, 1924; Holmes, 1965; Rast & Crimes, 1969
>247my PERMIAN (Late, Early)	HUNTER-BOWEN		GONDWANIAN		MELVILLIAN		SAALIAN
289my CARBONIFEROUS (Late, Early)	KANIMBLAN	ALICE SPRINGS	? ? ? BORCHGREVINK	ALLEGHANIAN; EARLY OUACHITA		Minor Episodes	VARISCAN (HERCYNIAN); (BRETONIC)
367my DEVONIAN (Late, Middle, Early)	TABBERABBERAN; BOWNING	Pertnjara	BORCHGREVINK	ACADIAN	ELLESMERIAN	LATE CALEDONIAN	
416my SILURIAN (Late, Early)	BENAMBRAN	Rodingan		SALINIC		CALEDONIAN	CALEDONIAN (CYMBRIAN)
446my ORDOVICIAN (Late, Early)	Haulage	DELAMERIAN	ROSS	TACONIAN			
509my CAMBRIAN (Late, Middle, Early)	TYENNAN; Jukesian			PENOBSCOT			GRAMPIAN (SARDINIAN)
575my LATEST PRECAMBRIAN	'D$_3$'	Duttonian & PETERMANN RA.	BEARDMORE	AVALONIAN			BAIKALIAN (CADOMIAN)

Figure 5 Principal episodes of late Precambrian and Paleozoic orogeny in Australia, Antarctica, eastern North America, arctic Canada, East Greenland, and Britain-Europe. Time scale after Armstrong and McDowall (1975).

Paper 9 by Paul E. Damon of the Laboratory of Isotope Geochemistry at the University of Arizona, who has been at the forefront of studies into the meter of Phanerozoic orogeny, demonstrates how geochronological and stratigraphic evidence may be integrated in the search for orogenic episodicity and periodicity. However, Schuchert's paleogeographic maps, which formed the basis for Damon's orogenic time scale, have been superseded by updated atlases (see Paper 21), and opinion is divided as to whether orogeny correlates with sea-level highstands or lowstands (see Part IV).

Although agreement is general that orogeny during the Phanerozoic was episodic, the questions remain as to whether such orogeny was *global* and *periodic* (sensu stricto). Intercontinental correlations of orogenic epochs have been proposed (e.g., Faul et al., 1963; Williams and Austin, 1973; Leonov, 1976), and a broad synchronism of certain Paleozoic orogenies in seven widely spaced regions is evident in Figure 5. Perhaps the best argument for the global nature of important orogeny is provided by the proposed link between patterns of sea-level fluctuation and orogenic episodicity (see Part IV). Global correlation of orogenic epochs is a current aim of the International Geological Correlation Program.

With regard to proposed periodicity (s.s.), recent studies suggest a relatively rapid rhythm of diastrophism in addition to the 150–200 m.y. spacing of the classical Caledonian, Variscan, and Alpine orogenies of Europe. The orogenic phases of Stille and Brinkmann, as dated by recent geological time scales (Table 3), have a mean period of 33 m.y. with a standard deviation of 19 m.y. If the dates of the Walachian and Attic, and the Erzgebirge and Sudetic phases are averaged, the mean period is 38±16 m.y. This mean is in close agreement with the orogenic pulse of 36±11 m.y. proposed by Damon (Paper 9). A mean period of 40–50 m.y. is suggested by the times of important Paleozoic orogeny shown in Figure 5. Moreover, Evernden and Kistler (1970) considered that

Table 3 Scale of the principal orogenic phases as defined by Stille and Brinkmann and dated by recent geological time-scales (in m.y.). From Roubault et al. (1967) and others

Alpine orogeny	~295 : Asturian phase
~2 : Walachian phase	~320 : Erzgebirge phase
~7 : Attic phase	~325 : Sudetic phase
~37 : Pyreneic phase	~345 : Bretonic phase
Laramide orogeny	*Caledonian orogeny*
~65 : Laramide phase	~395 : Ardennian phase
~80 : Subhercynian phase	~435 : Taconic phase
~100 : Austrian phase	~500 : Sardinian phase
~140 : Neocimmerian phase	
~195 : Cimmerian phase	*Baikalian orogeny*
Variscan orogeny	~570 : Cadomian phase
~225 : Palatine phase	
~260 : Saalian phase	

Mesozoic granites in the western United States were emplaced at ~30 m.y. intervals; Rubinshteyn (1967) postulated a global rhythm of 15–25 m.y. for Phanerozoic granitization and metamorphism; and Kaz'min (1975) considered that epochs of important rifting have repeated every 30–40 m.y. since Late Triassic time. The weight of evidence therefore suggests that Phanerozoic time has been punctuated by important continental diastrophism perhaps every 30–50 m.y., although the global nature and strict periodicity of such events remain suggestive but unproven. A comparable rhythm of diastrophism superimposed on the megacycles described above probably existed during the Precambrian, although it cannot be resolved by present dating methods.

Hypotheses advanced to explain the episodicity of Phanerozoic continental diastrophism propose discontinuous convection in the mantle (Walzer, 1974), periodic dissipation of the energy of the solid earth tides (Shaw et al., 1971), variation in the rate of the earth's rotation (Peyve, 1961; Williams and Austin, 1973; Kaz'min, 1975; Austin and Williams, 1978), and galactic control (Tamrazyan, 1954, 1967; Bingquan and Xinhua, 1976).

In Ocean Basins

In 1966 Langseth et al., to explain the distribution of heat-flow values through the ocean floor, suggested that sea-floor spreading was episodic. The idea of such episodicity was further advanced by Ewing and Ewing (1967) and Ewing et al. (1968), who drew attention to an abrupt change in sediment thickness between the crests and flanks of mid-ocean ridges; they pointed out that such changes in sediment thickness could result from major discontinuity in the rates either of sea-floor spreading or of sediment accumulation and chose to interpret the profiles as resulting from discontinuities in spreading rates. Their interpretation was supported by Le Pichon (1968), and numerous subsequent studies have indeed confirmed the episodicity of sea-floor spreading and diastrophism in the ocean basins. (However, the reasoning by Ewing and Ewing [1967] for discontinuous spreading has not withstood the test of drilling [Dr. J. I. Ewing, personal communication, April 1978].)

Peter R. Vogt and co-workers at the U.S. Naval Research Laboratory in Washington have played a leading role in the documentation of episodicity in sea-floor spreading (e.g., Schneider and Vogt, 1968; Vogt et al., Paper 10; Vogt et al., 1971). In 1969 Vogt et al. (Paper 10) proposed that discontinuities in the rate and direction of sea-floor spreading have occurred at intervals of 10–30 m.y. along much of the mid-ocean ridge. They illustrated five possible types of spreading discontinuity and discussed discontinuities known and surmised at that time. Although written soon after the formulation of plate tectonic theory, Paper 10 contains still-valid concepts. Dr. P. R. Vogt (personal communication, May 1978) comments as follows on this work:

> The paper is sufficiently basic and general that I would make no drastic revisions if I were to rewrite it to-day. *Factual* revisions would include the absolute dating of the older discontinuities, brought about by deep drilling after 1969. For example, the "Bermuda discontinuity" was formed ~110–115 m.y. B.P. (not 160 m.y. as tentatively given on p. 314), and the magnetic smooth-rough boundary at 155 m.y. B.P. (not 190 m.y.). The latter boundary, although still not completely understood, is now thought to be at least partly a result of changing geomagnetic parameters—reversal frequency and perhaps dipole intensity. Our 1969 hypothesis linking crustal structure to spreading rate (Figure 5) would also need revision; the oceanic layer (6.7 km/sec) is now known to thicken with crustal age out to about 40 m.y. B.P. Meanwhile, the second ("volcanic") layer thickness appears to be invariant of crustal age, and a spreading rate dependence has not been demonstrated. As to the origin of the plate reorganizations (spreading discontinuities), I would change the tone of the paper toward changes in the active driving forces (such as convection plumes), whereas others would no doubt continue to ascribe the discontinuities to changing interplate boundary forces, including the geometrical demand that three or more plates cannot continue to open about a fixed set of poles. Finally, our discussion of "mode-changes" (Figure 7) is still very much on target. Although the observed preference (even in wax models) for a "zed" geometry subsequently was analyzed in a more sophisticated fashion by Lachenbruch and Froideveaux, among others, the minimum-work concept still appears valid in general. Meanwhile, mode changes without significant changes in rotation pole or rate are well documented, and a convincing physical model for mode changes still lies in the future.

Subsequent work (e.g., Vine and Hess, 1970; Pitman and Talwani, 1972; Larson and Pitman, Paper 11; Vine, 1973; Lattimore et al., 1974; Vogt, 1975; Herz, 1977; Sclater et al., 1977) has confirmed the episodicity of diastrophic events in the ocean basins and has identified times at ~120–110, ~85–70, ~60–55, ~45–35, and ~10 m.y. ago when widespread changes took place in spreading rates and/or plate geometry. Paper 11 by Larson and Pitman is a particularly important contribution, since their extended magnetic reversal time scale allows revision of previous ideas on the timing and origin of sea-floor spreading events in the Atlantic

Ocean and identifies changes in average spreading rates for the Pacific. Average spreading rates in the Atlantic and Pacific oceans as a function of geological time are shown in their Figure 8. This work is further discussed by Berggren et al. (1975) and Larson and Pitman (1975).

A major advance in the past decade is the clearer picture obtained of processes in the ocean basins and how they may effect the continents. Numerous persons (e.g., Le Pichon, 1968; Vine and Hess, 1970; Brookfield, 1971; Coney, 1971; Frerichs and Shive, 1971; Larson and Pitman, Paper 11; Jahn et al., 1976) have suggested that diastrophic events in the ocean basins may be in consonance with post-Paleozoic continental tectonism. Correlation between times of relatively rapid spreading and of orogenesis (e.g., Coney, 1971; Larson and Pitman, Paper 11; Jahn et al., 1976) is plausible, "since periods of rapid plate accretion must correspond to periods of rapid plate consumption, and orogenic phenomena are mainly associated with plate consumption" (Turcotte and Burke, 1978, p. 341). Rice and Fairbridge (1975) postulated that the episodicity of sea-floor spreading itself may result from cyclic thermal runaway in mantle convection.

9

Reprinted from pages 15–23 and 31–33 of *The Late Cenozoic Glacial Ages,* K. K. Turekian, ed. New Haven, Conn.: Yale Univ. Press, 1971, 606 pp.

THE RELATIONSHIP BETWEEN LATE CENOZOIC VOLCANISM AND TECTONISM AND OROGENIC-EPEIROGENIC PERIODICITY

Paul E. Damon

First of all we have seen hardly anything of the earth's crust below a depth of 2 km; secondly, only one third of the globe is open to geological investigation —the rest is ocean; thirdly, a large portion of the continents is covered by shallow water or alluvial deposits, and of the remaining fraction only a very small portion is really well known.

L. U. De Sitter in Structural Geology, *1956, p. 483*

This digression from discussion of the familiar ocean basins to the mysterious continents may serve to emphasize that large elevated regions of the continents and the ocean basins may be produced by the same bulges of the mantle. The origin of rises may be determined by studying plateaus. Unfortunately we know even less about plateaus than about rises.

H. W. Menard in Marine Geology of the Pacific, *1964, p. 152*

These quotations from De Sitter in 1956 and Menard in 1964 serve to remind us of the giant steps—"the great leap forward"—taken by geology during the last decade or so. The once inaccessible ocean basins now seem familiar, and some of us must return to dry land once again and take a fresh look at the "mysterious" continents. Many questions remain to be answered, such as: Is orogeny periodic, episodic, or continuous? What is the relationship between continental volcanism and continental drift? What is the relationship between orogenic and epeirogenic movements on the continents and sea floor spreading? From my vantage point on the continental flank of the East Pacific rise here in Arizona, I would like to address myself to these questions, particularly, as they relate to the Late Cenozoic.

Is Continental Orogeny Periodic, Episódic, or Continuous?

On the subject of orogenic periodicity there are almost as many opinions as there are geologists. However, without attempting to list all plausible permutations and combinations, it will suffice to list only the point of view of certain major protagonists and their disciples.

1. During the Phanerozoic eon there were four major revolutions caused by periodic melting of the substratum. The inter-revolutionary periods were subject to minor disturbances resulting from partial melting and tidal creep of the substratum (Joly 1930).

2. Phanerozoic earth history consists of aperiodic, episodic, orogenic phases of very short duration (ca. 300,000 years) followed by anorogenic periods of very long duration (Stille 1924, 1940).

3. Superposed on a general tectonic unrest two major periodicities may be discerned—a longer periodicity of about 300 m.y. and a shorter periodicity of about 60 m.y. (Umbgrove 1947; Rutten 1949).

4. Phanerozoic earth history consists of essential uniformity in tectonism and volcanism with sporadic and not very great fluctuations in intensity. However, on a regional scale (e.g. western U.S.A.) plutonism may be catastrophic (Gilluly 1949, 1963).

It is safe to say that between them, Rutten and Gilluly demolished Stille's theory (item 3), which had reigned supreme for almost two decades. The existence of orogenic episodes of such short duration followed by long quiescent episodes is not a tenable point of view. When I arrived in Arizona in 1957, Gilluly's ideas (item 4) reigned supreme. The burden of proof fell on anyone who attempted to defend another proposition. Nevertheless, after seven years of geochronologic reseach in southwestern United States, I began, with the help of my students, to formulate an alternative point of view (Damon et al. 1964; Damon and Bikerman 1964; Damon and Mauger 1966).

The stimulus for the alternative point of view resulted from a statistical treatment of K-Ar dates for silicic volcanism and hypabyssal plutonism within the Basin and Range Province (Fig. 1). It can be seen that silicic magmatism during late Cretaceous and Cenozoic time, when plotted on a 5-m.y. class interval, resolves itself into two quasi-Gaussian distributions with a standard deviation of about 7.5 m.y. Furthermore, the two distributions peak almost exactly at the Cretaceous-Paleogene and Paleogene-Neogene boundaries. This result was not sought after, nor did it appear to be a mere coincidence, inasmuch as classical geologists had always sought to place boundaries at significant, observable "breaks" in the geologic record.

The suggested cause of these hypothetical discontinuities has been best expressed by Umbgrove (1947) in his classical treatise entitled *The Pulse of the Earth.* Briefly stated, diastrophism accompanied by silicic magmatism results in mountain building and a regression of the seas. Climatic deterioration thus ensues. Consequently, evolution is accelerated as a result of climatic stress and the

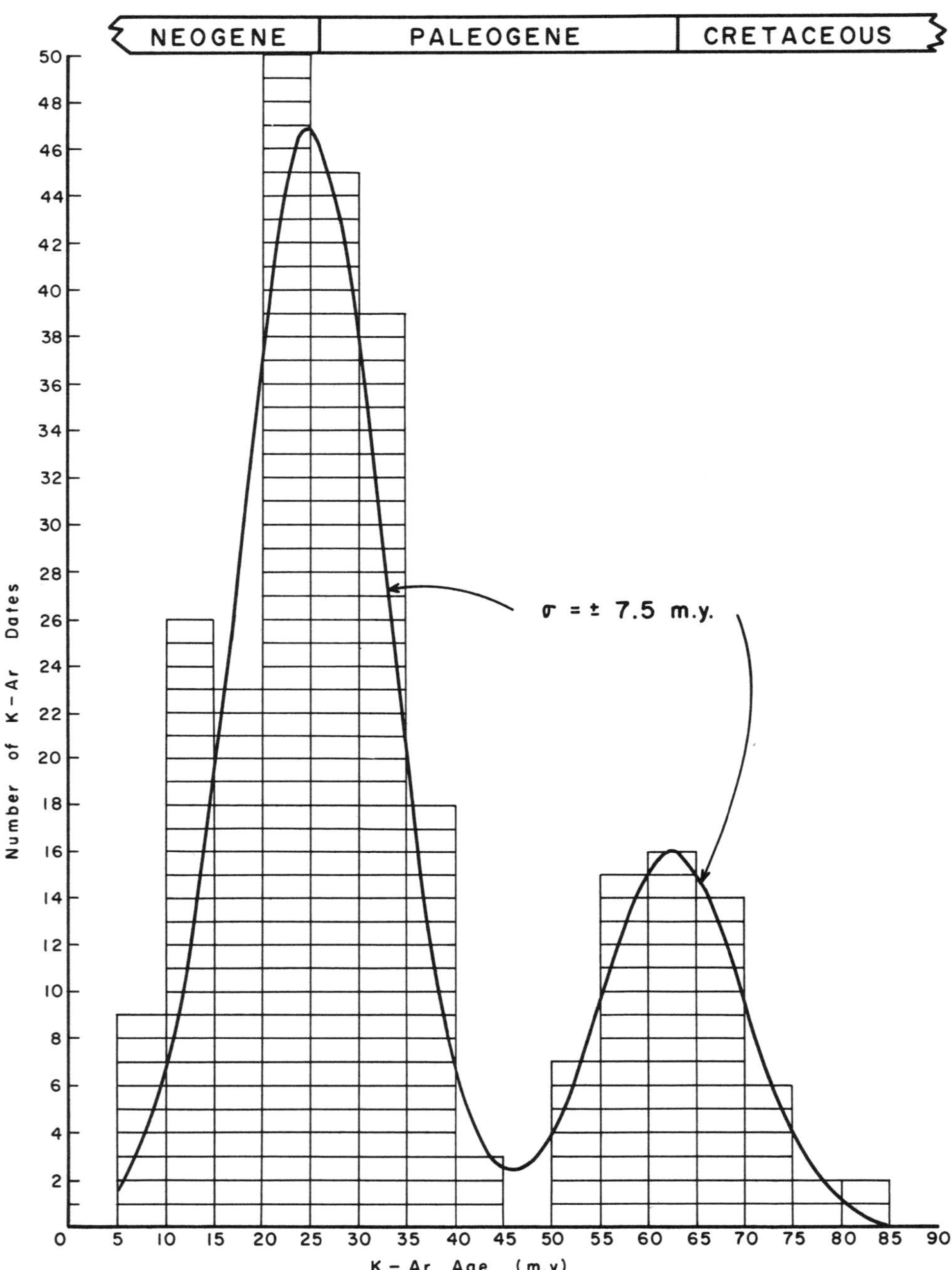

Fig. 1. Current version of histogram published previously by Damon and Mauger (1966) and Damon (1968b). Data are taken from Appendix I of Eastwood (1970) plus additional data from Armstrong (1970) and Damon et al. (1969).

destruction of floral and faunal habitats. Damon and Mauger (1966; see also Damon 1968a,b) suggested that the Laramide orogeny was a good example of Umbgrove's pulse of the earth. Recently, Kauffman and Kent (1968) have demonstrated a twofold increase in the rate of evolution of ammonites and other invertebrates during regressive, as compared to transgressive, stratigraphic sequences during the Cretaceous. Thus, even relatively minor fluctuations have measurable evolutionary effects.

At the time the original histogram was constructed, insufficient reliable data were available to extrapolate the K-Ar dating curve to earlier times and so, as an expedient, the area of transgression of the epicontinental seas onto the present outline of North America was used as an orogenic indicator. We obtained the areal data by planimeter measurements, using Schuchert's (1955) paleogeographic maps for want of a more up-to-date equivalent. The data for the Mesozoic and Cenozoic are presented graphically in Figure 2. Again, much to our surprise, each of the classical North American orogenies is represented by distinct regressions of the epicontinental seas. Subsequently, Evernden and Kistler (1970) have demonstrated the occurrence in California and western Nevada of five epochs of pre-Laramide, Mesozoic magma generation and emplacement (10–15 m.y. duration). Evernden and Kistler then correlated their magmatic episodes with the graph in Figure 2 and arrived at the following conclusion: "Our data require a refinement of this interpretation since the beginning of each Mesozoic plutonic episode is approximately coincident with a peak which marks the beginning of a temporary reversal of the general transgression curve" (p. 77 of preprint). These intrusive epochs and their ages are given in Table 1. A similar pattern has been observed for the Mesozoic batholithic rocks of British Columbia, Canada (Gabrielse and Reesor 1964). In order to explain the relationship in Figure 2, Grasty (1967) constructed a simple diastrophic model involving two "continents," one deformed and one undeformed. He concluded from this model (p. 5) that "an orogenic event is necessarily associated with a eustatic fall of sea level and that the depth of the epicontinental seas is a sensitive indicator of these events." It is now evident that the major features of Umbgrove's concept of the pulse of the earth are valid.

Encouraged by the success attained by the use of transgression of the epi-

Table 1. Intrusive Epochs in California and Western Nevada

(After Evernden and Kistler 1970)

M.Y.	*Geologic age*	*Intrusive epoch*
90–79	Late Cretaceous	Cathedral Range
117–104	Early Cretaceous	Huntington Lake
148–132	Late Jurassic	Yosemite
180–160	Early and middle Jurassic	Inyo Mountains
210–195	Middle and late Triassic	Lee Vining

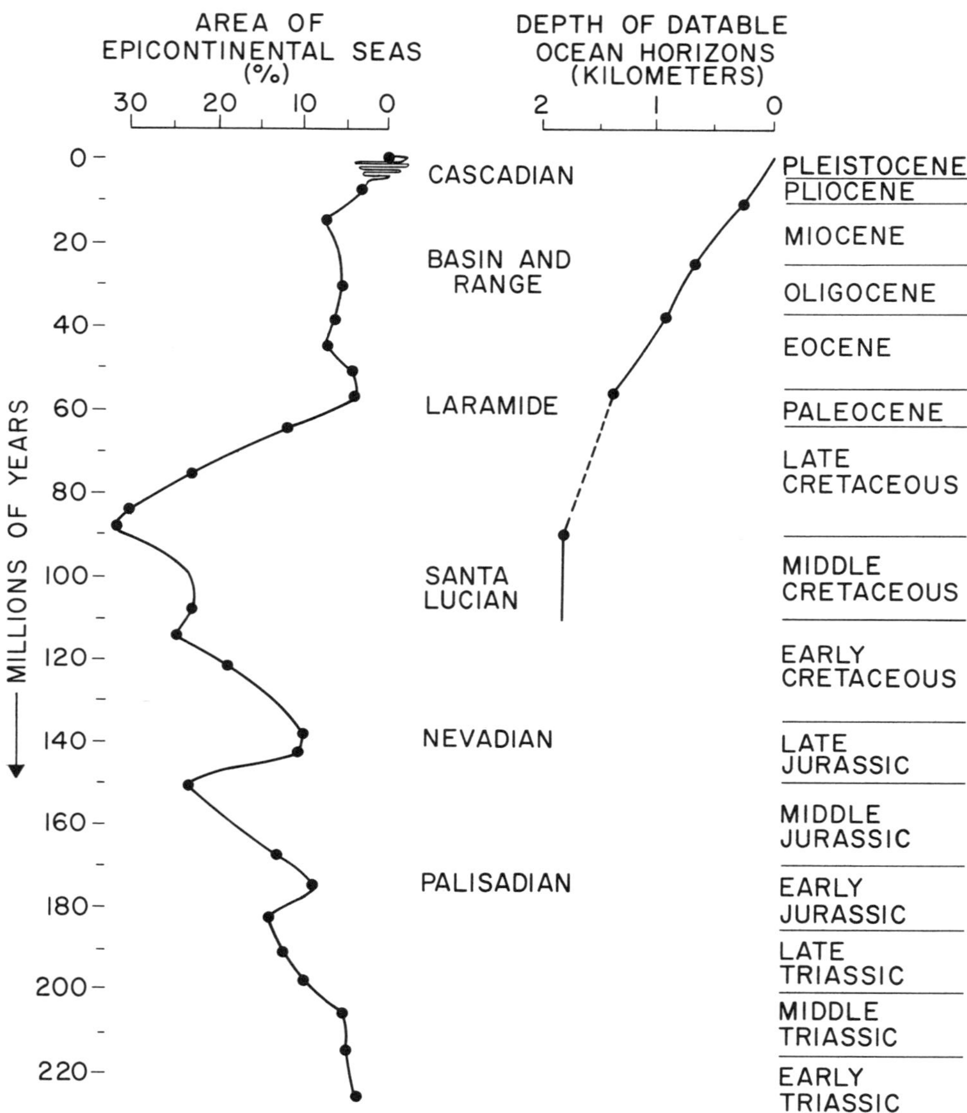

Fig. 2. Extent of the epicontinental seas and depth to datable ocean horizons. Area of epicontinental seas measured in units of percent of the present North American land area covered (including also Central America, northern Colombia, and Greenland, but excluding Hudson Bay). Depth of datable horizons data were from Menard and Ladd (1963); after Damon and Mauger (1966).

continental seas onto the present outline of North America as a measure of orogeny, I have made planimeter measurements for all of Schuchert's paleogeographic maps. The measurements were plotted as a histogram, with a 10-m.y. class interval to minimize correlation problems and overcome the sparsity of data which would result for class intervals of shorter duration (Fig. 3). In addition to the Mesozoic and Cenozoic orogenies previously noted (Fig. 2), the three classical Paleozoic North American orogenies (Taconic, Acadian, and Appalachian) show up as distinct regressions. Several other distinct regressions of nearly as great intensity are also indicated. These correspond to Caledonian and Hercynian movements in Europe.

In order to distinguish broad epeirogenic movements from the orogenic epochs, I passed a smooth curve through the transgression peaks and measured the extent of regression from the smoothed curve (Fig. 4). Stille's orogenies are

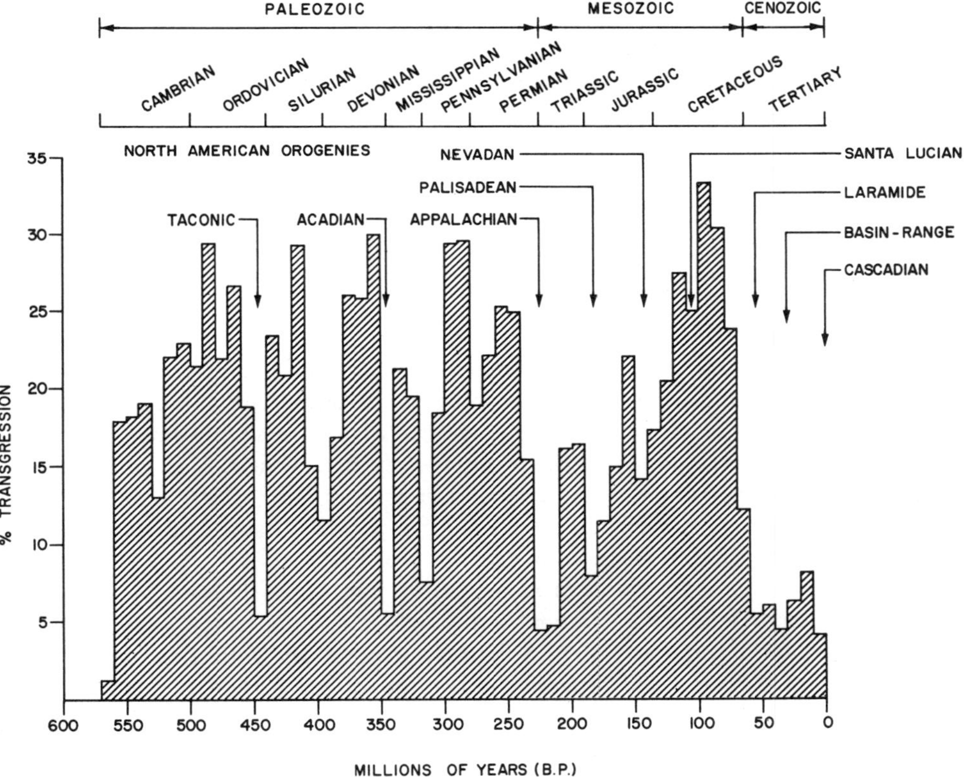

Fig. 3. Extent of epicontinental seas during the Phanerozoic as measured from Schuchert's (1955) paleographic maps. Area measured is the same as for Figure 2. Geological Society of London time scale (Harland et al. 1964) is used in this and following figures.

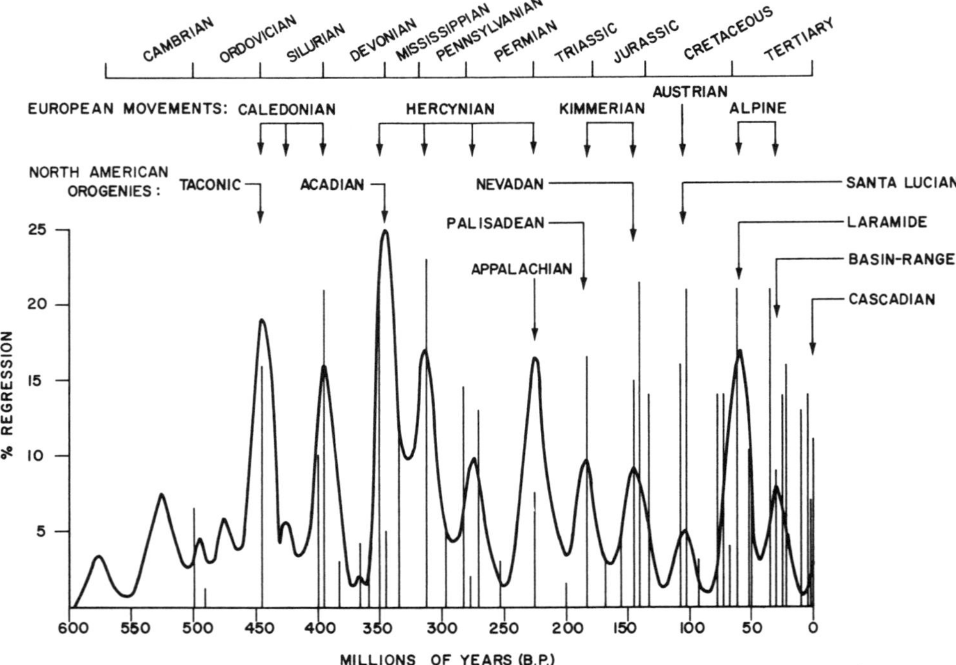

Fig. 4. Orogenic events measured as percent regression from a smooth curve passed through the peaks of marine transgression.

shown as straight vertical lines. Each of the main peaks can be correlated with classical major orogenies. Typically, several of Stille's orogenies fall under each peak but few occur during the quiescent periods. Apparently, most of Stille's orogenies are minor or regional episodes within more extensive events. The most pronounced peak corresponds to the Acadian orogeny. That this is no mere coincidence is evident from the analysis of Lyons and Faul (1968), of northern Appalachian geology in which they conclude (p. 315) that "The overridingly important event was the Acadian Orogeny."

Interestingly enough, if the orogenic pulses are approximated by Gaussian curves, the standard deviation is about 9 m.y. which is close to the value of ±7.5 m.y. observed for Laramide and mid-Tertiary magmatism within the Basin and Range Province. Clearly, the orogenic pulses are of more than regional significance and occur periodically. However, their apparent intensities differ by an order of magnitude. The relationship between apparent intensity and actual intensity of orogeny is a function of epeiric sea and continental shelf bathymetry, a subject which I intend to treat quantitatively in a subsequent paper.

The ages of the boundaries between geologic periods and the ages of regressive minima are presented in Table 2. There appears to be a good correspondence in most cases. The boundary between the Triassic and Jurassic is a notable ex-

Table 2. Period Boundaries and Regression Minima

Boundary between periods	*Age of boundary*	*Age of regression minima*
Precambrian–Cambrian	570	570–580
		520–530
Cambrian–Ordovician	500	490–500
		470–480
Ordovician–Silurian	430–440	440–450
		420–430
Silurian–Devonian	395	390–400
Devonian–Mississippian	345	340–350
Mississippian–Pennsylvanian	320	310–320
Pennsylvanian–Permian	280	270–280
Permian–Triassic	225	220–230
Triassic–Jurassic	190–195	175–185
Jurassic–Cretaceous	136	135–145
		100–110
Cretaceous–Paleogene	65	50–60
Paleogene–Neogene	26	20–30

Table 3. Duration of Periods and Orogenic Cycles

	Duration	*S.D.*	*Spread*
Periods	49.5	15.5	25–71
Orogenic cycles	35.9	10.7	20–50

ception. Where minima do not correspond to geologic boundaries, e.g. within the Cambrian, Ordovician, and Cretaceous, the geologic periods are more than 60 m.y. in duration. The Silurian is an exception to this rule.

The durations of periods and orogenic cycles are compared in Table 3. The orogenic cycles are 36 ± 11 (s.d.) m.y. in duration. The geologic periods are 50 ± 16 (s.d.) m.y. long. The three long periods account for the greater duration of geologic periods as compared to the orogenic cycles. Very recently, Reso (1969, p. 187) has suggested that one of these periods, the Cretaceous, should properly be divided into two: "The paleontologic record indicates that the mid-Cretaceous biological changes are as great as those which distinguish most systematic boundaries. Thus the evidence suggests that according to the methods employed to establish systems, the Cretaceous time interval should be regarded as comprising two Periods." Reso places this break at the base of the Cenomanian, which closely corresponds to the Santa Lucian regressive minimum in Figure 2.

Thus, to summarize my answer to the question posed at the beginning of this chapter: (1) Orogenies are periodic and can be approximated by Gaussian distributions for the intensities of diastrophism and magmatism; (2) the standard deviation of an orogenic pulse is about 8 m.y.; (3) the period of cycles is 36 ± 11 m.y.; (4) the apparent intensities of orogenic pulses vary by an order of magnitude; and (5) the rules used to define geologic boundaries typically resulted in these boundaries being fixed near the peak of orogenic pulses.

[*Editor's Note:* In pages 23 to 31 Damon discusses the relationship between epeirogeny and orogeny, evidence for the episodicity of sea-floor spreading, whether a relationship exists between continental orogeny and continental drift, and the relationship between rate of epeirogenic uplift and drift. His conclusions are included below.]

Conclusions

Proof of the power of a scientific theory, such as the atomic theory, is its effectiveness in helping the scientist correlate vast amounts of data and its potency in fathering new concepts, new relationships, and new tests of its validity. Certainly, the new global tectonics has proven its potency. So many ideas come to mind stimulated by this rapidly developing theory that one scarcely has the time or energy to explore them fully. Fortunately for the reader, most of the ideas which now come to this author's mind are beyond the scope of the paper and so I will conclude by briefly summarizing a few conclusions which must, as always in scientific work, remain tentative.

1. Orogenies are periodic and can be approximated by Gaussian distributions for the apparent intensities of diastrophism and magmatism. The standard deviation of an orogenic pulse is about 8 m.y. The period of cycles is 36 ± 11 m.y. The intensities of orogenic pulses vary by an order of magnitude.

2. The rules used to define geologic boundaries typically resulted in the boundaries being fixed near the peak of orogenic pulses. Certain long geologic periods, such as the Cretaceous, could appropriately be divided into two geologic periods.

3. The model of Mitchell and Reading (1969), in which continents oscillate between rises, provides an effective explanation of the main features of post-Paleozoic North American geology.

4. Epeirogenic uplift of continents is also a periodic phenomenon related to continental orogeny. During orogenic pulses the asthenosphere migrates upward into the lithosphere, and the resulting plastic crust allows continental drift to proceed more rapidly.

5. The rates of sea floor spreading and continental drift are measured in centimeters per year. Vertical epeirogenic movements proceed at centimeters per thousand years. The rate of vertical epeirogenic movement is determined by the rate of sea floor spreading and the slope of rises (approximately 1 meter per kilometer).

6. The late Cenozoic tectonic state of western North America is not typical of past states. It is at one extreme of an epeirogenic cycle during which it overrode the East Pacific rise. Consequently, it has been epeirogenically upwarped to an unusually high average altitude. Furthermore, the western states are entering a new period of orogeny, the Cascadian orogeny which will become more intense and spread eastward.

Summary. A statistical treatment of late Mezozoic and Cenozoic magmatic events coupled with analysis of paleogeographic data is used to demonstrate that orogenesis is periodic with a wavelength of 36 ± 11 m.y. The intensities of diastrophism and magmatism can be represented by a Gaussian distribution with a standard deviation of ± 8 m.y. Anorogenic periods are not of very long duration relative to orogenic periods, as suggested by Stille (1924, 1940). The intensities of orogenic pulses vary by an order of magnitude. The Acadian orogeny was the most intense event occurring during Phanerozoic time, as suggested by Lyons and Faul (1968).

Epeirogenic uplift of continents is also a periodic phenomenon related to continental orogeny. During orogenic pulses, the athenosphere migrates upward into the lithosphere, and the resulting plastic crust allows continental drift to proceed more rapidly. The rate of vertical epeirogenic displacement is determined by the rate of continental drift and the slope of rises (~1 meter per kilometer). Consequently, it may be measured in centimeters per thousand years as compared to centimeters per year for continental drift.

The Late Cenozoic tectonic state of western North America is not typical of its past history. It is at one extreme of an epeirogenic cycle and at an unusually high average altitude. Furthermore, the Cascadian orogeny is only in its beginning stages and, if the past is an adequate key to the future, it should become more intense and spread eastward.

Acknowledgments. It is a pleasure to thank Mr. William K. Smith and Professor John R. Sturgul for formulating the computer program used to obtain the graph in Figure 5. I am also grateful to Professor Donald E. Livingston for editing the manuscript.

This work was supported by U.S. Atomic Energy Commission contract AT(11-1)-689, and the State of Arizona.

REFERENCES

Armstrong, R. L. 1970. Geochronology of Tertiary igneous rocks, eastern Basin and Range Province, western Utah, eastern Nevada, and vicinity, U.S.A. *Geochim. et Cosmochim. Acta,* **34**:203.

Damon, P. E. 1968a. The relationship between terrestrial factors and climate. *Meteorol. Mon.* **8**:106.

Damon, P. E. 1968b. Potassium-argon dating of igneous and metamorphic rocks with applications to the Basin ranges of Arizona and Sonora. In *Radiometric Dating for Geologists,* E. I. Hamilton and R. M. Farquhar, eds. New York: Interscience, p. 1.

Damon, P. E., and associates. 1969. *Correlation and Chronology of Ore Deposits and Volcanic Rocks.* Annual Prog. Rep. No. COO-689-120 to Research Division, U.S. Atomic Energy Comm.

Damon, P. E., and M. Bikerman. 1964. Potassium-argon dating of post-Laramide plutonic and volcanic rocks within the Basin and Range Province of southeastern Arizona and adjacent areas. *Arizona Geol. Soc. Digest* **7**:63.

Damon, P. E., R. L. Mauger, and M. Bikerman. 1964. K-Ar dating of Lar-the Basin and Range Province. *Am. Inst. Min. Engineers Trans.* **235**:99.

Damon, P. E., R. L. Mauger, and M. Bikerman. (1964). K-Ar dating of Laramide plutonic and volcanic rocks within the Basin and Range Province of Arizona and Sonora. In *Cretaceous-Tertiary Boundary Including Volcanic Activity,* Internat. Geol. Congr., 22nd, Proc. **3**:45.

De Sitter, L. U. 1956. *Structural Geology.* New York: McGraw-Hill, 552 pp.

Eastwood, R. L. 1970. *A Geochemical-Petrological Study of Mid-Tertiary Volcanism in Parts of Pima and Pinal Counties, Arizona.* Ph.D. dissertation, Univ. Arizona, Tucson.

Evernden, J. F., and R. W. Kistler. 1970. Chronology of emplacement of Mesozoic batholithic complexes in California and western Nevada. Preprint of *U.S. Geol. Survey Prof. Paper* **623,** 91 pp.

Gabrielse, H., and J. E. Reesor. 1964. Geochronology of plutonic rocks in two areas of the Canadian Cordillera. In *Geochronology in Canada,* F. Fitz Osborne, ed. Toronto: Univ. Toronto Press, p. 96.

Gilluly, J. 1949. Distribution of mountain building in geologic time. *Geol. Soc. America Bull.* **60**:561.
Gilluly, J. 1963. The tectonic evolution of the western United States. *Geol. Soc. London Quart. Jour.* **119**:133.
Grasty, R. L. 1967. Orogeny, a cause of world-wide regression of the seas. *Nature* **216**:779.
Harland, W. B., et al., eds. 1964. *The Phanerozoic Time Scale,* Geol. Soc. London *Quart. Jour.* **120S**, 458 pp.
Joly, J. 1930. *The Surface-History of the Earth.* Oxford: Clarendon Press, 211 pp.
Kauffman, E. G., and H. C. Kent. 1968. Cretaceous biostratigraphy of western interior United States. *Annual Meeting, Geol. Soc. America, Mexico City,* Nov. 1968.
Lyons, J. B., and H. Faul. 1968. Isotope geochronology of the northern Appalachians. In *Studies of Appalachian Geology,* E. Zen, W. S. White, J. B. Hadley, and J. B. Thompson, Jr., eds. New York: Interscience, p. 305.
Menard, H. W. 1964. *Marine Geology of the Pacific.* New York: McGraw-Hill, 271 pp.
Menard, H. W., and H. S. Ladd. 1963. Oceanic islands, seamounts, guyots, and atolls. In *The Sea.* New York: Interscience, p. 365.
Mitchell, A. H. and H. C. Reading. 1969. Continental margins, geosynclines, and ocean floor spreading. *Jour. Geology* **77**:629.
Reso, A. 1969. The mid-Cretaceous paleontological break. *Annual Meeting Geol. Soc. America, Atlantic City, N.J.,* Nov. 1969.
Rutten, L. M. R. 1949. Frequency and periodicity of orogenetic movements. *Geol. Soc. America Bull.* **60**:1755.
Schuchert, C. 1955. *Atlas of Paleogeographic Maps of North America.* New York: Wiley, 177 pp.
Stille, H. 1924. *Grundfragen der vergleichenden Tektonic.* Berlin: Borntraeger, 443 pp.
Stille, H. 1940. *Einfuhrung in den Bau Amerikas.* Berlin: Borntraeger, 717 pp.
Umbgrove, J. H. F. 1947. *The Pulse of the Earth,* 2d ed. The Hague: Nijhoff, 358 pp.

10

Reprinted from pages 285–287, 288, 290, 291, 308 and 313–317 of *Tectonophysics*
8:285–317 (1969)

DISCONTINUITIES IN SEA-FLOOR SPREADING

PETER R. VOGT, OTIS E. AVERY, ERIC D. SCHNEIDER, CHARLES N. ANDERSON and DEWEY R. BRACEY

U.S. Naval Oceanographic Office, Washington, D.C. (U.S.A.)

SUMMARY

Time variations in the rate and direction of sea-floor spreading, though relatively slight, have occurred simultaneously along much of the Mid-Oceanic Ridge at intervals of 10–30 million years. Various criteria by which such discontinuities might be recognized are examined. The recent discovery that transform faults, at least those of small offset, may be created and destroyed repeatedly during the life of an ocean basin suggests several physical models for the creation and destruction of such faults. The shape of the Mid-Oceanic Ridge may represent a minimum-work configuration for a particular spreading rate and direction. When these quantities change the ridge shape adjusts itself to a new configuration if the energy barrier is not too great. The rotation of a ridge segment with respect to the spreading direction may require several million years and depends on spreading rate, ridge orientation, and fracture spacing. Observed rotation rates are about one third or one half the theoretical maximum.

INTRODUCTION

The JOIDES deep sea drilling program (Maxwell, 1969) has shown that, within the uncertainty of paleontological dating, the South Atlantic has been widening at a constant rate since 76 m.y. B.P. This result makes the relative magnetic reversal chronology of Heirtzler et al. (1968) approximately an absolute one, and at the same time precludes the major global stoppage of sea-floor spreading discussed by Ewing and Ewing (1967) and Schneider and Vogt (1968). Sea-floor spreading rates and directions have not been entirely time-invariant for all branches of the Mid-Oceanic Ridge, however, and in this paper we discuss the known and inferred time changes and their effects in a general way. On the assumption that the spreading pattern is relatively constant for a long time and then changes more or less abruptly, we may consider the changes as discontinuities.

TYPES OF DISCONTINUITIES AND THEIR CHARACTERISTICS

Several types of spreading discontinuities are conceivable, although they may not all be observed in nature. It will become apparent in this discussion that the discontinuities are interrelated.

(1) A total stoppage, followed by renewed spreading (Fig.1). Presumably, the longer the quiescent interval, the greater the depth to which the lithosphere can freeze, and the greater the probability that the revived rifting will not follow the old axis (Ewing and Ewing, 1967; Schneider and Vogt, 1968).

(2) A shift in the location of the rift axis. This will create sharp age boundaries where the young "window" of oceanic crust borders the older crust. The Iceland–Jan Mayen and Red Sea rifts are of this sort, except that the old crust is continental on one or both sides (Fig.2).

(3) A change in the rate of spreading (Schneider and Vogt, 1968) (Fig.3–5).

(4) A change in the direction of spreading, i.e., in the orientation of transform faults and hence also in the location of the pole of rotation (Menard and Atwater, 1968, Fig.6.).

(5) A change in the mode of spreading (Fig.7–10). In the "staircase mode" spreading is normal to a number of ridge segments separated by fracture zones. In the "oblique mode" the ridge segment *AB* is not perpendicular to the spreading direction.

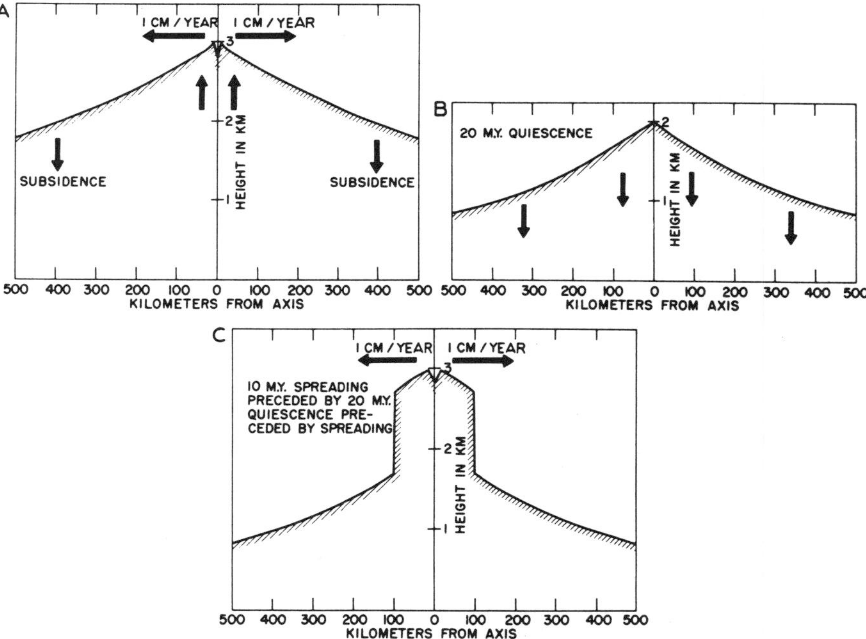

Fig.1. Ridge flank subsidence rates, estimated from Langseth et al. (1966) and assumed to be approximately a function of time only, suggest that regional bathymetry of a ridge could identify total stoppages. No such intermittent ridge has thus far been found.

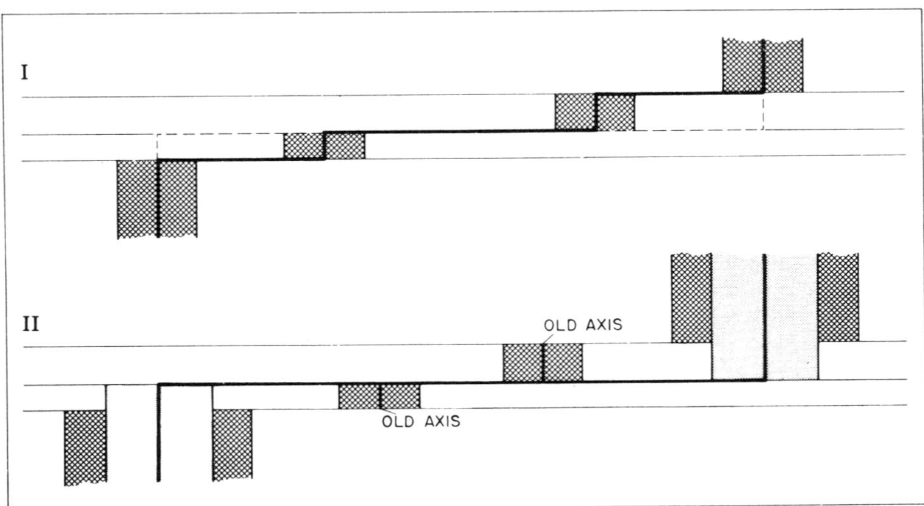

Fig.2. Hypothetical shift of ridge axis. In this example we suppose that changes in spreading rate or other parameters make many close-spaced fracture zones uneconomical, as discussed in text. New break tends to follow old lines of weakness parallel and perpendicular to rift.

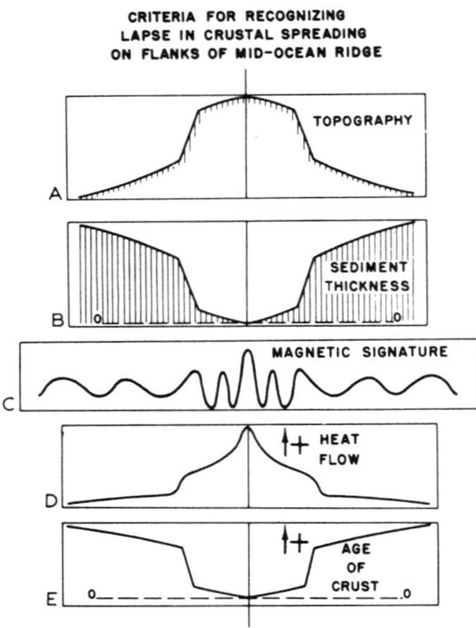

Fig.3. Various parameters conceivably useful in identifying slow-down or stoppage on segment of the Mid-Oceanic Ridge. Magnetic signature can be used only if anomalies can be dated absolutely or compared with anomalies on a ridge known to be spreading at a constant rate. The latter technique was used in identifying periods of slow spreading on Reykjanes Ridge between about 42 and 18 m.y. B.P. (Avery et al., 1969; Vogt et al., 1969c). High heat flow should correlate with high regional elevation.

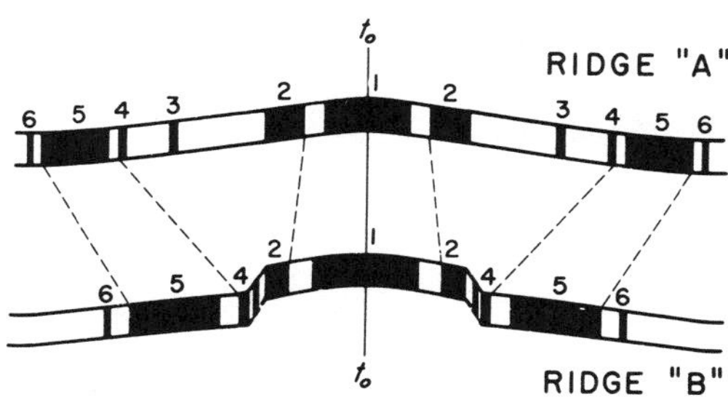

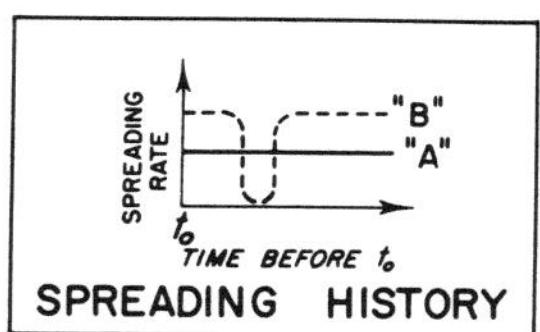

Fig. 4. Comparison of magnetic patterns between ridges will reveal discontinuities in spreading rate. If the reversal time scale is absolutely known (Maxwell, 1969), the discontinuity can be dated. Ridges do not have to be currentlv spreading for application of this technique.

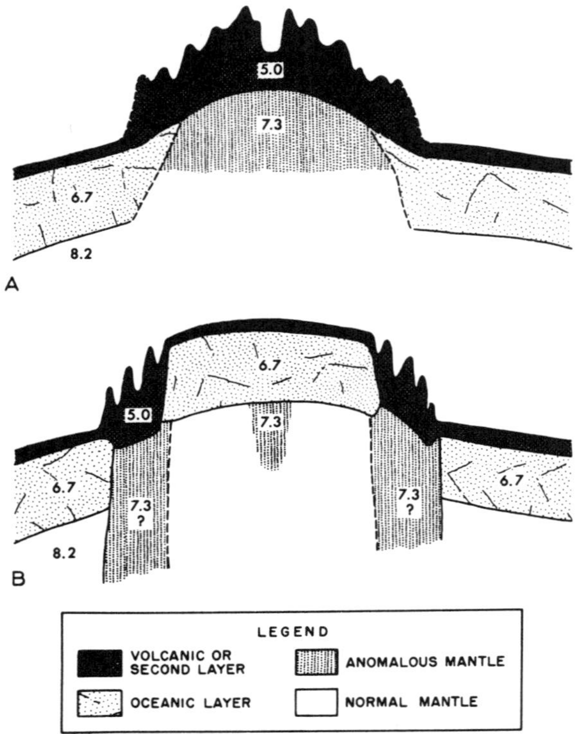

Fig.5. Crustal structure seems to depend on spreading rate (Talwani et al., 1965; Menard, 1967) suggesting an additional tool in identifying changes in rate. A. The ridge slowed down from a value greater than about 2 cm/year to a lesser value. B. Spreading rates fell below 2 cm/year for a period cf time (Schneider and Vogt, 1968).

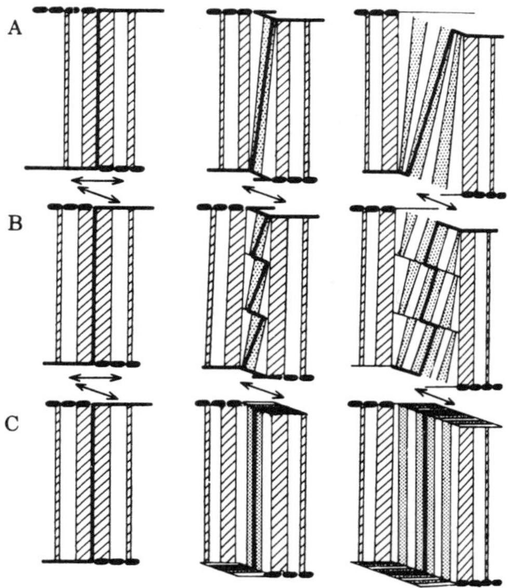

Fig.6. Changes in direction may give rise to transform faults. A–B. The observed tendency for ridges to be perpendicular to transform faults suggests that, whether or not new transform faults are created, the spreading axis will reorient itself perpendicular to the new direction in time (from Menard and Atwater, 1968). C. The new spreading axis becomes oblique, and the old transform fault also becomes a rhombus-shaped oblique rift provided rotation of spreading direction is clockwise for right-handed offsets of ridges by transform faults, or counterclockwise for left-handed offsets. Opposite direction changes require breaking corners of pre-existing crust before adjustment can be accomplished (Menard and Atwater, 1968).

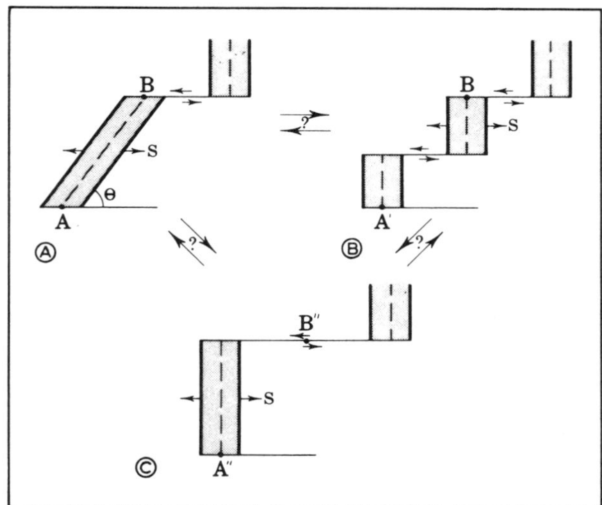

Fig.7. Changes in mode may occur in conjunction with changes of rate and direction. Transitions to a more efficient geometry can occur only if energy barrier between modes is not too great. Changes may be gradual or abrupt, in the latter case accompanied by axis shifts.

[*Editor's Note:* Figures 8, 9, and 10 have been omitted. In pages 288 to 308 Vogt et al. discuss stoppages and axis shifts, time-changes in spreading rate (plate motion rate), change of direction and mode, mode changes and the minimum work principle, and rotation of the spreading axis. Pages 309 to 312 are fold-out maps (Figures 16 and 17) and have been omitted.]

KNOWN AND SURMISED DISCONTINUITIES

As the age of a parcel of ocean floor tends to vary relatively linearly with distance from the axis of any particular segment of the Mid-Oceanic Ridge (Heirtzler et al., 1968), certain areas of the ocean will have to be covered by close-spaced grid surveys before the pattern of discontinuities is completely understood. In Fig.16 we have sketched the approximate isochrons corresponding to known spreading changes on a physiographic chart of the western North Atlantic (Heezen and Tharp, 1968). The positions of the isochrons presuppose a fairly constant spreading rate, now documented back to about 76 m.y. B.P. (Talwani et al., 1969). Only detailed surveying will establish whether the Mid-Atlantic Ridge has actually recorded some or all of the global changes in spreading pattern. Geomagnetic reversal chronologies have been established for about two thirds of the North Atlantic Basin time. The post- 76 m.y. B.P. sequence was derived by Heirtzler et al. (1968) in the South Atlantic and is now absolutely dated (Maxwell, 1969). A relative chronology derived from 40 east–west crossings of the Keathley sequence was reported by Anderson et al. (1969). There is some evidence that low amplitude lineations in the smooth zone represent sea-floor generated near the equator. If so, the reversal sequence can be extended back to the continental shelf. Dating of the Keathley sequence follows from the assumption of approximately constant spreading rates and the hypothesis that the smooth-rough transition reflects rapid polar wandering during Late Triassic times. No correlatable linear pattern has thus far been identified between the Bermuda discontinuity and a broad negative anomaly (Bracey, 1968, Fig.17) that is slightly below the bottom of the Heirtzler et al. (1968) reversal time scale. The pattern of anomalies and fracture zones in this region is purely speculative.

Atlantic discontinuities are speculative except for a change in spreading rate at about 4 m.y. B.P. (Phillips, 1967), a change in direction at about 10 m.y. B.P. on the 24°N transform fault (Fox et al., 1969), and the Chain fracture zone at 11°N (Van Andel et al., 1969), the Bermuda discontinuity (160 m.y. B.P.?), and possibly the discontinuity tentatively dated at 90 m.y. B.P. (Anderson et al., 1969).

Lesser transform faults are known to have disappeared between 12 and 18 m.y. B.P. between 56° and 61°N on Reykjanes Ridge (Avery et al., 1969; Vogt et al., 1969c) and prior to about 10 m.y. B.P. between 9° and 11°N on the Mid-Atlantic Ridge (Van Andel et al., 1969). In both areas the major fracture zone existed throughout. Apparently a major change in spreading rate, direction and mode of spreading occurred at least in the north Atlantic during the period 18–10 m.y. B.P.

The Bermuda discontinuity, tentatively dated at 160 m.y. B.P. appears to be of the same sort (Anderson et al., 1969), except that transform faults were created instead of destroyed.

There is mounting evidence that the discontinuities are global, although not everywhere clearly recognizable. Thus, the direction change between 7.5 and 3 m.y. B.P. in the northeast Pacific (Menard and Atwater, 1968) coincides approximately with a possible slowdown of spreading on the Mid-Atlantic Ridge (Phillips, 1967), the beginning of sea-floor spreading in the Red Sea, and the reorientation of spreading on the Chile Rise (Morgan et al., 1969).

The 18–10 m.y. B.P. discontinuity, discussed previously, seems to have affected at least the Atlantic branches of the Mid-Oceanic Ridge. Spreading rates changed in the south Pacific at this time (Heirtzler et al., 1968; Fig.13). World-wide geologic changes, including the slowdown of Alpine orogenic activity, may have effected changes in oceanic circulation and thus caused sedimentation rates on the ridge crest to decrease, thus explaining the scarcity of sediment on the crest (Ewing and Ewing, 1967) without resort to stoppages. A worldwide cooling trend that began at this time may have increased the vigor of currents and swept more sediment off the ridge crests, and decreased the carbonate fraction (hence the total sediment volume). Massive Miocene flood basalts were extruded during this period in the northwest United States.

The 45–35 m.y. B.P. discontinuity seems to have been sufficiently major to effect a permanent change in the geomagnetic reversal frequency (Heirtzler et al., 1968). Rapid sea floor spreading between Australia and Antarctica began then, while spreading on Reykjanes Ridge slowed down. Major orogenic activity along the Alps seems to have begun at the same time and lasted until the 18–10 m.y. B.P. discontinuity discussed above. When the magnitude and direction of spreading has been sufficiently well charted, it will be possible to apply the concepts of plate tectonics (Morgan, 1968, Isacks et al., 1968) to establish whether increased crustal destruction rates (calculated) can explain the increased orogenic activity along the Alpine tectonic front between about 40 and 15 m.y. B.P.

Between 60 and 40 m.y. B.P. the newly created Reykjanes Ridge was spreading at rates which declined relatively steadily from an initial value of 1.7 cm/year (60 m.y. B.P.) to 0.9 cm/year (48 m.y. B.P.). Possibly in adjustment to this rift, spreading directions changed in the North Pacific (Menard and Atwater, 1968). This correlation suggests that spreading direction may change on old ridges, in response to the development of new rifts elsewhere. This change is accomplished only gradually, being accompanied by direction changes of a few degrees per million years.

The 60 m.y. B.P. discontinuity seems to correspond with the Nevada orogeny of North America and was signalled by basic intrusives on Scotland and east and west Greenland (Grasty and Wilson, 1967), and a slowdown in the South Pacific (Heirtzler et al., 1968; Fig.13).

A discontinuity at the beginning of the Heirtzler et al. (1968) reversal chronology seems likely, because of a break-down in anomaly correlations at that time. In view of the uncertainties involved, this discontinuity may or may not be identical to the one labelled 90 m.y. B.P. in Fig.16. The broad negative anomaly near the Antilles (Fig.17) in any case does not occur within the Heirtzler time scale unless spreading rates exceeding 3 cm/year are assumed. Basic intrusives (85 m.y. B.P.) in Portugal may have signalled the first extension of the Atlantic Rift north of the Azores (Gasty and Wilson, 1967).

Nothing is known about the interval between this anomaly and the Bermuda discontinuity, but several discontinuities are likely to have

[*Editor's Note:* Figure 13, which shows spreading rates against time on four different parts of the mid-ocean ridge, has been omitted. Updated average spreading rates against time for the Atlantic and Pacific oceans are shown in Paper 11, Figure 8.]

occurred if the Cenozoic is typical with respect to the frequency of changes in spreading pattern.

The Bermuda discontinuity has been tentatively dated at about 160 m.y. B.P., as discussed previously, and this may coincide with the extrapolated time for the separation of Africa and South America (Maxwell, 1969).

The smooth-rough boundary (190 m.y. B.P.?) may either be a spreading discontinuity or an effect of rapid polar wandering. An axis shift early in the history of the Atlantic Basin is required to explain the greater width of the western smooth zone if it is to be maintained that these zones were created by sea-floor spreading.

The underlying causes for the observed changes are still poorly understood. The following observations must be explained by any theory:

(1) Changes in the spreading pattern seem to have occurred every 10–30 m.y., at least during the Cenozoic.

(2) Complete transitions from one spreading pattern to another seem to require 5 or 10 m.y. and involve trend change rates of several degrees per million years. Discontinuities are therefore not abrupt but may occupy strips of crust 100 km wide. Orientation changes may be completed 5 m.y. earlier in one region than another only 200 km away (Avery et al., 1969). The rotations are accomplished at about half or one third the theoretical maximum rate.

(3) Extensions of the rift into previously intact continental blocks seem to be associated with changes on preexisting ridges and occur during periods of relatively higher spreading rates (Fig.13).

(4) Discontinuities at 60, 45–35 and 18–10 m.y. B.P. were associated with orogenic changes. Basic intrusive activity signalled the development of new rifts at these times and also at 85 and 140–180 m.y. B.P. (Grasty and Wilson, 1967).

(5) The 45–35 m.y. B.P. discontinuity was also associated with a change of geomagnetic reversal frequency. This discontinuity is of the type proposed by Irving and Robertson (1969), because spreading, orogenic, and geomagnetic changes occurred in unison.

(5) The 45–35 m.y. B.P. discontinuity was also associated with a change of geomagnetic reversal frequency. This discontinuity is of the type proposed by by Irving and Robertson (1969), because spreading, orogenic, and geomagnetic changes occurred in unison.

(6) The boundary between 60 m.y. B.P. ocean floor south of Rockall Plateau (in the northeast Atlantic) and much older but as yet undated ocean floor immediately to the east is marked by a basement ridge (Avery et al., 1969; Johnson et al., 1969). This ridge rises 0.5 km above the average basement level on the young sea floor. There is a marked basement elevation discontinuity (3–4 km to the west, 5–6 km on the east), which would be consistent with the hypotheses illustrated in Fig.1 and/or 2.

(7) The Bermuda Discontinuity is marked by a 50 km-wide strip of high amplitude anomalies (Anderson et al., 1969), exceeded in amplitude only by the anomaly over the active ridge axis. Perhaps relatively thicker accumulations of highly magnetized pillow basalts were extruded as a result of the mechanics of a spreading direction change.

In this paper we do not attempt to propose theories for changes. We note only that the strong lithospheric plate concept (Morgan, 1968) implies that changes in the force distribution acting on the plates will be resisted until a certain threshold is exceeded. This threshold is determined by the relative weakness of the lithosphere along the ridge and trench boundaries compared to the strength in the interiors of the plates.

One possibility still viable is that a system of shallow convection cells supplies the net tractive forces on the bottom surfaces of the plates. There may be no relationship between ascending convection currents and ridges crests. The lithosphere is broken into plates as a response to these tractive forces and pre-existing lines of weakness, such as continent-ocean boundaries or earlier orogenic trends. The force distribution must also govern where new rifts form, else it is hard to see why the East African, Red Sea, Labrador and equatorial Atlantic rifts broke across presumably strong Precambrian shields. The strength of the plates is emphasized by the great length of the Americas plate (Morgan, 1968). This plate is remarkably narrow, yet unbroken, east of the Puerto Rico Trench - a tribute to the strength of the plates. Alternatively, the plate hypothesis has to be modified to include bending about vertical axes.

As the plates move, the deformation pattern perpetuates itself because continued rifting along existing plate boundaries is always easier than developing new breaks, all else being equal. Continued spreading changes the sizes and shapes of the plates until it becomes easier for the global tectonic pattern to change. This could occur even if the "basic" force distribution (convection currents?) remains fixed in space and time. Abrupt or gradual changes in force distribution would both produce relatively abrupt spreading changes because of the tendency for any given global plate-tectonic pattern to resist changes up to certain thresholds. The tendency (Heirtzler et al., 1968) for rotational poles to coincide with geographic poles suggests that episodes of rapid polar wandering should be accompanied by discontinuities in the spreading pattern.

An origin for discontinuities is suggested by the observation (Isack et al., 1968) that crustal destruction rates are slower along continent-continent plate interfaces (such as the Alpine-Himalayan belt) than where oceanic lithosphere is being destroyed. This probably results from the greater buoyancy of continents, which resist being forced into the denser mantle. The Atlantic Basin will not continue to grow in size indefinitely; a trench may develop, most likely along the continental margin where the lithosphere is already somewhat warped by thick sediment accumulation. Even without sediment loading the margin is a line across which physical properties in the upper mantle change and is therefore a likely locus of new deformation. The ocean basin then closes as ocean floor is consumed in the trench (provided the rate exceeds the rate new crust is generated at the ridge crest). Eventually the ridge crest is consumed (perhaps resulting in abrupt world-wide changes in the spreading pattern) and, finally, the last ocean floor disappears. After the continental plates begin to collide, destruction rates fall, resulting in decreases of spreading rates on other branches of the Mid-Oceanic Ridge. Thus, the correlation between slow spreading on Reykjanes Ridge and Alpine orogeny (40–20 m.y. B.P.) suggests that when the Tethyan ocean floor had been consumed, the relative movement of plates diminished in this area.

A simultaneous slowdown and spreading direction change may result if an ocean basin does not disappear coevally along its entire length. For example, if the northern part of a basin, closing along a north-south line of trenches, closed first, continent-continent collision begins there and retards the destruction rate. Meanwhile, oceanic lithosphere is still being consumed at a high rate in the south. As a consequence the plate being consumed turns counterclockwise, as do spreading directions on the other, accreting side of this plate.

Even though the spreading directions change relatively abruptly, several million years are required for the ridge crest to adjust itself to an optimum configuration (usually but not always perpendicular to adjacent transform faults).

REFERENCES

Anderson, C.N., Vogt, P.R. and Bracey, D.R., 1969. Magnetic anomaly trends between Bermuda and the Bahama–Antilles Arc. Trans. Am. Geophys. Union, 50: 189.

Avery, O.E., Vogt, P.R. and Higgs, R.H., 1969. Morphology, magnetic anomalies and evolution of the northeast Atlantic and Labrador Sea, 2. Magnetic anomalies. Trans. Am. Geophys. Union, 50: 184.

Bracey, D., 1968. Structural implications of magnetic anomalies north of the Bahama–Antilles islands. Geophysics, 33: 950–961.

De Boer, J., Schilling, J.G. and Krause, D.C., 1969. Magnetic polarity measurements on basalts from the Reykjanes Ridge. Trans. Am. Geophys. Union, 50: 190.

Ewing, J. and Ewing, M., 1967. Sediment distribution on the mid-ocean ridges with respect to spreading of the sea-floor. Science, 156: 1590–1592.

Fox, P.J., Pitman III, W.C. and Shephard, F., 1969. Evidence suggesting two poles of rotation for the central Atlantic crustal plate. Science, in press.

Godby, E.A., Hood, P.J. and Bower, M.E., 1968. Aeromagnetic profiles across the Reykjanes Ridge southwest of Iceland. J. Geophys. Res., 73: 7637–7649.

Grasty, R.L. and Wilson, J.T., 1967. Ages of Florida volcanics and of opening of the Atlantic. Trans. Am. Geophys. Union, 48: 212.

Harrison, C.G.A., 1968. Formation of magnetic anomaly patterns by dyke injection. J. Geophys. Res., 73: 2137–3142.

Heezen, B.C. and Tharp, M., 1968. Physiographic chart of the North Atlantic. Geol. Soc. Am.,

Heezen, B.C., Tharp, M. and Ewing, M., 1959. The floors of the oceans, 1. The North Atlantic. Geol. Soc. Am., Spec. Papers, 65: 122 pp.

Heirtzler, J.R., Dickson, G.O., Herron, E.M., Pitman, W. and Le Pichon, X., 1968. Marine magnetic anomalies, geomagnetic field reversals, and motions of the ocean floor and continents. J. Geophys. Res., 73: 2119–2136.

Irving, E. and Robertson, W.A., 1969. Test for polar wandering and some possible implications. J. Geophys. Res., 74(4): 1026–1036.

Isacks, B., Oliver, T. and Sykes, L.R., 1968. Seismology and the new global tectonics. J. Geophys. Res., 73: 5855–5900.

Johnson, G.L., 1967. North Atlantic fracture zones near 53°N. Earth Planetary Sci. Letters, 2: 445–448.

Johnson, G.L. and Heezen, B.C., 1967. The morphology and evolution of the Norwegian–Greenland Sea. Deep-Sea Res., 14: 755–771.

Johnson, G.L., Vogt, P.R. and Schneider, E.D., 1969. Morphology, magnetic anomalies and evolution of northeast Atlantic and Labrador Sea, 1. Morphology. Trans. Am. Geophys. Union, 50: 184.

Langseth, M.G., Le Pichon, X. and Ewing, M., 1966. Crustal structure of the mid-ocean ridges, 5. Heat flow through the Atlantic Ocean floor and convection currents. J. Geophys. Res., 71: 5321–5355.

Matthews, D.H. and Bath, J., 1967. Formation of magnetic anomaly pattern of Mid-Atlantic Ridge. Geophys. J., 13: 349–357.

Maxwell, A.E., 1969. Recent deep sea drilling results from the South Atlantic. Trans. Am. Geophys. Union, 50: 113.

McKenzie, D.P., 1967. Some remarks on heat flow and gravity anomalies. J. Geophys. Res., 72(24): 6261–6273.

Menard, H.W., 1967. Sea-floor spreading, topography and the second layer. Science, 157: 923–924.

Menard, H.W. and Atwater, T., 1968. Changes in direction of sea floor spreading. Nature, 219: 463–467.

Morgan, W.J., 1968. Rises, trenches, great faults and crustal blocks. J. Geophys. Res., 73: 1959–1982.

Morgan, W.J., Vogt, P.R. and Falls, D.F., 1969. Magnetic anomalies and sea-floor spreading on the Chile Rise. Nature, 222: 137–141.

Oliver, J. and Dorman, J., 1963. Exploration of sub-oceanic structure by use of seismic surface waves. In: The Sea. Interscience, New York, N.Y., 110–133.

Phillips, J.D., 1967. Magnetic anomalies over the Mid-Atlantic Ridge near 278N. Science, 157: 920–923.

Schneider, E.D. and Vogt, P.R., 1968. Discontinuities in the history of sea-floor spreading. Nature, 217: 1212–1222.

Schneider, E.D., Vogt, P.R. and Lowrie, A., 1969. Diapiric structures and magnetic anomalies of the pre-Cenozoic Atlantic Ocean. Trans. Am. Geophys. Union, 50: 212.

Sleep, N.H., 1969a. Heat flow, gravity, and sea-floor spreading. J. Geophys. Res., 74(2): 542–549.

Sleep, N.H., 1969b. Topography at intersection of mid-oceanic ridges and transform faults. Trans. Am. Geophys. Union, 50: 181.

Talwani, M.B., Le Pichon, X. and Ewing, M., 1965. Crustal structure of the mid-ocean ridges, 2. Computed model from gravity and seismic refraction data. J. Geophys. Res., 70(2): 341–352.

Talwani, M., Windisch, C., Langseth, M. and Heirtzler, J.R., 1969a. Recent geophysical studies on the Reykjanes ridge. Trans. Am. Geophys. Union, 50: 189.

Talwani, M., Pitman, W. and Heirtzler, J.R., 1969b. Magnetic anomalies in the north Atlantic. Trans. Am. Geophys. Union, 50: 189.

Van Andel, T.H. and Bowin, C.O., 1968. Mid-Atlantic Ridge between 22° and 23° north latitude and the tectonics of mid-ocean rises. J. Geophys. Res., 73: 1279–1298.

Van Andel, T.H., Phillips, J.D. and Von Herzen, R.P., 1969. Rifting origin for the Vema fracture zone in the North Atlantic. Science, in press.

Vogt, P.R., Schneider, E.D. and Johnson, G.L., 1969a. The crust and upper mantle beneath the sea. In: V.V. Beloussov and P.J. Hart (Editors), The Earth's Crust and Upper Mantle. Geophys. Monograph, 13, Am. Geophys. Union,

Vogt, P.R., Ostenso, N.A. and Johnson, G.L., 1969b. Magnetic anomalies and sea-floor spreading north of Iceland. J. Geophys. Res., in press.

Vogt, P.R., Avery, O.E., Morgan, W.J., Johnson, G.L., Schneider, E.D. and Higgs, R.H., 1969c. Morphology, magnetic anomalies and evolution of the northeast Atlantic and Labrador Sea, 3. Evolution. Trans. Am. Geophys. Union, 50: 184.

Wilson, J.T., 1965. A new class of faults and their bearing upon continental drift. Nature, 207: 343–347.

ERRATUM

Page 314, lines 34 through 37 are repeated in error.

11

Reprinted from pages 3645 and 3654–3660 of *Geol. Soc. America Bull.*
83:3645–3662 (1972)

World-Wide Correlation of Mesozoic Magnetic Anomalies, and Its Implications

ROGER L. LARSON
WALTER C. PITMAN III
Lamont-Doherty Geological Observatory of Columbia University, Palisades, New York 10964

ABSTRACT

In the course of correlating three sets of Mesozoic magnetic lineations in the western Pacific (the Phoenix, Japanese, and Hawaiian lineations), Larson and Chase (1972) determined a paleomagnetic pole for the Pacific plate for the Early Cretaceous. Using this pole we have derived a magnetic reversal model for the Hawaiian lineation set. We then have used this model to correlate the entire Hawaiian lineation set to the entire Keathley lineation set in the western North Atlantic. On the basis of these correlations and drill holes associated with the lineation patterns, we have extended the geomagnetic reversal time scale back to the base of the Late Jurassic (162 m.y. B.P.). A period of reversals occurred corresponding to the Hawaiian and Keathley lineations from 150 to 110 m.y. B.P., and these reversals are bracketed by long periods of dominantly normal polarity (the Cretaceous and Jurassic magnetic quiet zones).

This magnetic reversal time scale significantly alters previous notions of the timing and origin of sea-floor spreading features in the Atlantic Ocean. It implies that the Bay of Biscay opened sometime during the interval between 150 and 110 m.y. B.P.; that drift in the South Atlantic was initiated at sometime during the interval from 110 to 85 m.y. B.P. (probably close to 110 m.y. B.P.); and that the seaward portion of the marginal quiet zones of the eastern United States and northwestern Africa resulted from sea-floor spreading during the Late Jurassic period of dominantly normal magnetic polarity prior to 150 m.y. B.P.

In the Pacific during the late Mesozoic, spreading was occurring from at least five spreading centers joined at two triple points. The vast majority of the Pacific Basin today is occupied by only the Pacific-plate side of these spreading patterns. This implies that an area equal to most of the Pacific Basin has been subducted beneath the surrounding continents since the Early Cretaceous. Our magnetic reversal time scale calls for a rapid pulse of spreading from about 110 to 85 m.y. B.P. at all the spreading centers in both the Atlantic and Pacific Oceans. This implies a pulse of rapid subduction around the rim of the Pacific that we relate to episodes of large-scale plutonism in eastern Asia, western Antarctica, New Zealand, the southern Andes, and western North America during the Late Cretaceous.

[*Editor's Note:* In pages 3646 to 3654 Larson and Pitman discuss correlation and dating of the Hawaiian and Keathley lineations, timing of sea-floor spreading events in the Atlantic, and configuration of the ridges and continents at 110 m.y. B.P.]

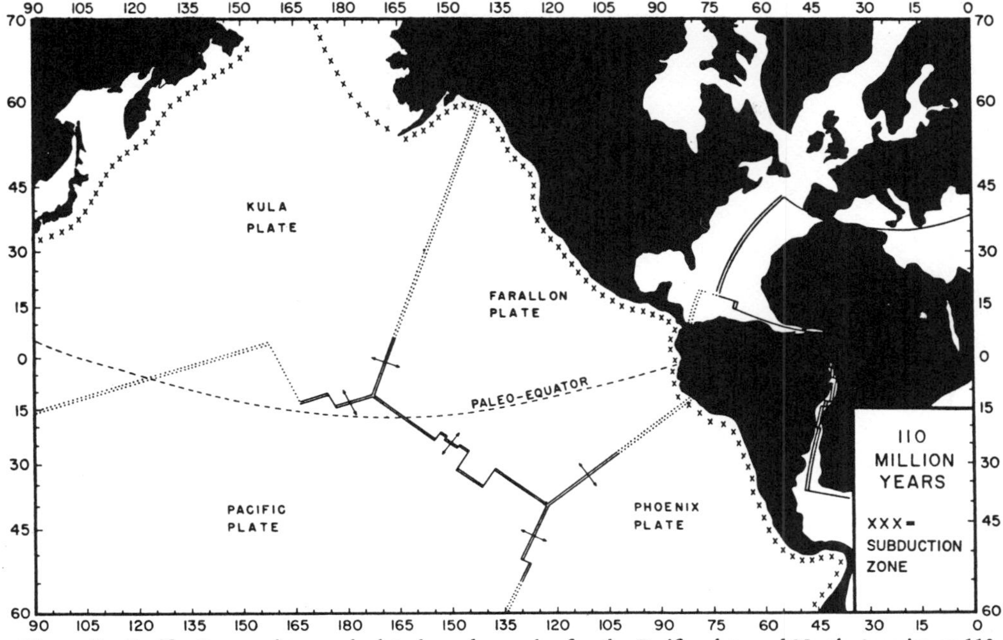

Figure 7. Pacific Ocean plates and plate boundaries at 110 m.y. B.P. after Larson and Chase (1972) plotted relative to the Atlantic Ocean continents at that time. Dotted plate boundaries show probable extensions of known features. Relative configuration accomplished by superimposing the paleomagnetic poles for the Pacific plate and North America at 110 m.y. B.P. (*see* text) and by reconstructing the Atlantic continents relative to North America. Plot construction based on the present-day location of North America, so the paleoequator appears as a curved line.

SPREADING RATES IN THE ATLANTIC AND PACIFIC OCEANS

Having derived a time scale of magnetic polarity events extending back through the Late Jurassic (Fig. 5), we computed average spreading rates for discrete time intervals for various oceanic plate boundaries. By spreading rate, we mean the half-rate of separation of two plates, and assume that the spreading has been bilaterally symmetric. Figure 8 shows the spreading-rate averages for the central Atlantic, South Atlantic, and Pacific Oceans. By central Atlantic, we mean the part between Africa and North America. Since almost all of the Indian Ocean appears to be younger than uppermost Cretaceous, we have not included it in the graph. The portion of the time scale that we wish to focus upon is the Mesozoic.

The spreading rates were determined by using the Heirtzler and others (1968) time scale and the time-scale extension proposed here. The data are plotted as average rates for the time span over which the average was made. This type of presentation gives the appearance of discontinuities in the spreading rates. In reality, transitions from one spreading rate to another probably follow smooth curves. However, it may be inferred from Figure 2 of Heirtzler and others (1968) that in the South Pacific, for example, there were long intervals during which the spreading rate was constant; yet, the spreading rates of adjacent intervals differed by at least a factor of two, and the transition from one interval to another was quite rapid.

The spreading rates for the central Atlantic were computed from the data of Pitman and Talwani (1972). Computations were made along an azimuth parallel to one of their synthetic fracture zones intersecting the ridge axis at 27° N. Since the quiet zone adjacent to North America is twice as wide as that adjacent to Africa we have taken an average and assumed that drift was initiated at 180 m.y. B.P.

[*Editor's Note:* Figure 5, which shows the geomagnetic reversal time scale from 162 m.y. B.P. to the present, has been omitted.]

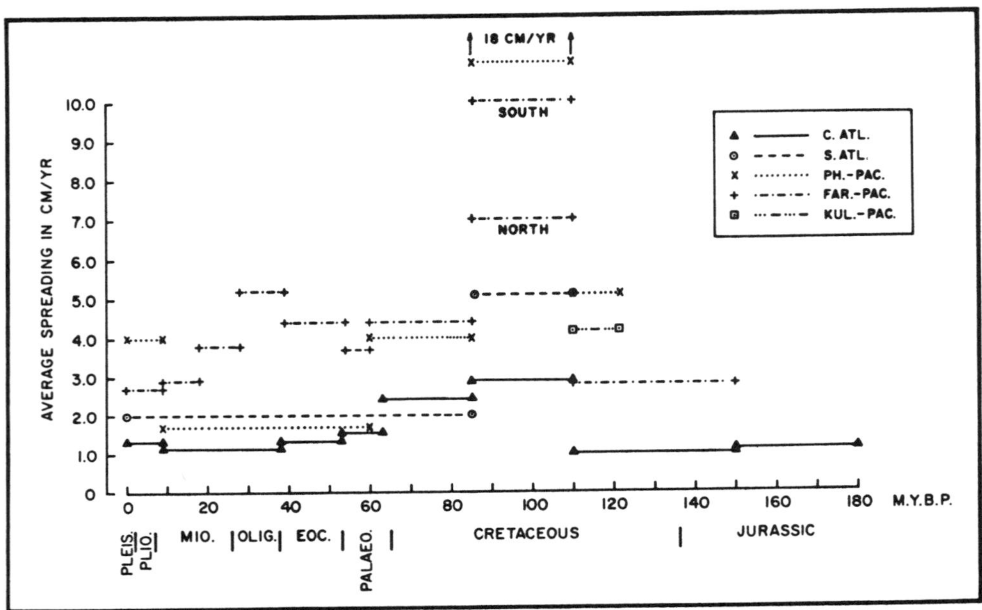

Figure 8. Average spreading rates in the Atlantic and Pacific Oceans as a function of geologic time. Averages were made over discrete time periods as discussed in the text. Note the marked increases in spreading rates of all spreading systems at 85 to 110 m.y. B.P.

The spreading rates for the South Atlantic were computed from the data of Dickson and others (1968) and Ladd and others (1972). We have also made the assumption based on previous arguments that the initiation of drift in the South Atlantic occurred at 110 m.y. B.P. We have computed the rates along an azimuth parallel to a synthetic fracture zone intersecting the ridge axis at 40° S. The azimuth of the fracture zone was determined by drawing a small circle about the pole used to fit South America against Africa (Bullard and others, 1965).

The spreading rates in the Pacific Ocean are described as motions across various plate boundaries. During the Late Cretaceous all of these spreading systems were located in the southern hemisphere (Fig. 7). Since then they have evolved into the present-day system of plate boundaries found in the eastern Pacific. For the Pacific-Phoenix plate boundary during the period from 85 m.y. B.P. to the present, the rates were computed from the data of Pitman and others (1968). Computations were made along an azimuth just north of and parallel to the Eltanin Fracture Zone. For the intervals of 110 to 85 m.y. and 125 to 110 m.y., the spreading rates are dependent upon the assumption that the fracture zone which marks the western boundary of the Phoenix lineations is a continuation of the Louisville Ridge–Eltanin Fracture Zone, and thus a continuation of the present-day system in the South Pacific (Larson and Chase, 1972, Fig. 13). The spreading rates were 18 cm/yr from 110 to 85 m.y. B.P. and 5 cm/yr for 125 to 110 m.y. B.P. The data used to compute the rates for the Pacific-Farallon plate boundary are from Heirtzler and others (1968) and Vine (1966). Computations have been made along an azimuth parallel to the Mendocino Fracture Zone. Two rates are given for the interval 110 to 85 m.y., one for the plate boundary north of Mendocino Fracture Zone and one for the plate boundary south of this offset. The rate across the Pacific-Kula plate boundary has been computed along an azimuth perpendicular to the trend of the anomalies.

LARGE-SCALE MOTIONS OF PLATES

We have actually determined two different types of plate motion. First, we have used trends of fracture zones and the age and location of magnetic lineations to determine the motions of the Kula, Farallon, and Phoenix plates with respect to the Pacific plate. Secondly, we may use the paleomagnetic data to determine the latitudinal motion of the Pacific plate with respect to the Eurasian and North American plates. As with all paleomagnetic data, the longitudinal motion is indeterminate.

As stated previously, we are somewhat skeptical about the determinations of the Cretaceous paleomagnetic pole for Eurasia because many of the measurements were made on rocks from within or adjacent to deformation belts of late Mesozoic or younger age. However, we note that the Eurasian data as presented by Van der Voo and Zijderveld (1969) displays good grouping. The mean pole is at 62° N., 171° E. The Cretaceous pole that we have used for North America is at 70° N., 170° W. (Larochelle, 1969). It is obvious that the mean Cretaceous pole for Eurasia is very nearly coincident with that for North America as plotted with respect to these land masses in their present positions. However, Van der Voo and Zijderveld pointed out that if the North Atlantic is closed according to the fit derived by Bullard and others (1965), then the Cretaceous paleomagnetic poles become separated by several thousand kilometers. This may be interpreted to infer that an error exists, either in the Cretaceous paleopole location for Eurasia, or in the fit of the European and North American continental margins as derived by Bullard and others (1965). However, we note that the separation is in an east-west direction so that the northward motion of the Pacific plate relative to Eurasia and North America is still at least 7,000 km.

As the Pacific plate was drifting northward relative to Eurasia and North America, the Kula plate was spreading away from the Pacific plate in roughly a northwesterly direction, and the Farallon plate was spreading away from the Pacific plate in a northeasterly direction. The entire Kula plate (including the accreting margin that was between the Kula and Pacific plates) and a northern part of the Pacific plate have been subducted under Eurasia and into a proto-Aleutian Trench. This means a minimum of 7,000 km of subduction under Eurasia. Most of the Farallon plate (with the exception of the Juan de Fuca and the Cocos plates) has been consumed under western North America. Again this is a minimum of 7,000 km.

The amount of subduction of the Phoenix plate under southern South America and western Antarctica is more difficult to estimate because the past spreading patterns are not as well established in the South Pacific. However, if accretion at the Phoenix-Pacific margin has been symmetric, we can deduce the amount of subduction by the present-day location of the ridge axis with respect to the continental margins of South America and Antarctica and the Phoenix lineations. Here the amount of crust consumed is equal to the length of the Eltanin-Louisville-Phoenix Fracture Zone complex, less the distance from the East Pacific Rise crest to Antarctica or South America. The amount of subduction measured in this way is about 5,000 km.

Another interesting facet of these large-scale motions was pointed out by Farrell (1968) on the basis of slightly different data. The present-day pole of relative motion for the Pacific plate and North American plate is at about 50° N., 80° W. (Chase, 1972). The present-day locations of the Cretaceous paleomagnetic poles for these two plates (50° N., 30° W. for the Pacific plate and 70° N., 170° W. for the North American plate) lie approximately along a single small circle about Chase's present-day pole of relative motion. This implies that Atwater's (1970) constant motion model for the Pacific and North American plates is grossly correct, not only for the late Tertiary, but for at least 6,000 km of relative motion in the last 110 m.y.

RELATION OF EPISODES OF FAST SPREADING TO CIRCUM-PACIFIC INTRUSIVE AND EXTRUSIVE ACTIVITY AND OROGENESIS

If there are significant increases in spreading rates at all ridge crests, and the earth is not expanding, then there must be a corresponding increase in the world-wide rate of crustal subduction. This might be reflected as increases in individual subduction rates at the various trenches around the margin of the Pacific Ocean (Fig. 7). It can be seen from Figure 8 that beginning at about 110 m.y. and lasting until about 85 m.y., there was a pulse of rapid spreading in the central and South Atlantic and in the Pacific at the Farallon-Pacific plate boundary, and probably the Phoenix-Pacific plate boundary. Whether this is true with regard to the Phoenix-Pacific plate boundary is dependent upon the validity of the previously discussed assumption of the Louisville Ridge–Eltanin Fracture Zone extending northwest to the Phoenix Fracture Zone. This implies the startlingly high rate of 18 cm/yr at the Pacific-Phoenix boundary from 110 to 85 m.y. B.P.

However, a rate of 9 cm/yr has been measured in the Indian Ocean (McKenzie and Sclater, 1972). The rate at the Pacific-Kula plate boundary was 4.2 cm/yr during the Early Cretaceous until 110 m.y. B.P. The rate since that time is unknown. It does not seem unwarranted to assume that spreading was quite rapid at this boundary during the 110 to 85 m.y. B.P. interval, since the evidence suggests that it was quite rapid at all the other boundaries.

This mid- to Late Cretaceous episode of rapid spreading appears to have been a world-wide event, and it is of great interest to relate it to major geologic events on the continents. The obvious candidates are major plutonic events that may be associated with a pulse of rapid marginal underthrusting. The increase in spreading rates in the Atlantic would obviously cause an increase in the rate of encroachment upon the Pacific-Tethys. The simultaneous increase in spreading rates in the Pacific, and probably in the Tethys, would probably be reflected by an increase in rates of underthrusting along all of the Pacific-Tethys subduction zones. We realize that rates of subduction do not necessarily increase at all margins, but intuitively feel that this is likely to be the case.

The mid- to early Late Cretaceous (115 to 85 m.y. B.P.) batholithic intrusion of the western United States is greater by an order of magnitude than any comparable period (Lindgren, 1915; Knopf, 1955; and Gilluly, 1963, 1965). We quote Gilluly (1965), ". . . the Middle Cretaceous was the time of emplacement of by far the greatest plutons of Phanerozoic time in the whole belt extending from Alaska to Baja California. The batholiths of the Peninsular Range, the Sierra Nevada, many of the Coast Range plutons of California, the Idaho batholith and many of its satellites in eastern Oregon and northern Idaho all belong to this episode. The plutons of northeastern Washington are outliers of the great Coast Range batholith of British Columbia." Gilluly (1963) attributed this massive plutonism to either subcrustal flow of great masses of differentiating mantle or to a smaller but still large mass of sialic crust. Hamilton (1969a, 1969b) has directly attributed this phase of batholithic intrusion to an accelerated rate of underthrusting at the west coast of North America, and he tentatively suggested a rate of 10 cm/yr, in good qualitative agreement with the rates of spreading at the Farallon-Pacific plate boundary. Another related phenomenon would appear to be the mid- to Late Cretaceous uplift (Ernst, 1970) and deformation of the Franciscan Formation which again may be ascribed to a sudden increase in spreading rates.

Although spreading in the Pacific may extend well back into the Paleozoic (Moores, 1970), spreading at the Pacific-Kula plate boundary began about 130 m.y. and continued at an average rate of 4.2 cm/yr until approximately 110 m.y. ago. At that time the rate may have accelerated as at other boundaries. Several orogenic events may correlate with the spreading history. Although there are granitic rocks older than 370 m.y. in Japan and some in the age range of 170 to 180 m.y., about 95 percent are younger than 120 m.y. (Kawano and Ueda, 1967). About 75 percent range in age from 50 to 100 m.y. The same pattern appears in Korea and eastern Asia where the mid- to Late Cretaceous was a time of extensive batholithic intrusion (Muyashiro, 1972, oral commun.).

Spreading on some part of the Pacific ridge system must have been occurring by at least 180 m.y. ago. This is demanded by the fact that drift between Laurasia and Gondwana began at about 180 m.y. ago. In fact, the Pacific spreading did not passively raft South America and Africa away from North America and Eurasia, but began to underthrust the Pacific margin of South America (James, 1971; Ian Dalziel, 1972, personal commun.). Paleozoic sediments were folded at this time in the southern Andes (Ian Dalziel, 1972, oral commun.) and andesite volcanism began in the Upper Triassic–Lower Jurassic (James, 1971). Although batholiths date from this time, the main period of batholithic intrusions appears to have been during the mid- to Late Cretaceous and lower Paleocene (Farrar and others, 1970; Hollingworth, 1964). "Folding in the west Andean geosyncline began in the Cenomanian and was continued by later Cretaceous and Early Tertiary phases," (Rutland, 1971). Radiometric ages determined on granite and granodiorite rocks from the southern Andes and western Antarctica (Halpern, 1970, 1971) indicate a major episode of intrusion from 70 m.y. to 120 m.y. with a peak at 90 to 100 m.y. A similar peak from 90 m.y. to 120 m.y. occurs in a histogram of radiometric ages of basement rocks of New Zealand (Landis and Coombs, 1967).

Spreading at the Pacific-Phoenix plate boundary began about 130 m.y. ago and continued at a rate of 5 cm/yr until 110 m.y. ago. This spreading period may well be the causal mechanism for the Early to mid- Cretaceous batholithic intrusion in the southern Andes, New Zealand, and western Antarctica. We speculate that from 110 m.y. to 85 m.y. there was rapid spreading at the Pacific-Phoenix plate boundary, perhaps as high as 18 cm/yr. We are more certain that spreading in the South Atlantic during this same interval was 5 cm/yr. This episode correlates remarkably well with the apparent main phase of batholith intrusion in the southern Andes, New Zealand, and western Antarctica and with initiation of folding of the west Andean geosyncline. Between 60 and 10 m.y. the spreading in the South Pacific was 1.75 cm/yr, but at about 10 m.y. the spreading rate increased to 4 cm/yr. In the South Atlantic the rate has been 2.0 cm/yr from about 85 m.y. to the present. The uplift that formed the present Andes began in the late Miocene with pulses in the mid-Pliocene and late Pliocene or Pleistocene (Hollingworth and Rutland, 1968). Ignimbrite volcanism dates from late Miocene and Pliocene (Rutland, 1971). Hamilton (1969a) regarded these volcanic fields as a modern analogue of the extensive Upper Cretaceous batholiths of western North America. Thus, the interval of slow (~2 cm/yr) spreading in the Pacific correlates with a period of relative quiescence in the Andes, and the rapid spreading in the Pacific that started in the late Miocene appears to have initiated the uplift and intrusion that are responsible for the modern Andes.

CONCLUSIONS

We have derived a magnetic reversal model for Late Jurassic–Early Cretaceous time based on the Hawaiian lineation sequence in the North Pacific. With the use of this model we have correlated the Hawaiian lineations to the Keathley lineations in the North Atlantic. This correlation, and the correlations of Larson and Chase (1972) of the Mesozoic-aged Phoenix, Japanese, and Hawaiian lineations establish that this period of reversals is bracketed by long periods of normal magnetic polarity. We have concluded that this period of reversals occurred from 150 to 110 m.y. B.P. on the basis of two drill holes that yield relatively precise basement ages, and are well located with respect to the lineation patterns. We do not expect that the reversal pattern will be altered significantly by future studies, as the only other Mesozoic lineation pattern not accounted for is the eastern counterpart of the Keathley lineations near the African margin, and possibly yet-undiscovered sequences at high latitudes. However, the time calibration of this pattern may be significantly changed by future drilling on the Mesozoic magnetic lineations.

If the calibration that we propose is approximately correct, it significantly alters the timing and explanation of various sea-floor spreading features of the Atlantic Ocean. The Bay of Biscay is restricted to opening between 150 and 110 m.y. B.P.; the South Atlantic opened between 110 and 85 m.y. B.P. (probably at 110 m.y. B.P.); and the seaward portions of the marginal quiet zones of the eastern United States and northwest Africa resulted from spreading during a period of no magnetic reversals prior to 150 m.y. B.P.

Thousands of kilometers of oceanic lithosphere have been subducted at the trench systems around the Pacific Basin since the Cretaceous. At least 7,000 km has been underthrust beneath North America, a similar amount beneath Eurasia, and probably about 5,000 km beneath South America and (or) western Antarctica.

Our time scale of geomagnetic reversals for the Mesozoic also calls for a pulse of relatively rapid spreading at all spreading centers in the Atlantic and Pacific Oceans from 110 to 85 m.y. B.P. We have related this pulse of rapid spreading to episodes of circum-Pacific intrusive and extrusive activity and orogenesis during this period. In particular, it appears to us that plutonism on a large scale characterized eastern Asia, western Antarctica, New Zealand, the southern Andes, and western North America during the mid- and early Late Cretaceous. This is best documented in western North America where more than 50 percent of the exposed batholiths date from 115 to 85 m.y. B.P. If the granodiorites and granites that make up these batholiths are derived from underthrust oceanic lithosphere, then these extensive plutons necessitate vast amounts of lithospheric subduction, which appears to be a consequence of rapid spreading in all the oceans during this period.

ACKNOWLEDGMENTS

We particularly wish to express our gratitude to the numerous persons that have contributed to the success of the Deep-Sea Drilling Project, for it is upon the scientific results of this project that this research largely has been based. This research has been sponsored by contract N-00014-67-A-0108-0004 from the Office of Naval Research and Grant GA-27281 from the Oceanography Section of the National Science Foundation to the Lamont-Doherty Geological Observatory.

REFERENCES CITED

Anonymous, 1964, Geological Society Phanerozoic time scale 1964: Geol. Soc. London Quart. Jour., v. 120, p. 260–262.

Atwater, T., 1970, Implications of plate tectonics for the Cenozoic tectonic evolution of western North America: Geol. Soc. America Bull., v. 81, p. 3513–3536.

Berggren, W. A., 1969, Cenozoic chronostratigraphy, planktonic foraminiferal zonation, and the radiometric time scale: Nature, v. 224, p. 1072–1075.

Bullard, E. C., Everett, J. E., and Smith, A. G., 1965, The fit of the continents around the Atlantic, *in* Symposium on continental drift: Royal Soc. London Philos. Trans., Ser. A, v. 258, p. 41–51.

Burek, P. J., 1970, Magnetic reversals: their application to stratigraphic problems: Am. Assoc. Petroleum Geologists Bull., v. 54, p. 1120–1139.

Chase, C. G., 1972, The N-plate problem of plate tectonics: Royal Astron. Soc. Geophys. Jour. (in press).

Christoffel, D., and Falconer, R.K.H., 1972, Marine magnetic measurements in the south-west Pacific Ocean and the identification of new tectonic features, *in* Hayes, D. E., ed., Antarctic oceanology II, The Australian-New Zealand sector: Washington, D. C., Am. Geophys. Union, Antarctic Research Series, p. 197–209.

Cox, A., 1969, Geomagnetic reversals: Science, v. 163, p. 237–245.

Dickson, G. O., Pitman, W. C., III, and Heirtzler, J. R., 1968, Magnetic anomalies in the South Atlantic and ocean-floor spreading: Jour. Geophys. Research, v. 73, p. 2087–2100.

Dietz, R. S., and Holden, J. C., 1970, Reconstruction of Pangea: breakup and dispersion of continents, Permian to present: Jour. Geophys. Research, v. 75, p. 4939–4956.

Emery, K. O., Uchupi, E., Phillips, J. D., Bowen, C. O., Bunce, E. T., and Knott, S. T., 1970, Continental rise off eastern North America: Am. Assoc. Petroleum Geologists Bull., v. 54, p. 44–108.

Ernst, W. G., 1970, Tectonic contract between the Franciscan Melange and the Great Valley Sequence—crustal expression of a late Mesozoic Benioff zone: Jour. Geophys. Research, v. 75, p. 886–901.

Farrar, E., Clark, A. H., Haynes, S. J., Quirt, G. S., Connand, H., and Zentilli, M., 1970, K-Ar evidence for the post Paleozoic migration of granitic intrusion faci in the Andes of Northern Chile: Earth and Planetary Sci. Letters, v. 10, p. 60–66.

Farrell, W. E., 1968, Has the Pacific basin moved as a rigid plate since the Cretaceous?: Nature, v. 217, p. 1034–1035.

Gilluly, J., 1963, The tectonic evolution of the western United States: Geol. Soc. London Quart. Jour., v. 119, p. 133–174.

—— 1965, Volcanism, tectonism and plutonism in the western United States: Geol. Soc. America Spec. Paper 80, 69 p.

Halpern, M., 1970, Rubidium-strontium dates and Sr^{87}/Sr^{86} initial ratios of rocks from Antarctica and South America: A progress report: Antarctic Jour. U.S., v. 5, no. 5, p. 160.

—— 1971, Evidence for Gondwanaland from a review of West Antarctic radiometric ages, *in* Research in the Antarctic; Washington, D. C., Am. Assoc. Adv. Sci., p. 717–730.

Hamilton, W., 1969a, The volcanic Central Andes—a modern model for the Cretaceous batholiths and tectonics of western North America, *in* McBirney, A. R., ed., Bulletin of Proceedings of the Andesite Conference: Oregon Dept. Geology and Mineral Industries Bull., v. 65, p. 175–184.

—— 1969b, Mesozoic California and the underflow of Pacific mantle: Geol. Soc. America Bull., v. 80, p. 2409–2430.

Hayes, D. E., and Pitman, W. C., III, 1970, Magnetic lineations in the North Pacific: Geol. Soc. America Mem. 126, p. 291–314.

Heezen, B. C., Tharp, M., and Ewing, M., 1959, The floors of the oceans, Part I: The North Atlantic: Geol. Soc. America Spec. Paper 65, 122 p.

Heirtzler, J. R., and Hayes, D. E., 1967, Magnetic boundaries in the North Atlantic Ocean: Science, v. 157, p. 185–187.

Heirtzler, J. R., Dickson, G. O., Herron, E. M., Pitman, W. C., III, and Le Pichon, X., 1968, Marine magnetic anomalies, geomagnetic field reversals, and motions of the ocean floor and continents: Jour. Geophys. Research, v. 73, p. 2119–2136.

Helsley, C. E., and Steiner, M., 1968, Evidence for long periods of normal magnetic polarity in the Cretaceous period: Earth and Planetary Sci. Letters, v. 5, p. 325–332.

Hollingworth, S. E., 1964, Dating the uplift of the Andes of Northern Chile: Nature, v. 201, p. 17–20.

Hollingworth, S. E., and Rutland, R.W.R., 1968, Studies of Andean uplift, Part I: Post-Cretaceous evolution of the San Bartolo area,

North Chile: Jour. Geology, v. 6, pt. 1, p. 49–62.

James, D. E., 1971, Plate tectonic model for the evolution of the Central Andes: Geol. Soc. America Bull., v. 82, p. 3325–3346.

Johnson, G. L., and Vogt, P. R., 1972, Morphology of the Bermuda rise: Deep-Sea Research (in press).

JOIDES, 1969, Deep-Sea Drilling Project, Leg 7: Geotimes, v. 14, no. 10, p. 12–14.

—— 1970, Deep-Sea Drilling Project, Leg 11: Geotimes, v. 15, no. 7, p. 14–16.

—— 1971a, Deep-Sea Drilling Project, Leg 14: Geotimes, v. 16, no. 2, p. 14–17.

—— 1971b, Deep-Sea Drilling Project, Leg 17: Geotimes, v. 16, no. 9, p. 12–14.

Kawano, V., and Ueda, Y., 1967, Periods of the igneous activities of the granitic rocks in Japan by K-Ar dating method: Tectonophysics, v. 4, p. 523–530.

Knopf, A., 1955, Batholiths in time, *in* Poldervaart, A., ed., Crust of the earth: Geol. Soc. America Spec. Paper 62, 762 p.

Ladd, J. W., Dickson, G. O., and Pitman, W. C., III, 1972, The age of the South Atlantic, *in* Nairn, A.E.M., and Stehli, F. G., eds., The ocean basins and margins: The South Atlantic: New York, Plenum Pub. Corp. (in press).

Landis, C. A., and Coombs, D. S., 1967, Metamorphic belts and orogenesis in southern New Zealand: Tectonophysics, v. 4, p. 501–518.

Larochelle, A., 1969, Paleomagnetism of the Monteregion Hills: further new results: Jour. Geophys. Research, v. 74, p. 2570–2575.

Larson, R. L., and Chase, C. G., 1972, Late Mesozoic evolution of the western Pacific Ocean: Geol. Soc. America Bull., v. 83, p. 3627–3644.

Larson, R. L., Smith, S. M., and Chase, C. G., 1972, Magnetic lineations of Early Cretaceous age in the western equatorial Pacific Ocean: Earth and Planetary Sci. Letters, v. 15, p. 315–319.

Lindgren, W., 1915, The igneous geology of the cordillera and its problems, *in* Problems of American geology: New Haven, Yale Univ. Press, p. 234–286.

Matthews, D. H., and Williams, C. A., 1968, Linear magnetic anomalies in the Bay of Biscay: a qualitative interpretation: Earth and Planetary Sci. Letters, v. 4, p. 315–320.

McElhinny, M. W., and Burek, P. J., 1971, Mesozoic paleomagnetic stratigraphy: Nature, v. 232, p. 98–101.

McKenzie, D., Sclater, J. G., 1972, The evolution of the Indian Ocean since the Late Cretaceous: Royal Astron. Soc. Geophys. Jour., v. 24, p. 437–528.

Moores, E., 1970, Ultramafics and orogeny, with models of the U.S. cordillera and the Tethys: Nature, v. 228, p. 837–842.

Morgan, W. J., 1972, Deep mantle convection plumes and plate motions: Am. Assoc. Petroleum Geologists Bull., v. 56, p. 203–213.

Opdyke, N. D., and Wensink, H., 1966, Paleomagnetism of rocks from the White Mountain plutonic-volcanic series in New Hampshire and Vermont: Jour. Geophys. Research, v. 71, p. 3045–3051.

Pimm, A. C., and Hayes, D. E., 1972, General synthesis: Initial Repts. Deep-Sea Drilling Project, v. 9 (in press).

Pitman, W. C., III, and Talwani, M., 1972, Sea-floor spreading in the North Atlantic: Geol. Soc. America Bull., v. 83, p. 619–646.

Pitman, W. C., III, Herron, E. M., and Heirtzler, J. R., 1968, Magnetic anomalies in the Pacific and sea-floor spreading: Jour. Geophys. Research, v. 73, p. 2069–2085.

Pitman, W. C., III, Talwani, M., and Heirtzler, J. R., 1971, Age of the North Atlantic Ocean from magnetic anomalies: Earth and Planetary Sci. Letters, v. 11, p. 195–200.

Raff, A. D., 1966, Boundaries on an area of very long magnetic anomalies in the northeast Pacific: Jour. Geophys. Research, v. 73, p. 3699–3705.

Rona, P. A., Brakl, J., and Heirtzler, J. R., 1970, Magnetic anomalies in the northeast Atlantic between the Canary and Cape Verde Islands: Jour. Geophys. Research, v. 75, p. 7412–7420.

Rutland, R.W.R., 1971, Andean orogeny and ocean-floor spreading: Nature, v. 233, p. 252–255.

Schouten, J. A., 1971, A fundamental analysis of magnetic anomalies and oceanic ridges: Marine Geophys. Research, v. 1, p. 111–144.

Schouten, H., and McCamy, K., 1972, Filtering marine magnetic anomalies: Jour. Geophys. Research (in press).

Siedner, G., and Miller, J., 1968, K-Ar determinations on basaltic rocks from southwest Africa and their bearing on continental drift: Earth and Planetary Sci. Letters, v. 4, p. 451–458.

Smith, A. G., and Hallam, A., 1970, The fit of the southern continents: Nature, v. 225, p. 139–144.

Talwani, M., and Eldholm, O., 1972, The boundary between continental and oceanic crust at the margin of rifted continents: Nature (in press).

Van der Voo, R., and Zijderveld, Z.D.A., 1969, Paleomagnetism in the western Mediterranean area: Koninkl. Nederlandse Akad. Wetensch. Verh., Gen. v. 26, p. 121–138.

Vine, F. J., 1966, Spreading of the ocean floor: new evidence: Science, v. 154, p. 1405–1415.

Vine, F. J., and Hess, H. H., 1970, Sea-floor spreading, *in* Maxwell, A., ed., The sea, Vol. 4, Part III: New York, John Wiley & Sons, p. 587–622.

Vogt, P. R., and Johnson, G. L., 1971, Cretaceous sea-floor spreading in the western North Atlantic: Nature, v. 234, p. 22–25.

Part II

GEOMAGNETIC EPOCHS

Editor's Comments on Papers 12 Through 15

12 **IRVING and PULLAIAH**
Excerpts from *Reversals of the Geomagnetic Field, Magnetostratigraphy, and Relative Magnitude of Paleosecular Variation in the Phanerozoic*

13 **IRVING and PARK**
Hairpins and Superintervals

14 **APARIN and VEDENKOV**
Periodic Variations in the Rate of Migration of Paleomagnetic Poles in the Phanerozoic

15 **CARMICHAEL**
An Outline of the Intensity of the Paleomagnetic Field of the Earth

Since the mid 1960s it has become evident that the paleomagnetic record is characterized by important long-term episodicity and fluctuation, which occur in three main fields represented by the four papers included here: polarity bias (Paper 12), apparent polar wander paths and rates (Papers 13 and 14), and paleointensity of the geomagnetic field (Paper 15).

POLARITY BIAS

The existence of long-term epochs of geomagnetic polarity bias was first clearly demonstrated by Steiner (1967), who recognized a tendency toward reversed polarity during mid-Tertiary, Permo-Carboniferous and latest Precambrian-Early Cambrian times and a tendency toward normal polarity in the mid-Mesozoic and Silurian (Figure 6). A period of ~300 m.y. is evident visually in Figure 6. Subsequent statistical analysis of the geomagnetic polarity record has identified long-term polarity variations of several periods, for example, ~300 m.y. and ~80 m.y. (Crain et al., 1969);

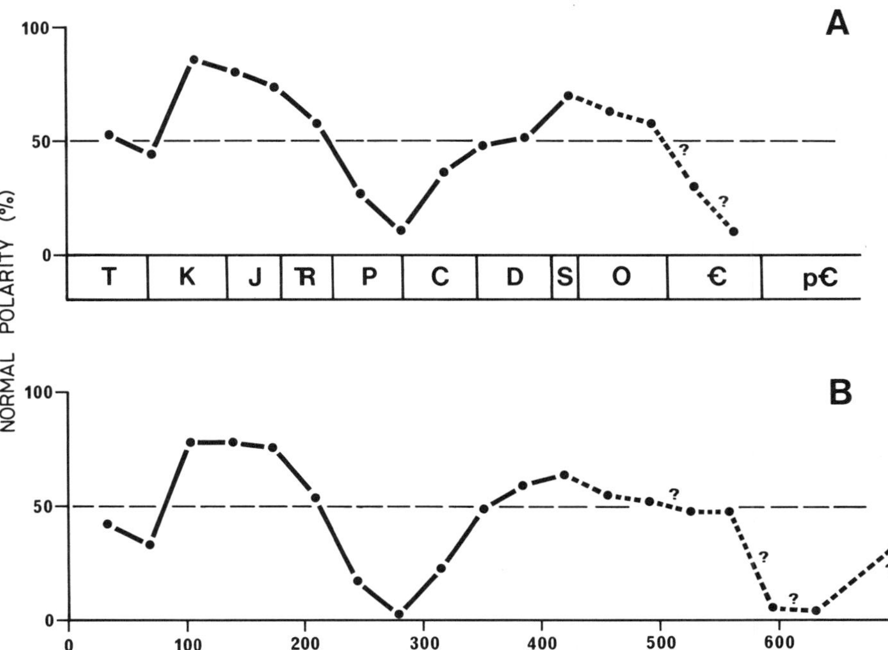

Figure 6 Geomagnetic polarity ratios according to Steiner (1967, Figure 10). A: Based on data compiled by Cox and Doell (1960). B: Based on continuous compilations of Irving (1964). Time scale in m.y.

~350 m.y. (McElhinny, 1971); 700±100 m.y. and 250±50 m.y. (Ulrych, 1972); ~300 m.y., ~113 m.y., and ~57 m.y. (Irving and Pullaiah, Paper 12); and 30–50 m.y. (Vogt, 1975; Jones, 1977). A period of ~300 m.y. is common to and dominant in most of these analyses and is discernible in Figure 13 of Paper 12. This period is evident also in Creer's (1975) plot of Phanerozoic polarity bias, in which he perceived marked discontinuities at 425, 375, 220, and 55 m.y. ago.

Paper 12 by Irving and Pullaiah provides a comprehensive review of polarity bias and paleosecular variation during the Phanerozoic. These authors suggested that such variations originated in processes near the core-mantle interface which may be expressed also in the time-scale of plate tectonics. Vogt (1975) indeed argued that major plate-tectonic changes at about 120–110, 80–70, and 45–42 m.y. ago correspond rather closely with changes in geomagnetic reversal frequency. Discontinuities in Phanerozoic polarity bias have been correlated also with orogeny (Khramov, 1977) and with apparent discontinuities in the rate of change of the earth's rotation (Creer, 1975; Whyte, Paper 34). Moreover, the postulated 700±100 m.y. period in polarity bias (although perhaps of doubtful validity; see Paper 12, p. 57) approximates the mean rhythm of geotectonic megacycles (see Part I), and the well established ~300 m.y. period in polarity bias is about twice the mean interval between certain major glaciations during the past billion years (see Papers 16 and 17). As outlined in the Introduction to this volume, perhaps a common thread is provided by the general similarity of such rhythms to galactic periods, which raises the question whether processes within and upon the earth might be influenced in some way by extraterrestrial events. Planetology ultimately may help resolve this problem of assessing the relative importance of endogenetic and exogenetic processes in earth history (see Part VII).

Numerous persons (e.g., Simpson, 1966; Crain, 1971; Hays, 1971) have proposed a correspondence between polarity bias and evolutionary events. McElhinny (1973, pp. 142–146) reviewed the evidence and concluded that "there appears to be no significant correlation."

APPARENT POLAR WANDER PATHS AND RATES

In 1972 Irving and Park (Paper 13) proposed that the apparent polar wander (APW) path relative to North America exhibited several sharp bends or "hairpins" by which post-Archean time was divisible into five long-term time units or "superintervals." Hairpins were ascribed to episodic major changes in the direction of motion of North America relative to the pole; they were accorded tectonic significance, since some appeared to correspond with times when new structural provinces developed. Sutton (1976b, p. 401) regarded the recognition of such hairpins in APW paths as "perhaps the most important advance" in the study of Precambrian global tectonics since the blending of ideas on orogenic megacycles with concepts of episodic disruption and reassembly of continents (see Papers 5–8).

Subsequent studies (e.g., Donaldson et al., 1973; Park et al., 1973; Piper, 1974, 1976; Irving and McGlynn, 1976) have confirmed the presence of major loops and hairpins in the APW paths of North America and other continents and have suggested correlation of certain hairpins with important tectonic events of the Proterozoic. A few new paleomagnetic poles can necessitate major revision of APW curves, however, and the tectonic significance of such data cannot yet be fully evaluated. This state of flux is emphasized by Dr. E. Irving (personal communication, 11 May 1978) in the following assessment of the concepts first presented in Paper 13:

> The picture of apparent polar wandering in the Precambrian of North America has changed considerably since 1972 and although the same individual poles are there, more have been produced and different paths drawn. For example, the pole path now generally favored for the interval 2200 to 1800 m.y. trends south from Alaska to Peru and is quite different from that given in our 1972 paper. In retrospect the least that can be said of our paper is that it focused attention on the potential geological importance of paleomagnetism in the Precambrian. The general idea of the correlation of hairpins with orogenic and magmatic episodes is rather vague and is not satisfactorily proved. Our data base is so flimsy that one should not expect any certainty in our ideas. However, the notion still gains some support from and perhaps even illuminates the data—there are Grenvillian and Keweenawan loops that are related in time and must surely be related causally with the Grenville Orogeny and the Keweenawan rift system. Just what this relationship is, is not at all clear. Similarly, there is a very good Coronation Loop with hairpin at about 1800 m.y. that must surely relate to the kinematics of the Hudsonian Orogeny. Although we are not very much closer to an understanding of what loops and hairpins really mean, their existence as a general feature of APW paths is not, I think, in doubt. When we have good Proterozoic APW paths for other Precambrian shields then we shall probably be in a position to begin constructing maps of Precambrian continental drift, but realistic maps are unlikely to be drawn for at least another decade. Then, we might know what loops and hairpins signify.

Morris et al. (1979) perceived cyclical APW paths of ~250 m.y. periodicity in global paleomagnetic data. They speculated (p. 96) that "the paths may be the signature of the precessional motion of all crustal units about some fixed axis of rotation."

The Russians V. P. Aparin and V. S. Vedenkov (Paper 14), in a study that complements that of APW paths, analysed the *rates* of APW for most continents during the Phanerozoic. They concluded that during this interval such rates fluctuated rhythmically with a period of 70 to 80 m.y. Thus, the overall rate of APW exhibits a pronounced discontinuity on average every ~35–40 m.y., which is in close agreement with the proposed meter of orogeny during the Phanerozoic (see Paper 9, and Part I). This apparent relationship is reinforced by Aparin and Vedenkov (1977), who perceived five episodes of increase or decrease in overall Paleozoic APW rates, which they correlated with important orogenic epochs. These data therefore suggest a connection between orogeny and changes in the *rates* of continental motions. Perhaps an integration of paths and rates of APW ultimately will clarify the tectonic significance of loops and hairpins since Archean time.

Paleomagnetic pole positions may be affected by small ($\lesssim 5°$) changes in the dipole offset of the geomagnetic field (Wilson and McElhinny, 1974; Hailwood, 1977). Since such data are available only for Tertiary time, it is uncertain whether the determined changes in offset constitute part of a very long-term fluctuation.

PALEOINTENSITY OF THE GEOMAGNETIC FIELD

Numerous studies (e.g., Briden, 1966; Carmichael, Paper 15, 1970; Smith, 1967a and b, 1970) have identified secular changes in the paleointensity of the geomagnetic field. McElhinny (1973, pp. 29–30), in reviewing such findings, pointed out the difficulties in determining paleointensities (see also Roy, 1977) and the wide scatter of data, but he concluded that "the measurements seem to indicate a clear decrease in the calculated geomagnetic dipole moment during the early Palaeozoic. A decrease to only ten per cent of the present value is indicated around 500 My ago." This decrease is illustrated by Carmichael in Paper 15, which outlines the paleomagnetic field intensity from 2500 m.y. ago to the present as obtained from igneous rocks of paleomagnetic stability. The data of Carmichael also indicate that the mean paleointensity ~1000–1200 m.y. ago was similar to the present value, which suggests that the minimum observed in the late Precambrian-early Paleozoic is a real effect and not due to a gradual decay of natural remanence with time.

Professor C. M. Carmichael (personal communication, 1 June 1978) comments as follows on Paper 15:

> Measures of the paleomagnetic field intensity over the past decade have confirmed the early view that such measurements are fraught with difficulty. In addition to the problems of alteration of the kind and amount of magnetic mineral, which was fully recognized in the article, it has become apparent that more formations have suffered partial remagnetization by chemical or thermal means than had previously been thought to be the case. At that time only the reversed and normal Keweenawan lavas were recognized as having been partly remagnetized. It is possible that much of the scatter of paleointensity values that was attributed to the four-fold variation in intensity, as determined from archeomagnetic data, may be due to partial remagnetization. With these reservations and still expecting paleointensity values to be less precise than paleodirections, I am confident that the general pattern

> of paleointensity is correct. This pattern is a strong field, at least as strong as now, from early in the history of the Earth with a much weaker field during most of the Paleozoic. In my view many of the difficulties encountered in unravelling the paleomagnetism of the Paleozoic are due to the inherent weakness of the field at that time. As to when the Earth's field originated, I am currently studying the magnetization of the banded iron-formations in the Isua supracrustal group of Greenland which are at least 3800 m.y. old. They are strongly but complexly magnetized and though not conclusive they indicate that the Earth had a substantial field at this early date.

The paleointensity of the geomagnetic field therefore appears to have undergone a long-term fluctuation since middle Proterozoic time. This fluctuation provides an observation that seems at odds with ideas on the origin of the geomagnetic field: proceeding back through the Phanerozoic the paleointensity of the geomagnetic field *weakens* while the overall rate of the earth's rotation *increases* (see Whyte, Paper 34). Factors other than the earth's spin-rate that exhibit long-term fluctuations measured in hundreds of millions of years therefore may greatly influence the intensity of the geomagnetic field. Perhaps fluctuations of geomagnetic paleointensity are related to geotectonic megacycles (Papers 1–7) or result from long-term changes in the earth's obliquity (see Paper 35 and Part VII).

Runcorn's hypothesis for the origin of geotectonic megacycles (Paper 5) may be tested by the determination of paleointensity values for early Precambrian rocks. McElhinny (1973, pp. 30–31) reviewed such data and concluded that they "all indicate that the earth's core must have been of sufficient size at least 2700 My ago to maintain a geomagnetic field equal to its present-day value. . . . The palaeointensity measurements from these very old rocks thus support the view that the formation of the core was essentially completed at an early stage in the earth's history." Together with Carmichael's suggestion (personal communication, 1978) that the earth had a substantial geomagnetic field at least 3800 m.y. ago, this would appear to invalidate Runcorn's hypothesis that the core has been growing gradually throughout most of geological time and that 3000 m.y. ago was of negligible size. The possibility of early formation of the earth's core is acknowledged by Professor Runcorn (see Part I).

12

Reprinted from pages 35–36, 37–38, 53 and 54–64 of *Earth-Sci. Rev.* **12**:35–64 (1976)

Reversals of the Geomagnetic Field, Magnetostratigraphy, and Relative Magnitude of Paleosecular Variation in the Phanerozoic

E. Irving and G. Pullaiah

ABSTRACT

The percentage of normal and reversed magnetization in land-based paleomagnetic studies of Phanerozoic rocks (0 to −570 m.y.) have been compiled in order to determine the long-term variation in polarity bias of the geomagnetic field. Where possible the results are compared with the record from marine magnetic anomalies. Only rarely is there an even balance between normal and reversed polarity. During the past 350 m.y. two quiet intervals can be recognized when few reversals occurred, the Cretaceous (KN about −81 to −110 m.y.) and Permo-Carboniferous (PCR about −227 to −313 m.y.). Less firmly established are two other quiet intervals, one in the Jurassic (JN about −145 to −165 m.y.), and one in the Triassic (TRN about −205 to −220 m.y.). Between these quiet intervals there are disturbed intervals when reversals were comparatively frequent. From −680 to −350 m.y. the paleomagnetic record is inadequate to delineate a succession of quiet and disturbed intervals although one is probably present. Maximum entropy spectral analysis reveals three periodicities, a dominant one at about 300 m.y. and others, less well-defined, at 113 and 57 m.y. The variations in polarity bias are compared with the paleosecular variation, and it is shown that the magnitude of the paleosecular variation is greater in disturbed than in quiet intervals. This indicates that the magnitude of paleosecular variation and polarity bias are governed by variations in the balance between non-dipole and dipole components of the field, and that these variations probably had their origin in processes near the core—mantle interface. The correspondence between the dominant periods of 300 m.y. and plate tectonics is noted and a causal relationship suggested.

INTRODUCTION

The main part of the geomagnetic field is of internal origin and is caused by dynamo action in the highly conducting fluid core. At any one place the field undergoes variations in direction of about 15° over periods of 10^2 to 10^4 years which are called secular variations. These are caused for the most part by irregular magnetic fields (non-dipole fields) which are thought to be generated by short-lived fluid eddies near the core—mantle interface. Over the longer term, however, the field has a simpler form. When averaged over 10^4 years or more it approximates to a geocentric axial dipole, the mean magnetic pole coinciding to an accuracy of a few degrees with the rotational pole. This dipole undergoes periodic reversals.

The cause of reversals is probably related to the interaction between the varying dipole (*D*) and non-dipole (*ND*) fields (Cox, 1968). In Cox's model there is a finite probability that the sum of the axial component of non-dipole contributions will overcome the dipole field causing a reversal. Hence the probability of reversal ought to be high if the ratio *ND/D* is large, and *vice versa.* Cox's model is set out conceptually in Fig. 1. If the dipole field is much larger than the sum of the axial components of the *ND* field, then the field is stable. As the *ND* fields increase, their axial components may overwhelm the dipole field, and the geomagnetic dynamo then amplifies the field in the opposite sense. Therefore the magnitude of paleosecular variation ought to be related to the frequency with which reversals occur; when reversals are frequent the paleosecular variation ought to be large, and when they are rare paleosecular variation ought to be small. Brock (1971) has shown

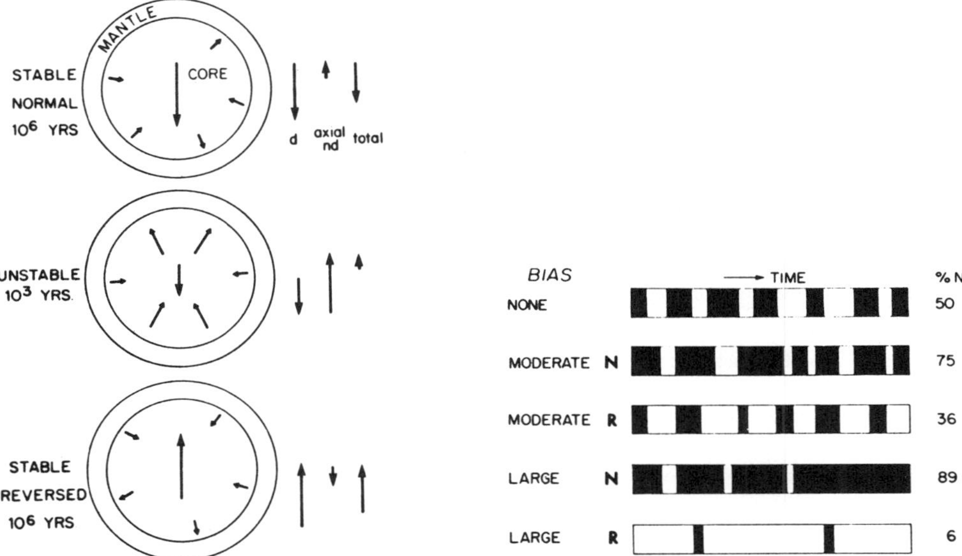

Fig. 1. Cox's model for reversals of the geomagnetic field (Cox, 1968). The radial dipoles are field sources imagined to be caused by eddies near the core—mantle boundary. The size of the core is exaggerated. On the right-hand side 'd' is the dipole component, and 'axial nd' the axial component of the radial dipoles.

Fig. 2. Polarity bias. The successive intervals of normal (black) and reversed (white) polarity are shown on strip charts. In the top chart normal and reversed polarity intervals are, on average, of equal length and there is no polarity bias. In the lower charts, normal (reversed) polarity intervals are on average much longer than the reversed (normal) ones. In the second and third charts moderate bias, and in the lower two charts large bias, are depicted. The corresponding polarity ratios are given on the right.

that these relationships are generally borne out by the paleomagnetic data. It needs to be emphasized that Cox's model is conceptual and its relation to any physical mechanism is not known. Nevertheless the general basis of the model can be tested paleomagnetically, and one of the purposes of this review is to make this test insofar as the present data allow.

It will be shown later that reversals are not a symmetrical phenomenon. By and large the length of time in which the field is normal is not equal to that in which the field is reversed. Only rarely, geologically speaking, are the two in balance, and it is more common for the field to be moderately or strongly biased (Fig. 2). Thus the polarity time scale, when viewed over short periods, consists of successive intervals of normal and reversed polarity (*polarity intervals*), but when viewed over longer periods, it consists of much longer intervals of normal or reversed bias. Times of strong polarity bias and infrequent reversals are referred to here as *quiet intervals.* Times of weak bias and frequent reversals are called *disturbed intervals.* As will be shown later, adjacent quiet and disturbed intervals usually have the same normal or reversed polarity bias, and bunches of them with the same bias are called *bias intervals.* The hierarchy of chronological terms is as follows: *polarity intervals* typically 10^6 years in length, *quiet* or *disturbed intervals* typically 10^7 years in duration, and *bias intervals* typically 10^8 years in duration. The corresponding time-rock terms are polarity zones, quiet and disturbed zones (in the oceans for example), and bias zones.

In this paper we attempt to delineate from the land-based paleomagnetic evidence the chronology of quiet, disturbed, and bias intervals for the Phanerozoic. We then compare this chronology with the record obtained from marine magnetic anomalies. We then compile the evidence of the magnitude of the paleosecular variation to determine what justification there is for the model of Fig. 1.

[*Editor's Note:* In pages 38 to 53 Irving and Pullaiah discuss the polarity data base, methods of erecting polarity time scales, and Cenozoic, Mesozoic and Paleozoic polarity.]

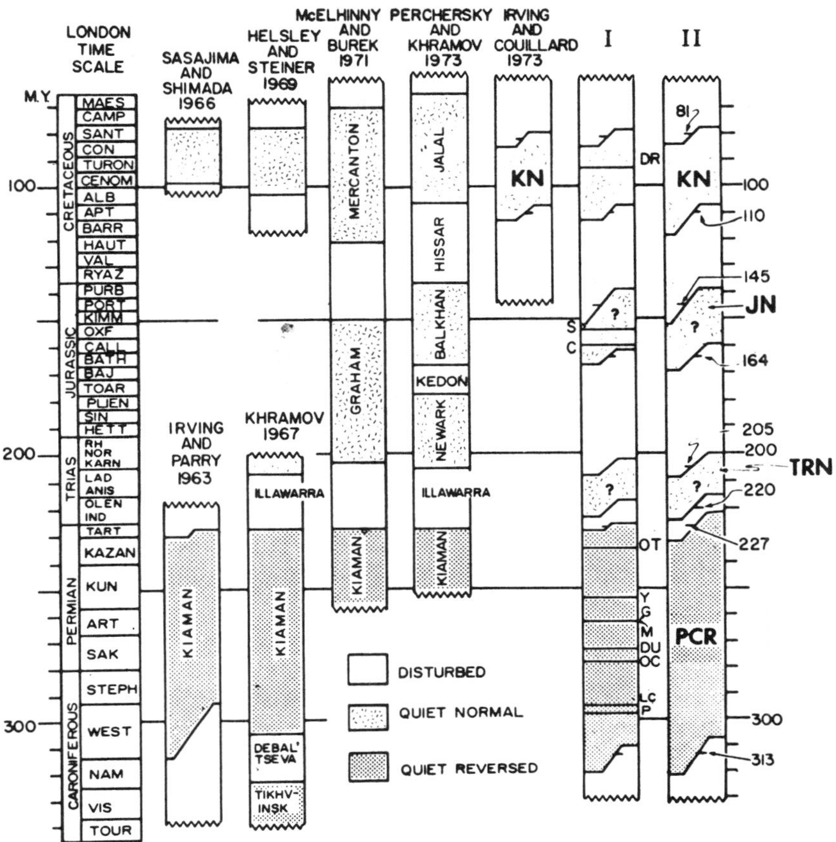

Fig. 12. Summary of chronologies of disturbed and quiet intervals. KN, JN, TRN and PCR are the Cretaceous, Jurassic, Triassic and Permo-Carboniferous quiet intervals. Two reversed polarity intervals within JN are shown: S, Somerville Formation (Steiner and Helsley, 1972); C, Upper Callovian sediments (Khramov, 1973). Eight records of normal polarity within PCR are shown: OT, Otpan Formation (Khramov, 1973); Y, Yates Formation (Peterson and Nairn, 1971); G, Garber Formation (Peterson and Nairn, 1971); M, Mendoza lavas (Creer et al., 1971); DU, Dunkard Series (Helsley, 1965); OC, Oak Creek Canyon, Supai Formation (Graham, 1955); LC, La Colina lavas (Thompson and Mitchell, 1972); P, Peterson Toscanite (Irving, 1966). The interrogation marks in JN and TRN indicate the uncertainty associated with them.

CHRONOLOGIES OF QUIET AND DISTURBED INTERVALS

Chronologies compiled by different workers are set out in Fig. 12. Errors in boundaries are given if available. Two scales based on present work are also given. Scale I refers to the London time-scale. Scale II is an attempt to give errors related to an absolute scale, by making a subjective assessment of the data adjacent to each boundary, and by taking account of errors in the London time-scale. Scale II is our best summary of the available data. KN is an interval of about 30 m.y. duration in the Upper Cretaceous, as Sasajima and Shimada (1966) and Helsley and Steiner (1969) originally contended. The assignments of McElhinny and Burek (1971) and Perchersky and Khramov (1973) seem to be too long, because of the presence of reversals toward the end of the Cretaceous, and in the later early Cretaceous, in both land and sea records. The record in the Jurassic and Triassic is rather unsatisfactory. Although JN is probably real, TRN is poorly known. McElhinny and Burek (1971) recognized one extended normal quiet interval, which they called the Graham. Perchersky and Khramov (1973) using U.S.S.R. data, recognised two quiet intervals, as we do. Their Balkhan interval is probably equivalent to JN. Their Newark interval is systematically younger than TRN. The reversed interval PCR has a duration of between 80 and 90 m.y.

Fig. 12 makes clear that there is no need to apply new names to quiet and disturbed intervals. The use of names such as Kiaman, Jalal or Mercanton has not assisted the discussion. The quiet intervals are set in the framework of the geological time-scale and can be referred to by established geological names, so the general reader and student knows immediately where they belong.

LONG-TERM POLARITY BIAS

Overlapping averages of polarity ratios for the Phanerozoic are shown in Fig. 13. The quiet intervals KN and PCR are discernible in the 10-m.y. averages, and a cyclical variation in polarity bias in the 50 and 100-m.y. averages. This cyclical change was noted previously by McElhinny (1971). The cyclical change is not due to the occurrence of quiet intervals superimposed on a general background of disturbed intervals in which the polarity is evenly balanced, because bias in the disturbed intervals themselves shows the same cyclical change (Fig. 14). The even balance between normal and reversed polarity during the Cenozoic reoccurs in the Lower Triassic and in the Late Silurian to Lower Devonian, but over the entire Phanerozoic it is comparatively rare. The 50-m.y. averages define four intervals, referred to as bias intervals, which alternately have a normal and reversed bias (Fig. 15). The late Cenozoic is apparently a time of transition and is referred to as bias interval 0.

Crain et al. (1969) have computed a Fourier power spectrum of normal polarity ratios using a compilation by Simpson (1966) with an effective

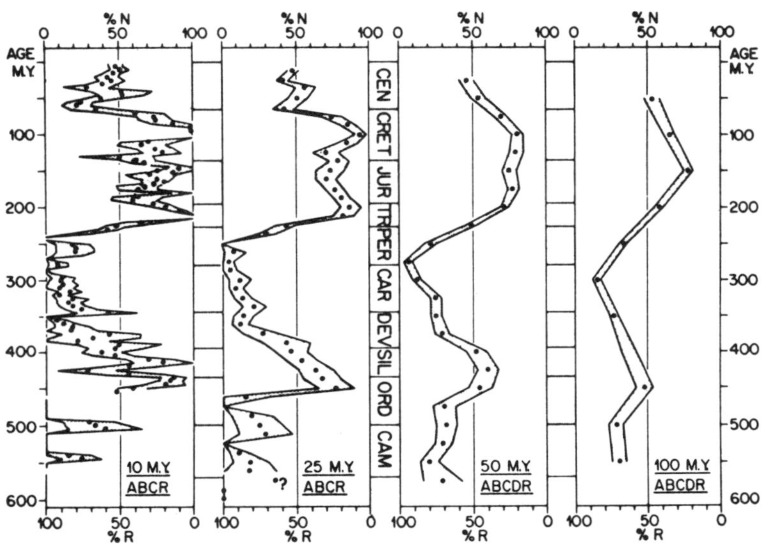

Fig. 13. Overlapping averages of polarity ratios. The means and limits of the standard errors are shown. The data used in these compilations are noted at the bottom of each column. *D* data are used in the 50 and 100-m.y. averages, but not in the shorter-term averages because their age control is inadequate.

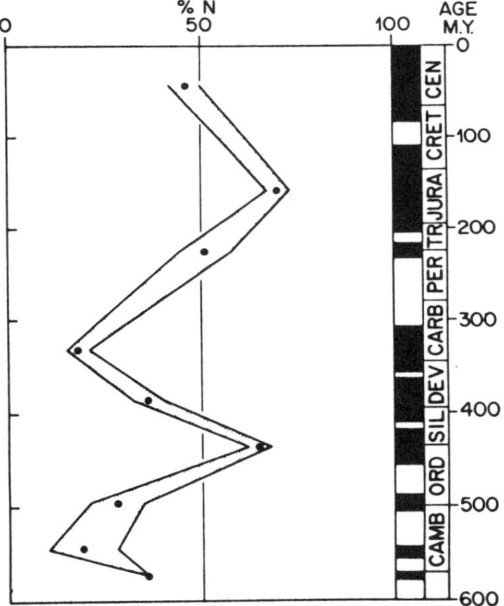

Fig. 14. Overlapping averages of polarity ratios for disturbed intervals. The means and standard errors are shown. The disturbed intervals are shown in black on the scale. The disturbed intervals in the earlier Paleozoic have been derived from the data of Figs. 9 and 13 and are only approximate.

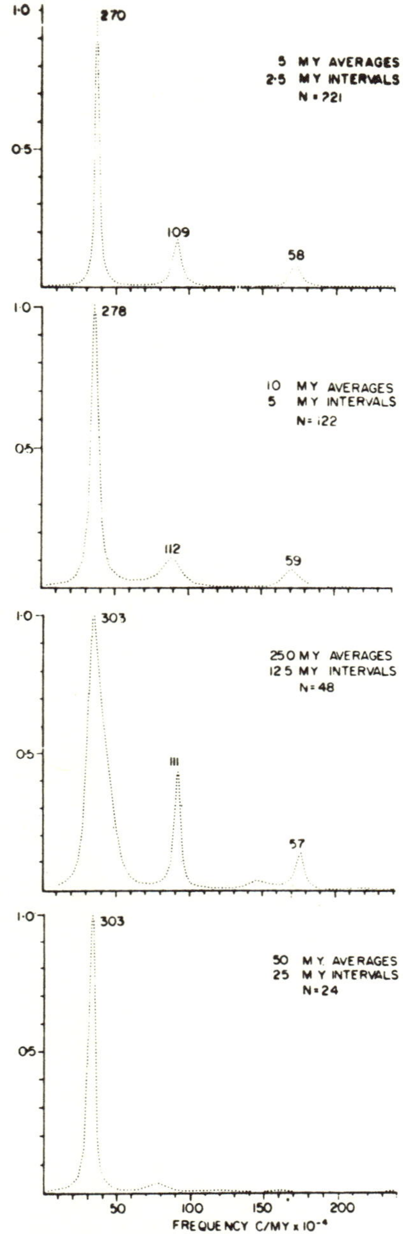

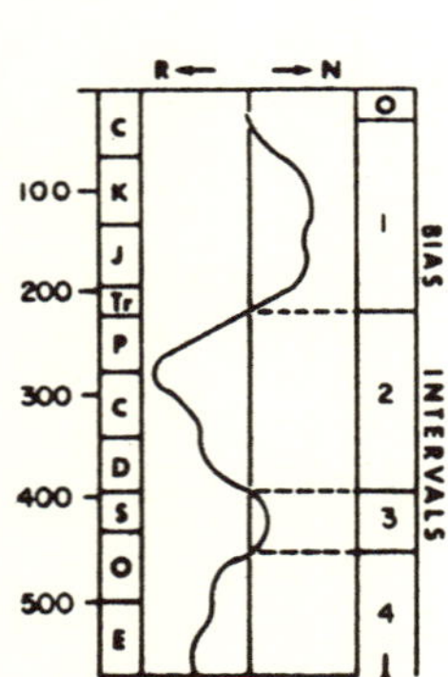

Fig. 15. Chronology of bias intervals for the Phanerozoic. The boundaries are points at which the polarity bias changes from normal to reverse. The mean curve is the 50-m.y. averages of Fig. 13.

Fig. 16. Maximum entropy power spectra for four averaging intervals. The ages of the peaks are given in m.y.

sampling interval of about 15 m.y. They observed periodicities of 300 m.y. and 80 m.y. Fourier spectral analysis is of limited usefulness in the treatment of short time-series. Maximum entropy spectral analysis (Burg, 1968) is a more powerful method, and Ulrych (1972) has applied it to the analysis of polarity ratios. He took the data compiled by McElhinny (1971), fitted it to a ninth-order polynomial, and interpolated with an effective sampling interval of 25 m.y. He obtained periodicities of 700 ± 100 m.y. and 250 ± 50 m.y.

We have carried out a maximum entropy spectral analysis using the overlapping averages of 5 m.y. (Fig. 9), 10 m.y., 25 m.y., 50 m.y. and 100 m.y. (Fig. 13). These have effective sampling intervals of 2.5 m.y., 5 m.y., 12.5 m.y., 25 m.y. and 50 m.y. respectively. The analysis was first done in each case with frequency intervals corresponding to the Nyquist frequencies of the respective sampling intervals with 240 estimates, and employing a range of prediction error filters. Later, to facilitate the comparison, we adopted a common frequency interval of $1 \cdot 10^{-4}$ c/m.y. This did not affect the resolution. All the spectra for different sampling intervals showed consistent peaks for prediction error filter lengths lying between 50 and 30% of the number of data samples. Fig. 16 gives the normalized power spectra for the various sampling intervals. There are three major components with mean periodicities of 297 ± 34 m.y., 113 ± 5 m.y. and 57 ± 1 m.y. The later two periodicities are not very evident in the spectrum from the 50-m.y. averages, although there is an indication of them. Small upward frequence shifts of the spectral peak are also observed in the latter averages and these effects may be due to the differences in the initial phase of the sinusoids, and on the smaller length of the data samples (Chen and Stegen, 1974). We have been unable to reproduce the 700-m.y. periodicity found by Ulrych (1972). We are somewhat sceptical about estimates of periods longer than the length of the record (600 m.y.). The periodicity of about 300 m.y. is common to the analysis of Crain et al. (1969), Ulrych (1972) and ourselves, and is evident visually in Fig. 13. The shorter periodicities of 113 and 57 m.y. (Fig. 16) are remarkably consistent, and they presumably correspond to the variations superimposed on the general trend that can be seen in the 25-m.y. averages (Fig. 13).

POLARITY BIAS AND RELATIVE MAGNITUDE OF PALEOSECULAR VARIATION

Brock (1971) has shown that the magnitude of the paleosecular variation in pre-Cenozoic times is, on average, about 15% less than during the Cenozoic. He suggested that the difference "presumably reflect(s) the reversal rate in the past". His suggestion is based on two arguments. First, his work suggested that the secular variation as observed paleomagnetically is caused largely by non-dipole components of the field, the effect of wobble of the dipole being small. Their contributions were in the ratio 6 : 1 respectively. Thus the magnitude of paleosecular variation may be expected to depend on

[*Editor's Note:* Figure 9, which shows data points for the earlier Mesozoic and Paleozoic and 5-m.y. averages of polarity ratios for the entire Phanerozoic, has been omitted.]

the ratio of the non-dipole (*ND*) to dipole (*D*) fields. Second, he appealed to Cox's model (Fig. 1) in which reversal frequency likewise varies with the ratio of *ND* to *D*. Referring now to the 50-m.y. averages in Fig. 13, then if this argument is correct, the paleosecular variation ought to be greater in the Cenozoic than in the Mesozoic and Late Paleozoic as Brock found. In particular, there ought to be a contrast between the paleosecular variation in the Cenozoic disturbed interval and the KN and PCR quiet intervals. We have made a study of paleosecular variation in the later Phanerozoic (0 to 350 m.y.) using a data base about twice as large as that available to Brock, and are able to determine whether Brock's suggestion is justified.

The rough measure of paleosecular variation used is the standard deviation s_e of the field directions at the equator, using the dipole perturbation model (Model A) of the geomagnetic field (Irving and Ward, 1964). The standard deviation $s_e = 46.8\ \sigma_e\ (1 + 3\ \sin^2\lambda)^{-1/4}$, where $\sigma_e = ND/D_e$ at the equator, and λ is the latitude. The dipole field *D* is imagined to be perturbed by a non-dipole field of constant magnitude but random in direction. The standard deviation s_e varies from pole to equator because of the latitude varia-

TABLE IV

Polarity bias and secular variation compared

		Polarity bias		Secular variation		
		n	% normal	N	σ_e	s°_e
1.	1945 field (Irving and Ward, 1964)	—	—	9	0.40 (0.01)	18.7 (0.5)
2.	-2000 to 0 m.y. (Irving and Ward, 1964)	—	—	34	0.38 (0.12)	17.8 (1.1)
3.	-2000 to 0 m.y. (Brock, 1971)	—	—	83	0.40 (0.01)	18.8 (0.5)
4.	-65 to 0 m.y., Cenozoic (Brock, 1971)	—	—	44	0.44 (0.02)	20.4 (0.7)
5.	-2000 to -65 m.y., pre-Cenozoic (Brock, 1971)	—	—	39	0.37 (0.02)	17.1 (0.7)
6.	-5 m.y. to present (5 or more sites)	128	63 (3)	74	0.39 (0.02)	18.4 (0.7)
7.	-5 m.y. to present (10 or more sites)	93	64 (4)	58	0.39 (0.02)	18.2 (0.8)
8.	-65 to -5 m.y., disturbed interval	150	48 (3)	53	0.42 (0.02)	19.5 (0.7)
9.	-80 to -110 m.y., KN quiet interval	17	100	5	0.31 (0.06)	14.7 (2.8)
10.	-307 to -237 m.y., PCR quiet interval	106	10 (3)	10	0.33 (0.04)	15.6 (1.6)
11.	-345 to 0 m.y., Carboniferous to Present	805	53 (2)	188	0.38 (0.01)	17.9 (0.43)

n is the number of polarity ratios; *N* is the number of paleosecular variation results; $\sigma_e = ND/D_e$; s°_e is the standard deviation of field variations reduced to the equator. Standard errors are given in brackets. The 63% normal magnetization in entry 6 is owing predominance of Bruhnes age date.

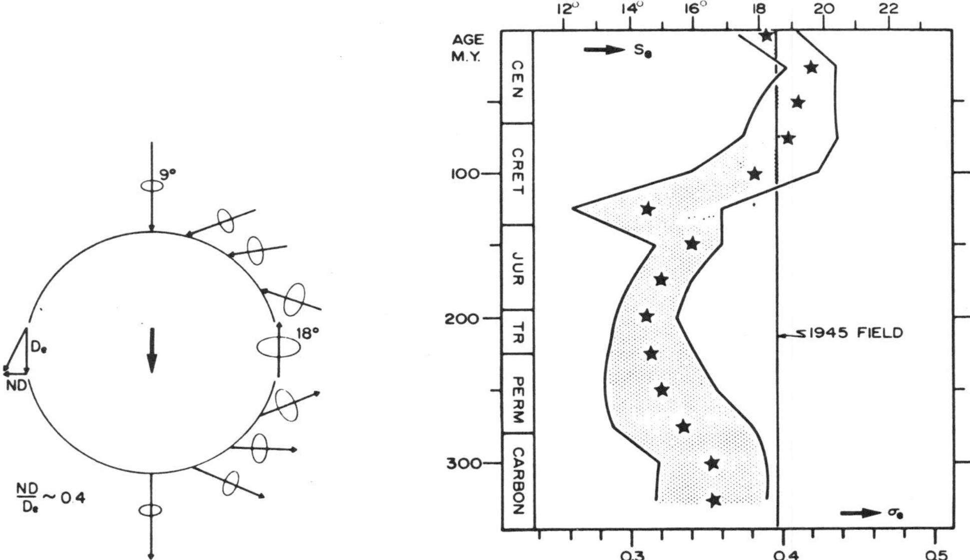

Fig. 17. Diagrammatic representation of secular variation. The arrows are the mean dipole field direction and the ovals represent the standard deviation of the field about this direction, the polar and equatorial values 9° and 18° being characteristic for the recent past.

Fig. 18. Paleosecular variation versus time. 50-m.y. overlapping averages with standard errors are plotted. The scale for standard deviation reduced to the equator (s_e) is shown above, and the ratio of non-dipole to dipole field at the equator (σ_e) below. There are insufficient data for analysis prior to —350 m.y.

tion of D (Fig. 17). The model is crude, and obviously is too simple to be very realistic, but Brock shows that s_e obtained in this way is a useful comparative measure of the magnitude of the paleosecular variation. Relatively large or small paleosecular variation can on average be detected in this way. Both σ_e and λ can be calculated from the paleomagnetic observations, since $\tan \lambda = \frac{1}{2} \tan I$ and $\sigma_e^2 = 3/k\ (1 + 3 \sin^2\lambda)$, where I is the mean paleomagnetic inclination, and k is Fisher's precision (Fisher, 1953). There are 188 suitable estimates of k in the literature for the past 350 m.y. We use selection criteria like those described by Brock, except that we admit results based on five or more sites whereas he used ten as his limit. The number of results based on five to nine sites is 56, and their inclusion makes no difference to the result, as is illustrated by a comparison of entries 6 and 7 in Table IV. Prior to the Middle Carboniferous, there are few paleomagnetic results suitable for paleosecular variation determinations, and no analysis is attempted. A check-list of the data used may be obtained from the authors upon request.

Polarity bias and the relative magnitude of the paleosecular variation are compared in Table IV entries 6 to 11. Other estimates of paleosecular variation are given, and there is excellent agreement between Brock's estimate for the Cenozoic and ours. The long quiet intervals, KN and PCR, have low paleosecular variation whereas the disturbed interval of the Cenozoic has a significantly higher paleosecular variation. There is a general and significant increase in the magnitude of paleosecular variation with time thus confirming Brock's result (Fig. 17). This result suggests that the mechanism responsible for paleosecular variation is also responsible for polarity bias, the important factor being the balance between the non-dipole and dipole fields.

DISCUSSION

The broad features of the reversal pattern for the Phanerozoic are made clear in Figs. 12 and 15. Normal and reversed polarities are rarely in balance, the field being either predominantly normal or predominantly reversed. At times of normal bias, normal polarity intervals are systematically longer than polarity intervals of reversed polarity, and vice versa. The presence of bias suggests that there are memory effects in the geomagnetic dynamo, because in the Mesozoic (−225 to −65 m.y.) a change from reversed to normal polarity usually occurs after a shorter lapse of time than the change from normal to reversed. In the Paleozoic (−550 to −225 m.y.) the opposite is generally true. There is a variation with a period of about 300 m.y. Although, during intervals of strong polarity bias, frequent short-lived reversals may occur, individual studies have shown that reversal frequency is least where the polarity bias is strongest, for example in the PCR and KN quiet intervals.

The quiet intervals are clearly of great significance stratigraphically, both for marine geology as Larson and Pitman (1972) have shown, and for continental stratigraphy. The boundaries of the quiet intervals are global time markers potentially observable in both the oceanic crust and terrestrial beds. The act of locating the upper and lower boundaries of KN in the Cretaceous sections of the world could, for example, solve many problems in the stratigraphy of that system. A full study of the upper boundary of the PCR could settle many of the outstanding problems relating to the Paleozoic—Mesozoic boundary, particularly those related to the correlation of marine and terrestrial strata. Moreover, there is little doubt that as more data are obtained, a chronology of quiet intervals will be found in the Early Paleozoic, and it is conceivable that their use could be extended into the Precambrian.

The comparison of the variations in the magnitude of paleosecular variation and polarity bias suggest that the chronology of reversals derives from variations in the balance between non-dipole and dipole fields. The source of the non-dipole field is generally ascribed to eddies near the surface of the outer core, and the form of these is likely to be sensitive to the changes in temperature and/or topography of the lower mantle (Hide, 1967). Hence it is reasonable to suppose that the processes at the base of the mantle are responsible for polarity bias, reversal frequency variations, and the magnitude of the secular variation. The fundamental periodicity associated with these processes is 300 m.y.; there appear to be slow processes at the core—mantle boundary, which affect the balance between the non-dipole and dipole fields, and which also either inhibit reversals, or allow them to happen more frequently. Crain et al. (1969) compared this 300-m.y. period with that of galactic rotation and suggested a cause in galactic processes. Niitsuma (1973) has extended this discussion and suggests that there is a correlation between incidence of few reversals and the times when the solar system passed through the spiral arms of the galaxy. These times he gives as −50, −80, −180, −270, −370, −440, −510, and −560 m.y. There does not seem to be any clear-cut correspondence with the results presented here (see for example Fig. 13) and these times, but the periods of 113 and 57 m.y. evident in the spectral analysis could conceivably have some relationship to such a process. It seems to us, however, that the comparison between paleosecular variation and polarity bias demonstrates that the cause is not extraterrestrial, but is within the earth, and of more relevance therefore is the comparison between the 300 m.y. and the time-scale of plate tectonics. It can hardly be fortuitous that perhaps the most general features of geomagnetic history — the change from reversed to normal bias coincident with the Paleozoic—Mesozoic boundary — is also one of the greatest breaks in the geological record. Therefore it seems to us that mantle processes which affect the core—mantle boundary, and hence the relative magnitudes of the non-dipole and dipole fields, are the most likely cause of this dominant periodicity in the geomagnetic field. There are several possible causes: (1) periodic changes of the temperature distribution over the core—mantle interface caused by through-mantle convection, mantle plumes for example; (2) convection in the lower mantle; (3) slow chemical evolution involving exchange of material between mantle and core; (4) slow changes in the position of the core—mantle interface due to alternate freezing and melting of the outer core and lower mantle; (5) polar wander, which will move the mantle relative to the core and hence change the temperature distribution at the core—mantle interface relative to the axis of rotation; (6) evolution of the inner core (we are indebted to A. Brook for suggesting this possibility).

Whatever the cause, the evidence clearly supports the contention that the timetable of reversals is governed by mantle behaviour (Irving, 1966; Hide, 1967; Cox, 1969). The geomagnetic clock appears not to be a phenomenon separate and independent of geological processes, but may have common cause with geology, in the earth's mantle. During the last decade it has often been argued that through-mantle convection cannot occur, but recently there has been a tendency to rehabilitate the idea, and this seems to us to be the most reasonable way of accounting for the long-period changes in the bias of the field and its relationship to geological history. There is apparently an inner tectonic evolutionary history of the lower mantle governing the main features of the geomagnetic field, and an outer tectonic evolution of the upper mantle governing geological history. It seems to us reasonable to assume that they are two parts of the same global process, namely through-mantle convection.

ACKNOWLEDGEMENTS

We are very grateful to Roger Larson for discussion of many aspects of the marine geomagnetic record, and to Jean Roy for discussion of Fig. 10. We are particularly grateful to J.C. Gupta and A. Nandi for their help with numerical analyses.

REFERENCES

Baldwin, B., Coney, P.J. and Dickinson, W.R., 1974. Dilemma of a Cretaceous time scale and rates of sea-floor spreading. Geology, 2: 267—270.

Brock, A., 1971. An experimental study of paleosecular variation. Geophys. J. R. Astron. Soc., 24: 303—317.

Burg, J.P., 1968. A new analysis technique for time series data. Paper presented at NATO advanced Study Institute on Signal Processing, Enschede, Netherlands.

Chen, W.Y. and Stegen, G.R., 1974. Experiments with maximum entropy power spectra of sinusoids. J. Geophys. Res., 79: 3019—3022.

Cox, A., 1968. Lengths of geomagnetic polarity intervals. J. Geophys. Res., 73: 3247—3260.

Cox, A., 1969. Geomagnetic reversals. Science, 163: 237—245.

Cox, A., 1973. Plate Tectonics and Geomagnetic Reversals. Freeman, San Francisco, Calif., 702 pp.

Crain, I.K., Crain, P.L. and Plaut, M.G., 1969. Long period Fourier spectrum of geomagnetic reversals. Nature, 223: 283.

Creer, K.M., 1955. A Preliminary Palaeomagnetic Survey of Certain Rocks in England and Wales. Thesis, Univ. Cambridge, U.K. 203 pp.

Creer, K.M., Mitchell, J.G. and Valencio, D.A., 1971. Evidence for a normal geomagnetic field polarity event at 262 ± 5 m.y. within the late Palaeozoic reversed interval. Nature, 233: 87—89.

Dickinson, W.R. and Rich, E.I., 1972. Petrologic intervals and petrofacies in the Great Valley sequence, Sacramento Valley, California. Geol. Soc. Am. Bull., 83: 3007—3024.

Doell, R.R., 1955. Palaeomagnetic study of rocks from the Grand Canyon of the Colorado River. Nature, 176: 1167.

Fairbairn, H.W., Faure, G., Pinson, W.H., Hurley, P.M. and Powell, J.L., 1963. Initial ratio of strontium 87 to strontium 86, whole-rock age and discordant biotite in the Monteregian igneous province, Quebec. J. Geophys. Res., 68: 6515—6522.

Fisher, R.A., 1953. Dispersion on a sphere. Proc. R. Soc. Lond., A217: 295—305.

Graham, J.W., 1955. Evidence of polar shift since Triassic time. J. Geophys. Res., 60: 329—347.

Harland, W.B., Smith, A.G. and Wilcock, B., 1964. Geological Society of London Phanerozoic time-scale. Quart. J. Geol. Soc. Lond., 120: 260—262.

Heirtzler, J.R. and Hayes, D.E., 1967. Magnetic boundaries in the North Atlantic Ocean. Science, 157: 185—187.

Heirtzler, J.R., Dickson, G.O., Herron, E.M., Pitman, W.C. and Le Pichon, X., 1968. Marine magnetic anomalies, geomagnetic field reversals, and motions of the ocean floors and continents. J. Geophys. Res., 73: 2119—2136.

Helsley, C.E., 1965. Paleomagnetic results from the Lower Permian Dunkard Series of West Virginia. J. Geophys. Res., 70: 413—424.

Helsley, C.E. and Steiner, M.C., 1968. Evidence for long intervals of normal polarity during the Cretaceous period. Earth Planet. Sci. Lett., 5: 325—332.

Hicken, A., Irving, E., Law, L.K. and Hastie, J., 1972. Catalogue of paleomagnetic directions and poles, first issue. Publ. Earth Phys. Branch, Dept. Energy, Mines and Resour. Ottawa, 45: 1—135.

Hide, R., 1967. Motions of the earth's core and mantle, and variations in the main geomagnetic field. Science, 177: 55—56.

Irving, E., 1966. Paleomagnetism of some Carboniferous rocks of New South Wales and its relation to geological events. J. Geophys. Res., 71: 6025—6051.

Irving, E., 1970. The Mid-Atlantic Ridge at 45° N, XVI. Oxidation and magnetic properties of basalt; review and discussion. Can. J. Earth Sci., 7: 1528—1538.

Irving, E. and Couillard, R., 1973. Cretaceous normal polarity interval. Nature, 244: 10—11.

Irving, E. and Hastie, J., 1975. Catalogue of paleomagnetic directions and poles, second issue. Publ. Earth Phys. Branch, Dept. Energy Mines and Resour., Ottawa, in preparation.

Irving, E. and Parry, L.G., 1963. The magnetism of some Permian rocks from New South Wales. Geophys. J. R. Astron. Soc., 7: 395—411.

Irving, E. and Ward, M.A., 1964. A statistical model of the geomagnetic field. Pure Appl. Geophys., 57: 47—52.

Khramov, A.N., 1958. Paleomagnetism and Stratigraphic Correlation. Gostoptechizdat, Leningrad, 218 pp. (English transl. published by School of Earth Sciences, Australian National University, Canberra, Australia.)

Khramov, A.N., 1967. The earth's magnetic field in the late Paleozoic. Akad. Nauk S.S.R. Izv. Earth Phys. Ser., 86—108.

Khramov, A.N. (compiler), 1973. Paleomagnetic Directions and Paleomagnetic Poles. Akad. Sci. U.S.S.R., Sov. Geophys. Comm., Mosc., 2: 88 pp.

Larochelle, A., 1969. Paleomagnetism of the Monteregion Hills: further new results. J. Geophys. Res., 74: 2570—2575.

Larson, R.L. and Hilde, T.W.C., 1975. A revised time scale of magnetic reversals for the Early Cretaceous and Late Jurassic. J. Geophys. Res., 80: 2586—2594.

Larson, R.L. and Pitman, W.C., 1972. World-wide correlation of Mesozoic magnetic anomalies and its implications. Bull. Geol. Soc. Am., 83: 3645—3662.

McElhinny, M.W., 1971. Geomagnetic reversals during the Phanerozoic. Science, 172: 157—159.

McElhinny, M.W. and Burek, P.J., 1971. Mesozoic palaeomagnetic stratigraphy. Nature, 232: 98—102.

McMahon, B.E. and Strangway, D.W., 1968. Investigation of Kiaman Magnetic Division in Colorado redbeds. Geophys. J. R. Astron. Soc., 15: 265—285.

Morley, L.W., 1963. See reprinted extract and discussion in Cox (1973), pp. 224—225.

Niitsuma, N., 1973. Galactic Rotation and Geomagnetic Reversals. Rock Magnetism and Paleogeophysics. Publ. Rock Magn. Paleogeophys. Res. Group, Tokyo Univ., 1: 130—133.

Obradavich, J.D. and Cobban, W.A., 1974. A Time-scale for the Late Cretaceous of the Western Interior of North America. Spec. Publ. Geol. Assoc. Can., in press.

Opdyke, N.D., Glass, B., Hays, J.D. and Foster, J., 1966, Paleomagnetic study of deep-sea cores. Science, 154: 349—357.

Perchevsky, D.M. and Khramov, A.N., 1973. Mesozoic palaeomagnetic scale of the U.S.S.R. Nature, 244: 499—501.

Peterson, D.N. and Nairn, A.E.M. 1971. Palaeomagnetism of Permian redbeds from the SW U.S.A. Geophys. J. R. Astron. Soc., 23: 191—207.

Roy, J.L., 1969. Paleomagnetism of the Cumberland Group and other Paleozoic formations, Can. J. Earth Sci., 6: 663—669.

Roy, J.L. and Park, J.K., 1969. Paleomagnetism of the Hopewell Group, New Brunswick, J. Geophys. Res., 74: 594—604.

Roy, J.L. and Park, J.K., 1974. The magnetization process of certain red beds: vector analysis of chemical and thermal results. Can. J. Earth Sci., 11: 437—471.

Sasajima, S. and Shimada, M., 1966. Paleomagnetic studies of the Cretaceous volcanic rocks in SW Japan. J. Geol. Soc. Jap., 72: 503—514.

Simpson, J.F., 1966. Evolutionary pulsations and geomagnetic polarity. Bull. Geol. Soc. Am., 77: 197—203.

Steiner, M.B. and Helsley, C.E., 1972. Jurassic polar movement relative to North America. J. Geophys. Res., 77: 4981—4993.

Taylor, P.T., Zietz, I. and Dennis, L.S., 1968. The geological implications of paleomagnetic data from eastern margin of the United States. Geophys., 33: 755—780.

Theyer, F. and Hammond, S.R., 1974. Cenozoic magnetic time-scale in deep sea cores: completion of the Neogene. Geol., 2: 487—492.

Thompson, R. and Mitchell, J.G., 1972. Paleomagnetic and radiometric evidence for the age of the lower boundary of the Kiaman magnetic interval in South America. Geophys. J. R. Astron. Soc., 27: 207—214. lrych, T., 1972. Maximum entropy power spectrum of long period geomagnetic reversals. Nature, 235: 218—219.

Vine, F.J. and Matthews, D.H., 1963. Magnetic anomalies over oceanic ridges. Nature, 199: 947—949.

13

Reprinted from *Canadian Jour. Earth Sci.* **9**:1318–1324 (1972)

Hairpins and Superintervals

E. IRVING AND J. K. PARK
Earth Physics Branch, Department of Energy, Mines and Resources, Ottawa, Canada

The path of apparent polar wandering relative to North America has several sharp turning points or hairpins. These hairpins record rapid changes in the direction of motion of North America relative to the pole. The youngest of these hairpins (in the Cretaceous) corresponds approximately in time to a major "tectonic reorganization" suggested by Coney on other grounds. The older hairpins are also supposed to represent first-order tectonic breaks, and they are therefore used to divide post-Archean time into 5 superintervals. The diastrophic significance of these superintervals is briefly discussed.

Introduction

Chronologies of polarity and magnetic intervals[1] which have durations of 10^6 or 10^7 years are now being established for the past few hundred million years. In this paper a method of subdividing time into much longer units (called 'superintervals') is introduced. Chronologies of polarity and magnetic intervals are based on observations of the *polarity* of remanent magnetization, and since reversals of the geomagnetic field are a global phenomenon, reflecting changes in the core of the earth, they provide time scales applicable to the entire earth. The chronology of superintervals described below is based on observation of the *directions* of remanent magnetization. Directions of magnetization are expressed as corresponding poles, and variations in direction as paths of apparent polar-wandering. Sharp turning points (referred to here as 'hairpins') occur in these paths, and it is the ages of these hairpins that are used to define the limits of superintervals.

The polar path described the horizontal movements of a tectonic unit (or lithospheric plate) relative to the pole, and the hairpins therefore correspond to periodic changes in the direction of these movements. If such changes of motion affect each plate independently, then the time at which hairpins occur will differ from plate to plate, and the chronologies of superintervals will have a regional, but not a global, application. If, however, the periodic reorganizations of plate motions are a global phenomenon (and this would be the expectation from plate theory) the *time* at which changes occur (although their form may differ) will be essentially identical; that is, the chronology of superintervals will be global in its applications. There are therefore two parts to this discussion: firstly the description of such chronologies for each continental shield, and secondly the comparison of these chronologies from shield to shield. In this paper we describe, as a beginning, a *very* tentative chronology of superintervals for the Canadian Shield. The comparison with results from other shields will be made later.

[1]A polarity interval is a length of time during which (aside from very short-lived reversals) the geomagnetic field had constant polarity; generally they have durations of millions of years or less. A magnetic interval typically has a duration of several tens of millions of years, and is characterized by a certain reversal frequency; the distinction made is between lengths of time during which the field reversed frequently and the polarity intervals were short, and lengths of time during which the field rarely reversed and the polarity intervals were longer. (See Fig. 4.)

Chronology of the Canadian Shield

The chronologies of Precambrian time in Canada have developed from a knowledge of the structure and radiometric ages of the Canadian Shield. About 30 years ago M. E. Wilson (1941) recognized various geological provinces. Later, Gill (1949) developed the idea of structural provinces which he thought of as regions in which the structural trends, and hence presumably the tectonic stresses operative during crustal stabilization, were approximately uniform in direction. At the same time J. T. Wilson (1949) noted that, in addition to uniformity of structure, each province had characteristic radiometric ages. Making use of developments in mapping, and especially of the age determinations of R. K. Wanless and colleagues, Stockwell (1963, 1970) refined this subdivision into structural provinces, and calculated their mean radiometric (based mostly on K–Ar) ages. These ages Stockwell considered to define the orogenies, each of which he thought of as a period of mountain building and folding, which may, or may not, be accompanied by metamorphism and granitic intrusion. That is, the radiometric ages correspond to the later phases in the classical cycle of geosynclinal theory – subsidence, deposition, folding, metamorphism, mountain building – a concept dominated by vertical motions. In view of the great success of plate theory, which seeks to explain the evolution of the lithosphere in terms of dominantly horizontal motions, it is reasonable to ask what Stockwell's chronology might mean in terms of possible long-term horizontal motion of the Canadian Shield. The movements of the Shield relative to the pole are therefore now reviewed. Readers are referred to Du Bois (1962), Black (1963), Spall (1971), and Robertson and Fahrig (1971) for earlier reviews of these movements.

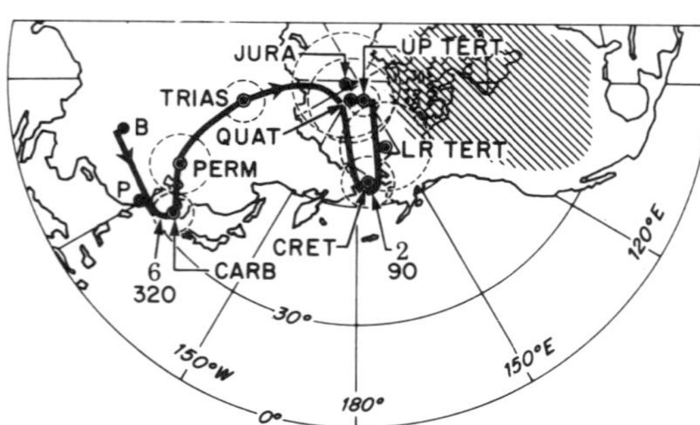

FIG. 1. Apparent polar wandering relative to North America since the late Paleozoic. Mean poles for Carboniferous and later periods (listed in Table 1) are given with their error circles ($P = 0.05$). P is the pole for the Perry Formation (Upper Devonian) and B that for the Bloomsburg Formation (Upper Silurian) which are the most reliable poles for the Silurian and Devonian of North America. Hairpins 2 and 6 and their approximate ages in millions of years are indicated. Polar equal-area projection.

Apparent Polar Wandering

Apparent polar movement relative to North America is shown in Figs. 1 to 3[2] in the usual way; the field is assumed to be a geocentric axial dipole and North America is fixed. During the later Cenozoic the pole is near the present geographic pole. In the Lower Cenozoic it moves towards Alaska, and during much of the Cretaceous it remains in the vicinity of what is nowadays the Bering Sea. The pole then doubles back upon itself during the Jurassic, reaching central Siberia in the Triassic. In the Upper Palaeozoic it is in North China, where there is apparently another hairpin bend.

The few results that are available from the early Paleozoic suggest that the pole moved into what is nowadays southeast Asia and the eastern Indian Ocean. This reconstruction of the early Paleozoic polar path given in Fig. 2 is *very* tentative. The pole appears to move across what is nowadays Australia into the central Pacific, and then executes a series of loops, first moving towards North America,

[2]The results reviewed in this paper (see footnote to Table 1 and legend to Fig. 2) are referenced by a list number. This list (Hicken *et al.* 1972, in press) will be published later this year and is in computer format. It contains the relevant statistics and bibliography and notes as to geological age etc., and it is divided into 12 sections, corresponding to the geological time units, as follows: 01 Precambrian, 02 Cambrian, 03 Ordovician, 04 Silurian, 05 Devonian, 06 Carboniferous, 07 Permian, 08 Triassic, 09 Jurassic, 10 Cretaceous, 11 Tertiary, 12 Plio-Pleistocene and Quaternary. This time number is used in Table 1 and elsewhere where necessary, but if the geological age is clear from the context (*e.g.* in legend of Fig. 2) it is omitted. Within these time sections each result is identified by a three digit number which is given, for example, in Fig. 2. Sometimes the results were obtained by averaging several determinations, often values of different authors (*e.g.* 010305 Portage Lake Lavas Fig. 2) and the details of the averaging procedures are explained in the computer listing.

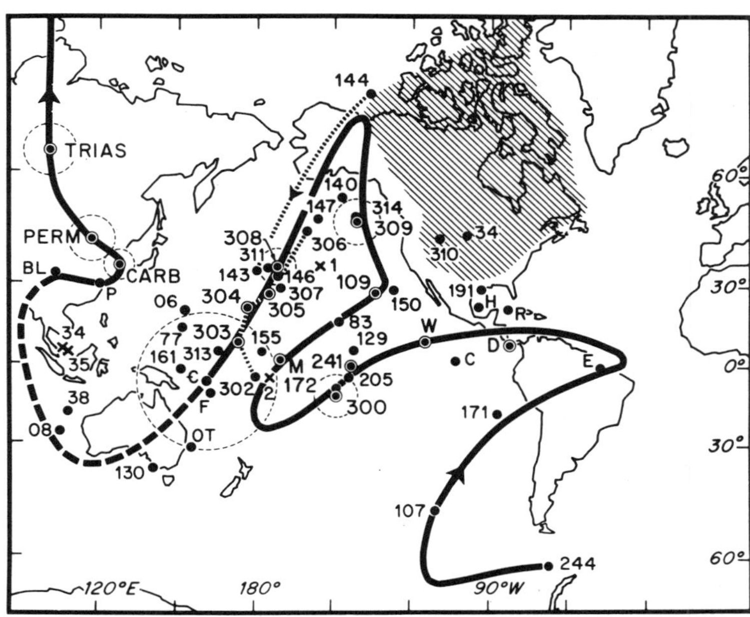

FIG. 2. Apparent polar-wandering relative to Canadian Shield Archean to Triassic. Mercator's projection is used. The exposed and probable subsurface extent of the shield is shaded. Control points that are considered to be of high reliability (they are commonly averages of several determinations) are marked by dot and exscribed circle. Crosses are results from Scotland, obtained by rotation about the pivot point (North America to Europe) of Bullard *et al.* (1965). The finely broken lines connect results whose order is known from superimposition and cross-cutting relationships. The results are identified by the directory number of Hicken *et al.* (1972, to be published) who list the numerical details and references to originals. *Devonian*: 038 Onondaga Limestone; *P* (115) Perry Formation combined. *Silurian*: *B* (027) Bloomsburg Formation; 034 and 035, Arrochar and Carabal Hill complexes 417 m.y. *Ordovician*: 006 Juniata Formation; 008 Trenton Group. *Cambrian*: mean of 019, 023, 045, 066, 078. *Precambrian*: 001 Stoer Group 935 m.y.; 002 Torridon Group 751 m.y.; 034 Baraga County dikes; 077 Eileen Sandstone; 083 Sibley Series; 107 Matachewan dikes 2500 m.y., Ontario; 109 Abitibi dikes 1230 m.y.; 129 Crocker Island Alkali Complex 1475 m.y.; 130 Allard Lake anorthosite; 140 Mamainse Point Lavas (normal) 1076 m.y., Ontario; 143 Upper Gargantua lavas, Ontario; 144 Lower Gargantua, Ontario; 146 North Shore volcanics (normal) 1115 m.y., Minnesota; 147 Osler Formation, Ontario; 150 Lower Keweenawan Lavas, Ironwood, Michigan; 155 Pike's Peak Granite 1030 m.y., Colorado; 161 Grenville dikes, Ontario; 171 Nipissing diabase 2100 m.y., Ontario; 172 Sherman Granite and associated rocks 1410 m.y.; 191 Gunflint Formation 1700–2300 m.y., Ontario; 205 Michikamau anorthosite 1400 m.y.; 241 igneous rocks of St. Francois Mountains 1300–1550 m.y.; 244 Stillwater Complex 2700 m.y., Montana; 300 Belt Series combined; 302 Jacobsville Sandstone combined; 303 Freda Sandstone and Nonesuch Shale combined 1046 m.y.; 304 Copper Harbor Lavas combined 1046–1200 m.y.; 305 Portage Lake Lavas combined 100 m.y.; 306 North Shore volcanics (reversed) combined 1115 m.y., Minnesota; 307 Beaver Bay complex combined, Minnesota; 308 Keweenawan Intrusions combined 1100 m.y.; 309 Keweenawan (reversed) combined 1100 m.y.; 310 Alona Bay Lavas combined; 311 Mamainse Point Lavas (normal) combined 1176 m.y., Ontario; 313 Franklin diabases combined 675 m.y.; 314 Mugford Volcanic Series 950 m.y. More recent data, in process of publication and not yet indexed in the directory of Hickens *et al.* (1972), are denoted by letters and referenced in the usual way: *C* Cameron Bay 1740 m.y. (Irving *et al.* 1972*a*); *D* Dubawnt Group 1725 m.y. (Park *et al.* 1972); *E* Et-Then Group 1250–1845 m.y. (Irving *et al.* 1972*b*); *F* Frontenac dikes, NW trend (Park and Irving 1972); *H* Hornby Bay Group sediments (Irving *et al.* 1972*a*); *M* Mackenzie diabases combined 1250 m.y. (Fahrig and Jones 1969, Irving *et al.* 1972*b*); *OT* Ottawa basic rocks (Irving *et al.* 1972*c*); *R* Port Radium Sill (Irving *et al.* 1972*a*); *W* Western Channel diabase 1500–1750 m.y. (Irving *et al.* 1972*a*). Data points *D*, *E*, *F*, *M*, *OT*, and *W* are based on detailed demagnetization studies and many sampling localities. Data points *C*, *H*, and *R* are based on demagnetization studies, but single sites only.

then back into the central Pacific, turning eastward across the Pacific to northern South America, and finally passing across South America into the southeast Pacific. Since the Archean the pole path seems to have undergone about two and one-half roughly sinusoidal oscillations, with periods of several hundreds of millions of years, and peak-to-peak amplitudes of the order of 90°.

In Fig. 2 poles from the Silurian (no. 34 and 35) and Precambrian (no. 1, 935 m.y. and no. 2, 751 m.y.) are plotted after rotation for closure of the Atlantic. The former fall near the pole for the Bloomsburg Formation (BL), and the latter are near the descending limb of the late Proterozoic polar path (track 2 see below). These results provide rough spot-checks of the reconstruction of the path of polar wandering relative to the Canadian Shield.

Chronology of Superintervals

The later Phanerozoic hairpins are numbered 2 and 6 (Fig. 1) because their peaks coincide with Magnetic Intervals 2 (the late Mesozoic Normal Interval) and 6 (the late Paleozoic Reversed Interval Fig. 4). In Fig. 3 the earlier hairpins are numbered in decades 10, 20, 30, 40, and 50 to allow indexing of intermediate features that will doubtless be found. Hairpin ages have been calculated assuming a constant rate of polar movement and are given in Figs. 3 and 4. The assumption of constant polar movement may in future be proved inadequate, but the present data do not warrant more detailed treatment. The polar paths between the hairpins are referred to as 'tracks', and are numbered 1 to 5. The time corresponding to each track is called a superinterval. Their limits are given in Fig. 4.

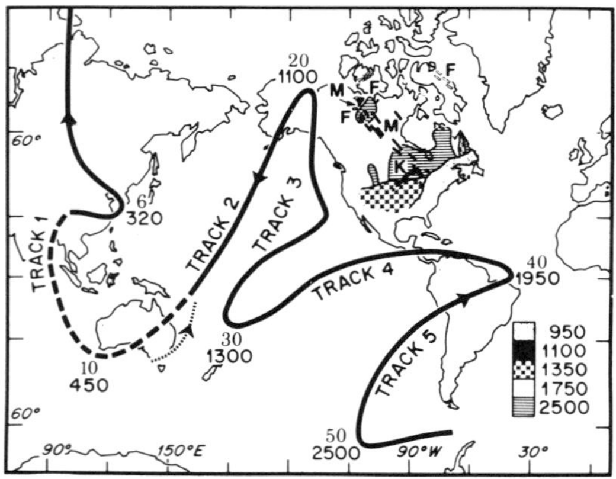

FIG. 3. Generalized polar path relative to the Canadian Shield. Hairpins 10, 20, 30, 40, and 50 are marked with their approximate ages in millions of years. The polar tracks between them are numbered. Note the speculative nature of the early Paleozoic curve. The finely broken line is the path obtained from Grenville rocks. The provinces of the Canadian Shield and its subsurface extension in the USA are shown schematically, the approximate dates of crustal stabilization are shown in the legend. *M* refers to the Mackenzie igneous rocks (1250 m.y. Fahrig and Jones 1969), *K* to the Keweenawan rocks (1100 m.y.) and *F* to the Franklin igneous rocks (675 m.y. Fahrig *et al.* 1971).

The earlier hairpins appear to be much bolder features, and to occur less frequently than their later Phanerozoic counterparts. This may reflect the evolution of diastrophic processes, or it may be a consequence of the lower density (per unit time) of the observations for earlier time. Hairpins 10 and 50 are only poorly defined, but present evidence suggests that they correspond very approximately to the Taconic and Kenoran orogenies (Fig. 4). Hairpin 30 corresponds closely to the Elsonian orogeny of Stockwell (Fig. 4). Hairpins 20 and 40 are older than the Grenvillian and Hudsonian orogenies (as defined by the peaks in the K–Ar histograms, Fig. 4) by 150 and 250 m.y. respectively. Hairpins 6 and 2 correspond well to the Appalachian and the Columbian–Laramide orogenies of the western Cordillera, and predate the Paleozoic–Mesozoic and Mesozoic–Cenozoic boundaries by 50 and 20 m.y. respectively. The opening of the Laurasian Atlantic follows hairpin 2. It is noteworthy also that hairpins 2 and 6 are centered in magnetic intervals when few reversals occurred (Fig. 4), but the data are insufficient to determine if there is any similar correspondence for earlier time.

Hairpins 2 and 6 may be considered to subdivide track 1, and perhaps several superintervals should be inserted in the Phanerozoic. The nature of the early Paleozoic pole path needs clarifying before further subdivision is attempted.

Discussion

The hairpins shown in Figs. 1 to 3 correspond to very large changes in the direction of movement of the Canadian Shield relative to

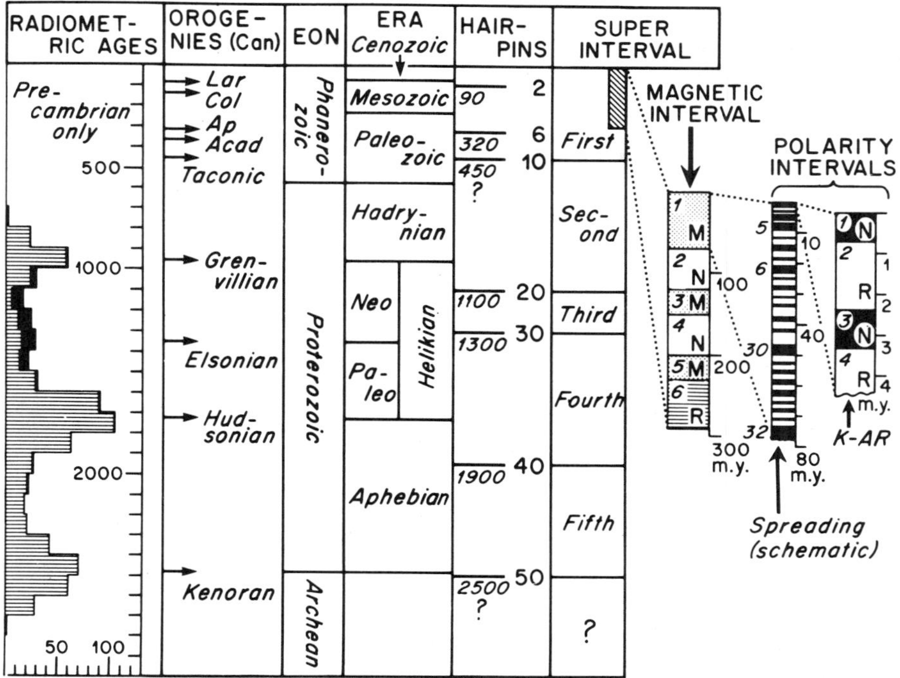

FIG. 4. Summary of paleomagnetic time-scales. On the right the scale of Polarity Intervals is given – the scale determined by K–Ar dating (recent summary by Doell *et al.* 1971) and its extension from sea-floor studies (Heirtzler *et al.* 1968). *N* denotes predominantly normal polarity, and *R* predominantly reversed polarity. The scale of Magnetic Intervals (McElhinny and Burek 1971, Irving 1972) is set out (*M* denoting mixed polarity) and further to the left the superintervals and the hairpins that define them. The geological eras and Stockwell's (1970) time scale are given in the center, and the corresponding frequency histogram of radiometric ages on the left. This histogram has been compiled from Stockwell (1970) for the exposed shield in Canada (shaded part), and from Muelberger *et al.* (1967) for the subsurface and exposed shield in the USA.

the pole. Following a time (superinterval) of motion in approximately the same direction (the speed may of course vary) the Canadian Shield has been momentarily halted (hairpin) and then the motion has been renewed in some very different direction. This seems to have occurred several times since the Archean.

Within each superinterval the approximately uniform direction of motion and the stress field that this maintains may be responsible for the approximately consistent structural trends found within each of the provinces of the shield (Stockwell 1970, Fig. IV-2), and for the linear nature of some province boundaries. It is noteworthy that tracks 2 and 3 are essentially parallel to the Grenville Front (stretches from southeastern United States to Labrador) and track 4 is parallel to the boundary between crust stabilized at 1350 m.y. and older crust to the north (Fig. 3).

Hairpins 6 and 2 correspond to late Paleozoic and late Mesozoic orogenies and are followed by widespread (indeed worldwide) restriction of the continental seas that mark the commencement of the Mesozoic and Cenozoic. This is reminiscent of the feature, already noted, that hairpins 20 and 40 occur earlier than Grenvillian and Hudsonian orogenies. The inference is that the restriction of the epicontinental seas, and the peaks in the K–Ar histograms (Fig. 4) are a consequence of uplift caused by the reorganization of plate motions.

It is noteworthy that Coney (1971) has suggested, on the basis of entirely different evidence, that 80 m.y. ago (roughly coincident with hairpin 2) a large change in the 'absolute'

TABLE 1. North American Phanerozoic mean poles

No.	Period	N	Pole	K	A_{95}
12A	Plio-pleistocene	6	84.3 N, 154.5 E	57	8.9
12B	Plio-pleistocene	9	87.4 N, 162.3 E	17	12.7
11	Upper Tertiary	13	86.9 N, 192.9 E	256	2.6
11	Lower Tertiary	5	73.8 N, 154.3 W	45	11.5
10	Cretaceous	11	65.9 N, 174.7 W	63	5.8
9	Jurassic	4	86.8 N, 096.4 E	56	12.4
8A	Triassic	11	61.2 N, 099.8 E	102	4.5
8B	Triassic	12	63.2 N, 100.0 E	61	5.6
7	Permian	9	42.7 N, 115.9 E	68	6.3
6	Carboniferous	14	35.6 N, 126.8 E	75	4.6
5	Devonian (Perry)	4	29.4 N, 119.0 E	93	9.6
4	Silurian (Bloomsburg)	1	32.0 N, 102.0 E	—	—
2	Cambrian	5	07.4 S, 160.4 E	9	27.6

Note: The mean poles have been obtained by averaging the several individual results that are identified by the reference numbers in Hicken *et al.* (1972) through which the original descriptions may be traced: 12(A) 12047, 069, 094, 095, 136, 161; 12B 12047, 069, 094, 095, 136, 161, 098, 115, 160; 11 (Upper) 11104, 106, 107, 151, 190, 216, 217, 231, 232, 295, 320, 361, 362; 11 (Lower) 11101, 165, 212, 238, 312; 10 10017, 028, 029, 030, 035, 049, 050, 071, 072, 073, 090; 9 09034, 072, 103, 104; 8A 08040, 042, 044, 068, 089, 111, 113, 135, 165, 166, 167; 8B 08040, 042, 044, 068, 089, 111, 113, 135, 165, 166, 167, 071; 7 07049, 050, 051, 073, 094, 095, 149, 205, 206; 6 06063, 066, 087, 088, 116, 135, 136, 137, 138, 140, 170, 171, 172, 174; 2 02019, 023, 045, 066, 078.

motion of the North American plate occurred as determined by the 'hot spot' method. (Coney also suggested that there was a similar period of "tectonic reorganization" 40 m.y. ago. His maps show only a small change in direction at this later time, and there is no corresponding hairpin; it is doubtful whether the small changes are detectable paleomagnetically.) Making use of Coney's ideas regarding the interrelationship between orogenies and changes in the direction of plate motion, the following qualitative model is proposed. After a time of free-plate motion a momentary 'plate-jam' occurs, brought about either through locking of a major convergent juncture by continental lithosphere (Mackenzie 1969), or by locking of major transcurrent junctures through the development of marginal irregularities (Donaldson and Irving 1972). The forces at depth causing plate motions are still operative and the plates soon free themselves, but not without reorganization of their margins and changes in directions of motion. This freeing process is followed by the restriction of the epicontinental seas, due in part to the creation by rifting of new deep ocean, to the uplift of rifted margins, and to the rebound of deceased subduction zones – all processes that may be expected to reduce sea-level. Since most organisms live either in epicontinental seas, or upon the lowland areas that depend to some extent on such seas for their rainfall, this tectonic sequence may have set in train wholesale extinctions of life. This explains, qualitatively, the lag between the hairpins and the Phanerozoic era boundaries, and the lag between the older hairpins and the K–Ar peaks. The Precambrian–Cambrian boundary is marked by the sudden production of varied and abundant life, whereas later era boundaries correspond to wholesale extinctions, suggesting that it has other causes. The absence of a hairpin at the end of the Precambrian (Fig. 3) is therefore not unexpected. This model does not explain the correspondence between the ages of hairpins 50 and 30 and the Kenoran and Elsonian orogenies (Fig. 4). However, hairpin 50 is not well-defined, and the nature of the Elsonian orogeny is only poorly understood (the K–Ar peak, for example, is comparatively feeble, Fig. 4).

Finally, it is noted that the poles obtained from Grenville rocks do not conform to the main polar-wandering curve (Fig. 3). This feature is discussed elsewhere (Irving *et al.* 1972*c*; Palmer and Carmichael 1972).

Acknowledgments

Many fruitful discussions with J. A. Donaldson, J. C. McGlynn, and J. L. Roy, and criticism of the manuscript by W. A. Robertson are gratefully acknowledged.

BLACK, R. F. 1963. Palaeomagnetism of part of the Purcell system in southwestern Alberta and southeastern British Columbia. Geol. Surv. Can., Bull. 83, pp. 1–31.

BULLARD, E. C., EVERETT, J. E., and SMITH, A. G. 1965. The fit of the continents around the Atlantic Phil. Trans. Roy. Soc. **258**, pp. 41–51.

CONEY, P. J. 1971. Cordilleran tectonic transitions and motion of the North American plate. Nature **233**, pp. 462–465.

DOELL, R. R., GROMME, G. S., DALRYMPLE, G. B., and COX, A. V. 1971. Geomagnetic polarity epoch time scale. Upper Mantle Project, U.S., Program Final Report, Nat. Acad. Sci., Washington, 128 p.

DONALDSON, J. A. 1972 (In press). Grenville Front and rifting of the Canadian Shield. Nature.

DU BOIS, P. M. 1962. Palaeomagnetism and correlation of Keweenawan rocks. Geol. Surv. Can., Bull. 71, pp. 1–75.

FAHRIG, W. F., IRVING, E., and JACKSON, G. D. 1971. Paleomagnetism of the Franklin diabases. Can. J. Earth Sci. **8**, pp. 455–467.

FAHRIG, W. F. and JONES, D. L. 1969. Paleomagnetic evidence for the extent of Mackenzie igneous events. Can. J. Earth Sci. **6**, pp. 679–688.

GILL, J. E. 1949. Natural divisions of the Canadian Shield. Trans. Roy. Soc. Can. **43**, pp. 61–91.

HEIRTZLER, J. R., DICKSON, G. D., HERRON, E. M., PITMAN, W. C., and LE PICHON, X. 1968. Marine magnetic anomalies, geomagnetic field reversals, and motions of the ocean floor and continents. J. Geophys. Res. **73**, pp. 2119–2136.

HICKEN, A., IRVING, E., LAW, L. K., and HASTIE, J. 1972. Directory of paleomagnetic directions and poles: first issue. Publ. Earth Phys. Br., Ottawa. (*In press.*)

IRVING, E. 1972. Paleomagnetic stratigraphy – names or numbers. Comments on Earth Sci., Geophysics **2**. (*In press*), pp. 125–130 cf.

IRVING, E., DONALDSON, J. A., and PARK, J. K. 1972*a*. Paleomagnetism of the Western Channel diabase and associated rocks, Northwest Territories. Can. J. Earth Sci. **9**, pp. 960–971.

IRVING, E., PARK, J. K., and MCGLYNN, J. C. 1972*b*. Paleomagnetism of the Et-Then Group and Mackenzie diabase in the Great Slave Lake area. Can. J. Earth Sci. **9**, pp. 744–755.

IRVING, E., PARK, J. K., and ROY, J. L. 1972*c*. Palaeomagnetism and the origin of the Grenville Front. Nature **236**, pp. 344–346.

MACKENZIE, D. P. 1969. Speculations on the consequences and causes of plate motions. Geophys. J. Roy. Astron. Soc. **18**, pp. 1–32.

MCELHINNY, M. W. and BUREK, P. J. 1971. Mesozoic paleomagnetic stratigraphy. Nature **232**, pp. 89–102.

MUELBERGER, W. R., DENISON, R. E., and LIDIAK, E. G. 1967. Basement rocks in continental interior of United States. Bull. Am. Assoc. Petrol. Geol. **51**, pp. 2351–2380.

PALMER, H. C. and CARMICHAEL, C. M. 1972. Grenville paleomagnetism and possible tectonic implications. (Abstr.). Trans. Am. Geophys. Un. **53**, 357 p.

PARK, J. K. and IRVING, E. 1972. Magnetism of dikes of the Frontenac axis. Can. J. Earth Sci. **9**, pp. 763–765.

PARK, J. K. IRVING, E., and DONALDSON, J. A. 1972 (In press). Paleomagnetism of the Dubawnt Group. Geol. Soc. Am. Bull.

ROBERTSON, W. A. and FAHRIG, W. F. 1971. The great Logan paleomagnetic loop – the polar wandering path from Canadian Shield rocks during the Neo-Helikian Era. Can. J. Earth Sci. **8**, pp. 1355–1364.

SPALL, H. 1971. Precambrian apparent polar wandering; evidence from North America. Earth Planet. Sci., Lett. **10**, pp. 273–280.

STOCKWELL, C. H. 1963. Second report on structural provinces, orogenies and time-classification of rocks of the Canadian Precambrian Shield. *In*: Age determinations and geological studies. Geol. Surv. Can., Pap. 62-17, pp. 123–133.

——— 1970. Geology and economic minerals of Canada. R. J. W. Douglas (*Ed.*). Geol. Surv. Can., Econ. Geol. Rep. **1**, Fifth ed., 44 p.

WILSON, J. T. 1949. Origin of continents and Precambrian history. Trans. Roy. Soc. Can. **43**, pp. 157–184.

WILSON, M. E. 1941. The Precambrian. Geol. Soc. Am., 50th Anniv. Vol., pp. 269–305.

14

Reprinted from *Akad. Nauk SSSR Doklady Earth Sci. Sec.* **222**:49–50 (1975)

PERIODIC VARIATIONS IN THE RATE OF MIGRATION OF PALEOMAGNETIC POLES IN THE PHANEROZOIC[1]

V. P. Aparin and V. S. Vedenkov

L. V. Kirenskiy Institute of Physics, Siberian Division, USSR Academy of Sciences, Krasnoyarsk

(Presented by Academician A. V. Peyve, July 23, 1974)

The movement of the lithosphere relative to the geomagnetic dipole, the orientation of which is "tied" to the axis of the Earth's rotation, is indirectly reflected in the migration of virtual geomagnetic poles. Therefore, evaluation of the rate of pole migration will give a direct estimate of the rate of movement and rotation of lithospheric plates. Moreover, comparison of curves of the pole migration rates with corresponding geomagnetic reversals will provide interesting information, especially on the interaction of the mantle and the core [1].

[1]Translated from: Periodicheskiye izmeneniya skorosti peremeshcheniya paleomagneticheskikh polyusov v fanerozoye. Doklady Akademii Nauk SSSR, 1975, Vol. 222, No. 2, pp. 415–416.

Using our previously proposed method [2], we calculated on a M-222 computer the migration rates of virtual geomagnetic poles, the data being paleomagnetic sections of various regions of all continents except Antarctica. In all, we used 921 pole-position determinations, and were thus able to calculate the migration rates. From these we then plotted the migration rate of virtual geomagnetic poles as a function of time for various major regions. The curves for these regions were plotted for the following time intervals: European part of the USSR (Russian platform), Early Ordovician (O_1) to Early Cretaceous (Cr_1); Siberia, Early Cambrian (Cm_1) to Early Cretaceous (Cr_1); northeastern USSR, Late Permian (P_2) to late Neogene (Ng_2); Urals and Ural region, Early Cambrian (Cm_1) to Late Triassic (T_3); southern and southwestern USSR,

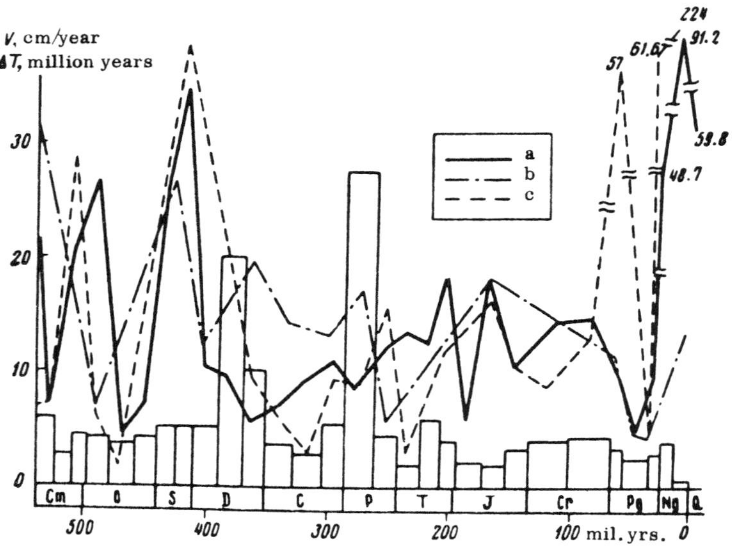

Fig. 1

Late Jurassic (J_3) to Quaternary (Q); southern Europe and Near East, Late Carboniferous (C_3) to Late Cretaceous (Cr_2); western Europe, Early Cambrian (Cm_1) to Quaternary (Q); Africa (few data), Middle Cambrian (Cm_2) to Early Permian (P_1); America, Middle Cambrian (Cm_2) to Neogene (Ng); South America (few data), Late Cambrian (Cm_3) to Late Permian (P_2); Australia, Early Cambrian (Cm_1) to Quaternary (Q); Japan, early Paleogene (Pg_1) to Quaternary (Q).

All our curves are broken lines, on which peaks and troughs occur at a relatively similar age for any given continent. In view of this fact, as well as the limited age range of these regional curves and the nonuniform distribution of original data over the age range, especially in the Mesozoic, we averaged regional data over the continents, lumping for this purpose the figures for North and South America.

As can be seen from Fig. 1, each curve consists of several periodically recurring rhythms of increasing and decreasing rate of pole migration. The duration of these rhythms ranges from 70 to 80 million years; their amplitudes are highest for the early Paleozoic and the Cenozoic. Moreover, although the migration rates calculated from Quaternary magnetic sections may have been overestimated as a result of secular field variations, this overestimation was eliminated from early Paleozoic data by averaging.

The chronologic coincidence of early Paleozoic peaks of migration-rate curves, especially in the Silurian, indicates a synchronous, uniform displacement of continents relative to the pole in the early Paleozoic. One explanation of this coincidence could be the existence of larger continents in the geologic past, such as Gondwana, Laurasia and Pangea. Indeed, this is specifically confirmed by analysis of pole migration paths. The good agreement of paleomagnetic directions on models of Gondwana and Laurasia [3] also supports this explanation. However a mismatch of behavior of continental curves of pole migration rates is already clearly apparent for the Carboniferous and in Australia, in the Devonian. It indicates the fracturing of the old continent into plates, which then moved at different rates relative to the pole.

We have plotted histograms of variation of the duration of polarity epochs (ΔT) in Fig. 1 for Eurasia (a), Australia (b) and America (c). Data on the distribution of geomagnetic reversals in the Phanerozoic were taken from the article of Braginsky et al. [4]; the age scale was taken from the article of Afanas'yev [5]. Comparison of this histogram with curves of pole migration rates shows that peaks of reversal frequencies (troughs on histogram) generally lead those of migration rates by several score million years. According to Kalinin [6], higher frequency of geomagnetic reversals is due to growth of the Earth's core as a result of fusion and differentiation of the lower mantle. This in turn leads to gravity-induced convection, a powerful mechanism of tectogenesis [7]. Thus, periodic variations in the rate of movement of the lithosphere are functions of the increase in the core volume and the frequency of geomagnetic reversals. In this connection we should note the work of Irving [8], Kravchinskiy [9] and Sheynmann [10], which shows that periods of rapid migration of virtual geomagnetic poles were related to major tectonic events.

We thank A. N. Khramov for discussing our data.

REFERENCES

1. Stacey, F. Physics of the Earth, Russian transl., 1972.
2. Anaparin, V. P. and V. S. Vedenkov. Geomagnetizm i aeronomiya, 13, No. 3, 501, 1973.
3. Petrova, G. N. and A. N. Khramov. Fizika Zemli, No. 4, 65, 1970.
4. Braginsky, S. I., G. N. Petrova and A. N. Khramov. Report on XV IUGG General Assembly, p. 261, 1971.
5. Afanas'yev, G. D. Problemy geokhimii i kosmologii (Problems of Geochemistry and Cosmology), p. 61, Nauka Press, 1968.
6. Kalinin, Yu. D. Geomagnetizm i aeronomiya, 11, No. 1, 187, 1971.
7. Artyushkov, Ye. V. Fizika Zemli, No. 5, 18, 1970.
8. Irving, E. J. Geophys. Res., 71, No. 24, 6025, 1966.
9. Kravchinskiy, A. Ya. Geotektonika, No. 6, 34, 1973.
10. Sheynmann, Yu. M. Byull. Moscovsk. obshch. ispyt. prirody, No. 5, 5, 1973.

15

Reprinted from *Earth and Planetary Sci. Letters* **3**:351–354 (1967)

AN OUTLINE OF THE INTENSITY OF THE PALEOMAGNETIC FIELD OF THE EARTH

C. M. CARMICHAEL
Department of Geophysics, University of Western Ontario, London, Canada

Received 11 December 1967

An outline of the paleomagnetic field intensity from 2.5×10^{-9} yr to the present has been obtained from igneous rocks, mainly basaltic lavas, that have been shown to have paleomagnetic stability. The field had an intensity equal to or larger than at present during the Pre-Cambrian. Near the end of the Pre-Cambrian and the early part of the Paleozoic it was quite weak then increased during the Mesozoic and Tertiary to present values. The scatter in the values is consistent with a short term variation in intensity of a factor of four being superimposed on the long term trend.

1. INTRODUCTION

Paleomagnetic studies have been very successful in delineating the history of the direction of the magnetic field of the earth, but relatively little work has been done on the history of the field's intensity over a comparable span of time. This preoccupation with direction measurements is due in part to the relative difficulty of intensity determinations, but the main reason for it is the desire to chart the drifting of continents and the wandering of poles.

The natural remanent magnetization (NRM) of lavas, dikes and other igneous bodies is produced by the earth's magnetic field as these bodies cool to temperatures below the Curie point of their magnetic minerals. Laboratory studies have shown that for weak fields of the order of that of the earth, the magnetization produced is directly proportional to the intensity of the magnetizing field. This fact provides a means of measuring the intensity of the paleomagnetic field in which an igneous rock was formed by comparing its NRM to the thermoremanent magnetization (TRM) produced in the laboratory in a magnetic field of known intensity. Comparisons of this sort for some historic lavas have produced a TRM in the present field about equal to the NRM as would be expected. For older lavas however the NRM is found to be considerably less than the TRM produced in the present field, which has been interpreted as being due to spontaneous demagnetization of the NRM with time since the rock was formed. For want of conclusive evidence to the contrary it has generally been considered, up to now, that the intensity of the paleomagnetic field has been similar to that of the present field.

The intensity of the field over the past few thousand years has been determined from pottery and bricks by a method due to Thellier and described in a review paper, Thellier and Thellier [1]. It has been found that the intensity has varied from one half to about twice the present intensity. Using baked contacts and some lavas with a modified Thellier method, variations in intensity of a similar magnitude have been found for Miocene time by Coe [2] and for the Tertiary by Smith [3]. Briden [4] has reported on a number of intensity determinations of his own and others showing weak fields in the middle Paleozoic and stronger fields, though still less than the present, during the Mesozoic. Nesbitt [5] has used the ratio of intensity of magnetization of red sandstones to their susceptibility to get a measure of the field. This is a completely independent method which shows low values in the Paleozoic and large values for two Pre-Cambrian samples.

2. METHOD

This report gives a summary of the results and the method used in determining the paleomagnetic field intensity for twenty time periods from 25×10^8 yr to the present. The rock types used were mostly basaltic lavas but some were andesites and welded tuffs. All had been shown to possess paleomagnetic stability by measurements of their remanent directions.

With the possible exception of some welded tuffs, rocks of this type cannot survive the repeated heatings of the Thellier method without alterations of their magnetic minerals. Even the single heating to above the Curie temperature to produce a TRM can alter existing minerals and produce marked changes in their magnetic properties. It is essential that a procedure be used which minimizes these changes and, what is more important, which is able to detect them if they occur. The most definitive test for alteration used was a comparison of the curve of thermal demagnetization of the NRM, measured in field-free space as the sample was being heated, with the similar curve for the TRM produced in the same sample by cooling in 0.5 oersteds. Differences in the shape of the curve or in the zero-field Curie temperature indicate alteration of the magnetic minerals and the sample must be discarded. A more insidious alteration, which can not be detected by this technique, is one in which the nature of the magnetic mineral does not change but the amount does. Conversion of a strong magnetic mineral to a weak or nonmagnetic one, which decreases the amount, and precipitation of new magnetic material from the silicates, which increases it, are examples of this type of alteration. Such changes in amount of magnetic mineral can be detected by comparing the saturation remanence before and after heating.

Unwanted viscous magnetizations, low temperature partial thermoremanences and changes due to spontaneous demagnetization are preferentially contained in the low coercive force and low Curie temperature components of a rock. For this reason the NRM must be 'washed' by a combination of thermal and AC demagnetization. The procedure used was to heat the sample to 300°C and cool it in field-free space followed by AC demagnetization in steps of about 75 oersteds peak. So that comparisons could be made between the various remanences, all four, namely natural remanent magnetizations (NRM), thermoremanent magnetization produced in 0.5 oersteds (TRM 0.5 Oe), saturation remanence before heating (Sat. RM) and saturation remanence after heating (Sat. RM heated) were given the same magnetic washing treatment.

The intensity of the paleomagnetic field in units of % of the present field was calculated from the following expression,

$$\frac{\text{NRM}}{\text{sat. RM}} \Big/ \frac{\text{TRM (0.5 Oe)}}{\text{sat. RM (heated)}} \times \text{latitude factor} .$$

The latitude factor corrects for the variation of the magnetic field intensity with magnetic latitude using the relative variation of the present field combined with the paleolatitude of the sample. The calculation was carried out using the successive values of the remanences after the same stage of magnetic washing. In this way values of the paleomagnetic field intensity based on progressively harder magnetic components can be compared.

With the procedure described it is not possible to make the thermal demagnetization test for alteration on the same sample that is used for all four of the remanences. The TRM and the heated saturation remanence can be measured using the alteration test sample but a second sample of the same rock must be used for the NRM and unheated saturation remanence. Alternatively, when heating tests on a sample show no alteration, a fresh sample from the same rock can be used for all four remanences. Both methods were used in this study but most intensities were obtained using the first one.

3. RESULTS

The values of paleomagnetic field intensity are plotted in fig. 1 in units of % of the present field. Results for samples from different parts of the same formation are plotted on the same vertical line. The N or R in the symbol denotes normal or reversed polarity relative to the present field with the symbol for nearly horizontal samples being left blank. The vertical dotted line at about 10×10^8 yr joins determinations from a sequence of reversed and normal Keweenawan lavas from the Lake Superior region. A paleomagnetic report on these will be published shortly by H. C. Palmer of the University of Western

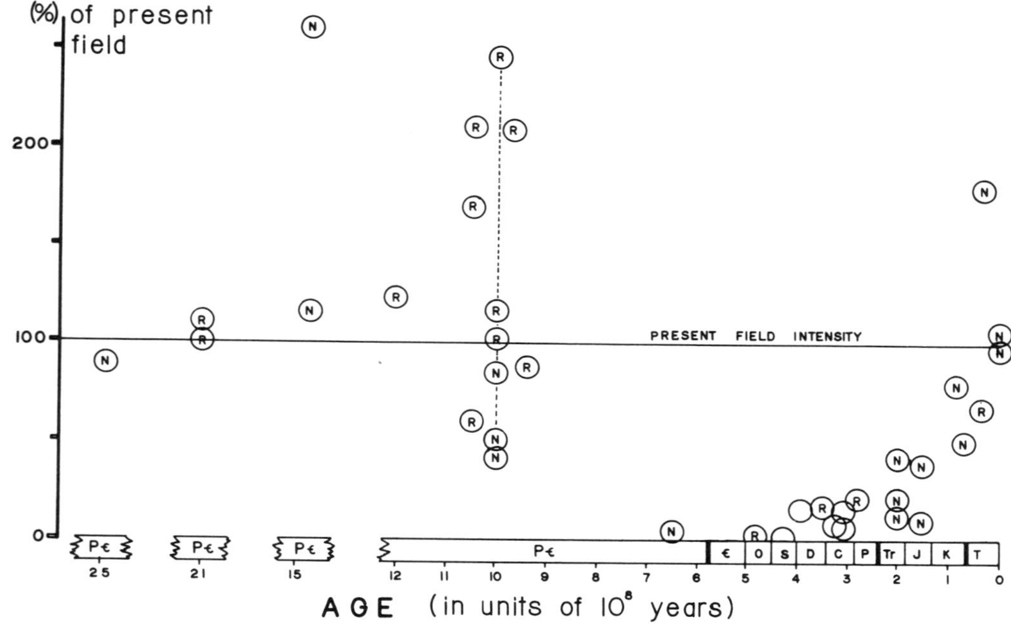

Fig. 1. Paleomagnetic field intensity at various ages as a percentage of the present field.

Pє – Pre-Cambrian	D – Devonian	J – Jurassic
є – Cambrian	C – Carboniferous	K – Cretaceous
O – Ordovician	P – Permian	T – Tertiary
S – Silurian	Tr – Triassic	

Ontario. The intensities and vector directions of the magnetization of these flows indicate that part of their magnetization has been added subsequent to emplacement. This added component has had the effect of increasing the intensity of the reversed flows and decreasing that of the normal ones. The original magnetizing field must have been approximately 110 to 120% of the present. As a check on the method it was applied to four recent lavas from Mount Etna, the Azores, Hawaii and Mount Mihara. Of these the Etna and Hawaii samples had to be discarded because of mineral changes on heating but the Azores sample gave a value of 107% and the Mihara one of 98% of the present field at these sites.

4. DISCUSSION

This outline of the paleomagnetic field intensity increases the coverage from the Paleozoic to the present and extends it well back into the Pre-Cambrian. It has the advantage that internal comparisons of relative intensity from one time to another should be reasonably valid since all determinations were made by the same method, using the same equipment operated by the same person. It suffers from the same shortcoming as other intensity studies, with the exception of Smith's work on the Tertiary, in that too few determinations were made on any one formation to give a reliable measure of the field. This is a particularly bad fault if the field has always had a short term four fold variation in intensity as it seems to have had for the past 60 million years. The scatter of points would suggest that some such variation has always been a feature of the field. There are also large time gaps in the Pre-Cambrian including the crucial one from 7 to 9×10^8 yr. (The author would be very pleased to learn about any dated basalts that fall within this range.) After all of these reservations have been noted however, there is a definite pattern apparent.

I am persuaded that the paleomagnetic field intensity was as strong or stronger than at present during the Pre-Cambrian. Near the end of the Pre-Cambrian and during the early part of the Paleozoic it was quite weak, probably less than 10% of the present. From the latter part of the Paleozoic through the Mesozoic and the Tertiary the intensity increased to present values. Superimposed on this trend there seems to be a short term variation similar in magnitude to the four fold one found for the recent field. In evaluating this intensity pattern it must be born in mind that no allowance has been made for spontaneous demagnetization with time. Should such demagnetization have been significant it will not change the general shape of the distribution but will mean that progressively older values are systematically too low.

If this pattern remains as more intensity data accumulates it puts an additional boundary condition on theories of the origin and history of the earth's magnetic field. In particular the low intensity in the late Pre-Cambrian to middle Paleozoic and the subsequent increase must be accounted for by any theory.

REFERENCES

[1] E. Thellier and O. Thellier, Ann. Geophys. 15 (1959) 285.
[2] R. S. Coe, J. Geophys. Res., in press.
[3] P. J. Smith, Geophys. J. Astr. Soc. 12 (1967) 239.
[4] J. C. Briden, Nature 212 (1966) 264.
[5] J. D. Nesbitt, Nature 210 (1966) 618.

Part III

EPISODICITY OF GLACIATION

Editor's Comments on Papers 16 Through 20

Variations of terrestrial climate during geological time are most convincingly revealed by the apparent long-term episodicity of major glaciation. Early this century Schuçhert (1914) discussed evidence for glaciation in the Lower Huronian, the late Proterozoic, the Late Silurian (the Pakhuis Formation tillites, South Africa, now regarded as Late Ordovician to Early Silurian), the Permian, and the Pleistocene (see Figure 1 in the Introduction, and C. A. Ross, ed., *Paleobiogeography*, Benchmark Papers in Geology, vol. 31, 1976, pp. 367–377). Schuchert (p. 286) concluded that "the earth since the beginning of geologic history has periodically undergone more or less widespread glaciation and that the cold climates have been of short geologic duration." Subsequent work, as summarized by Harland and Herod (1975), Frakes (1979), and Papers 16 and 17, has confirmed that at least seven major episodes of terrestrial glaciation occurred during the past 2300 m.y. and perhaps six in the past 1000 m.y. As discussed in Papers 16 and 17, these glaciations seem best regarded as discrete episodes separated by long intervals marked by substantially warmer climates and the absence of ice sheets. Such episodic glaciation appears to be superimposed on very long-term (~600 m.y.) changes in facies associations, areal extent, and paleolatitudes of deposition of glaciogenic rocks (see Paper 35 and Part VII).

Climatic change may result from both endogenetic and exogenetic processes. Nearly all recent papers that discuss the possible causes of long-term episodicity of glaciation consider an exogenetic control, and the five papers included here fall within this category. The former idea of mountain building as an important factor in glaciation (e.g., Huntington, 1914; Umbgrove, 1947, pp. 268–269) cannot be sustained, since orogeny occurred more frequently than glaciation during the Phanerozoic (see Figure 3 in the Introduction and Part I). Recent proponents of an endogenetic control of glaciation (e.g., Crowell and Frakes, 1970; Donn and Shaw, 1977; Crowell, 1978) maintain that glaciation may occur at any time, and results primarily from the movement of continents into latitudes and arrangements of potential glaciation. Paleomagnetic data (e.g., Smith et al., 1973; Smith and Briden, 1977) demonstrate, however, that continental land masses have at certain times during the Phanerozoic occupied very high paleolatitudes without incurring continental glaciation, thus suggesting that endogenetic processes are not first-order controls of global climate. On this argument Professor R. W. Fairbridge (personal communication, 28 August 1978) comments that "To me a key point in favor of exogenetic control is the fact that the South Pole reached Antarctica in the Permian, and *stayed there* from Triassic to Tertiary time without glaciation. Clearly, polar continental masses are not enough." The debate nevertheless is very much alive (see Tarling, 1978), and as discussed in Part VII planetology may help identify the relative importance of endogenetic and exogenetic processes to terrestrial climatic change.

The five papers included here are divisible into two groups: galactic models (Papers 16–18) and solar models (Papers 19 and 20). The Milankovitch astronomical theory of ice ages relates to short-term climatic changes, which are not considered here.

GALACTIC MODELS

Forbes (1931) was the first to postulate a link between the apparent rhythm of major glaciations (taken as 210 m.y.) and the period of revolution of the sun about the galactic center (then estimated at 230 m.y.), although Shapley (1921) previously had proposed a cosmic influence on global climate. The possibility of a connection between the presumed rhythm of glaciation and the sun's galactic period, each taken to be ~200–250 m.y., was raised also by Umbgrove (1942, 1947) and Shapley (1949). Umbgrove (1942, p. 154) noted that "it is difficult to decide for the moment whether this coincidence is a mere question of chance, but it is obvious that this point will have to be taken into consideration in the future development of science." The idea of a galactic influence in climatic change has indeed been pursued by numerous geologists and astronomers since Umbgrove, led by the Russians G. F. Lungershausen and G. P. Tamrazyan and the Canadian J. Steiner.

Lungershausen (1957) perceived a glacial rhythm of ~190–200 m.y. in then available geochronological data, which he equated with contemporary estimates of the sun's galactic period or "cosmic year." He envisaged a strongly eccentric solar orbit around the galactic center and proposed that during "cosmic winters" when the sun was in the outer, more rarified regions of the Galaxy, overall planetary cooling and terrestrial glaciation ensued. Tamrazyan (1959), in contrast, equated postulated times of glaciation with the periodic passage of the solar system through the galactic plane where absorbing galactic nebulae or dust clouds are located. Tamrazyan (1967) subsequently presented an elaborate scheme in which proposed periodic changes in terrestrial tectonism, volcanism, paleoclimates, metallogenesis, and organic evolution were ascribed to a varying gravitational potential along an eccentric solar orbit with an anomalistic period of 176 m.y. (The anomalistic period from perigal to perigal is shorter than the full period of 212 m.y. given by Tamrazyan due to the known retrograde rotation of the solar galactic orbit.) The Phanerozoic glacial chronologies of Forbes (1931) and Lungershausen (1957) are noteworthy for the inclusion of a Late Ordovician glaciation in addition to those of the Permo-Carboniferous and Cenozoic; their chronologies anticipated Fairbridge's (1960) illustration of a comparable glacial scheme and were proposed well before a Late Ordovician glaciation became widely accepted in the early 1970s.

Johann Steiner of the University of Alberta deserves much credit for the renewed attention given by earth scientists in the West to possible galactic control of episodic and periodic glaciation. In 1967 Steiner presented a wide-ranging hypothesis by which events in the stratigraphic, tectonic, biological, and geomagnetic records were ascribed to a systematic variation of the gravitational parameter *G* resulting from the slight eccentricity of the solar galactic orbit; Steiner also estimated a duration of ~280 m.y. for the cosmic year. In 1973 Steiner and Grillmair (Paper 16) applied this model to episodic and periodic glaciation, which they attributed to successive passages of the sun through apsides of an eccentric solar orbit. This paper also reviews other works on the same topic, and the evidence for and ages of ancient glaciations. Steiner (1973, 1978) has expanded his galactic model to encompass other episodic events in the geological record. In 1975 Williams (Paper 17) further reviewed and updated the known age limits of major glaciations during the past billion years, summarized the present view of galactic structure and dynamics and, based on a reestimated length of the cosmic year of ~303 m.y., postulated that major glaciation occurs during passage of the sun through diametrically opposite, little-warped regions of the flexed galactic disk. Another proposal for galactic control of periodic glaciation, advanced in 1975 by astronomer W. H. McCrea (Paper 18) and also advocated by Gribbin (1978), attributes glaciation to an *increase* in solar radiation consequent upon passage of the sun through dust lanes bordering galactic spiral arms. Gribbin considered that a double-arm spiral structure for the Galaxy would lead to two so-induced glaciations per cosmic year. In 1977 Whyte (Paper 34) correlated times of glaciation with episodes of apparent deceleration of the earth's rotation; Fairbridge (1978) pointed out that these apparent spin minima have occurred on a galactic time scale.

Since the equation by Forbes, Umbgrove, Shapley, and Lungershausen of one glaciation per cosmic year of ~200–250 m.y., important refinements of glacial and galactic periods have occurred. Available geochronological data, as summarized in Papers 16 and 17, indicate that the mean or modal times of major glaciations have recurred about every 140–170 m.y. during the past billion years, apart from an apparent double cycle spanning the Mesozoic. (Unfortunately, some confusion recently has arisen concerning the apparent periodicity of glaciation. Tarling [1978], in refuting any such rhythm, mistook the galactic time-scale of Steiner and Grillmair [Paper 16] for their glacial chronology. Furthermore, the 250 m.y. period of glaciation employed by McCrea [Paper 18] and Gribbin [1978] is obsolete, having been obtained from Allen [1973], whose principal source in turn was Holmes [1965].) The estimated duration of the cosmic year, in contrast, has been increased to a little more than 300 m.y. (Paper 17). The former idea of one terrestrial glaciation for each galactic revolution by the sun therefore has been replaced by the concept that *two* glaciations usually occur on average every cosmic year.

Despite these refinements, recent hypotheses concerning possible galactic causes of glaciation are not without their problems. The physical basis of Steiner and Grillmair's (Paper 16) galactic model was criticized by Bond (1974), and the model does not use the anomalistic cosmic year, which would appear to invalidate the correlation of glaciation with apsides. The scheme of Williams (Paper 17) has an apparently missing glacial event in the Late Jurassic ~145 m.y. ago. The model of McCrea (Paper 18) and Gribbin (1978) seems invalidated by the apparent lack of a nearby cosmic dust cloud that could have caused the last Cenozoic glaciation (Dennison and Mansfield, 1976) and by recent maps of the Galaxy (Paper 17, Figure 2; French and Osborne, 1976; de Vaucouleurs and Pence, 1978) that indicate a multi-arm spiral structure rather than the double-arm structure assumed by Gribbin (1978); moreover, it is difficult to see how an ice age could be triggered by an *increase* in solar luminosity (see Paper 19 and Öpik, 1977). Nonetheless, Lindsay and Srnka (Paper 37) considered that an apparent cyclical variation of micrometeoroid flux with depth in the lunar soil (time) accords with solar passage through galactic dust lanes with a period of ~10^8 years.

Apparently, therefore, despite a tantalizing relation between glacial and galactic time scales, no such hypothesis can as yet provide a completely satisfactory explanation of the seemingly near-periodic fluctuations of terrestrial climate.

SOLAR MODELS

In a pioneering series of more than thirty papers spanning forty years, the Estonian astronomer E. J. Öpik (now at the Armagh Observatory, Northern Ireland) developed the hypothesis that past variations in terrestrial climate, as indicated by the apparent near-periodicity of ice ages, have resulted from postulated fluctuations of solar radiation due to periodic convective overturn of material within the sun. Paper 19 comprises excerpts from a recent review of Öpik's model. The time-scale and apparent period of major glaciations employed by Öpik have been superseded by more recent data (see Papers 16 and 17), but this does not affect Öpik's argument. A more important objection might be raised at the calculated secular warming of global climate since early Precambrian time (Paper 19, Figure 10 and Table 5), which is contrary to the secular decrease of mean global temperature suggested by biological (Hoyle, 1972) and isotopic (Knauth and Epstein, 1976) evidence (see also Part V).

Support for Öpik's contention that glaciation results from decrease in solar radiation may be provided by the solar neutrino "problem," which is reviewed by Cameron (Paper 20) and further discussed by Hartmann (Paper 38). They speculate that the failure to detect the predicted level of neutrino emission from the sun might indicate that the sun's luminosity is at present anomalously low, due to a temporarily expanded state of the solar core. This idea has prompted correlation of the apparently low neutrino flux with the late Cenozoic ice age, and the thought that a mechanism for intermittent mixing might exist within the sun that ultimately has controlled the timing of terrestrial glaciation. However, the solar neutrino problem is far from resolved: several astronomers have outlined possible steady-state solar models with very low neutrino flux (e.g., Auman and McCrea, 1976; Beaudet et al., 1977), and others have advocated caution in the interpretation of results and the desirability of further research (e.g., Ulrich, 1975; Freedman et al., 1976). Evidence for past variations in solar activity is discussed also in Part VII.

16

Reprinted from *Geol. Soc. America Bull.* **84**:1003–1018 (1973)

Possible Galactic Causes for Periodic and Episodic Glaciations

J. STEINER *Department of Geology, University of Alberta, Edmonton 7, Alberta, Canada*
E. GRILLMAIR *Calgary, Alberta, Canada*

ABSTRACT

Present knowledge of the geochronology of Phanerozoic and Precambrian glaciations from all continents is summarized. Late Precambrian glaciations appear to group into three age clusters. Seven major glacial episodes are well documented and dated in the geological record. The duration of glacial episodes varies from 60 to 100 m.y.

A partly speculative galactic model has been derived from present general knowledge of the Milky Way galaxy. This simple model is shown to be approximately 80 percent valid in terms of Newtonian gravitational theory. A model applying Newtonian gravitation more strictly would be more complex. According to the model, the solar system experiences a periodically and episodically varying galactic parameter during its orbits as far back as 3,000 m.y. Minima of the varying galactic parameter correlate with Phanerozoic and Precambrian glaciations well. The reason for these apparent correlations is not known, but they suggest the possibility of a causal relation.

This study enhances, within limits, the possibility of valid intercontinental Precambrian time-stratigraphic correlations on the basis of glaciogene deposits. The model appears to point toward multilateral causes of periodic ice ages in which the various causes may be assigned an approximate degree of importance. The contributing causes include major ice-age hypotheses so far proposed, plus the supposed galactic variable as the trigger mechanism. During the Phanerozoic, the variation of the galactic parameter appears to agree with other paleoclimatic criteria, which are not discussed in detail.

Apart from the galactic-geologic correlations, an attempt is made to evaluate the degree of validity of the galactic model, and it is concluded that only future work, together with continuously improving geological and galactic knowledge, may eventually answer this clearly fundamental question.

INTRODUCTION

Umbgrove (1947) noted that the length of a cosmic year corresponds approximately to cyclic occurrences of Phanerozoic to late Precambrian glaciations and major orogenic cycles. Lungershauzen (1957) further emphasized this correlation by postulating an alleged Ordovician glaciation for which there was some Russian evidence (as reviewed by Schwarzbach, 1963). Tamrazjan (1967) attempted to correlate a variety of geological phenomena with the length of the anomalistic cosmic year, but he had to neglect paleoclimatic criteria, including glaciations, since in his model the cosmic year is foreshortened by the retrograde rotation of the Kepler ellipse of the solar galactic orbit. Machado (1967) postulated pulsating gravitation of periods similar to Tamrazjan, but offered no physical explanation. Steiner (1967) compared major geological phenomena, including glaciations, to a controversial gravitational model of the galaxy for the past 1¼ cosmic yrs. Schwarzbach (1968), in reviewing new ice-age hypotheses, extended Steiner's model to 2 cosmic yrs and noted how convincing such an extension appears. Gidon (1970a, 1970b) discussed briefly the possibility of a correlation between solar galactic revolutions and glaciations. All these comparisons, with the exception of Steiner's and Schwarzbach's, were based on the long since outdated circa-200-m.y. duration of a cosmic year. The presently accepted duration of the cosmic year has also been compared to major paleomagnetic polarity epochs (Crain and Crain, 1970; Crain and others, 1969), biological phenomena (Hatfield and Camp, 1970), and occurrences of carbonatite (Macintyre, 1971).

This study is based on an up-to-date galactic

Phanerozoic Time-Scale, London Geological Society

	Harland and others, 1964	Lambert, 1971	I.U.G.S. (Rast, 1971)
Permian-Carboniferous boundary	280 m.y.	280 m.y.	285 ± 10 m.y.
Silurian-Ordovician boundary	430 to 440 m.y.	445 (?) m.y.	440 ± 10 m.y.
Base of Cambrian	570 m.y.	590 (?) m.y.	570 ± 15 m.y.

model extending back 3 b.y., and assesses its correlation with adequately dated ancient ice ages. In this model, the duration of the cosmic year is assumed to be decreasing from 400 m.y., in the early Precambrian, to its present duration of 274 m.y.

ANCIENT ICE AGES

Numerous compilations of ancient glacial deposits have been made (Coleman, 1926; Schwarzbach, 1963; Cahen, 1963; Harland, 1964a, b; Holmes, 1965). Some authors have suggested nearly continuous glaciation at high latitudes from late Precambrian to the end of the Paleozoic (Crowell and Frakes, 1970; Crawford and Daily, 1971). We consider glaciations as separate episodes since successive glaciogene deposits in many parts of the world are stratigraphically separated by rock sequences of the order of thousands of meters, which show evidence of prolonged periods of very warm climate.

Seven major glacial episodes[1] are well documented on several or all continents (Table 1). One middle Precambrian and three Phanerozoic glaciations are well dated. Unquestionably, late Precambrian glaciogene deposits appear to group into three age clusters considering available radiometric dates and stratigraphic data. Therefore, three major late Precambrian glacial episodes are tentatively defined below. In addition eight, more problematic older Precambrian glacial deposits occur in rock sequences of which the age is known within several hundred millions of years. Some of these older Precambrian glacial deposits may be mutually correlative. For the Phanerozoic era, this study employs the time-scale of the London Geological Society (Harland and others, 1964; Harland and Francis, 1971). The supplement to the Phanerozoic time-scale of the London Geological Society (Harland and Francis, 1971) does not appreciably differ from the 1964 time-scale for this purpose. Lambert's (1971) recommended ages and the composite time-scale of the I.U.G.S. (Rast, 1971) compare with the 1964 time-scale as shown above.

Phanerozoic Glaciations

Late Cenozoic Glaciation. As far as the Pleistocene epoch is concerned, evidence of glaciation goes back approximately 14 m.y. into the Miocene epoch in the high latitudes of Alaska and Antarctica (Denton and Armstrong, 1969; Denton and others, 1970).

Permian-Carboniferous Glaciation. The Permian-Carboniferous glaciation of the Gondwanaland continents (Wanless and Cannon, 1966; Hamilton and Krinsley, 1967), is best known next to the Pleistocene glaciation. Girdler (1964), Irving (1964), Creer (1970), and McElhinny and Luck (1970) review the high paleolatitudes obtained from these continents toward the end of the Paleozoic era. Crowell and Frakes (1970) cite evidence of late Paleozoic Gondwanaland glaciation ranging from mid-lower Carboniferous to mid-Permian time (approximately from 255 to 340 m.y.). Meyerhoff and Meyerhoff (1972) briefly summarized the recent Russian evidence of five Permian-Carboniferous glacial centers of northern Asia. Evidence in Russia of Permian-Carboniferous glaciation extends into the Upper Permian (Mikhaylov and others, 1970) and may range from 235 to 320 m.y. Siberia was in the middle paleolatitudes according to Creer's (1970) Permian-Carboniferous Eurasian pole positions.

Silurian-Ordovician Glaciation. Recently a Silurian-Ordovician glaciation has been disclosed in northern Africa (Beuf and others, 1968; Fairbridge, 1969, 1970, 1971; Bennacef and others, 1971). The Late Ordovician Table Mountain tillite of South Africa has been known for some time (Du Toit, 1954), but has been dated only recently (Cocks and others,

[1] The inferred geochronologic unit *glacial episode* is applied here to specific periods of the entire geological record, rather than only to climatic episodes of the Quaternary, as implied by the Code of Stratigraphic Nomenclature (Am. Comm. Strat. Nomenclature, 1961).

TABLE 1. ADEQUATELY DATED EPISODES OF GLACIATION

Glacial episode	Absolute age	Location
Late Cenozoic*	0 to 14 m.y.	Affected all continents and sea-floor sediments (Evans, 1971)
Permian-Carboniferous*	255 to 340 m.y.	Australia
		South Africa
		Madagascar
		South America
		Falkland Islands
		Antarctica
		India—Pakistan
	235 to 320 m.y.	Siberia
Silurian-Ordovician*	410 to 470 m.y.	North and South Africa
		Argentina, Brazil (?), Bolivia (?)
		Spain, Normandy, Russia (?)
		Nova Scotia (?), Yukon Territory
Eocambrian	650 ± 50 m.y.	Australia
	600 ± 30 m.y.	Norway*, Sweden*
	600 ± 30 m.y.	East Greenland*
	600 to 640 m.y.(?)	Spitzbergen*
600 to 640 m.y.(?) or	660 to 680 m.y.(?)	European Russia†
	560 to 630 m.y.	Normandy
	570 to 600 m.y.	Newfoundland
	600 m.y. plus	Brazil
	620 to 650 m.y.	Algeria*
	570, 600, 650 m.y.	China*
Infracambrian I	740 to 750 ± 40 m.y.	Australia
	715 m.y.(?), 810 m.y.(?)	European Russia†
	747 to 810 m.y.	Siberia
	750 m.y.	China*
	800 ± 50 m.y.	British Columbia
	820 m.y.	Southern Appalachians
	825 m.y.	Washington
	750 ± 50 m.y.	Katanga
Infracambrian II	950 ± 50 m.y.	Katanga and Lower Congo (Gabon, ?Angola)
	950 m.y.	Russian Tien Shan*
	950 m.y.	China*
Gowganda Glaciation	2,288 ± 87 m.y.	Eastern Canada
		North-central United States and Wyoming

* These age assignments are based on stratigraphic data mainly, but the strata have been correlated with radiometrically dated sequences mostly on the same continent or the age is based on the fossil record of overlying or interbedded strata.

† For details of these uncertain ages, see text.

1970). Paleozoic glacial rocks in Ethiopia have not been dated (Dow and others, 1971). Lower Paleozoic pole positions indicate high paleolatitude for Africa (McElhinny and others, 1968; McElhinny and Luck, 1970; Creer, 1970). Lower Paleozoic high paleolatitude for South America (Creer, 1970) appears to correlate with evidence of Ordovician mountain glaciation from Argentina (Harrington, 1956) and Silurian (?) tilloids of Brazil and Bolivia (De Oliveira, 1956; Schwarzbach, 1963; Beurlen, 1970). Schenk (1971, 1972), who suggested the possibility of a Late Ordovician to Early Silurian glaciation for Nova Scotia, also summarized the French language literature regarding Ordovician glacial evidence from northern Africa, Spain, and Normandy. Ziegler (1959) reported an Ordovician tillite from the Yukon Territory. The original Russian evidence for a Silurian-Ordovician glaciation (Schwarzbach, 1963) has apparently not been reviewed recently, but some of these deposits (for example, Novaya Zemlaya) may have been assigned to the Permian-Carboniferous. The range of Silurian-Ordovician glaciations is estimated as 410 to 470 m.y.

Late Precambrian Glaciations (600 to 1,000 m.y.)

Evidence of late Precambrian glaciation has been reported from all continents except Antarctica (Harland, 1964a; Crawford and Daily, 1971). Many of the deposits are adequately dated; others are not. Although the I.U.G.S. Subcommission on Precambrian Stratigraphy recommends that the terms "Eocambrian" and "Infracambrian" be dropped (De Villiers, 1970), they are retained here (Table 1) because the galactic model proposed below predicts three late Precambrian glaciations. The term "Eocambrian" is used for glaciations immediately preceding the Cambrian of 600 ± 50 m.y., and "Infracambrian I" is used for evidence of an older glaciation which preceded the Cambrian by approximately 160 m.y. (760 ± 50 m.y.). "Infracambrian II" is defined for possible glacial evidence centering around 940 ± 50 m.y.[2] Such terminology is advisable (*see* Rankama, 1970) because rapidly accumulating age determinations suggest that there was more than one late Precambrian glacial episode. In addition, present Rb/Sr methods on shale and K/Ar determinations on glauconite are often inconsistent with known stratigraphy which may be due, in part, to metamorphism, and thus a liberal error of ±50 m.y. has been used in defining the late Precambrian glacial episodes. Some K/Ar dates on related igneous rocks have also been put in doubt by subsequent Rb/Sr determinations. Late Precambrian paleomagnetic data have not been reviewed and critically evaluated on a world-wide basis and are difficult to interpret, due to scarcity of data and incomplete knowledge of stratigraphic relations in some areas.

Australia. The stratigraphy of the late Precambrian Australian tillite is summarized by Dunn and others (1971), who assigned an age of 650 m.y. to the Egan-Marinoan glaciation, and 740 to 750 m.y. to the Moonlight Valley–Sturtian glaciation, on the basis of extensive Rb/Sr determinations on shale (Compston and others, 1966; Compston and Arriens, 1968; Compston and Taylor, 1969; Cooper and others, 1971). The latter authors however, also reported seemingly contradictory Rb/Sr isochrons of 790 m.y., approximately 600 m.y., and 950 to 1,200 m.y. from Australian shale of uncertain stratigraphic positions and associated with glaciogene deposits.

Europe and Greenland. Banks and others (1969) suggest an age of 600 m.y. for the the Mortensnes (Varangian; equivalent term, Varegian, see Geijer, 1963) tillite of Finnmark, Norway (Reading and Walker, 1966), on the basis of sedimentation rates of the overlying Cambrian and Ordovician strata. The clearly equivalent tillite in Sweden is discussed by Geijer (1963). Spencer (1969) applied Banks' method to the upper tillite of the East Greenland succession (Berthelsen and Noe-Nygaard, 1965), and also estimated an age of 600 m.y. In both conformable successions, the base of the Cambrian is taken as 570 m.y., and the error should therefore be less than 30 m.y., including the uncertainty of the base of Cambrian. The age of the Spitzbergen tillite has not been radiometrically determined, but it has been correlated to the Norwegian and Greenland tillite (Winsnes, 1965). Chumakov (1968) assigns the Spitzbergen tillite to the middle Vendian (also spelled Wendian, Am. Geol. Inst.). No age dates are available for the Dalradian tillite in Scotland (Spencer, 1971); however, the tillite has been shown to be younger than 1,000 m.y. (Leggo and Spencer, 1969). In Normandy the upper Brioverian, which includes the Granville tillite, has been dated as 630 to 560 m.y. (Leutwein and Sonet, 1966; Leutwein, 1968).

Russia. The age, and partly the stratigraphic subdivisions and correlations of the various late Precambrian glaciogene deposits of Russia, appear to be in a state of flux (Zubtsov and Zubtsova, 1966; Bessonova and Chumakov, 1967; Kirsanov, 1968; Furduy, 1968; Zhuravleva and Chumakov, 1968; Postnikova and Revenko, 1969). Keller and others (1968) assign most late Precambrian tillite of European and Asiatic Russia to the Vendian (675 ± 25 to 570 ± 10 m.y.) and many authors note that tillite and tilloid occur in the middle of the succession. Salop (1968) cited the following

[2] The definitions of Eocambrian and Infracambrian I and II are patterned after present French usage.

glauconite ages for tillite and tilloid: Rybachy of Kola Peninsula, 715 m.y.; Volhynian of Russian Platform, 810 m.y.; Chingasan of Siberia, 747 to 810 m.y. Chumakov and Cailleux (1971) assigned an age of approximately 660 to 680 m.y. to the glaciation of European Russia on the basis of glauconite dates in the Poljud Range (glauconitic rocks underlying and overlying tillite yield ages of 680 to 686 m.y. and 620 m.y., respectively) and a maximum age of 640 to 660 m.y. of post-tillite dolerite and basalt of the northern Ukraine and Belorussia. The lower tillite of Tien Shan is near the base of the late Riphean succession. The latter has been assigned an age range of 950 ± 50 to 675 ± 25 m.y. (Keller and others, 1968).

China. Saito (1969) listed five glaciations for the late Precambrian of China: Huishan glaciation, 950 m.y. (Infracambrian II); Hsiho glaciation, 750 m.y. (Infracambrian I); Wuhsingsham glaciation, 650 m.y. (Eocambrian); Nantou glaciation, 600 m.y. (Eocambrian); and Cambrian glaciation, 570 m.y. (Eocambrian). Although these ages are not confirmed by radiometric dating, they seem to fit the definitions of late Precambrian glaciations used in this study which are based on data from all other continents. The Siberian and Russian Tien Shan glaciations appear to correlate with some of Saito's age estimates.

North America. In Canada, the glaciogene Toby Conglomerate (Aalto, 1971) comprises the base of the Windermere system and may be assigned an age of 800 ± 50 m.y. Harrison (1972) concluded that the Toby Conglomerate could not be older than 850 m.y., which marks the end of Belt time. The Belt and Windermere systems are separated by a major unconformity, and Douglas and others (1970, Chart 3) considered the Toby Conglomerate to be younger than 764 m.y. This date has been invalidated by Ryan and Blenkinsop's (1971) Rb/Sr determinations on the Hellroaring granite.

Hughes and Brückner (1971) tentatively assigned an age of 600 to 570 m.y. to the late Precambrian rocks, including the Conception Group of southern Newfoundland, on the basis of K/Ar radiometric dating of underlying granite. The Conception Group includes tillite and glacio-marine deposits (Brückner and Anderson, 1971).

Crittenden (1972, written commun.) advises of two more relevant age determinations. Volcanic rocks associated with diamictite in northeast Washington yield an age of 825 m.y. Tillite of the southern Appalachians unconformably overlies rhyolitic volcanic rocks which have been shown to be 820 m.y. old (Rankin and others, 1969). Crittenden and others (1972) used the upper Precambrian diamictites in correlating western North American strata.

Africa. The "Grand Conglomérat" of the Katanga succession (Cahen and Lepersonne, 1967) is thought to be of true glacial origin, and Cahen (1970, Fig. 2) estimated its age as 950 (±50) m.y. on the basis of numerous radiometric ages of igneous rocks and detailed correlations. The "Petit Conglomérat," which is considered to be a tillitic facies of the Middle Kundelungu, 2,000 m above the "Grand Conglomérat," may be estimated as approximately 750 (±50) m.y. on the basis of the same correlations. Cahen and Snelling (1966) indicated that the upper tillite (Tillite Supérieur) of Gabon, Congo (Braz), Lower Congo (Léo), and Angola is older than 740 m.y. Clifford (1970) correlated the lower Congo succession with the Katanga system.

The base of the Eocambrian tillite in Algeria 1,000 m stratigraphically below the Silurian-Ordovician glaciogene deposits, is tentatively estimated as 620 to 650 m.y. (Biju-Duval and Gariel, 1969).

South America. In Brazil, tillite and fluvioglacial deposits of the Lavras series are separated from the overlying Bambui Group by an erosional surface (De Oliveira, 1956). These latter, practically unmetamorphosed sediments have been shown to be 600 m.y. old by Rb/Sr whole-rock determinations, supported by K/Ar and common lead ages (Cordani and others, 1968). The Lavras glaciogene sediments are thus older than 600 m.y. and younger than 750 m.y. (Grabert, 1967).

Older Precambrian Glaciations

Gowganda Glaciation. The best documented middle Precambrian glaciation is the Huronian Gowganda tillite (Schenk, 1965; Casshyap, 1969; Young, 1970) of Canada and the United States, which is dated as 2,288 ± 87 m.y. (Fairbairn and others, 1969). Huronian paleolatitude determinations range from 50° to 87° (Irving, 1964; Girdler, 1964; Symons, 1966, 1967).

Other Precambrian Glaciations. The lower tillite (Tillite Inférieur or Tillite des Monts Bamba) of Western Congo, Northern Angola, and Southern Gabon (Cahen and Lepersonne, 1967), may be older than 1,000 m.y. and younger than 1,740 ± 75 m.y. (Bonhomme and others, 1966). This tillite disconformably to unconformably overlies the Sansikwa rocks. The argillite strata of the supposed lateral equivalents of the Sansikwa yield the 1,740 m.y. isochron (Appendix, Cahen and Snelling, 1966). The Tillite Inférieur and Tillite Supérieur are stratigraphically separated by the 1,500 to 2,000 m Haut Shiloango–Louila succession.

Holmes (1965) lists additional and much more problematic Precambrian glaciations as follows: Cudapah, India (?); Urundi (NE. of L. Tanganyika); Botnian Group, Finland; Transvaal Group, Pretoria; Ketelidian Group, southeast Greenland; Witwatersrand Group, Transvaal; and Timiskaming Group, Michigan. The corresponding orogenic cycles have been assigned the following ages: Cudapah, 500 to 1,600 m.y. (Aswathanarayana, 1968); Burundian, 1,300 to 2,100 m.y. (Cahen, 1970); Ketelidian Group, prior to 1,500 to 1,600 m.y. (Larsen and Møller, 1968); Sveco-Fennian, 1,650 to 2,300 m.y. (Kratz and others, 1968); Transvaal system, 1,950 to 2,350 m.y. (Clifford, 1970); Witwatersrand Group, 2,350 to 2,600 m.y. (Clifford, 1970); and Timiskaming, older than 2,560 m.y. (Wanless and others, 1972).

MILKY WAY GALAXY

The Milky Way Stellar System is a spiral galaxy approximately 30 kiloparsec (kpc) in diameter (1 kiloparsec = 3,258 light yrs). The solar system is situated within the spiral arms at 10 kpc from the galactic center and revolves around this center and close to the galactic plane in an elliptical orbit. The circular orbital velocity of the sun is taken as 250 km/sec. The eccentricity of the solar orbit is such that the galactocentric distance varies from approximately 10 kpc to 11.7 kpc. One complete revolution is commonly termed a cosmic year. The orbital positions nearest to and farthest from the galactic center are called perigalacticum and apogalacticum, respectively (that is, apsides). The sun is now approaching its perigalactical position and will reach it within 8 ± 4 m.y. Estimates of the sun's position with regard to its perigalacticum range from 4 m.y. to over 12 m.y. The length of the present cosmic year is approximately 274 m.y. While revolving around the galactic center, the sun vibrates perpendicular to the galactic plane with a periodicity of 77 m.y., but the distance of the sun to the galactic plane never exceeds 0.1 kpc. The circular orbital velocities of hydrogen clouds as a function of galactocentric distance are known within certain limits from radio-astronomical observation. These astrophysical observations are mainly restricted to galactocentric distances of less than 10 kpc. Present galactic knowledge indicates that the solar system originated outside of its spiral arm and probably condensed from a dust cloud near the edge of the spiral arms of the galaxy. The original solar orbit was most likely characterized by high eccentricity and a steep inclination to the galactic plane, as indicated by orbital parameters of young stars. On the basis of continuously improving astrophysical observations and astronomical knowledge, various mass models of the galaxy have been proposed. The quantitative aspects of this summary are based largely on Innanen (1966a, b, c, d, e; Innanen and Fox, 1967). In this study it is necessary to identify the various cosmic years and an open-ended terminology is proposed whereby the present cosmic year is termed Pr, the one preceding it Pr-1, and the one before it Pr-2, and so forth (Table 2).

GALACTIC MODEL

On the basis of the observed circular orbital velocities given by Innanen (1966c), the central galactic force per unit mass (*see* Appendix) within the galactic plane and as a function of the galactocentric distance is calculated (Fig. 1). The central galactic force is at a maximum near the edge of the spiral arms and tends toward minimal values at the galactic center and intergalactic space. The degree of validity of this curve (Fig. 1) is discussed in the Appendix. The central galactic force experienced by the solar system at the present time is taken as unity.

In this model, due to the interaction of the elliptical orbit of the sun and the circular distribution of the central galactic force within the galactic plane, the solar system experiences a varying central galactic force during its galactic orbit.

A central galactic force curve versus time for the solar system is calculated on the basis of the Kepler-Newtonian equations of motion and the following model:

1. The central galactic force per unit mass in the galactic plane is varying as a function of

TABLE 2. GALACTIC MODEL DURATION OF COSMIC YEARS

Cosmic year	Decrease of duration (approximately)	Duration (approximately)	End of cosmic year–Perigalacticum
Present cosmic year (Pr)	35 m.y.	280 m.y.	minus 8 ± 4 m.y.
Pr-1	27 m.y.	315 m.y.	280 m.y.
Pr-2	21 m.y.	342 m.y.	595 m.y.
Pr-3	15 m.y.	363 m.y.	937 m.y.
Pr-4	11 m.y.	378 m.y.	1,300 m.y.
Pr-5	6 m.y.	389 m.y.	1,678 m.y.
Pr-6	3 m.y.	395 m.y.	2,067 m.y.
Pr-7	1 m.y.	398 m.y.	2,462 m.y.

Error at least ± 0.1 cosmic yrs.

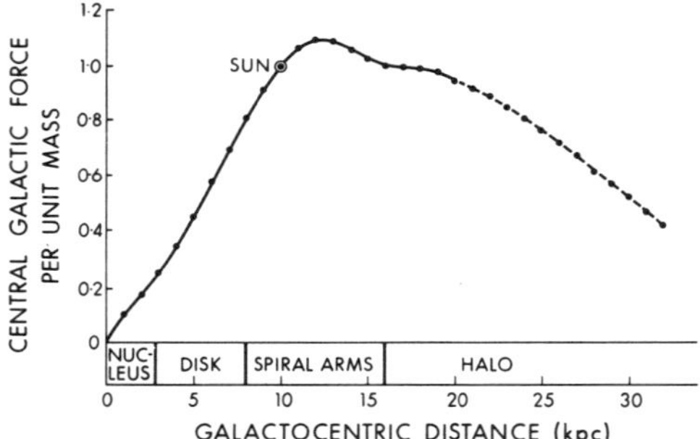

Figure 1. Central galactic force per unit mass plotted as a function of galactocentric distance. The degree of validity of this function is discussed in the Appendix. 1 kiloparsec = 3,258 light yrs. Dashed curve = extrapolation of Innanen model.

galactocentric distance as shown in Figure 1 (*see* Appendix). The model utilizes only the portion of the curve with a galactocentric distance of greater than 10 kpc.

2. The duration of the present cosmic year is approximated to 280 m.y. and the present time is equated to the perigalactical position.

3. In addition to rotating around the galactic center in an elliptical orbit, the sun is assumed to be spiralling inward with a radial contraction velocity component of 1.4 km/sec.

4. The inward spiralling necessitates that the length of the cosmic years decreased with the passage of time. From Pr-7 to Pr the duration of the cosmic years is assumed to have decreased from 400 m.y. to 280 m.y., according to a log-log relation (Table 2).

5. The eccentricity of the solar orbit is also assumed to have decreased from a maximal value in a straight line function (Table 3).

The solar system is assumed to have originated outside of the maximum (Fig. 1) of the central galactic force. Since the sun is spiralling inward, it crossed this maximum, and the relations of the apsides and minima of the central galactic force are as follows (Fig. 2; Table 4): zero to 280 m.y., perigalactica are minima; 280 to 1,120 m.y., both perigalactica and apogalactica are minima; and 1,120 to 3,060 m.y., apogalactica are minima. Astronomical data permit the construction of a series of galactic models, but models employing other and different variables do not compare as well with present knowledge of glacial episodes. The present model neglects the solar orbital motion perpendicular to the galactic plane. This motion would conceivably result in superimposed, minor oscillations of approximately 40 m.y. on the curve shown in Figure 2.

TABLE 3. GALACTIC MODEL ECCENTRICITY OF SOLAR ORBIT

Cosmic year	Eccentricity
Pr	0.080
Pr-1	0.133
Pr-2	0.186
Pr-3	0.239
Pr-4	0.239

This model is not based on quantitative astronomical and astrophysical data prior to approximately 350 m.y., but it is thought to approximate some qualitative, though speculative, considerations of galactic dynamics. Thus this galactic model must be considered as unsubstantiated prior to 350 m.y. (Fig. 2).

COMPARISON OF THE GALACTIC MODEL WITH GLACIAL EPISODES

The galactic model predicts twelve minima of the central galactic force experienced by the

TABLE 4. COMPARISON OF ESTABLISHED GLACIAL EPISODES AND CENTRAL GALACTIC FORCE MINIMA OF THE GALACTIC MODEL

Glacial episodes established in geological history		Galactic model (probable error at least ± 0.1 cosmic yrs)	
Name	Geological time-scale	Galactic time-scale	Apsides
Late Cenozoic	0 to 14 m.y.	minus 8 ± 4 m.y.	Perigalacticum
Permian-Carboniferous	235 to 340 m.y.	280 m.y.	Perigalacticum
Silurian-Ordovician	410 to 470 m.y.	437 m.y.	Apogalacticum
Eocambrian	mean: 616 m.y., σ = ± 30 m.y.	595 m.y.	Perigalacticum
Infracambrian I	mean: 777 m.y., σ = ± 40 m.y.	766 m.y.	Apogalacticum
Infracambrian II	circa: 950 ± 50 m.y.	937 m.y.	Perigalacticum
		1,119 m.y.	Apogalacticum
		1,489 m.y.	Apogalacticum
		1,873 m.y.	Apogalacticum
Gowganda	2,288 ± 87 m.y.	2,265 m.y.	Apogalacticum
		2,660 m.y.	Apogalacticum
		3,060 m.y.	Apogalacticum

σ = standard deviation.

solar system for the past 3 b.y. (Table 4; Fig. 2). Seven of these minima correlate with the mid-point or mean age of seven glacial episodes with errors of a few million years to a maximum of 75 m.y. in late Precambrian time. The degree of clustering of dated glacial evidence around central galactic force minima is illustrated in Table 5 and is apparent on Figure 2. The presently documented duration of glacial episodes is also summarized (Table 5). An average misfit of ±13 m.y. is indicated between the mid-point of glacial episodes and seven apsides of the galactic model. Such an average error is much less than the combined probable error of the galactic model and the geological time-scale, plus the fact that glaciation could occur at near minimal values of the central galactic force as implied by the definition of glacial episodes. During the late Cenozoic glaciation, high-latitude glaciation started in the Miocene 14 m.y. ago, and the sun is presently 8 ± 4 m.y. from the perigalacticum.

The poorest correlations are exhibited by Chumakov and Cailleux's (1971) age assignment of the late Precambrian glaciation of European Russia (misfit: 75 m.y.) and the Australian Egan-Marinoan glaciation (misfit: 55 m.y.). Out of thirty dated glacial localities only the two above-mentioned glaciations correlate with near maxima of the central galactic force (Table 1 and Fig. 2).

The galactic model predicts five more Precambrian glaciations as follows (Table 4; Fig. 2): 1,120 m.y., 1,490 m.y., 1,870 m.y., 2,660 m.y., and 3,060 m.y., all at least ±0.1 cosmic yrs. This model can thus be further tested by obtaining radiometric dates from the older Precambrian glaciogene deposits listed by Holmes (1965).

DISCUSSION

Although the comparison of the galactic model with glaciations of geological history appears promising, it is advisable to attempt to discuss the degree of validity of the galactic model. Quantitative knowledge of the Milky Way and other galaxies is limited. Subject to the limitations discussed in the Appendix, and considering the range and magnitude of galactic phenomena hereto not completely understood, it may be concluded that the degree of validity of the galactic model is of the same order of magnitude as that of late Precambrian geochronology. The paleoclimatic history of the past 1,000 m.y. is incompletely known and the solar galactic parameters are probably in doubt by at least 10 percent, as indicated by the even figures of 10 kpc and 250 km/sec. In other words, two very fuzzy curves are being compared and the degrees of agreement between them is surprising, although within certain limits the model has been chosen in such a way as to maximize

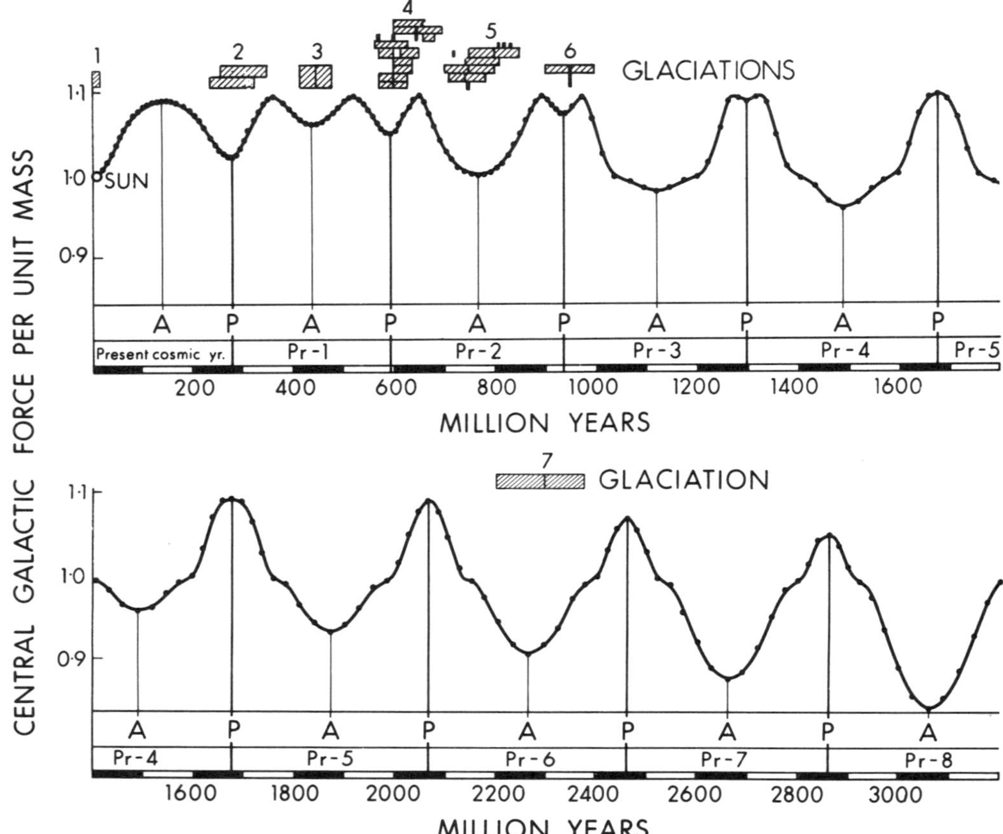

Figure 2. Comparison of galactic model with well-documented glacial episodes. Central galactic force per unit mass is plotted as a function of time, as experienced by the solar system. A = Apogalacticum. P = Perigalacticum. Cosmic years are designated as Pr (present cosmic yr), and Pr-1, Pr-2, and so forth. Glacial Episodes: 1. Late Cenozoic; affected all continents and seafloor sediments. 2. Permo-Carboniferous; Australia, South Africa, Madagascar, South America, Falkland Islands, Antarctica, India-Pakistan, Siberia. 3. Siluro-Ordovician; Africa, South America, parts of Europe, Northern Canada?. 4. Eocambrian; Australia, Europe, Greenland, Spitzbergen, Newfoundland, South America, North Africa, China. 5. Infracambrian I; Australia, European Russia and Siberia, China, North America, Africa. 6. Infracambrian II; Africa, Asiatic Russia, China. 7. Gowganda; North America.

positive correlation. The comparison is semi-quantitative since only maxima and minima are being compared. The fact that the Infracambrian II glacial episode correlates with a comparatively minor minimum of the galactic curve (Fig. 2) may be due to details of the present galactic model, since a slight increase of the eccentricity or a minor decrease of the radial contraction velocity component would rectify this point.

The reason for the apparent correlation of this galactic model with geological glaciations is not understood, but a causal relation appears plausible. Steiner (1967) argued that the central galactic force per unit mass may somehow be proportional to the Newtonian gravitational "constant" (G). A systematic and periodic variation of G would result in proportionally varying energy output of the sun. However, a spacial or spacio-temporal variation of G is irreconcilable with generally accepted physical theory. According to Bell (1953) and Willet (1949) a varying energy output of the sun offers the only hope for a satisfactory explanation of world-wide changes of climate, whatever their length and magnitude. Hoyle and Narlikar (1971, 1972) discussed a possible spacio-temporal variation of proper mass and also touched upon some of the resulting geophysical and astrophysical consequences. According to them there appears to be a need for such a theory, due to contradictory extragalactic observations and their interpretation.

This study appears to clarify somewhat the thorny problem of world-wide late Precambrian glaciations and within perhaps ±50 m.y. the glaciogene deposits may indeed facilitate valid intercontinental time-stratigraphic correlations, as suggested by Harland (1964a, b), Dunn and others (1971), and Crittenden (1971). Within a given glacial episode, Crowell and Frakes's (1970) model may be valid, and glacial centers on different continents may vary in time due to changing paleolatitude which is thought to be caused by continental drift. In the future, this model may be utilized to estimate a probable error of possible intercontinental correlations of older Precambrian glaciations.

Assuming that Figure 2 reflects a valid paleoclimatic curve, the galactic winters straddle the perigalactical positions in the Phanerozoic and center around the apogalactica in the Precambrian, according to this model. The Silurian-Ordovician and Infracambrian II glacial episodes are exceptions to this rule and are transitional consequences of the model (Fig. 2; Table 4). The Precambrian-Phanerozoic boundary (Eocambrian glacial episode) marks the midpoint of this change of regime. According to Kepler's second law, the orbital velocity of the sun is fastest at the perigalactica and slowest at the apogalactica. In this model, the Phanerozoic is thus characterized by long warm periods representing the near apogalactical positions (galactic summers) and comparatively shorter, cool periods representing the perigalactical positions (galactic winters) (Fig. 2). This is in good agreement with general paleoclimatic data. The reverse is true for the Precambrian when shorter, warm periods alternate with longer, cool periods according to this model.

In conclusion, we feel that the above facts and uncertainties should encourage geologists to pursue possible galactic-geological correlations further, until a clearer picture of their validity emerges.

TOWARD A SCHEME OF MULTILATERAL CAUSES OF ICE AGES

The galactic model is thought to provide the basic favorable conditions for glacial episodes in intervals of 160 to 400 m.y. At these periodic and partly episodic intervals when radiation from the sun may be at a minimum due to the galactic variables, other causes which operate continuously become effective and reinforce the galactic trigger mechanism.

TABLE 5. TIME DIFFERENCE BETWEEN APSIDES OF THE GALACTIC MODEL AND MIDPOINT OF GLACIAL EPISODES

	Time difference	Duration of glacial episode
Late Cenozoic	9 m.y.	14 m.y. plus
Permian-Carboniferous	8 m.y.	105 m.y.
Silurian-Ordovician	3 m.y.	60 m.y.
Eocambrian	21 m.y.	110 m.y.
Infracambrian I	11 m.y.	110 m.y.
Infracambrian II	13 m.y.	unknown
Gowganda	23 m.y.	unknown
Mean = 12.6 m.y.		

It is not the purpose of this study to review the numerous ice-age hypotheses which have been proposed in the past. However, most of them rely on the changing distribution of continents (including mountains) and oceans with regard to the rotational poles which, in turn, supposedly affect precipitation and (or) atmospheric and oceanic circulation. Such mechanisms probably control only the size, thickness, and rate of spreading of ice sheets.

The basic trigger mechanism should have a periodicity of the order of 10^7 to 10^8 yrs and the galactic variables discussed above represent the only known external periodic phenomena of this magnitude. The solar system variables proposed by Milankovitch (1938) rank next in order of magnitude (10^4 to 10^5 yrs). The correlation of the summer insolation curve with repeated Pleistocene ice advances appears to be well established (Evans, 1971; Mitchell, 1965). Periodicities of 40,000, and possibly 80,000 yrs, are documented and correlated. During a given glacial advance within a glacial episode, high paleolatitude and (or) high paleoaltitude must also be an important factor, as suggested by Crowell and Frakes (1970) in somewhat different context. Climatic variations of the order of 10^2 and 10^3 yrs have recently been reported by Bray (1972) for the past 20,000 yrs, but their causes are uncertain (also see Willet, 1949). Table 6 tabulates a possible scheme of multilateral causes of ice ages listed in order of importance. These causes may have to coincide in order to initiate a glacial episode.

TABLE 6. SUGGESTED MULTILATERAL CAUSES FOR THE INITIATION OF GLACIAL EPISODES

(1) Galactic variables due to solar orbital motions parallel to the galactic plane	Provides periodicities of 1.6 to 4 $\times$ 10^8 yrs	Basic trigger mechanism (glacial episode)
(2) Galactic variables due to solar orbital motions perpendicular to the galactic plane	Provides periodicities of 4 $\times$ 10^7 yrs	Reinforcing mechanism which operates all the time, but which by itself does not cause glaciations (subdivision of glacial episode)
(3) Solar system variables	Provides periodicities of 4 to 8 $\times$ 10^4 and possibly 10^5 yrs	Reinforcing mechanism which operates all the time, but which by itself does not cause glaciations (glacial advance)
(4) Changing distribution of continents (including mountains) and oceans affecting precipitation and (or) atmospheric and oceanic circulation	Probably mainly controls size, thickness, and rate of spreading of ice sheets	Contributing mechanism which is mainly of importance if mean annual temperatures are below a certain threshold value
(5) High paleolatitude and (or) high paleoaltitude	Provides locality of glaciation	Due to maximum development of glaciogene deposits in these localities provides maximum chance for such deposits to be preserved in the geological record
(6) Varied, unknown causes	Variations of the order of 10^2 and 10^3 yrs	Minor superimposed climatic variations

Listed in order of importance.

ACKNOWLEDGMENTS

Helpful critical comments were provided by H.A.K. Charlesworth, M. D. Crittenden, J. C. Crowell, and J. A. Westgate. Steiner also wishes to acknowledge fruitful conversations with R. A. Burwash and R. E. Folinsbee. Partial financial support for this study was provided by the National Research Council of Canada.

APPENDIX. GALACTIC MODEL (THE TWO-BODY PROBLEM)

In the simplest model (the two-body problem), the centripetal force per unit mass F' is given by $F' = \omega^2 r$, where ω is angular circular velocity; r is orbital radius. In the case of the galactic model (Fig. 1), the central galactic force per unit mass $K'(R)$ which is a function of the galactocentric distance R is given by

$$K'(R) = V_c^2 R$$

where V_c is the circular orbital velocity observed at a given galactocentric distance and which is reconciled with and extended by mass models of the galaxy.

The following vector equation is assumed to be valid:

$$K(R) = L = \frac{d}{dt}(IV_c)$$

where L is torque about the galactic rotational axis; I is rotational inertia, and $K(R)$ is the central galactic force. The utilization of the two-body problem, including the above-mentioned assumption, represents the major shortcoming of this model, which has been adopted for reasons of simplicity. Some physicists object to it, but the model may not be as invalid as it appears.

In Newtonian gravitational theory, the two-body problem is applicable only outside of the galaxy since it assumes the galactic mass to be concentrated in the center. Within the galaxy a much more complex relation applies and the central part of Figure 1 with galactocentric distances of less than 9 kpc is invalid. The distribution of the central galactic force per unit mass in the nucleus and the disc must be different than shown in Figure 1. In the spiral arms and the halo, the two-body problem becomes increasingly more valid with increasing galactocentric distance, as outlined below.

According to Innanen (1966e) the sun was never closer to the galactic center than 9.5 kpc. According to this model the sun is postulated to have been 20 kpc from the galactic center in the early Precambrian and migrated to its present 10 kpc galactocen-

TABLE 7. MULTIPLES OF SOLAR MASS IN A COLUMN OF UNIT AREA PERPENDICULAR TO THE GALACTIC PLANE OF THE GALAXY AND AS A FUNCTION OF GALACTOCENTRIC DISTANCE

Galactocentric distance (kpc)		Solar masses/parsec^2
0.2		13,782.0
0.4		7,470.0
0.5		5,612.0
0.6		4,245.0
0.8		2,707.0
1.0		2,174.0
1.5		1,320.0
2.0		960.0
2.5		792.0
3.0		679.0
3.5		597.0
4.0		527.0
5.0		412.0
6.0		321.0
7.0		247.0
8.0		180.0
9.0		135.0
10.0	Solar Neighborhood	93.2
11.0		59.3
12.0		32.8
13.0		14.3
14.0		6.6
15.0		3.4
16.0		2.2
18.0		0.6
20.0		0.06

Data from Innanen, 1966c.

tric distance. The galactic model is a two-dimensional model restricted to the galactic plane since the solar orbital motions perpendicular to the plane have been neglected for the time being for reasons of simplicity. The immense central concentration of the mass of the galaxy near the galactic plane and within galactocentric radii of less than 9 kpc is illustrated by Innanen's (1966c) data (Table 7), which justifies the utilization of the two-body problem. It may be stated that in the two-dimensional galactic model and the present solar galactocentric distance, the two-body problem is valid by more than 90 percent. In three dimensions, the halo contains 31 percent of the mass of the galaxy (Innanen, 1966c) and thus, within the spiral arms the validity of a three-dimensional galactic model, would be somewhat less than 70 percent.

On the basis of the galactic-mass distribution, the validity of the galactic model proposed herein may therefore be almost 80 percent so far as the utilization of the two-body problem is concerned. This estimate is the same as that given by Ogorodnikov (1965) for equivalent equations applicable to the solar neighborhood of 10 kpc galactocentric distance. The complete galactic model, as used herein, cannot be verified quantitatively prior to 350 m.y. by means of presently available astronomical and astrophysical data. The model stands and falls depending on detailed correlations with geological and other earth science data.

REFERENCES CITED

Aalto, K. R., 1971, Glacial marine sedimentation and stratigraphy of the Toby Conglomerate (Upper Proterozoic), southeastern British Columbia, northwestern Idaho and northeastern Washington: Canadian Jour. Earth Sci., v. 8, p. 753–787.

American Commission of Stratigraphic Nomenclature, 1961, Code of stratigraphic nomenclature: Am. Assoc. Petroleum Geologists Bull., v. 45, p. 645–665.

Aswathanarayana, U., 1968, Metamorphic chronology of the Precambrian provinces of South India: Canadian Jour. Earth Sci., v. 5, p. 591–600.

Banks, N. L., Edwards, M. B., and Reading, H. G., 1969, Geol. Soc. London Proc., no. 1657, p. 191–192.

Bell, B., 1953, Solar variation as an explanation of climate change, *in* Shapley, H., ed., Climatic change: Cambridge, Harvard Univ. Press, p. 123–136.

Bennacef, A., Beuf, S., Biju-Duval, B., De Charpal, O., Gariel, O., and Rognon, P., 1971, Example of cratonic sedimentation: Lower Paleozoic of Algerian Sahara: Am. Assoc. Petroleum Geologists Bull., v. 55, no. 12, p. 2225–2245.

Berthelsen, A., and Noe-Nygaard, A., 1965, The Precambrian of Greenland, *in* Rankama, K., ed., The Precambrian, v. 2: New York, John Wiley & Sons, Inc., 454 p.

Bessonova, V. Ya., and Chumakov, N. M., 1967, Glacial sediments of the upper Precambrian of Belorussia: Akad. Nauk SSSR Doklady, v. 178, no. 4, p. 53–56 (English trans. by Am. Geol. Inst.).

Beuf, S., Bennacef, A., Biju-Duval, B., De Charpal, O., Gariel, O., and Rognon, P., 1968, Les grands ensembles sédimentaires du Paléozoique inférieur du Sahara: Soc. Géol. France Compte Rendu, fasc. 8, p. 260–263.

Beurlen, K., 1970, Geologie von Brasilien: Berlin, Stuttgart, Gebrüder Borntraeger, 444 p.

Biju-Duval, B., and Gariel, O., 1969, Nouvelles observations sur les phénomènes Glaciaires "Eocambriens" de la Bordure Nord de la Synéclise de Taoudeni, entre le Hank et le Tanezrouft, Sahara Occidental: Palaeogeography, Palaeoclimatology, Palaeoecology, v. 6, p. 283–315.

Bonhomme, M., Weber, F., and Favre-Mercuret, R., 1966, Age par la méthode rubidium-strontium des sédiments du Bassin de Franceville (République Gabonaise): Alsace Lorraine (Strasbourg), Bull. Serv. Carte Géol., v. 18, no. 4, p. 243–252.

Bray, J. R., 1972, Cyclic temperature oscillations from 0–20,300 yr BP: Nature, v. 237, p. 277–279.

Brückner, W. D., and Anderson, M. M., 1971, Late Precambrian glacial deposits in southeastern Newfoundland—A preliminary note: Geol. Assoc. Canada Proc., v. 24, no. 1, p. 95–102.

Cahen, L., 1963, Glaciations Anciennes et Derive des Continents: Soc. Géol. Belgique Annales, T. 86, p. 21–30.

—— 1970, Igneous activity and mineralisation episodes in the evolution of the Kibaride and Katangide orogenic belts of central Africa, *in* Clifford, T. N., and Gass, I. G., eds., African magmatism and tectonics: Edinburgh, Oliver & Boyd, 461 p.

Cahen, L., and Lepersonne, J., 1967, The Precambrian of the Congo, Rwanda, and Burundi, *in* Rankama, K., ed., The Precambrian, v. 3: New York, Interscience Publishers, 325 p.

Cahen, L., and Snelling, N. J., 1966, The geochronology of Equatorial Africa: Amsterdam, North-Holland Publishing Co., 195 p.

Casshyap, S. M., 1969, Petrology of the Bruce and Gowganda Formations and its bearing on the evolution of Huronian sedimentation in the Espanola-Willisville area, Ontario (Canada): Palaeogeography, Palaeoclimatology, Palaeoecology, v. 6, p. 5–36.

Chumakov, N. M., 1968, Late Precambrian glaciation of Spitsbergen: Akad. Nauk SSSR Doklady, v. 180, no. 6, p. 115–118 (English trans. by Am. Geol. Inst.).

Chumakov, N. M., and Cailleux, A., 1971, Glaciation et éolisation dans l'est et le nord de l'Europe à l'éocambrian: Rev. Géomorphologie Dynam., v. 20, no. 1, p. 1–4.

Clifford, T. N., 1970, The structural framework of Africa, *in* Clifford, T. N., and Gass, I. G., eds., African magmatism and tectonics: Edinburgh, Oliver & Boyd, 461 p.

Cocks, L.R.M., Brunton, C.H.C., Rowell, A. J., Rust, I. C., 1970, The first Lower Palaeozoic fauna proved from South Africa: Geol. Soc. London Quart. Jour., v. 125, p. 583–603.

Coleman, A. P., 1926, Ice ages, recent and ancient: London, Macmillan, 296 p.

Compston, W., and Arriens, P. A., 1968, The Precambrian geochronology of Australia: Canadian Jour. Earth Sci., v. 5, p. 561–584.

Compston, W., and Taylor, S. R., 1969, Rb-Sr study of impact glass and country rocks from the Henbury Meteorite crater field: Geochim. et Cosmochim. Acta, v. 33, p. 1037–1043.

Compston, W., Crawford, A. R., and Bofinger, V. M., 1966, A radiometric estimate of the duration of sedimentation in the Adelaide Geosyncline, South Australia: Geol. Soc. Australia Jour., v. 13 (1), p. 229–276.

Cooper, J. A., Wells, A. T., and Nicholas, T., 1971, Dating of glauconite from the Ngalia Basin, Northern Territory, Australia: Geol. Soc. Australia Jour., v. 18, pt. 2, p. 97–106.

Cordani, U. G., Melcher, G. C., and de Almeida, F.F.M., 1968, Outline of the Precambrian geochronology of South America: Canadian Jour. Earth Sci., v. 5, p. 629–632.

Crain, I. K., and Crain, P. L., 1970, New stochastic model for geomagnetic reversals: Nature, v. 228, p. 39.

Crain, I. K., Crain, P. L., and Plaut, M. G., 1969, Long period Fourier spectrum of geomagnetic reversals: Nature, v. 223, p. 283.

Crawford, A. R., and Daily, B., 1971, Probable non-synchroneity of Late Precambrian glaciations: Nature, v. 230, p. 111–112.

Creer, K. M., 1970, A review of palaeomagnetism: Earth-Sci. Rev., v. 6, p. 369–466.

Crittenden, M. D., Jr., 1971, Tillites as potential time lines in the Precambrian: Geol. Soc. America, Abs. with Programs (Ann. Mtg.), v. 3, no. 7, p. 534.

Crittenden, M. D., Jr., Stewart, J. H., and Wallace, C. A., 1972, Regional correlation of upper Precambrian strata in western North America: 24th Internat. Geol. Cong., v. 1, p. 334–341.

Crowell, J. C., and Frakes, L. A., 1970, Phanerozoic glaciation and the causes of ice ages: Am. Jour. Sci., v. 268, p. 193–224.

Denton, G. H., and Armstrong, R. L., 1969, Miocene-Pliocene glaciation in southern Alaska: Am. Jour. Sci., v. 267, p. 1121–1142.

Denton, G. H., and others, 1970, Late Cenozoic glaciation in Antarctica: Antarctic Jour. U. S., v. 5, p. 15–21.

De Oliveira, A. I., 1956, Brazil, *in* Jenks, W. F., ed., Handbook of South American geology: Geol. Soc. America Mem. 65, 378 p.

De Villiers, J., 1970, Report of the I.U.G.S. subcommission on Precambrian stratigraphy: I.U.G.S. Geol. Newsletter, v. 1969, no. 4, p. 317–320.

Douglas, R.J.W., Gabrielse, H., Wheeler, J. O., Scott, D. F., and Belyea, H. R., 1970, Geotectonic correlation chart for Western Canada, Chart III, *in* Douglas, R.J.W., ed., Geology and economic minerals of Canada: Canada Geol. Survey Econ. Geology Rept. no. 1.

Dow, D. B., Beyth, M., and Hailu, T., 1971, Palaeozoic glacial rocks recently discovered in Northern Ethiopia: Geol. Mag. 108 (1), p. 53–60.

Dunn, P. R., Thomson, B. P., and Rankama, K., 1971, Late Precambrian glaciation in Australia as a stratigraphic boundary: Nature, v. 231, p. 498–502.

Du Toit, A. L., 1954, The geology of South Africa (3rd ed.): Edinburgh, S. H. Haughton.

Evans, P., 1971, Towards a Pleistocene time-scale, *in* The Phanerozoic time-scale, a supplement, Pt. 2: Geol. Soc. London Spec. Pub. no. 5, 356 p.

Fairbairn, H. W., Hurley, P. M., Card, K. D., and Knight, C. J., 1969, Correlation of radiometric ages of Nipissing diabase and Huronian metasediments with Proterozoic orogenic events in Ontario: Canadian Jour. Earth Sci., v. 6, p. 489.

Fairbridge, R. W., 1969, Early Paleozoic South Pole in Northwest Africa: Geol. Soc. America Bull., v. 80, p. 113–114.

—— 1970, South Pole reaches the Sahara: Science, v. 168, p. 878–881.

—— 1971, Upper Ordovician glaciation in Northwest Africa? Reply: Geol. Soc. America Bull., v. 82, p. 269–274.

Furduy, R. S., 1968, Upper Precambrian tillite of the Kolyma region: Akad. Nauk SSSR Doklady, v. 180, no. 4, p. 72–75 (English trans. by Am. Geol. Inst.).

Geijer, P., 1963, The Precambrian of Sweden, *in* Rankama, K., ed., The Precambrian, v. 1: New York, Interscience Publishers, 279 p.

Gidon, P., 1970a, Glaciations Majeures et révolution galactique du système solaire: Acad. Sci. Comptes Rendus, Sér. D, v. 271, no. 4, p. 385–387.

—— 1970b, L'altername glaciaire-interglaciaire au cours d'une glaciation majeure: Acad. Sci. Comptes Rendus, Sér. D, v. 271, no. 17, p. 1493–1494.

Girdler, R. W., 1964, The paleomagnetic latitudes of possible ancient glaciations, *in* Nairn, A.E.M., ed., Problems of paleoclimatology: London, Interscience Publishers, p. 113–115.

Grabert, H., 1967, Ergebnis und Ausdeutung radiometrischer Untersuchungen an Graniten des Brasilianischen Schildes: Neues Jahrb. Geologie u. Paläontologie Monatsh. 5, p. 268–281.

Hamilton, W., and Krinsley, D., 1967, Upper Paleozoic glacial deposits of South Africa and Southern Australia: Geol. Soc. America Bull., v. 78, p. 783–800.

Harland, W. B., 1964a, Evidence of Late Precambrian glaciation and its significance, *in* Nairn, A.E.M., ed., Problems of paleoclimatology: London, Interscience Publishers, p. 150–155.

—— 1964b, Critical evidence for a great infra-Cambrian glaciation: Geol. Rundschau, v. 54, p. 45–61.

Harland, W. B., and Francis, E. H., eds., 1971, The Phanerozoic time-scale, a supplement: Geol. Soc. London Spec. Pub., no. 5, 356 p.

Harland, W. B., Smith, A. G., and Wilcock, B., eds., 1964, The Phanerozoic time-scale: Geol. Soc. London, 458 p.

Harrington, H. J., 1956, Argentina, *in* Jenks, W. F., ed., Handbook of South American geology: Geol. Soc. America Mem. 65, 378 p.

Harrison, J. E., 1972, Precambrian belt basin of northwestern United States: Its geometry, sedimentation and copper occurrences: Geol. Soc. America Bull., v. 83, no. 5, p. 1215–1240.

Hatfield, C. B., and Camp, M. J., 1970, Mass extinctions correlated with periodic galactic events: Geol. Soc. America Bull., v. 81, p. 911–914.

Holmes, A., 1965, Principles of physical geology (2d ed.): London, Nelson, 1288 p.

Hoyle, F., and Narlikar, J. V., 1971, On the nature of mass: Nature, v. 233, p. 41–44.

—— 1972, Cosmological models in a conformably invariant gravitational theory—II, a new model: Royal Astron. Soc. Monthly Notices, v. 155, p. 323–335.

Hughes, C. J., and Brückner, W. D., 1971, Late Precambrian rocks of eastern Avalon Peninsula, Newfoundland—A volcanic island complex: Canadian Jour. Earth Sci., v. 8, p. 899–915.

Innanen, K. A., 1966a, The angular momentum distribution of mass models of the galactic system: Jour. Astrophysics, v. 143, p. 150–152.

—— 1966b, Mass models of the galactic system: Jour. Astrophysics, v. 143, p. 153–167.

—— 1966c, A mass model of the galactic system: Zeitschr. Astrophysik, v. 64, p. 158–164.

—— 1966d, Orbits of globular clusters and high-velocity stars in a mass model of the galactic system: Zeitschr. Astrophysik, v. 64, p. 445–456.

—— 1966e, The sun's orbit in a mass model of the galactic system: Zeitschr. Astrophysik, v. 64, p. 457–459.

Innanen, K. A., and Fox, D. R., 1967, Velocity dispersions in a mass model of the galactic system: Zeitschr. Astrophysik, v. 66, p. 308–313.

Irving, E., 1964, Paleomagnetism: New York, John Wiley & Sons, Inc., 399 p.

Keller, B. M., Korolev, V. G., Semikhatov, M. A., and Chumakov, N. M., 1968, The main features of the Late Proterozoic paleogeography of the USSR: 23rd Internat. Geol. Congress, Prague, Academia, v. 4, p. 189–202.

Kirsanov, V. V., 1968, Precambrian stratigraphy of the axial part of the Moscow syneclise: Akad. Nauk SSSR Doklady, v. 178, no. 5, p. 78–82 (English trans. by Am. Geol. Inst.).

Kratz, K. O., Gerling, E. K., and Lobach-Zhuchenko, S. B., 1968, The isotope geology of the Precambrian of the Baltic Shield: Canadian Jour. Earth Sci., v. 5, p. 657–660.

Lambert, R. St.-J., 1971, The pre-Pleistocene Phanerozoic time-scale—A review, *in* Harland, W. B., and Francis, E. H., eds., The Phanerozoic time-scale, a supplement: Geol. Soc. London Spec. Pub. no. 5, 356 p.

Larsen, O., and Møller, J., 1968, Potassium-argon age studies in West Greenland: Canadian Jour. Earth Sci., v. 5, p. 683–691.

Leggo, P. J., and Spencer, A. M., 1969, Rb/Sr radiometric analysis of granitic clasts from the Scottish Dalradian tillite [abs.]: Am. Geophys. Union Trans., v. 50, no. 4, p. 331.

Leutwein, F., 1968, Contribution à la connaissance du précambrien récent en Europe Occidentale et développement géochronologique du Briovérien en Bretagne (France): Canadian Jour. Earth Sci., v. 5, p. 673–682.

Leutwein, F., and Sonet, J., 1966, Contribution à la connaissance de l'évolution géochronologique de la partie nord-est du Massive Armoricain Francais: *in* Interpretation Géologique des Mesures Effectuées au Spectromètre de Masse dans le Domaine de la Géochronologie Absolut: Centre National de la Recherche Scientifique, Colloques Internat., no. 151, 583 p.

Lungershauzen, G. F., 1957, The periodic change of climate and the earth's gigantic glaciation: Sovetskaya Geologiya, Sbornik statei, v. 59, p. 88–115.

Machado, F., 1967, Geological evidence for a pulsating gravitation: Nature, v. 214, p. 1317–1318.

Macintyre, R. M., 1971, Apparent periodicity of carbonatite emplacement in Canada: Nature, Phys. Sci., v. 230, no. 1, p. 23–24.

McElhinny, M. W., and Luck, G. R., 1970, Paleomagnetism and Gondwanaland: Science, v. 168 (3933), p. 830–832.

McElhinny, M. W., Briden, J. C., Jones, D. L., and Brock, A., 1968, Geological and geophysical implications of paleomagnetic results from Africa: Rev. Geophysics, v. 6, p. 201–238.

Meyerhoff, A. A., and Meyerhoff, H. A., 1972, The new global tectonics: Major inconsistencies: Am. Assoc. Petroleum Geologists Bull., v. 56, no. 2, p. 269–336.

Mikhaylov, Y. A., Ustritskiy, V. I., Chernyak, G. Y., and Youshitz, G. P., 1970, Upper Permian glaciomarine sediments of the northeastern USSR: Akad. Nauk SSSR Doklady, v. 190, p. 100–102 (English trans. by Am. Geol. Inst.).

Milankovitch, M., 1938, Astronomische Mittel zue Erforschung der erdgeschichtlichen Klimate: Geophys. Handb., v. 9, p. 593–698.

Mitchell, J. M., 1965, Causes of climatic change: Boston, Am. Meteorol. Soc.

Ogorodnikov, K. F., 1965, Dynamics of stellar systems: Oxford, Pergamon.

Postnikova, I. Y., and Revenko, E. A., 1969, New data on the Wendian Complex of the Volga-Ural Region: Akad. Nauk SSSR Doklady, v. 188, no. 5, p. 107-109 (English trans. by Am. Geol. Inst.).

Rankama, K., 1970, Proterozoic, Archean and other weeds in the Precambrian rock garden: Geol. Soc. Finland Bull., v. 42, p. 211–222.

Rankin, D. W., Stern, T. W., Reed, J. C., Jr., and Newell, M. F., 1969, Zircon ages of felsic volcanic rocks in the upper Precambrian of the Blue Ridge, Appalachian Mountains: Science, v. 166, p. 741-744.

Rast, N., 1971, Isotope dating in the USSR—An essay review, *in* Harland, W. B., and Francis, E. H., eds., The Phanerozoic time-scale, a supplement: Geol. Soc. London Spec. Pub. no. 5, 356 p.

Reading, H. G., and Walker, R. G., 1966, Sedimentation of Eocambrian tillites and associated sediments in Finnmark, Northern Norway: Palaeogeography, Palaeoclimatology, Palaeoecology, v. 2, p. 177–212.

Ryan, B. D., and Blenkinsop, J., 1971, Geology and geochronology of the Hellroaring Creek Stock, British Columbia: Canadian Jour. Earth Sci., v. 8, p. 85–95.

Saito, R., 1969, Glacier problems of late Precambrian eon: Kumamoto Jour. Sci. Ser. B, sec. 1, v. 8, no. 1, p. 7–44.

Salop, L. I., 1968, Pre-Cambrian of the USSR: Internat. Geol. Cong., 23rd, Prague, Academia, v. 4, p. 61–73.

Schenk, P. E., 1965, Depositional environment of the Gowganda Formation (Precambrian) at the south end of Lake Timagami, Ontario: Jour. Sed. Petrology, v. 35, p. 309–318.

—— 1971, Southeastern Atlantic Canada, northwestern Africa, and continental drift: Canadian Jour. Earth Sci., v. 8, no. 10, p. 1218–1251.

—— 1972, Possible Late Ordovician glaciation of Nova Scotia: Canadian Jour. Earth Sci., v. 9, no. 1, p. 95–107.

Schwarzbach, M., 1963, Climates of the past: London, D. Van Nostrand, 328 p.

—— 1968, Neuere Eiszeithypothesen: Eiszeitalter u. Gegenwart, v. 19, p. 250–261.

Spencer, A. M., 1969, Discussion—Author's written reply: Geol. Soc. London Proc., no. 1657, p. 192.

—— 1971, Late Precambrian glaciation in Scotland: Geol. Soc. London Mem. 6, 98 p.

Steiner, J., 1967, The sequence of geological events and the dynamics of the Milky Way Galaxy: Geol. Soc. Australia Jour., v. 14, pt. 1, p. 99–132.

Symons, D.T.A., 1966, A paleomagnetic study on the Gunflint, Mesabi and Cuyuna iron ranges in the Lake Superior region: Econ. Geology, v. 61, p. 1336–1361.

—— 1967, Paleomagnetism of Precambrian rocks near Cobalt, Ontario: Canadian Jour. Earth

Sci., v. 4, p. 1161–1170.
Tamrazjan, G. P., 1967, The global historical and geological regularities of the earth's development as a reflection of its cosmic origin (as a sequence of interaction in the course of galactic movement of the solar system): Ostrava, Vysoka Skola Banska, Sbornık, v. 13, p. 5–24.
Umbgrove, J.H.F., 1947, The pulse of the earth: The Hague, Martinus Nujhoff, 358 p.
Wanless, H. R., and Cannon, J. R., 1966, Late Paleozoic glaciation: Earth-Science Rev., v. 1, p. 247–286.
Wanless, R. K., Stevens, R. D., Lachance, G. R., and Delabio, R. N., 1972, Age determinations and geological studies: Geol. Survey Canada, Paper 71–2, Erratum, p. 5.
Willet, H. C., 1949, Long period fluctuations of the general circulation of the atmosphere: Jour. Meteor., v. 6, p. 34–50.
Winsnes, T. S., 1965, The Precambrian of Spitzbergen and Bjørnøya, *in* Rankama, K., ed., The Precambrian, v. 2: New York, Interscience Publishers, 454 p.
Young, G. M., 1970, An extensive early Proterozoic glaciation in North America?: Palaeogeography, Palaeoclimatology, Palaeoecology, v. 7, p. 85–101.
Zhuravleva, Z. A., and Chumakov, N. M., 1968, Catagraphs, oncoliths and stromatoliths from the Upper Precambrian of Eastern Belorussia: Akad. Nauk SSSR Doklady, v. 178, no. 3, p. 48–50 (English trans. by Am. Geol. Inst.).
Ziegler, P. A., 1959, Frühpaläozoische tillite im östlichen Yukon-Territorium (Kanada): Eclogae Geol. Helvetiae, v. 52, p. 735–741.
Zubtsov, Y. I., and Zubtsova, Y. I., 1966, Precambrian tillites of the Tien Shan: Akad. Nauk SSSR Doklady, v. 169, no. 1, p. 21–24 (English trans. by Am. Geol. Inst.).

Manuscript Received by the Society February 15, 1972
Revised Manuscript Received June 29, 1972

17

Reprinted from *Earth and Planetary Sci. Letters* **26**:361–369 (1975)

POSSIBLE RELATION BETWEEN PERIODIC GLACIATION AND THE FLEXURE OF THE GALAXY

G.E. WILLIAMS

Received January 8, 1975
Revised version received April 4, 1975

At least five discrete episodes of widespread Earth glaciation or global cooling have occurred since late Precambrian time. A notable feature of such episodes is the ~155-m.y. mean period of their mean ages, key dates being ~770, ~615, ~445, ~295, ~145? and minus ~10? m.y. B.P.

Employing the following values for parameters of the Sun's galactic orbit, present circular velocity = 225 ± 25 km sec^{-1}, present galactocentric distance = 9.0 ± 0.5 kiloparsec, and eccentricity = 0.1 ± 0.01, it is calculated as a first approximation that the period of the Sun's revolution around the galactic centre is 303^{+65}_{-51} m.y. Assuming that the two galactic regions which appear to be tidally flexed by the Magellanic Clouds are stationary with respect to the galactic centre, then the mean period of solar passage between regions of given, diametrically-opposite galactic flexure is 152^{+33}_{-26} m.y. Since the solar system now lies within a little-flexed region of the galactic disc, and the Earth is presently subjected to a climatic regime of high-latitude glaciation, it is suggested that passage through such regions may have resulted in the ~155-m.y. rhythm of Earth glaciation since late Precambrian time.

1. Introduction

With the wide acceptance of important Earth glaciations during the late Precambrian (e.g. [1–4]) and Late Ordovician (e.g. [5–7]), and the firm establishment of Permo-Carboniferous and late Cenozoic glaciations, a picture has formed of repeated major fluctuations in global climate since late Precambrian time. According to Fairbridge [8, p. 505], an "approximate cyclicity" is suggested by this chronology, whose "coincidence with the possible timing of the galactic cycle is intriguing but not proven". The present paper demonstrates that times of glaciation since the late Precambrian correlate closely with a hitherto unrecognized periodic parameter of the Sun's galactic revolution, suggesting that such cyclicity of glaciation may indeed reflect galactic motions of the solar system.

2. Periodicity of glaciation

Ages of late Precambrian glaciogenic sequences, as determined isotopically or estimated stratigraphically, are listed in Table 1 and plotted as a histogram in Fig. 1. The ages are divisible into three groups: the two younger, better-dated age groups (here designated late Precambrian I and II) have ages of ~615 ± 40 m.y. and $\sim 770^{+50}_{-30}$ m.y. respectively, and the oldest age group (late Precambrian III) a more tentative age of $\sim 940^{+60}_{-90}$ m.y. Age groups I and II represent the well-known "pair" of late Precambrian tillites.

Table 2 lists the estimated ages of late Precambrian glaciations together with those of *important* Phanerozoic

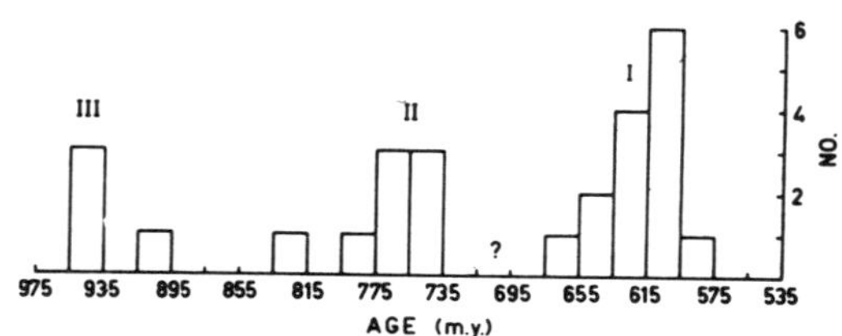

Fig. 1. Frequency histogram of published ages of late Precambrian glaciogenic sequences as listed in Table 1. Mean age plotted for each location; maximum and minimum ages not included.

TABLE 1

Published ages of late Precambrian glaciogenic sequences

Glacial episode	Age (m.y.)	Location	References
Late Precambrian I	<668 ± 23*	Norway	[9]
	~600	Norway	[10]
	~600	East Greenland	[11]
	~620	Spitsbergen	[12]
	>572 ± 20	Scotland	[13]
	~630–560	Normandy	[14]
	680*–660*	European Russia	[15]
	640–620	European Russia	[16]
	{ 715–670	Newfoundland	[17]
	{ 600–570	Newfoundland	[18]
	~650–570	Algeria, Mauritania	[19, 20]
	$\gtrsim$620*	Ghana	[21]
	653 ± 70–570	Southern Africa	[22]
	650 ± 50*	Australia	[3, 23]
	$\gtrsim$600 ± 25	Brazil	[24]
	675–570	Siberia	[25]
	650–570	China	[26]
Late Precambrian II	810 ± 90–668 ± 23*	Norway	[9]
	810*, 715*	European Russia	[27]
	<820	Appalachians	[28, 29]
	~850–800	Western North America	[30]
	>740 ± 50	Gabon, Congo, Angola	[31]
	~800–750	Katanga	[32, 33]
	840 ± 40–710	Zaire	[33]
	<850, <719 ± 28	Southern Africa	[22]
	750 ± 40*	Australia	[3, 23]
	810*–747*	Siberia	[27]
	~750	China	[26]
Late Precambrian III	~950–850	Finnmark	[34]
	$\lesssim$950 ± 50	Russian Tien Shan	[25]
	~945 ± 50	Katanga, Lower Congo	[33]
	~950	China	[26]

The upper boundary of the Precambrian is here placed at 570 m.y. B.P., after [35].
* Isotopic age determination of glaciogenic sequence.

glaciations. Each age group is best regarded as reflecting a *separate* glacial episode, because successive glaciogenic sequences in many parts of the world are stratigraphically separated by thick sequences exhibiting evidence of prolonged periods of markedly warmer climate. Moreover, since palaeomagnetic studies (e.g. [40–42]) demonstrate that major portions of Gondwanaland occupied very high palaeolatitudes continuously during the Palaeozoic, the deglaciations of that supercontinent during the Silurian and Late Permian would seem to reflect real climatic change. An *external* control of global climate appears likely.

A notable feature of the glacial episodes listed in Table 2 is the periodicity of their mean ages. A mean period of ~155 m.y. best fits the data, taking the mid-point of the late Cenozoic glaciation at 10 m.y. in the future; if it is assumed that the Earth is now at the glacial midpoint, a mean period of ~152 m.y. provides a reasonable fit. Although there is an apparent gap in the sequence of glaciations at ~145 m.y. B.P. (Late

TABLE 2

Ages of major glaciations during past billion years

Glacial episode	Absolute age (m.y.)	Mean age (m.y.)	155-m.y. periodicity (m.y.B.P.)
Late Cenozoic	20–0[1]	minus 10? (±30)	minus 10
Nil (Late Jurassic)[2]			145
Permian–Carboniferous	350–240 (Gondwanaland)[3] ~280 (Siberia)[4]	295 (±55)	300
Late Ordovician	450–440[5]	445 (±5)	455
Late Precambrian I	See Table 1, Fig. 1	~615 (±40)	610
Late Precambrian II		~770($^{+50}_{-30}$)	765
Late Precambrian III		~940($^{+60}_{-90}$)	920

[1] Commencement of *extensive* late Cenozoic glaciation ~20 m.y. B.P. [36].
[2] See text for discussion.
[3] After [37–39].
[4] According to Keller [16], *reliable* indications of Permo-Carboniferous glaciation in the USSR appear to be restricted to the Late Carboniferous–Early Permian, that is, around 280 m.y. B.P.
[5] After [7,8]. According to Fairbridge [8], there is no reliable evidence of important glaciation during the Siluro-Devonian.
Time scale after [35].

Jurassic, according to the time scale of Harland et al. [35]), relatively cool air-temperature on land at that time was suggested by Arkell [43, p. 616] on the evidence of dwarf Purbeckian fauna. Furthermore, possible glaciogenic deposits of about this age occur in Antarctica (Late Jurassic? "Mawson Tillite" in Victoria Land [44], and tillite in the Queen Maud Mountains [45]) and in western Victoria (Early Cretaceous of the Otway Basin: E.D. Gill, personal communication, 1974). That any mid-Mesozoic cooling was moderate, and did not lead to widespread glaciation, may well be attributable to changes in polar albedo and global atmospheric circulation accompanying the *known* poleward expansion of vegetation during Mesozoic–early Cenozoic time (e.g. [46–48]).

Numerous persons have sought a relation between times of major glaciation and the duration of the Sun's orbit around the galactic centre, termed the cosmic year (see review by Steiner and Grillmair [49]). These attempts are not completely convincing, since:

(a) The duration of the cosmic year has been estimated variously at ~200 to ~280 m.y. (e.g. [49–51]).

(b) Correlation of glaciation with perigalacticum and apogalacticum (the apsides) of an eccentric solar orbit [49,50] appears doubtful. Since the orbit's major axis slowly rotates in the opposite sense to the Sun's orbital motion [52, p. 598], a consequence of point (a) above is that the average time taken by the Sun to move between apsides is *less than* 100–140 m.y., and thus markedly below the sought-for value of ~155 m.y. Indeed, according to Innanen [53], the Sun's average perigal-apogal orbital time is only 110 m.y.

Periodic glaciation since at least late Precambrian time may, however, relate to a previously unrecognized parameter of the solar galactic orbit having a mean period of ~155 m.y. and prevailing today. This paper proposes such a parameter.

3. The Milky Way Galaxy

The Milky Way Galaxy is a spiral stellar system probably ~50 kiloparsec (kpc) in diameter (1 kpc = 3258 light years) undergoing differential clockwise revolution viewed from the galactic north (Fig. 2A). The Sun moves around the galactic centre in an orbit of eccentricity 0.1 ± 0.01 ([54, p. 574], and S.P. Wyatt, personal communication, 1974; see Appendix) and present galactocentric distance 9.0 ± 0.5 kpc [55–57]

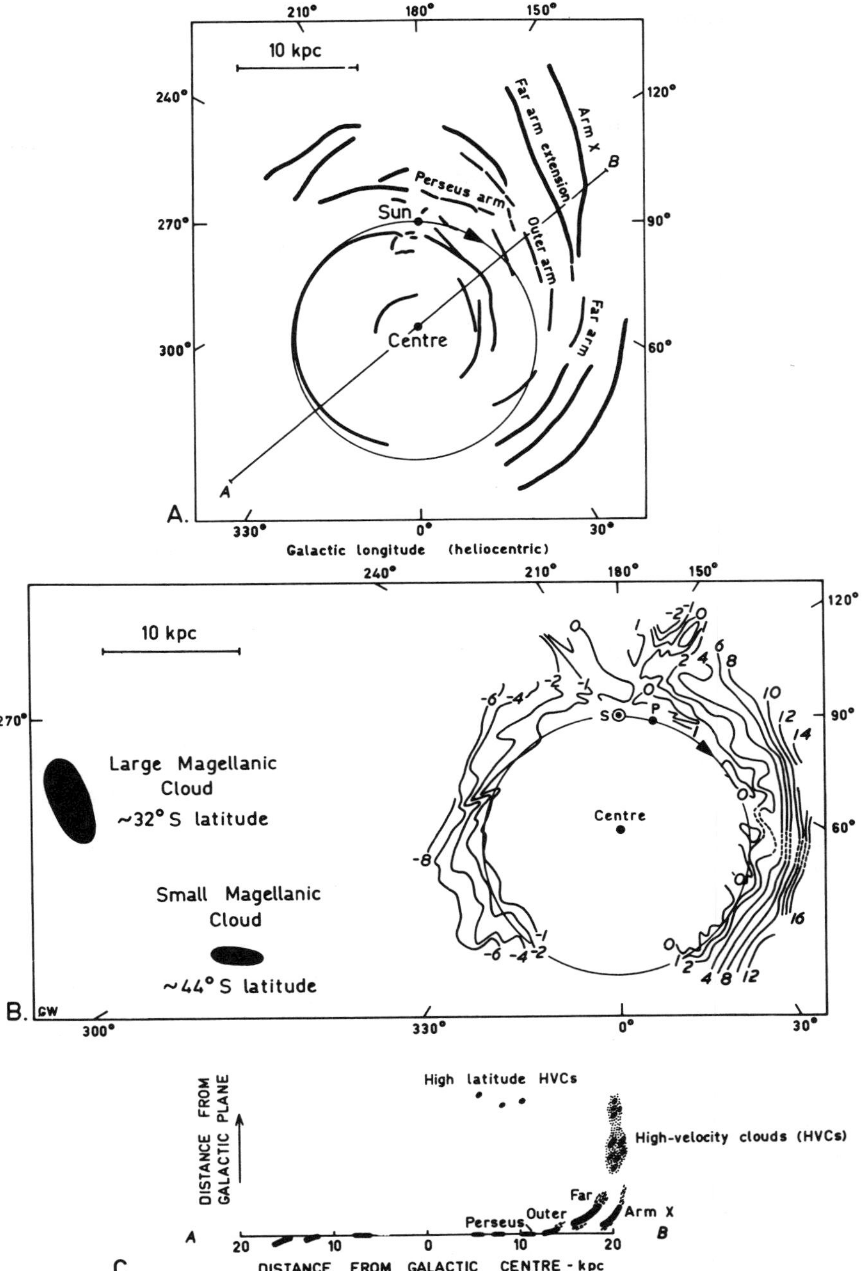
210°
180°
150°
10 kpc
240°
120°
Far arm extension
Arm X
B
Perseus arm
Sun
270°
90°
Outer arm
Centre
Far arm
300°
60°
A
A.
330°
0°
30°
Galactic longitude (heliocentric)
240°
210°
180°
150°
120°
10 kpc
270°
90°
Large Magellanic Cloud
~32° S latitude
Centre
60°
Small Magellanic Cloud
~44° S latitude
B.
GW
300°
330°
0°
30°
High latitude HVCs
High-velocity clouds (HVCs)
DISTANCE FROM GALACTIC PLANE
Far
Arm X
Outer
Perseus
A
B
20
10
0
10
20
C.
DISTANCE FROM GALACTIC CENTRE - kpc

with a present circular velocity of 225 ± 25 km sec^{-1} [57,58]. In addition, the Sun vibrates up to ~100 parsec (pc) north and south of the galactic plane with a mean period of ~70 m.y., presently lying ~15 pc north of (or "above") the plane [54].

A relief map of the Galaxy is shown in Fig. 2B, and a galactic cross-section in Fig. 2C. A remarkable structure is revealed: there is an upward flexure of the plane of neutral hydrogen in the outer spiral arms between galactic longitudes 20° and 140°, and a reciprocal downward flexure on the opposite side of the Galaxy. Near the Sun's orbital path the deviation from the galactic plane is as much as 100–200 pc ([59,60] and Fig. 2B). There is evidence that stars too participate in the bending of the Galaxy [61].

The Galaxy possesses two companion or satellite galaxies, the Magellanic Clouds, which lie ~50–60 kpc from the Sun at galactic latitudes ~32° and ~44° south ([62,63] and Fig. 2B). The Clouds are connected by a bridge of neutral hydrogen, suggesting that they form one system [58,64]. The orientation of the Clouds' orbit is unknown [65].

The downward flexure of the Galaxy is oriented toward the Large Magellanic Cloud or LMC (Fig. 2B), and numerous astronomers now consider that such flexure is a tidal distortion caused by the Clouds, particularly the LMC (e.g. [59,65–69]). That bending originated in this manner is strongly supported by the comparable flexure of external spiral galaxies with nearby companions [69]. Doubts that the mass of the Clouds is sufficient to tidally distort the Galaxy at the Clouds' present distance [67] may be allayed by the recent claim [71] that the masses of galaxies have been underestimated by an order of magnitude due to the presence of invisible matter.

The solar system at present lies almost midway between the upward- and downward-flexed regions of the galactic disc (Fig. 2B).

4. Galactic model for periodic glaciation

It is evident from Fig. 2B that the Sun's movement around the galactic centre produces these four significant events: twice during each revolution the solar system moves through the inner margins of those regions where the Galaxy is tidally flexed, and twice moves through those regions where tidal flexure is minimal or absent. The solar system is now within one such latter region.

Employing the following best estimates of the Sun's orbital parameters (with probable limits of uncertainty in parentheses), present circular velocity $V_0 = 225$ (±25) km sec^{-1}, present galactocentric distance $R_0 = 9.0$ (±0.5) kpc, and eccentricity $e = 0.1$ (±0.01), it is calculated (Appendix) as a first approximation that the duration of the present cosmic year P is 303^{+65}_{-51} m.y. It follows that for stationary flexed regions, the mean period of solar passage between regions of given, diametrically-opposite galactic flexure ($P/2$) is 152^{+33}_{-26} m.y. The absolute durations of such passages probably vary up to ~10% from the mean, due to the retrograde rotation of the major axis of the Sun's orbit.

The best estimate of $P/2$ (~152 m.y.) compares favourably with the proposed ~155-m.y. mean period of Earth glaciation (Table 2), raising the question of a causative link. Since the solar system now lies within a region of minimal galactic flexure, and the Earth is presently subjected to a climatic regime of high-latitude glaciation, it is suggested that the ultimate cause of Earth glaciation, at least since late Precambrian time, may be passage of the solar system through little-flexed regions of the Galaxy. The immediate cause of Earth glaciation may be variation in energy received from the Sun (e.g. [72–75]); possibly the properties of the interstellar medium [72] and/or convective processes in the peripheral region of the Sun (e.g. [73]) are

---

Fig. 2. Galactic structure. A. Spiral structure as viewed from galactic north. Sun moves in clockwise direction (arrow) in orbit of eccentricity 0.1 [54]. Line *AB* refers to Fig. 2C. Modified after [69]. B. Relief map of outer parts of Galaxy as viewed from galactic north, showing: height (in hundreds of parsec) of centroid of hydrogen density relative to galactic plane; Magellanic Clouds; Sun (*S*); and perigalacticum (*P*). Modified after [59,70]. C. Section perpendicular to galactic plane along line *AB* shown in Fig. 2A. No vertical exaggeration of scale. After [69].

Note: The apparent asymmetry of structures in Fig. 2A and C is largely observational, since most radio-mapping of the Galaxy has been conducted from Earth's northern hemisphere and Earth's north pole is oriented along galactic longitude 123°.

influenced by the action of the Magellanic Clouds. Significantly, there is strong astronomical evidence that the Sun's luminosity is now anomalously low (e.g. [74–76]).

If Earth glaciation is indeed a response to periodic fluctuations of solar luminosity, evidence of correlative climatic changes might exist in the crustal records of the other terrestrial planets, particularly Mars. In this regard it may be significant that, according to Hartmann [77], Mars has experienced major climatic variations with time scales of the order of 10^8 years.

5. Conclusions

The present computation of the period of the Sun's revolution around the galactic centre (303^{+65}_{-51} m.y.) suggests a link between times of Earth glaciation since the late Precambrian, and probable times of passage of the Sun through galactic regions least distorted by the tidal action of the Magellanic Clouds. Possibly the properties of the interstellar medium and/or the energy output of the Sun are influenced sufficiently by the action of the Clouds to have important climatic consequences for the Earth and other terrestrial planets.

Appendix

Duration of the cosmic year, and period of Sun's passage between given regions of tidally-flexed Galaxy

From Van de Kamp [78, p. 47], for an eccentric solar orbit:

$$R_p = R_a(1-e) \qquad (1)$$

where R_p is the Sun's perigalactic distance, R_a is the Sun's mean galactocentric distance, and e is the eccentricity of the Sun's orbit.

Since the Sun is now only ~15–20° from perigalacticum ([54] and Fig. 2B), $R_0 \simeq R_p$, where R_0 is the Sun's present galactocentric distance. Hence:

$$R_0 \simeq R_a(1-e) \qquad (2)$$

Taking $R_0 = 9.0 \pm 0.5$ kpc [55–57] and $e = 0.1 \pm 0.01$ (Wyatt [54] recently has given e at 0.1. According to Professor Wyatt (personal communication, 1974), "The best estimate of e remains very close to 0.1. There is an uncertainty of roughly 10% because our observational knowledge of the components of the Sun's velocity relative to circular velocity at R_0 is ±1 km/sec or so."), then

$$R_a \simeq 10.01 \pm 0.67 \text{ kpc} \qquad (3)$$

Because of the Sun's near-perigalactic position, from Kepler's second law the Sun's present circular velocity V_0 (225 ± 25 km sec^{-1} [57,58]) is maximal for the current cosmic year. In conserving angular momentum:

$$mR_0V_0 = mR_aV_a \qquad (4)$$

from whence:

$$R_0 = R_aV_a/V_0 \qquad (4a)$$

where V_a is the mean solar orbital velocity, and m the Sun's mass. Substituting eq. 4a in eq. 2:

$$V_a \simeq V_0(1-e) \qquad (5)$$

$$= (225 \pm 25)(0.9 \pm 0.01)$$

$$= 202.8 \pm 24.8 \text{ km sec}^{-1} \qquad (6)$$

The duration of the current cosmic year P is approximated by:

$$P = 2\pi R_a/V_a \qquad (7)$$

Substituting eq. 3 and eq. 6 in eq. 7:

$$P = \frac{2\pi \times (10.01 \pm 0.67) \times 3258 \times 9.463 \times 10^{12}}{(202.8 \pm 24.8) \times 3.156 \times 10^{13}} \qquad (8)$$

$$= 303^{+65}_{-51} \text{ (or } 310 \pm 58) \text{ m.y.}$$

since 1 kpc = 3258 light yrs; 1 light yr = 9.463×10^{12} km; and 1 m.y. = 3.156×10^{13} sec.

Employing 303^{+65}_{-51} m.y. for the value of P, it follows that for stationary flexed galactic regions the mean period of solar passage between regions of given, diametrically-opposite flexure ($P/2$) is 152^{+33}_{-26} m.y.

Substitution of galactic constants accepted by the International Astronomical Union in 1966, namely $V_0 = 250$ km sec^{-1} and $R_0 = 10$ kpc, does not alter the computed magnitude of P, since these values give the same angular velocity of galactic rotation (25 km sec^{-1} kpc^{-1}) as those adopted here. Steiner's [50] widely-cited estimate of ~280 m.y. for P was obtained by computation similar to that above, employing V_0 = 250 km sec^{-1} and R_0 = 10 kpc (the IAU "constants"), and $e = 0.07$ after Trumpler and Weaver [52, p. 601].

It is thus the differing assumed values of e which distinguish Steiner's from the present calculation of the solar revolution period. Wyatt's [54] computation of e supersedes that by Trumpler and Weaver [52], being based on Delhaye's [79] proposed elements of the Sun's peculiar motion, which have not since been improved significantly in accuracy (S.P. Wyatt, personal communication, 1974). Steiner's [50] estimate of P thus should be abandoned.

Acknowledgements

I thank those who provided helpful discussion and correspondence, and referees for constructive comments.

References

1 W.B. Harland, Evidence of late Precambrian glaciation and its significance, in: Problems in Palaeoclimatology, A.E.M. Nairn, ed. (Interscience, London, 1964) 119.
2 H.G. Reading and R.G. Walker, Sedimentation of Eocambrian tillites and associated sediments in Finnmark, northern Norway, Palaeogeogr., Palaeoclimatol., Palaeoecol. 2 (1966) 177.
3 P.R. Dunn, B.P. Thomson and K. Rankama, Late Pre-Cambrian glaciation in Australia as a stratigraphic boundary, Nature 231 (1971) 498.
4 A.M. Spencer, Late Pre-Cambrian glaciation in Scotland, Geol. Soc. London Mem. 6 (1971) 100 pp.
5 R.W. Fairbridge, South Pole reaches the Sahara, Science 168 (1970) 878.
6 S. Beuf, B. Biju-Duval, O. de Charpal, P. Rognon, O. Gariel and A. Bennacef, Les Grès du Palèozoïque Inférieur au Sahara (Technip, Paris, 1971) 464 pp.
7 W.B. Harland, The Ordovician ice age, Geol. Mag. 109 (1972) 451.
8 R. W. Fairbridge, Glaciation and plate migration, in: Implications of Continental Drift to the Earth Sciences, 1, D.H. Tarling and S.K. Runcorn, eds. (Academic Press, London, 1973) 503.
9 I.R. Pringle, Rb–Sr age determinations on shales associated with the Varanger Ice Age, Geol. Mag. 109 (1973) 465.
10 N.L. Banks, M.B. Edwards and H.G. Reading, Late Pre-Cambrian glaciation in Scotland. Discussion, Proc. Geol. Soc. London 1657 (1969) 191.
11 A.M. Spencer, Late Pre-Cambrian glaciation in Scotland. Abstract and discussion, Proc. Geol. Soc. London 1657 (1969) 177.
12 N.M. Chumakov, Late Precambrian glaciation of Spitsbergen, Akad. Nauk S.S.S.R. Dokl. 180 (1968) 115 (A.G.I. translation).
13 A.M. Spencer and M.O. Spencer, The late Precambrian/ Lower Cambrian Bonahaven Dolomite of Islay and its stromatolites, Scott. J. Geol. 8 (1972) 269.
14 F. Leutwein, Contribution à la connaissance du précambrien récent en Europe Occidentale et développement géochronologique du Briovérien en Bretagne (France), Can. J. Earth Sci. 5 (1968) 673.
15 N. Chumakov and A. Cailleux, Glaciation et éolisation dans l'est et le nord de l'Europe à l'éocambrien, Rev. Géomorphol. Dyn. 20 (1971) 1.
16 B.M. Keller, Great glaciations in history of the Earth, Int. Geol. Rev. 15 (1973) 1067.
17 M.M. Anderson, A possible time span for the late Precambrian of the Avalon Peninsula, southeastern Newfoundland in the light of worldwide correlation of fossils, tillites, and rock units within the succession, Can. J. Earth Sci. 9 (1972) 1710.
18 C.J. Hughes and W.D. Brückner, Late Precambrian rocks of eastern Avalon Peninsula, Newfoundland – a volcanic island complex, Can. J. Earth Sci. 8 (1971) 899.
19 B. Biju-Duval and O. Gariel, Nouvelles observations sur les phénomènes Glaciaires "Eocambriens" de la Bordure Nord de la Synéclise de Taoudeni, entre le Hank et le Tanezrouft, Sahara Occidental, Palaeogeogr., Palaeoclimatol., Palaeoecol. 6 (1969) 283.
20 N. Clauer, Utilisation de la méthode rubidium-strontium pour la datation de niveaux sédimentaires du Précambrien supérieur de l'Adrar mauritanien (Sahara occidental) et la mise en évidence de transformations précoces des minéraux argileux, Geochim. Cosmochim. Acta 37 (1973) 2243.
21 N.A. Bozhko, G.A. Kazakov, D.M. Trofimov, K.G. Knoppe and Y.A. Gatinskiy, New absolute dating of west African glauconites, Akad. Nauk S.S.S.R. Dokl. 198 (1972) 138 (A.G.I. translation).
22 J.D.A. Piper, Latitudinal extent of late Precambrian glaciations, Nature 244 (1973) 342.
23 W. Compston and P.A. Arriens, The Precambrian geochronology of Australia, Can. J. Earth Sci. 5 (1968) 561.
24 C.A.L. Isotta, A.C. Rocha-Campos and R. Yoshida, Striated pavement of the upper Precambrian glaciation in Brazil, Nature 222 (1969) 466.
25 B.M. Keller, V.G. Korolev, M.A. Semikhatov and N.M. Chumakov, The main features of the late Proterozoic paleogeography of the U.S.S.R., 23rd Int. Geol. Congr. 4 (1968) 189.
26 R. Saito, Glacier problems of late Pre-Cambrian eon, Kumamoto J. Sci. Ser. B(1) 8 (1969) 7.
27 L.I. Salop, Pre-Cambrian of the U.S.S.R., 23rd Int. Geol. Congr. 4 (1968) 61.
28 D.W. Rankin, T.W.Stern, J.C. Reed and M.F. Newell, Zircon ages of felsic volcanic rocks in the upper Precambrian of the Blue Ridge, Appalachian Mountains, Science 166 (1969) 741.
29 D.W. Rankin, Late Precambrian and early Paleozoic paleogeography in western Virginia and North Carolina, Geol. Soc. Am. Abstr. with Progr. Annu. Meet. 5 (1973) 209.
30 F.K. Miller, E.H. McKee and R.G. Yates, Age and correlation of the Windermere Group in northeastern Washington, Geol. Soc. Am. Bull. 84 (1973) 3723.

31 L. Cahen and N.J. Snelling, The Geochronology of Equatorial Africa (North-Holland, Amsterdam, 1966) 196 pp.
32 L. Cahen and J. Lepersonne, The Precambrian of the Congo, Rwanda, and Burundi, in: The Precambrian, 3, K. Rankama, ed., (Interscience, New York, 1967) 143.
33 L. Cahen, Igneous activity and mineralisation episodes in the evolution of the Kibaride and Katangide Orogenic Belts of central Africa, in: African Magmatism and Tectonics, T.N. Clifford and I.G. Gass, eds. (Oliver and Boyd, Edinburgh, 1970) 97.
34 A. Siedlecka and D. Roberts, A late Precambrian tilloid from Varangerhalvøya – evidence of both glaciation and subaqueous mass movement, Årbok Nor. Geol. Unders. (1972) 135.
35 W.B. Harland, A.G. Smith and B. Wilcock eds., The Phanerozoic Time-scale (Geol. Soc., London, 1964) 458 pp.
36 D.E. Hayes et al., Leg 28 deep-sea drilling in the Southern Ocean, Geotimes 18 (6) (1973) 19.
37 L.A. Frakes and J.C. Crowell, Late Paleozoic glaciation: II, Africa exclusive of the Karroo Basin, Geol. Soc. Am. Bull. 81 (1970) 2261.
38 J.C. Crowell and L.A. Frakes, Late Paleozoic glaciation: IV, Australia, Geol. Soc. Am. Bull. 82 (1971) 2515.
39 J.C. Crowell and L.A. Frakes, Late Paleozoic glaciation: V, Karroo Basin, South Africa, Geol. Soc. Am. Bull. 83 (1972) 2887.
40 J.C. Briden, Palaeomagnetic polar wander curve for Africa, in: Palaeogeophysics, S.K. Runcorn, ed. (Academic Press, London, 1970) 277.
41 M.W. McElhinny and G.R. Luck, Palaeomagnetism and Gondwanaland, Science 168 (1970) 830.
42 A.G. Smith, J.C. Briden and G.E. Drewry, Phanerozoic world maps, Spec. Pap. Palaeontol. 12 (1973) 1.
43 W.J. Arkell, Jurassic Geology of the World (Oliver and Boyd, Edinburgh, 1956) 806 pp.
44 G. Warren, Mawson Tillite, in: Geology of Victoria Land between the Mawson and Mulock Glaciers, Antarctica, B.M. Gunn and G. Warren, eds., Geol. Surv. N.Z. Bull. 71 (1962) 117.
45 G.A. Doumani and V.H. Minshew, General geology of the Mount Weaver area, Queen Maud Mountains, Antarctica, in: Geology and Paleontology of the Antarctic, J.B. Hadley, ed., Am. Geophys. Union Antarct. Res. Ser. 6 (1965) 127.
46 J.C. Briden and E. Irving, Palaeolatitude spectra of sedimentary palaeoclimate indicators, in: Problems in Palaeoclimatology, A.E.M. Nairn, ed. (Interscience, London, 1964) 199.
47 J.A. Wolfe, Paleogene floras from the Gulf of Alaska region, U.S. Geol. Surv. Open-file Rep. 374 (1969) 114 pp.
48 P.D.W. Barnard, Mesozoic floras, Spec. Pap. Palaeontol. 12 (1973) 175.
49 J. Steiner and E. Grillmair, Possible galactic causes of periodic and episodic glaciations, Geol. Soc. Am. Bull. 84 (1973) 1003.
50 J. Steiner, The sequence of geological events and the dynamics of the Milky Way Galaxy, J. Geol. Soc. Aust. 14 (1967) 99.
51 G.M. Clemence, Astronomical reference systems, in: Basic Astronimical Data, K.A. Strand, ed. (Univ. Chicago Press, Chicago, 1963) 1.
52 R.J. Trumpler and H.F. Weaver, Statistical Astronomy (University of California Press, Berkeley, Calif., 1953) 644 pp.
53 K.A. Innanen, The Sun's orbit in a mass model of the galactic system, Z. Astrophys. 64 (1966) 457.
54 S.P. Wyatt, Principles of Astronomy, 2nd ed. (Allyn and Bacon, Boston, Mass., 1971) 686 pp.
55 L.A. Balona and M.W. Feast, A new determination from OB stars of the galactic rotation constants and the distance to the galactic centre, Mon. Not. R. Astron. Soc. 167 (1974) 621.
56 G. Rybicki, M. Lecar and M. Schaefer, R_0 from galactic disk models, Bull. Am. Astron. Soc. 6 (1974) 453.
57 C. Cruz-González, Local density gradient and galactic parameters determined from nearby stars, Mon. Not. R. Astron. Soc. 168 (1974) 41.
58 D.S. Mathewson, M.N. Cleary and J.D. Murray, The Magellanic Stream, Astrophys. J. 190 (1974) 291.
59 G. Elwert and D. Hablick, The gravitational effect of the Magellanic Clouds on the HI-layer of the Galaxy, Z. Astrophys. 61 (1965) 273.
60 G.L. Verschuur, Bending of the galactic plane and the nature of the high velocity clouds, Astron, Astrophys. 27 (1973) 407.
61 D. Sher, Statistical significance of some optical evidence for the bending of the galactic plane, Astrophys. Space Sci. 18 (1972) 468.
62 A.D. Thackeray, Survey of principal characteristics of the Magellanic Clouds, in: The Magellanic Clouds, A.B. Muller, ed., Astrophys. Space Sci. Library 23 (Reidel, Dordrecht, 1971) 3.
63 D.S. Mathewson, Radio continuum observations of the Magellanic Clouds, in: The Magellanic Clouds, A.B. Muller, ed., Astrophys. Space Sci. Library 23 (Reidel, Dordrecht, 1971) 98.
64 F.J. Kerr, Properties of the neutral hydrogen in the Magellanic Clouds, in: The Magellanic Clouds, A.B. Muller, ed., Astrophys. Space Sci. Library 23 (Reidel, Dordrecht, 1971) 50.
65 E.S. Anver and I.R. King, The Influence of the Magellanic Clouds on the Milky Way, Astron. J. 72 (1967) 650.
66 H.J. Habing and H.C.D. Visser, Discussion, in: Radio Astronomy and the Galactic System, H. van Woerden, ed. (Academic Press, London, 1967) 159.
67 C. Hunter and A. Toomre, Dynamics of the bending of the Galaxy, Astrophys. J. 155 (1969) 747.
68 A. Toomre, V_0 from Magellanic Cloud orbits, Q. J. R. Astron. Soc. 13 (1972) 266.
69 R.D. Davies, Observations of the outer spiral structure of the Milky Way and its relation to the high velocity clouds, Mon. Not. R. Astron. Soc. 160 (1972) 381.
70 F.J. Kerr, The large-scale distribution of hydrogen in the Galaxy, Annu. Rev. Astron. Astrophys. 7 (1969) 39.
71 J. Einasto, A. Kaasik and E. Saar, Dynamic evidence on

massive coronas of galaxies, Nature 250 (1974) 309.
72 M. Krook, Interstellar matter and the solar constant, in: Climatic Change, H. Shapley, ed., (Harvard University Press, Cambridge, Mass., 1965) 143.
73 E.J. Öpik, Climatic change in cosmic perspective, Icarus 4 (1965) 289.
74 F.W. Dilke and D.O. Gough, The solar spoon, Nature 240 (1972) 262.
75 A.G.W. Cameron, Major variations in solar activity? Rev. Geophys. Space Phys. 11 (1973) 505.
76 W.A. Fowler, What cooks with solar neutrinos? Nature 238 (1972) 24.
77 W.K. Hartmann, Geological observations of Martian arroyos, J. Geophys. Res. 79 (1974) 3951.
78 P. van de Kamp, Elements of Astromechanics (Freeman, San Francisco, Calif., 1964) 140 pp.
79 J. Delhaye, Solar motion and velocity distribution of common stars, in: Galactic Structure, A. Blaauw and M. Schmidt, eds. (University of Chicago Press, Chicago, Ill., 1965) 61.

18

Reprinted from *Nature* **255**:607–609 (1975)

Ice ages and the Galaxy

W. H. McCrea

University of Sussex, UK

The passage of the Solar System through a dust lane bordering a spiral arm of the Galaxy may cause a temporary variation of the Sun's radiation and so lead to an ice epoch on Earth.

RECENT contributions to *Nature*[1,2] indicate that there still exists no consensus as to what agency triggers the onset of an ice age on Earth. Most discussions make passing mention of some possible astronomical agency, but without any critical up-to-date examination. The purpose of this paper is to show that one particular astronomical hypothesis gains much in plausibility in the light of recent far-reaching advances in our knowledge of the entire Milky Way Galaxy—and, indeed, of the behaviour of galaxies in general. Certain processes which we could previously say might occur, we can now say almost certainly must occur and must do so in ways that seem to account for several characteristic features of ice ages.

In a remarkable paper[3] published in 1939 on "The effect of interstellar matter on climatic variation", Hoyle and Lyttleton suggested that the passage of the Solar System through a cloud of interstellar matter could result in an ice age on Earth. Sir George Simpson had inferred that a slight increase in the flux of solar radiation reaching the Earth is needed to initiate an ice age by leading to an increase in the rate of precipitation. Hoyle and Lyttleton showed that matter might fall into the Sun from an interstellar cloud at a rate such that the release of gravitational energy would produce a temporary brightening of the Sun by the requisite amount. They were able to infer that the hypothesis would account qualitatively for several associated properties. Their work has been quoted repeatedly since, although conclusive comments have been lacking. On the geological and meteorological side, the inconclusiveness has been associated with the absence of positive predictions; the work had shown rather what could occur than what had to occur. On the astronomical side, it required cloud densities much greater than, until very recently, astronomers deemed credible; also the general astronomical significance of accretion came to be largely discredited and unfortunately this particular application suffered unduly in consequence.

Energy-release in accretion

Hoyle and Lyttleton inaugurated extensive studies of the accretion process in many astronomical contexts. For the present purpose the simple theory in their paper will still suffice; using so far as possible the original notation, it is:

Let the Sun of mass M and radius R move with uniform velocity v through a cloud having uniform (undisturbed) density ρ. The gravitational attraction of the Sun deflects material into the wake of the Sun where its transverse motion is destroyed by mutual collisions. Out to a certain distance λ, say, such material is left with less than escape speed. It will fall into the Sun, thereby releasing energy GM/R per unit mass, where G is the gravitational constant, since it falls from distances large compared with R. Simple dynamics shows that in this way the Sun sweeps up all the cloud material within distance σ of its path, where

$$\sigma = 2\,GM/v^2 \tag{1}$$

and in fact $\lambda = \sigma$. The rate of accretion is therefore $dM/dt = \pi\sigma^2\rho v$. Thus the resulting rate of energy release is a fraction ε of the normal luminosity of the Sun L, where

$$\varepsilon = \frac{GM}{RL}\frac{dM}{dt} = 4\pi\frac{G^3M^3\rho}{RLv^3} \tag{2}$$

If the parameters are initially in cgs units, it happens to be convenient to write $\rho = nH_2$, where H_2 is the mass of the hydrogen molecule, and n is the density in equivalent hydrogen molecules per cubic centimetre, and also to write $v = 10^5 V$, where V is then the speed in km s^{-1}. Figure 1 is a logarithmic plot of n against V for $\varepsilon = 1, 10, 100\%$, using solar values of M, R, L. It also shows σ in astronomical units plotted against V.

This simple theory may be criticised for treating the cloud-material as composed of independent particles rather than as a compressible fluid. But later elaborations of accretion theory[4,5] tend to show that the simple theory should give valid orders of magnitude. As regards the energy released, all that signifies is that the computed mass finds its way on to the Sun from distances much greater than the solar radius. Then the stated amount of gravitational energy must be liberated; the detailed mechanism need not concern us here. Hoyle and Lyttleton showed also that ancillary effects like the gravitational attraction of the cloud as a whole need not be considered in this connection. The resistance of the cloud[6] is probably unimportant in cases of interest here but any effect it has would be favourable to the processes under discussion.

Astronomical relevancies

What is seen as a spiral arm of a galaxy like our own is a region where there are (short lived) very bright stars illuminating clouds of interstellar gas and dust. Stars and gas must move through the arms; otherwise, because of the observed differential rotation, the arms would have become tightly coiled. Observation thus shows that the spiral structure is some wave-pattern moving through the disk of stars and gas; or, what is the same thing, it is a pattern through which these objects move. Observation leads to the further inference that these objects enter an arm at its inner (concave) edge, except perhaps in the outermost parts of the system. Furthermore, observation often shows a dust lane along this edge. This is interpreted as a shock-wave effect whereby the clouds become temporarily compressed as they cross this lane. It may be inferred that the compressed clouds are the sites of formation of clusters of new stars, the most massive of which are the very bright stars that illuminate the arm as they proceed on their way but burn themselves out before emerging into the interarm region. At each passage through an arm, only a small fraction of the total diffuse matter can be formed into stars; otherwise very little could still remain. All this interpretation is supported by many other observations; it depends on no particular theory of spiral arms although, as is well-known, a density-wave theory has been elaborated.

On this picture, the Sun and its planetary system were formed in a compressed cloud some 5×10^9 yr ago; since then it has circulated round the Galaxy some 20 times. It can be estimated that it crosses a spiral arm at intervals of the order of 10^8 yr, spending the order of 10^7 yr crossing the main part of the arm and the order of 10^6 yr crossing the compression lane. So it has passed through an arm some 50 times in all. At present, astronomers plot its position as just inside, or just about to enter, the inner edge of the so-called 'Orion arm'; consequently we infer that it entered the associated compression lane a million or more years ago and left it only comparatively recently.

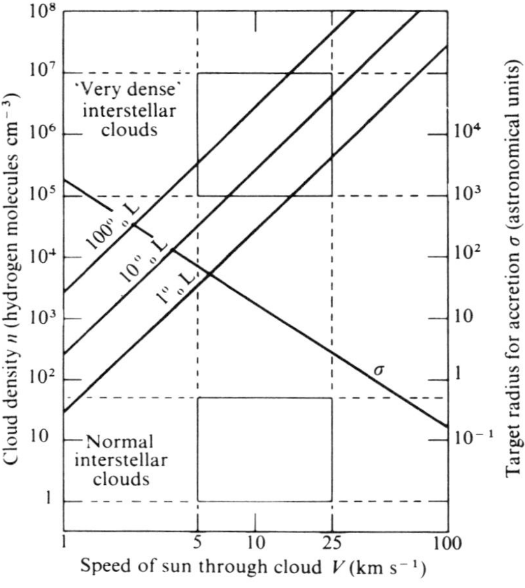

Fig. 1 If the Sun has velocity V through a cloud of density n, the material within distance σ from its path given by equation (1) falls into it, thereby releasing energy at rate εL given by equation (2), L being the Sun's luminosity.

Knowledge of interstellar clouds has also advanced greatly in recent years; the discovery and study of an unexpected variety of molecules therein has led to remarkable inferences about their densities in certain cases. In 1972 Lequeux[7] reviewed some of the pioneering recent developments. In particular, he described three 'very dense' clouds: one with a mean density of about 10^5 hydrogen molecules per cm^3 over a distance of 10–15 pc, one for which he estimates about 4×10^5 molecules cm^{-3} over about 0.33 pc, and a third with a density "probably much larger than 10^7 cm^{-3} over an extent of at least 1.5 pc". Since the 'compression lanes' are apparently the densest extended regions of interstellar matter in a galaxy, and since densities in the range just quoted are actually inferred, we can only conclude that the densities in the lanes reach at least this range of values. For 'normal clouds' Lequeux quotes a density range which in the same units is 1–50 cm^{-3}. These ranges are indicated in Fig. 1.

Ice ages

According to a summary given by Allen[8], ice epochs occur on Earth with a period of about 250 Myr, each epoch having a duration of a few million years and consisting of several glaciations with period (irregular) of glaciation and inter-glaciation about 0.25 Myr, the glaciation itself lasting about a fifth of the period. The most recent epoch was the Quaternary, which is inferred to have begun about a million years ago, and the latest glaciation ended about 11,000 yr ago. In the geological literature there seems to be a good deal of uncertainty about the whole subject; nevertheless it seems to be fairly generally agreed that there is a characteristic time of the order of 10^8 yr for the occurrence of an ice epoch, and a characteristic duration of the order 10^6 yr, each epoch consisting of several glaciations. ('Ice age' seems to be used sometimes for 'ice epoch' and sometimes for 'glaciation'; it seems best to speak of the whole set of phenomena as 'ice ages'.)

Implications of repeated dust lane encounters

Since the speed of the Sun relative to the nearer stars is about 20 km s^{-1} and since there is no known systematic difference between the motions of stars like the Sun and of interstellar material in the same region of the Galaxy, this must be also about the Sun's speed relative to the nearer interstellar clouds. Further, Allen[8] quotes for the r.m.s. random velocity of the clouds in the line of sight a value 9 km s^{-1}. So the speed of the Sun V relative to an individual cloud which it encounters may be taken to be in about the range 5–25 km s^{-1}; this seems to be the sort of range usually considered in the literature in discussing related problems; it is indicated in Fig. 1. The combined ranges of V and n for 'normal' interstellar clouds are shown by the lower rectangle in the figure. We see that any accretion in such conditions can have only a negligible effect on the Sun's luminosity. Thus the passage of the Solar System through a normal interstellar cloud has no significance in the present content.

In Fig. 1 the combined range of V and n for 'very dense' clouds is shown by the upper rectangle. It is remarkable how this is at once seen to produce conditions of crucial interest. Throughout practically the whole range, accretion could produce a temporary brightening of the Sun by over 1%, and it may exceed 100% in part of the range. The elementary theory may somewhat overestimate the possibilities, nevertheless there is ample margin for asserting that something significant can and must occur.

Lequeux's figures suggest also that, if the Sun enters a dense cloud, it is likely to travel a distance of the order of a parsec through it. At, say, 20 km s^{-1} this would take about 50,000 yr.

For the proportion of space near the galactic plane occupied by clouds Allen[8] quotes a figure of 4%. This implies that a star moving through such a region would spend 4% of its time inside some cloud. Were the same proportion to apply in a region of dense clouds, then the time just mentioned would imply an average interval of the order of 10^6 yr between immersions. But the appearance of the dark lanes suggests that clouds may be much closer together there; so a million years may be an overestimate for these regions.

If now we accept the general line of reasoning of Hoyle and Lyttleton about the effect of a temporary increase in the solar radiation and if we combine the present results with the foregoing inferences regarding the encounters of the Solar System with compressed clouds, we then infer: at intervals of the order of 10^8 yr—when the Solar System enters a compression lane—there must occur on Earth an ice epoch. A glaciation must last on average the order of 50,000 yr—the time of travel through a particular cloud. The Solar System would take a few million years to cross the lane, and several glaciations would be expected to occur within that time. Since the Solar System is near the inner edge of the Orion arm, it has presumably 'recently' crossed an associated lane; so there should have been an ice epoch during the past few million years extending up to near the present time. When the Solar System is not in a region of very dense clouds, no such effects would occur.

All this agrees remarkably well with what is known about ice ages as summarised in the preceding section. On the one hand, there is naturally much more to be done on both the astronomical and the meteorological aspects. In particular, the way in which a change in the Sun's radiation might trigger an ice-age is still conjectural, and if one has been triggered it could be that some ebb and flow of ice could persist for quite a long time before that ice age could be said to be past. On the other hand, there are compulsive features; very dense clouds do exist and the Sun must go through them and when it does its radiation must be affected. If therefore this can be accepted as the explanation of ice ages, other features in the overall picture may be regarded as confirmed, such as the behaviour of the Sun relative to the Orion arm of the Galaxy. Of course, if ice ages are satisfactorily accounted for in this way, there is less support for an explanation[9] depending on intrinsic variability of the Sun. Also it is now worth looking for other effects that may follow from the causes discussed here, as I have attempted to do elsewhere (Halley Lecture, Oxford 1975).

Recently my attention has been called by Professor Kiang of Dunsink Observatory to the fact that the late Harlow Shapley[10,11] long ago anticipated in principle the basic ideas of both Hoyle and Lyttleton and the present work. Shapley wrote at times when relevant astronomical knowledge was more tentative and restricted, but his insight into its possible significance was astonishing. I regret not knowing of his suggestions at an earlier stage.

Received April 8; accepted May 6, 1975.

1 Gribbin, J., *Nature*, **254**, 14, 20 (1975).
2 Sellars, A., and Meadows, A. J., *Nature*, **254**, 44 (1975).
3 Hoyle, F., and Lyttleton, R. A., *Proc. Camb. Phil. Soc.*, **35**, 405–415 (1939).
4 McCrea, W. H., *Gas dynamics of cosmic clouds in IAU Symp. No. 2* Chapter 36 (North Holland, 1955).
5 Ruderman, M. A., and Spiegel, E. A., *Astrophys. J.*, **165**, 1–15 (1971).
6 McCrea, W. H., *Mon. Not. R. astr. Soc.*, **113**, 162–179 (1953).
7 Lequeux, J., *On the Origin of the Solar System* (edit. by Reeves, H.), 118–134 (Paris, 1972).
8 Allen, C. W., *Astrophysical Quantities* (third ed.) (London, 1973).
9 Dilke, F. W. W., and Gough, D. O., *Nature*, **240**, 262–4; 293–4 (1972).
10 Shapley, H., *J. Geol.*, **29**, 502–4 (1921).
11 Shapley, H., *Sky Telesc.*, **9**, 36–7 (1949).

19

Reprinted from pages 253, 257, 259–260, 263–269, 273–274, and 276 of *Irish Astron. Jour.* **12**:253–276 (1976)

SOLAR STRUCTURE, VARIABILITY, AND THE ICE AGES

(*SOLAR VARIABILITY AND CLIMATE*)*

E. J. Öpik

1. *Introduction*

In geologic time, terrestrial climate has undergone considerable variations of which the most prominent ones are those revealed in the ice ages. Setting aside minor or short-lived fluctuations (on a time scale of one hundred thousand to thousands of years or less), the Major Ice Ages, or Cold Epochs as we could call them, interrupted the relatively warm periods at intervals of about 250 million years. These Cold Epochs were, with the chronology in round figures: the present Quaternary (we are still in it); the Permo-Carboniferous, 250 million years ago; the Eocambrian, 500 m.y. ago; the Algonkian, 750 m.y. ago; the Huronian, about 1000 m.y. ago. There were earlier Cold Epochs which, however, are not so well dated. They lasted for only a few million years each, being of insignificant duration as compared to the long warm epochs in between (as was the Tertiary with the preceding Mesozoic) which thus are to be regarded as representing the normal climatic conditions of our planet. During the Cold Epochs, there were secondary fluctuations of climate known to geologists as the Ice Ages proper, increasing glaciations and warmer interglacials, on a time scale of 10^4 - 10^5 years. We are at present in a Quaternary interglacial which started 10,000 years ago with the melting of extended icecaps.

[*Editor's Note:* In pages 253 to 256 Opik discusses Quaternary climatic changes.]

Attempts at explanation of the Ice Ages have been numerous—the number of hypotheses is not much different from that of the authors. Most favoured have been "local", terrestrial explanations which, however, are unable to account for the periodic recurrence, global extent and amplitude of the variations. Thus, secular variations of the earth's orbital elements (Milankovitch) is still a favorite idea with some authors, although it ignores the fact that the mean insolation (amount of heat received) of the earth depends only on the orbital semi-major axis or the mean distance from the sun and that, by considering only the seasons (and even these being affected but slightly) and neglecting convective heat exchange between the latitudes or climatic zones (an artificial assumption which would yield almost absolute zero temperature in the polar winter), theory has been constructed on paper with little semblance to reality.

[*Editor's Note:* In pages 257 to 259 Öpik argues that neither variation of the earth's orbital elements nor of atmospheric CO_2 can account for Quaternary climatic changes.]

Variations in the solar radiative output seem to offer the most likely explanation. Yet here, too, strange proposals have been made. Thus, some very distinguished authors have suggested that ice ages were caused by an *increase* of solar radiation received by the earth: the increased mean temperature would cause more moisture to be carried by the air, leading to increased atmospheric precipitation and purportedly to more snow in polar regions. The latter statement is contrary to the logics of physical processes. Indeed, snow can be deposited only when the temperature falls below the freezing point and this would happen, on a warmer earth, farther out toward the poles, on a smaller area: thus meaning less or no ice around the poles which, at the freezing point, cannot feel the warming up and greater amount of moisture over the rest of the globe. The excess moisture would be deposited as rain, not as snow, before the freezing point is reached.

Of course, increased cloudiness could lead to some cooling because of its high reflectivity even despite an increased solar input.

With 0.5 as nearly the mean global cloudiness and a corresponding global mean temperature near +15.6°C, an increase of the cloud cover to 0.6 would cause the global mean to drop to a glacial low of +9.0°C, while at 0.4 of the earth's surface covered by clouds the mean temperature would rise to a Tertiary optimum of +21.4°C. Here seems to be a simple means of varying the climate without varying the sun, a real *deus ex machina.*

However, cloudiness, especially in the tropical and temperate zones of the earth, is mainly caused by cooling and condensation of adiabatically expanding rising atmospheric currents (cumulus clouds and cyclonic barometric lows). These can cover only one-half of the earth's surface, the rest being reserved for their counterpart of descending currents (clear sky between the cumuli and anticyclonic areas). The theoretical fraction of overcast sky is thus 50%, which actually at present is close to the global mean and hardly could change much whatever the insolation (except for some circumpolar and nocturnal fog which is of secondary importance and, little affecting the global reflectivity, works in the opposite direction as a protective screen). As with other purely terrestrial causes which have been proposed (and which it would take us too far to enumerate them all), increased cloudiness as triggered by an increase in the moisture content of the atmosphere is extremely unlikely to play a role in the long-range fluctuations of terrestrial climate.

3. *Variations of Solar Radiation*

The most straightforward cause of the Ice Ages is then to be sought in a virtual decrease of solar radiation. This was the guiding idea in a series of researches which stretched over four decades and resulted in the publication of over 30 papers and monographs (a total of more than 700 printed pages) directly concerned with the subject, as well as many supporting publications (*see* Listings in *Refs.* 1, 2, and 3). As the result of a *morphological* or many-sided approach, a consistent model has now emerged which not only claims a high probability of reality as suggested by the convergence of several otherwise unconnected lines of research, but also leads to unexpected conclusions on other problems, such as star formation or solar neutrino flux.

Only a general outline of this vast subject can be given in the present brief review.

The interpretation of the Ice Epochs as the consequence of long-range variability of the sun, resulting in recurrent temporary dimming after intervals of a few hundred million years, rests on a few very plausible initial assumptions or hypotheses, from which the conclusions then unquestionably follow through the application of well known laws of nuclear and general physics and mechanics. Calculations of the internal structure of *unmixed* stellar models are playing here a dominant role. "Unmixed" means here a sustained difference in composition between the central and outer parts of a star; this quality is not one of the hypothetical ones, but follows already from the working of nuclear reactions, hydrogen being converted into helium in the central hot regions, while preserved outside. Von Zeipel-Eddington's rotational currents are utterly inefficient with respect to mixing at the solar rate of rotation, requiring intervals of the order of 10^{14} - 10^{15} years and thus incapable of interfering with the very much faster nuclear reactions (timescale 10^{10} years) (*Ref.* 4). The success of the unmixed models in explaining the structure of giant stars lends solid support to our assumption.

With this, only one hypothetical assumption is needed to make our model work. Namely a content in the sun of elements heavier than helium, or "the metals" as they are called in the jargon of stellar structure, of not less than 1 - 2%, and probably 3% by weight must be assumed. This will pinpoint the carbon cycle, and not the p - p (proton-proton) reaction as the main supply of solar energy, conditioning a higher central temperature and convective instability in the core. A depression of the neutrino flux during our Quaternary Ice Age would then also be indicated, qualitatively at least in agreement with experiments. Important is the metal content in the central regions of the sun, which may be higher than observed in stellar atmospheres. Of the two modes of star formation, either gravitational collapse of a vast cloud, or gradual accretion of dust and gas around an initial kernel of meteoritic matter, the latter would be more likely to produce the initial conditions as just set forth here. The difficulty of angular momentum as an obstacle to contraction would then be removed, because accreting particles from a stream which is not connected with the nucleus would rather slow down instead of accelerating the rotation; and a preference for small, not large stellar masses would result, in agreement with stellar statistics. Jupiter may be an example of such an initial nucleus, unmixed and with heavier elements around the centre, which did not grow to stellar dimensions simply because of the limited supply of material—all having been swallowed up by the sun in the making.

[*Editor's Note:* In pages 261 to 262 Öpik first discusses metal content and convective transport in the sun. He then calculates that the mean global temperature varies directly, and the extent of polar ice sheets inversely, with the solar luminosity *L*.]

6. *Disturbance of Nuclear Energy Production and Solar Luminosity*

It has been shown (*Ref.* 5), analytically and with supporting integrations of stellar structure that an increase ΔL of the nuclear energy production in a stellar (solar) core above the equilibrium "thermostatic" value L_0 leads to a *decrease* of the external radiative output L or the luminosity proper by an almost equal amount, so that

$$L = L_0 - \Delta L \tag{4}$$

This seemingly paradoxical statement follows from the physically-mechanically (and numerically) defined radiative-equilibrium structures, which in static equilibrium can transport only a narrowly determined amount of radiation,

$$Q_0 = L_0 \tag{5}$$

When the energy production in the convective core is increased to $Q_i = L_0 + \Delta L$, the conditions of radiative equilibrium in an intermediate zone of whatever extent impose on the model a definitive average value of the radiative flux Q_m, little differing from the equilibrium value L_0 and intermediate between the internal energy generation, Q_i, and the external output, $Q_e = L$, so that quite closely (for 1st order small variations)

$$Q_m \approx \tfrac{1}{2}(Q_i + Q_e) = L_0 \tag{6}$$

which evidently is satisfied by

$$Q_e = L = L_0 - \Delta L \tag{4}$$

Thus an amount $Q_i - Q_e = 2\,\Delta L$ must be absorbed in the intermediate zone, leading to expansion while the excess is converted into gravitational potential energy. With expansion, the temperature in the core decreases inversely as the radius, and with a high-power dependence of energy generation, ϵ, on temperature (for the carbon cycle),

$$\epsilon \sim T^s \tag{7}$$

at $s \geqslant 15$, the energy generation rapidly decreases and an equilibrium configuration with $Q_i = L_0 = L$ is restored, on a time scale of the order of 5×10^5 years for the sun [about $2(s + 3) = 36$ times faster than the Helmholtz-Kelvin time scale of gravitational relaxation]. Besides, this thermostatic mechanism with nuclear energy generation has been predicted by Öpik in a very early, "pre-Eddingtonian", publication (*Ref.* 6).

Actually, re-adjustments of stellar structure, accompanied by internal transport of thermal and potential energy, lead to a delayed and softened response of the external radiative output to an increase in the energy generation in the core. In Fig. 4 an imaginary instantaneous response (B) to a sudden increase of the core output is compared with a calculated (by integrations) delayed response (A) (*Ref.* 5); not only is the minimum delayed, but its depth—that of a shallow minimum—is reduced to 76% of the "instantaneous" value. A further smoothing agent must be seen in a gradual, not instantaneous, onset of the central disturbance of energy generation.

7. *Gas Diffusion and Opacity as the Trigger of the Disturbance*

The most difficult problem is presented by the physical cause or the processes

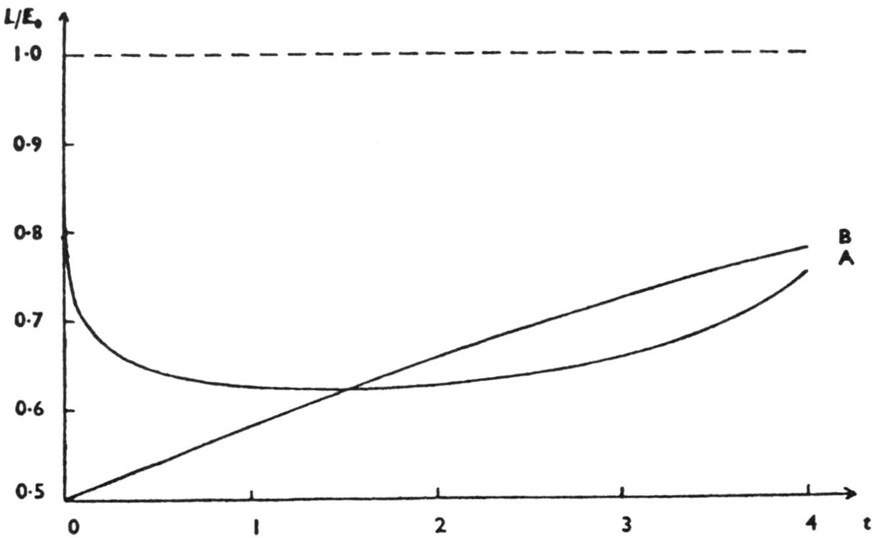

FIG. 4—*Variation of luminosity caused by a 1·5-fold sudden increase in the energy output.*
Abscissae : time in 10^8 years, for a solar mass.
Ordinates : relative luminosity.
A : complex decay (second approximation).
B : instantaneous response (first approximation).
Broken line : original (equilibrium) luminosity.

leading to a rapid (10^6 as against 10^8 or 10^9 years) increase in the nuclear energy output of the core. Yet, without introducing additional or *ad hoc* hypotheses, by considering logically and physically all the important processes occurring inside a star, a model has emerged which unequivocally leads to the events sought for: gas diffusion (of hydrogen toward the depleted core), usually neglected in theories of stellar structure because of its slowness, offers here the solution. The time scale of diffusion is proportional to the square of the linear dimension and, while the diffusion time scale over the entire solar radius is of the order of 3×10^{12} years, on a certain *a priori* constructed model and over the small distances in the critical regions around the solar core, it was found that gas diffusion (as described below) could lead to the required repeated pulses of solar luminosity at intervals of about 600 million years. With "good astrophysical accuracy", this is close enough to the 250 million years as registered for the recurrence of the Ice Epochs. The figures in such estimates can be expected to reflect only a close order of magnitude because of the lack of knowledge of the precise chemical composition and layering of the solar interior.

8. *Propagation and Decay of Disturbance*

The course of events then can be described as it follows from integrations of evolutionary sequences of stellar models generating nuclear energy by conversion of hydrogen into helium, typically assumed to vary with the 15th power of temperature (the exact power is not so important, provided it is high; the introduced simplifications could be revised with the more exact expression for the carbon cycle, but no *qualitative* change in the conclusions can be expected). In dwarf stars, of the

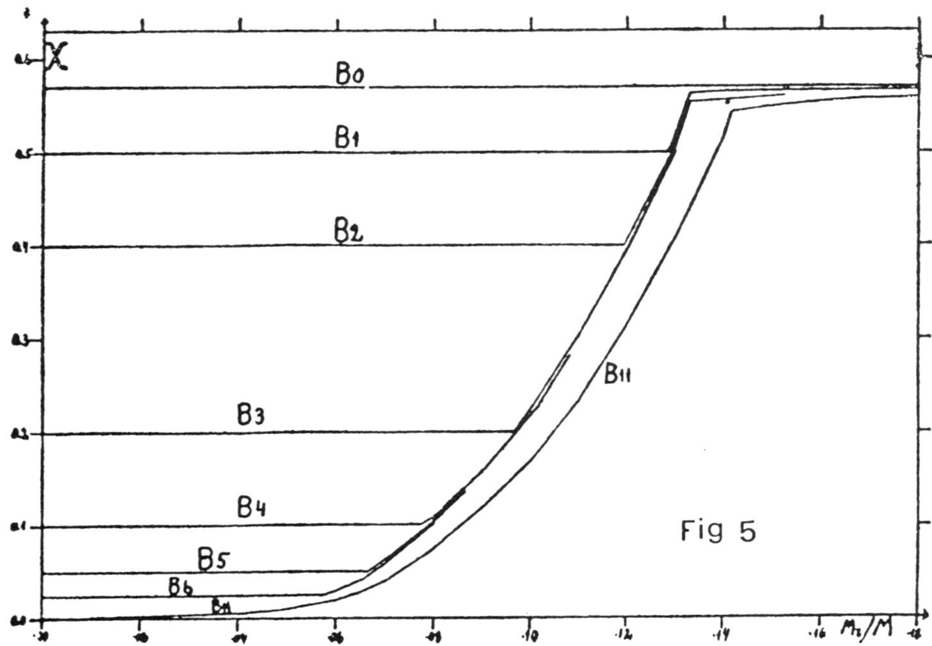

Radial distribution of hydrogen and shrinkage of the convective core with depletion of hydrogen in a sequence $B_0 \rightarrow B_{11}$ of evolving stellar models. Abscissae, fractional mass of the core; ordinates, hydrogen fraction by weight. At B_{11} the convective core virtually ceases to exist.

order of solar mass, radiation pressure can be neglected for a good approximation. As a non-dimensional criterion of convective instability in a spherically symmetric stellar model, the local polytropic index,

$$n = d\,(\log p)/d\,(\log T) - 1 \tag{8}$$

(p = pressure, T = temperature)

can be used. When this falls below the adiabatic value (or the temperature gradient exceeds the adiabatic gradient for monoatomic gas),

$$n < n_a = 1.5 \tag{9}$$

convection starts. In a stratified stellar interior, with a radial gradient of molecular weight (decreasing outwards for stability), n may fall below n_a. In such a *superadiabatic* layer of stratified molecular weight μ, the critical value of the index for onset of convection is

$$n < n_c = 2.5/[1 + 2.5\ \mathrm{d}\,(\log \mu)/\mathrm{d}\,(\log p)] - 1 \tag{10}$$

As soon as convection has started in the superadiabatic layer ($n < 1.5$), the gradient of molecular weight is wiped out and convection proceeds vigorously at a finite negative excess, $\Delta n = n - 1.5$, with a finite excess of temperature ΔT over the adiabatic value [see Equ. (1)]. Through a "surf" effect, convection then spreads over the entire superadiabatic layer. This is also essentially the process leading to the disturbance of energy generation, ΔL. In the course of evolution of a stellar core, while hydrogen is converted into helium, the convective core (at $n = 1.5$) shrinks (*see* Fig. 5), while outside, in the surrounding radiative-equilibrium layers,

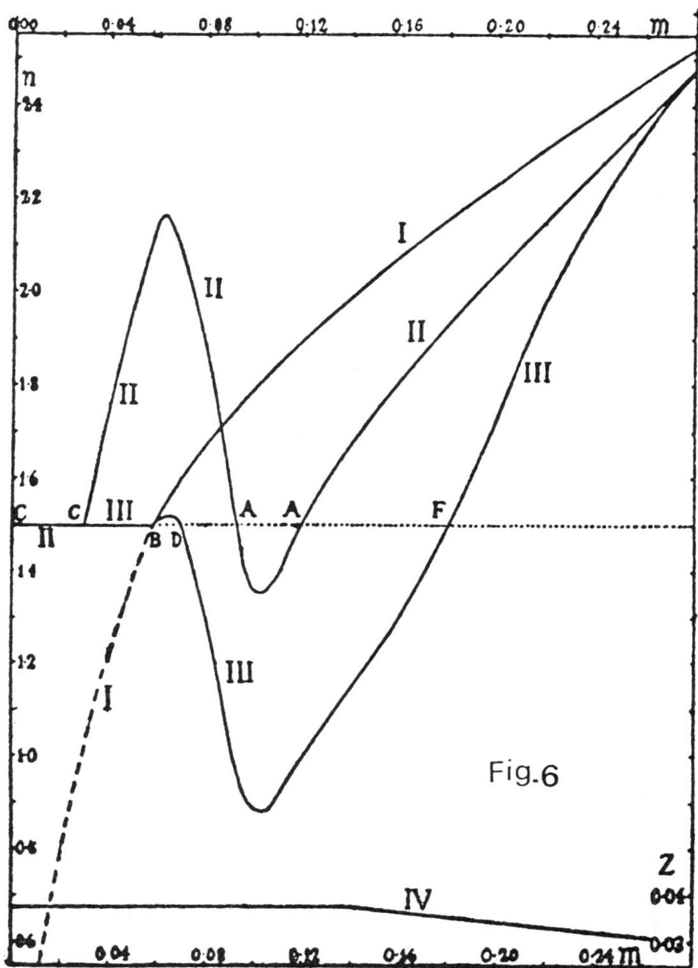

I-III, radial variation of polytropic index in an evolved typical stellar model (n, ordinates on the left; m, abscissae, fractional mass): I, without diffusion; II, with diffusion; III, with diffusion and a slightly increased central "metal" content, Z from 0.03 to 0.04 (curve IV, with ordinates on the right).

the polytropic index varies characteristically, with a tendency to drop toward the critical value at some distance from the boundary of the convective region, especially when the original metal content even slightly increases inwards (*see* Fig. 6). This is due to increased opacity (metals are its main source), requiring a steeper temperature gradient to cope with the radiative flux, and a depressed polytropic index according to the radiative equilibrium formula

$$n = (M_r/Q_r)\ T^4/(Bk\ p) - 1 \qquad (11)$$

(M_r = mass, Q_r = radiative flux from inside of radius r, k = opacity, B = $3/(16\ \pi\ a\ c\ G)$, a = radiation constant, c = velocity of light, G = gravitation constant).

Fig. 7 also shows the role of diffusion in supplying hydrogen to the exhausted core, so that the regions in radiative equilibrium outside the core also are participating in the energy generation despite their lower temperature.

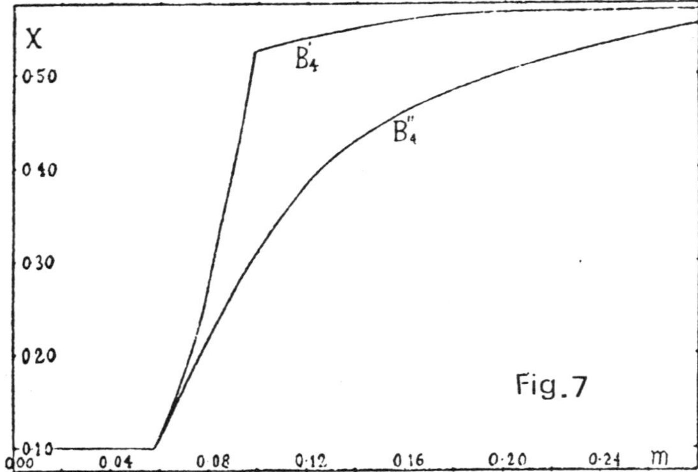

Radial distribution of hydrogen in an evolved stellar model: B'_4, by neglecting diffusion; B''_4, by allowing for diffusion of hydrogen during 3×10^9 years. Ordinates, hydrogen fraction by weight ; abscissae, fractional inner mass.

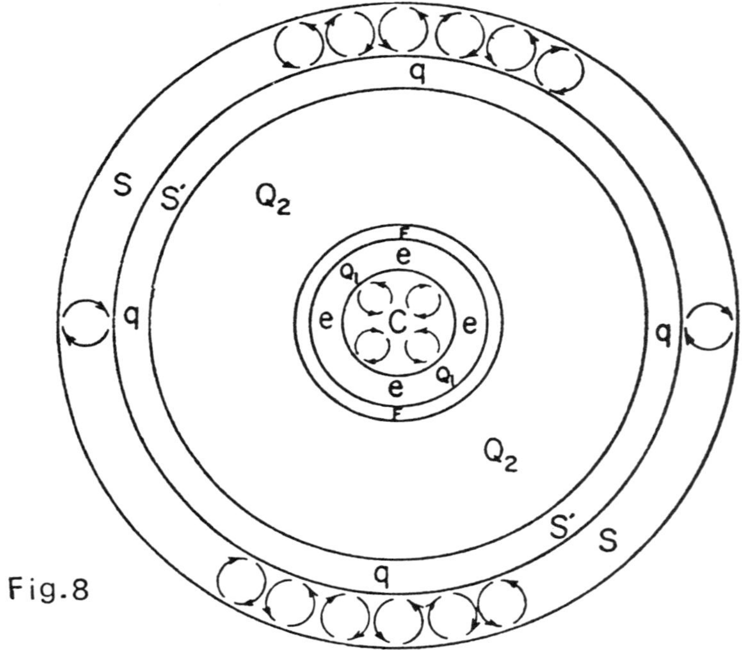

Cross section of evolved solar model, with inner convective core C, radiative-equilibrium zones Q_1, Q_2 and outer convective layers *S'*, *S*. In *F*, "metals" are accumulated and the opacity increased.

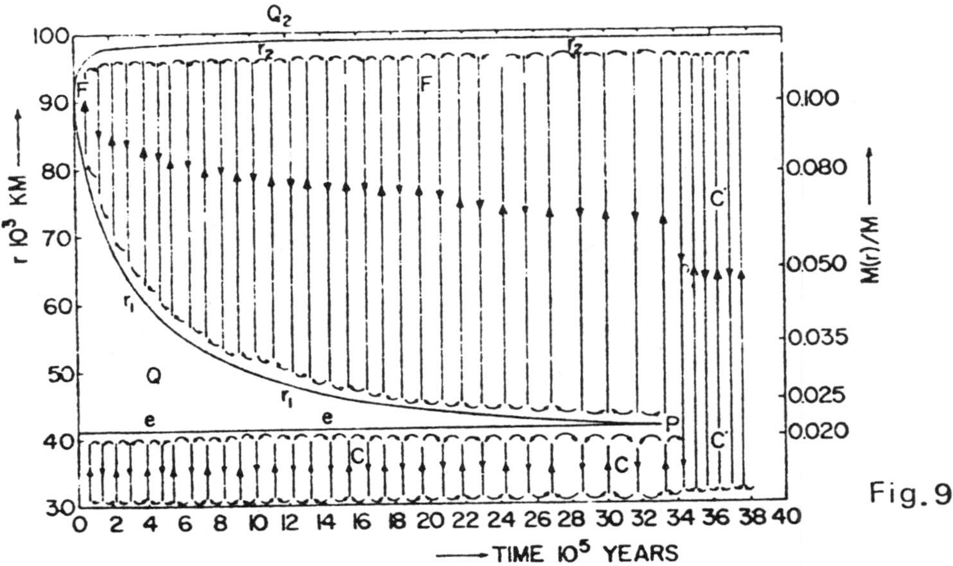

Calculated propagation of convective disturbance in a typical evolved model of solar mass. The arrows symbolize convection, spreading from outside (F) to join ultimately the convective core.

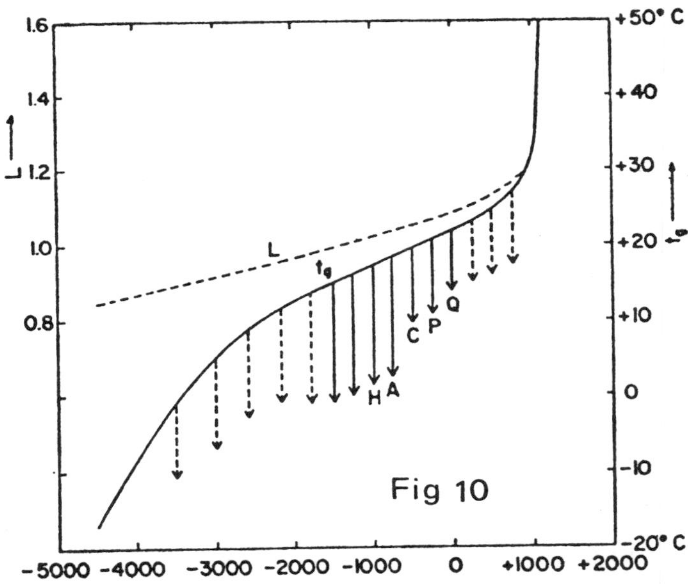

Theoretical variation of terrestrial climate (t_g, mean global temperature) and solar luminosity (L) as depending on the internal structure of the sun. Abscissae, time, millions of years before or after present. The arrows symbolize ice ages: Quaternary (Q), Permo-Carboniferous (P), Eocambrian (C), Algonkian (A), Huronian (H).

9. *Luminosity and Disturbances in Evolving Stellar Models*

Integrations of evolving stellar models thus picture the internal structure of the sun (Fig. 8) as consisting of a central convective region (C), the actual seat of the nuclear reactions, shrinking in size as the hydrogen content (X_c) decreases (Fig. 5), being replaced by helium. Outside is a vast region (Q_1, Q_2) in radiative equilibrium containing most of the solar mass, topped at the surface by an outer ("subphotospheric") convective region (S). As the convective core shrinks, hydrogen is there depleted to a degree leading to *inward* diffusion of this light constituent (Fig. 7), contrary to its natural tendency of it to diffuse outwards. The "metals", however, stay put, in a region (F) of depressed polytropic index (II or III in Fig. 6). The depression is enhanced in the shell F not far out from the inner core, where the departing hydrogen has caused the metal content and the opacity to increase. When the critical limit of n_c (Equ. 10) is reached, convection breaks through, erasing the stratification of molecular weight and, on a time scale of 3 — 4 million years as estimated from the quantitative theory of convective transport (*Ref.* 7), reaches and joins the central core: a sudden influx of atomic fuel (hydrogen) from outside then triggers off the increase ΔL. The sequence of events is described in Fig. 9.

The outcome is a Major Ice Age, with its deep minimum starting at the time when the outer convective superadiabatic region from F reaches the original core. Even before the two convective regions come into contact, the inward transport of more hydrogen (*cf.* Fig. 7) into the hotter inner regions would already increase the nuclear energy output and cause pre-dimming several million years before the final glacial outbreak, as actually is suggested by the palaeoclimatic data (*see* Fig. 3).

Table 5 and Fig. 10 show the calculated variation of solar luminosity and its effect on terrestrial climate over the past 4500 m.y., including a future outlook, for a start already with some deficiency of hydrogen in the core: 33%, instead of the average hydrogen content of 53% in the entire sun. Such initial deficiency could have been the consequence of accretion, the sun being built around an initial condensation rich in meteoric material. A higher mass fraction of "metals" in the core, favouring our model of the recurrent disturbances (*see* Fig. 6) and leading to the Ice Ages, could then be better understood.

Table 5

Evolution of the Sun, Starting with $X_c = 33\%$ (Central Hydrogen Content)*

10^6 yrs. (B.P.)	—4500	—3000	—1500	0	+600	+920	+1040	+1092	+1114	+1118	+1122
X_c%	33	28	20	10	5	2.5	1.25	0.62	0.31	0.10	0.08
Aver. X%	53.0	51.7	50.2	48.8	47.9	47.6	47.5	47.4	47.4	47.4	47.4
Luminosity (present = 1)	0.85	0.92	1.00	1.09[a]	1.15	1.20	1.26[b]	1.41	1.67	2.11	3.25
Earth temp. °C	—17**	+5†	+15	+22[a]	+26	+30	+34[b]	+43	+59	+80	+125

*Average 53% hydrogen, 44% helium, 3% heavier elements at the start.
B.P. time, millions of years after (+) or before (—) present.
**Global ice age. †Polar ice.
[a] Normal luminosity and temperature during end of Tertiary (not now, in our ice age).
[b] Heat peak due to changes in the internal structure of the sun.

[*Editor's Note:* Figure 3, which shows gradually falling temperature in Middle Europe during the Late Tertiary, has been omitted.

In pages 270 to 273 Öpik claims that the postulated secular warming of global climate since the formation of the earth (Figure 10) is supported by available paleontological and geological data. (This is further discussed in the Editor's Comments.) He then proposes that a solar "flickering" can account for the succession of Quaternary glaciations and argues forcibly that solar passage through interstellar clouds would cause planetary *warming* and not glaciation (cf. Paper 18).]

12. *Summary*

In summarizing we may say that, under the sole condition that the sun, or at least its central regions, would contain not less than 2 - 3% of the elements heavier than helium with Kramers opacity, the interplay of nuclear reactions, diffusion of hydrogen, and forced turbulence spreading from a super-adiabatic shell inwards toward a convective core, must necessarily lead to temporary decreases of solar luminosity, each lasting for a few million years and repeated within a few hundred million years. These dimmings are to be identified with the major Ice Epochs in the earth's geological past. A similar interplay of diffusion (or solar wind) with turbulence and opacity stratification in the subphotospheric shell of the sun would lead to a kind of "flickering", to advances and retreats of glaciers, with a periodicity of the order of 10^5 years and known to us as the glacials and interglacials of the Quaternary (ours is at present an interglacial, with the solar luminosity still depressed by 9% as compared to its normal value during the preceding warm Tertiary). A morphological analysis (*Ref.* 2) of the terrestrial climatic heat balance by zones of latitude and seasons, as well as for local conditions, with empirically determined absorptivity and emissivity of the atmosphere-surface complex, and with allowance for convective exchange of heat between different parts of the globe (the coefficient of proportionality in the general transport formula being determined empirically for each location and season while the convective transport is given as the difference between insolation and radiative loss), is used as a basis in judging the variations of solar luminosity during the Quaternary.

[*Editor's Note:* In pages 274 to 276 Öpik answers certain criticisms of his solar model and discusses atmospheric planetary greenhouse effects.]

References

(1) E. J. Öpik, "Climatic Change in Cosmic Perspective", *Icarus*, 4, pp. 289-307 (1965). *Armagh Obs. Contrib.* No. 51.

(2) E. J. Öpik, "A Climatological and Astronomical Interpretation of the Ice Ages and of the Past Variations of Terrestrial Climate". *Armagh Obs. Contrib.* No. 9, pp. 1-79 (1953).

(3) *Irish Astron. J.*, Vol. 10 (1972), *Special Issue*, pp. 17-22.

(4) E. J. Öpik, "Rotational Currents". *Monthly Not. Roy. Astron. Soc.*, 111, pp. 278-288 (1951). *Armagh Obs. Contrib.* No. 8.

(5) E. J. Öpik, "Secular Changes of Stellar Structure and the Ice Ages". *Monthly Not. Roy. Astron. Soc.*, 110, pp. 49-68 (1950); *Armagh Obs. Contrib.* No. 5.

(6) E. J. Öpik, "Notes on Stellar Statistics and Stellar Evolution". *Acta et Comm. Univ. Tartu*, IV (3), pp. 1-46 (1922); *Tartu Obs. Publ.*, 25, No. 2.

(7) E. J. Öpik, "Transport of Heat and Matter by Convection in Stars". *Monthly Not. Roy. Astron. Soc.*, 110, pp. 559-589 (1950). *Armagh Obs. Contrib.* No. 7.

(8) E. J. Öpik, "Convective Transport in the Problem of Climate". *Geophysical Bulletin* No. 8 (1953), *Dublin Inst. of Adv. Studies*, 14 pp.

(9) F. W. W. Dilke and D. O. Gough, "The Solar Spoon". *Nature*, 240, pp. 262-264 and 293-294 (1972).

(10) E. J. Öpik, "Mars—the Intermediate between Earth and Moon". *Irish Astron. J.*, 11, pp. 85-99 (1973); *Armagh Obs. Contrib.* No. 85.

(11) E. J. Öpik, "Rilles and Water on the Moon?" *Irish Astron. J.*, 9, pp. 79-80 (1969); *Armagh Obs. Leaflet* No. 99.

(12) E. J. Öpik, "Planetary Climatology". *Irish Astron. J.*, 9, pp. 173-210 (1970). *Armagh Obs. Leaflet* No. 105.

20

Reprinted from *Rev. Geophysics and Space Phys.* **11**:505–510 (1973)

Major Variations in Solar Luminosity?

A. G. W. Cameron

Belfer Graduate School of Science
Yeshiva University, New York, New York 10033

There has been much discussion as to whether small variations in the effective solar luminosity, associated with small variations in the orbit of the earth, may be responsible for major climatic changes [see, for example, *Broecker*, 1968]. However, the actual luminosity of the sun has appeared to be remarkably constant during the present century, in which precise measurements have been made, and the astrophysical theory of stellar structure and evolution suggests that such constancy is to be expected. Consequently, to the extent that very small variations in solar luminosity have been connected with possible causes of climatic variation, it has necessarily been assumed that the effects are greatly magnified in the response of the earth.

However, the failure of the solar neutrino experiment to detect neutrinos emitted from the solar interior indicates that something is badly wrong with our understanding of the way in which the sun operates. This problem is still not understood, but among the remedies recently suggested that show some promise of accounting for the problem, a major variation in the solar luminosity would have marked terrestrial consequences. This review gives a brief account of recent developments in this area.

As of the time of preparation of this manuscript, this whole area of study is less than one year old. Therefore it seems appropriate to tell the story chronologically.

The solar neutrino problem has existed for a considerably longer period of time. For a recent review of the problem prior to the episodes described here, see the article by *Bahcall and Sears* [1972]. The sun generates energy by converting hydrogen into helium by a complex series of nuclear reactions; the nuclear physics of these processes is now believed to be well understood. Following many of these reactions, neutrinos are emitted that should escape readily from the interior of the sun; most of those that encounter the earth pass right through without any interaction, because the interaction cross section for neutrinos is exceedingly small. Nevertheless, the cross section for inducing nuclear reactions by these neutrinos is not zero, and a major attempt to detect them has been under way for the last several years by Raymond Davis, Jr. He has established a large tank deep in a mine in South Dakota which is filled with C_2Cl_4, commercial cleaning fluid, in the amount of 10^5 gallons. Many of the solar neutrinos should be energetic enough to cause ^{37}Cl nuclei within the tank to be transformed into ^{37}Ar, which decays back to ^{37}Cl with a half-life of 35 days. The radioactive argon is periodically swept out of the tank and concentrated with argon carrier into a very small volume that can be introduced into a well-shielded counter, so that the induced radioactivity can be detected.

For convenience in describing the results of this counting, a new unit has been invented, equal to 10^{-36} reactions per second per ^{37}Cl target nucleus, which is called the solar neutrino unit, or SNU. The best-evolved models of the sun currently predict that the Davis experiment should detect 7 SNU. For some time Davis's upper limit has been below this level and has gradually crept downward, until now it is placed at 1 SNU. In fact, a year ago he was quoting 0.3 ± 0.6 SNU, with the possibility that the positive indicated count was due to cosmic ray background, so that the whole amount should be taken as an upper limit.

The whole problem was discussed at some length at the Solar Neutrino Conference, held February 25–26, 1972, in Irvine and San Clemente, California (*Reines and Trimble* [1973]; full proceedings of the conference are available from the editors at the University of California, Irvine). Although a number of suggestions were made at that time, I shall mention only two and shall concentrate on only one of these.

In the course of this conference, W. A. Fowler suggested two 'desperate' remedies for the situation. The first piece of desperation postulates the location of a nuclear resonance in the ^{6}Be compound nucleus, formed when two ^{3}He nuclei fuse together, precisely at the optimum place with respect to the thermonuclear bombarding energy, but this resonance would have most unusual and improbable nuclear properties. A specific search for this hypothetical resonance has since been carried out, with negative results [*Parker et al.*, 1972], and I shall not mention it further.

Fowler's other desperate suggestion was that the core of the sun might be in a temporarily expanded condition, so that much of the internal nuclear fire was banked down, with a consequent drastic lowering in the emitted neutrino flux from the center of the sun. At that time he did not have any specific suggestion for the mechanism that would produce this core expansion. In giving the conference summary [see *Reines and Trimble*, 1972], I suggested a possible mechanism that could provide the core expansion. The nucleus ^{3}He is an intermediate product in the nuclear reactions that convert hydrogen into helium. Its abundance in the solar core is a steady state equilibrium set up between formation and destruction. This equilibrium abundance decreases as the temperature is raised, so that the abundance of ^{3}He is expected to increase with increasing radial distance in the core of the sun. If something should occur that would mix up the solar core, this would bring ^{3}He toward the center of the sun, which would produce an increased rate of nuclear energy generation there, so that the additional heat generated would cause an expansion of the core and a decrease of all the nuclear reaction rates there. This would continue until the heat had had a chance to be transported out of the core, which would require a long time because of the radiative diffusion required to do this.

In July Fowler published his two desperation remedies [*Fowler*, 1972]. Here he not only suggested that a sudden mixing within the solar interior could greatly reduce the neutrino flux from the sun but pointed out that accompanying this decrease in the neutrino flux there would necessarily be a decrease in the solar luminosity. He then suggested that, since we are now at a time at which the solar neutrino flux is abnormally low, we should also be at a time when the solar luminosity is abnormally low, and this might be related to the present epoch of glaciation.

Meanwhile, three different calculations were being carried out to determine what would happen if a sudden mixing should occur within the core of the sun, and the reports of these three investigations appeared in *Nature* during December 1972.

The first of these investigations was by *Dilke and Gough* [1972]. These authors carried out a stability analysis of the structure of the solar core, crudely approximated by a plane parallel stratified fluid layer, in which the abundance of ^{3}He increases away from the center. This analysis indicated that the fluid layer should become overstable, so that oscillations of increasing amplitude would occur in the layer, eventually leading to a mixing within the core of the sun. Dilke and Gough carried out an evolutionary study of the consequences of such a mixing and found that a decrease in solar luminosity should occur. They discussed the effect that this might have on terrestrial ice ages. However, the Dilke and Gough mixing mechanism has subsequently been criticized on a number of technical grounds by P. K. Ulrich and R. T. Rood (unpublished manuscript, 1973).

The second of the studies published in December was that of *Rood* [1972], who arbitrarily mixed the solar interior by varying amounts and studied the effects on the solar neutrino flux and luminosity. He found that if a large enough fraction of the solar interior were suddenly mixed, the solar neutrino flux would drop sufficiently to reduce the expected Davis counting rate below 1 SNU, and there would be a corresponding decrease in the solar luminosity.

Similar results were published at the same time by *Ezer and Cameron* [1972]. In their work, a model of the sun that had been evolved for 4.6×10^9 years had the central 0.56 of its mass suddenly mixed. This was considered to be an atypical mixing, because it brings large amounts of fresh hydrogen toward the center of the sun, which allows the center of the sun to expand and relax toward a new equilibrium state. After this relaxation had occurred, Ezer and Cameron then mixed the central region of the sun again, producing a result that should be more typical of periodic mixing in the interior of the sun. This brought the solar neutrino flux down to a counting rate equivalent to about 0.5 SNU. The effects on the solar luminosity are shown in Figure 1; the result of the second mixing should be considered more typical. It may be seen that the solar luminosity drops to about $\frac{2}{3}$ of the present value. Since the solar neutrinos are below the Davis upper limit only while the solar luminosity is near its minimum value, it is apparent that this explanation would require us to identify the present time as being near

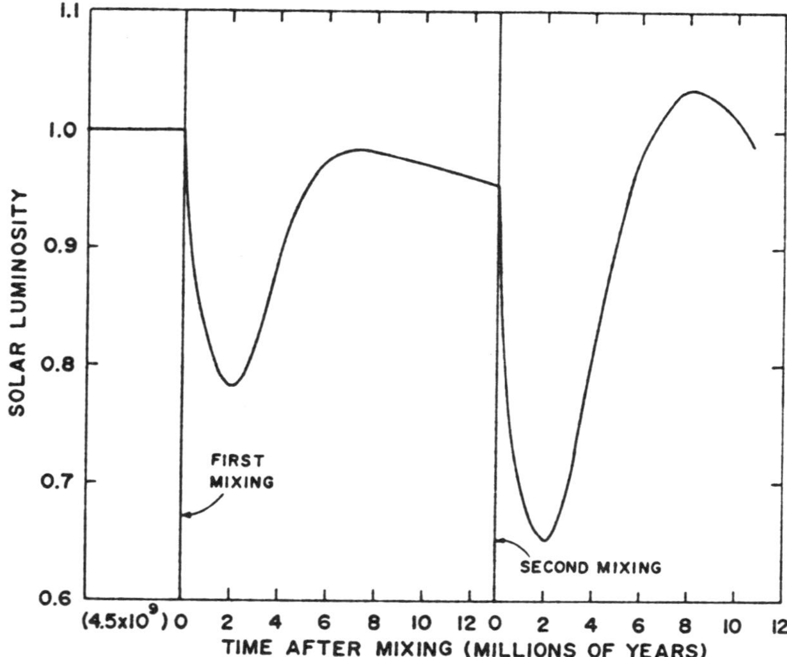

Fig. 1. Variation of the solar luminosity during the numerical experiments carried out by *Ezer and Cameron* [1972].

the minimum in the solar luminosity, so that a normal solar model should be considerably more luminous than has been assumed in these calculations, and the luminosity should then come down to the present observed luminosity of the sun, or perhaps a little bit lower. Such a change in the normal luminosity of the sun can be produced if one changes the ratio of hydrogen to helium in the solar interior.

Although, like Rood, Ezer and Cameron arbitrarily mixed the solar composition in the core of the sun to bring about the sudden core expansion, they suggested as a possible mixing mechanism the recent results of *Sakurai* [1972], who found that Eddington-Sweet circulation currents in the solar interior have a tendency to pump angular momentum toward the center of the sun, provided that angular momentum is being drained out of the solar surface layers by a solar wind torque. This might have the effect of promoting a slow mixing in the inner layers of the sun, but it is not clear that the mixing could be sudden.

We are probably at the beginning of a prolonged debate over whether or not subtle hydrodynamic effects can produce sudden or not-too-slow mixing in the core of the sun. I do not think it necessary that the student of the earth should wait until this debate is settled before inquiring whether such remarkably large excursions in the solar luminosity may have some interesting terrestrial effects that he can try to calculate or to discern in the geological record. If the large-scale glaciation of the earth appears to require a major drop in solar luminosity, then let him make that assumption and not be intimidated by previous astronomical dogma.

If we define a major period of glaciation as one in which the polar caps are ice covered, there are conflicting estimates of the duration of such periods of glaciation, ranging from 10 to 50 m.y. [see articles in *Fairbridge*, 1967]. I have not found any critical discussion of this duration in the literature. The major decrease in solar luminosity indicated by the calculations shown in Figure 1 lasts only some 6 m.y. This is essentially the Kelvin-Helmholtz contraction time for the solar core, and it cannot be changed very greatly if the mixing event in the solar core is a sudden one. However, the duration of the low-luminosity period of the sun could be lengthened if the mixing process in the solar core were one that drained the ^{3}He reservoir only gradually toward the center, over a period of many millions of years, with the transport perhaps being limited by angular momentum diffusion rates or something of the sort. However, if the solar core mixing takes place on such an intermediate time scale, the excursion downward in the solar luminosity should be correspondingly shallower. In any case, the luminosity decrease shown in Figure 1 depends on the assumption of the amount of mass mixed in the sun, and it has no particular significance other than to indicate how large the luminosity excursion could be. A recent estimate of temperature variations in the terrestrial oceans has been made by *Fenyves et al.* [1973], who indicate that the ocean temperatures may have been significantly depressed from the normal values during the last 50 m.y. If this relatively long time scale should be correct, then the mixing in the solar core would have to take place on such an intermediate time scale, but a very large amount of mass would have to be mixed in order that the solar neutrino flux could fall below the Davis upper limit.

The geological record appears to indicate that major periods of glaciation occur about every 2 or 3 $\times$ 10^8 years [*Schwarzbach*, 1963]. This should be a useful boundary condition for those attempting to determine a mixing mechanism for the sun; the mixing process must operate in a major intermittent manner.

If the normal luminosity of the sun is much higher than at the present time, a number of major astronomical problems are posed. The normal main sequence mass-luminosity relation for stars must be incorrect, and the calculated main sequence hydrogen-burning lifetimes must also be incorrect, but it is not clear what the sign of the error is. I have discussed this problem with several observational astronomers and have not found anyone who believes that such errors cannot exist. The calculations show that when the solar luminosity decreases, the solar radius also decreases, with the result that the sun moves parallel to the main sequence. Thus the sun would be temporarily mimicking a star of somewhat lower mass, and it would not be possible to detect these luminosity excursions by examining the distribution of main sequence stars in Hertzsprung-Russell diagrams.

The prospect of a normally increased solar luminosity also poses a great challenge to those who attempt to calculate the global behavior of the atmosphere and oceans. If the solar luminosity increases, then the atmospheric temperature will rise, and so will the atmospheric humidity. This may increase the terrestrial cloud cover and hence the planetary albedo. Thus there may be a partial shielding of the effects of the increased luminosity from conditions at the planetary surface, and the temperature in the tropical regions may rise considerably less than would the temperature of a blackbody in space at the earth's distance from the sun. This may be due in part to the increased planetary albedo, but it may also be due in part to the increased rate of energy transport from the equator to the poles that might occur with the removal of the polar ice caps. Most hydrodynamic programs that attempt to simulate the global behavior of the atmosphere carry as inputs the observed temperature of the ocean surface. But what should be the temperature of the ocean surface in an interglacial epoch? Thus some fundamental questions in the area of the earth-sea interaction must be answered before it will be possible to simulate the interglacial climate and to determine the response of the earth to major changes in the solar luminosity.

The possibility of major changes in the solar luminosity may also have had dramatic effects on biological evolution on the earth. Presumably there are relatively gentle evolutionary pressures existing when the earth has ice-free poles and a hot and humid climate. Certainly the environmental pressures on the evolutionary process are very much greater during a glacial period when the ice sheets are advancing from and retreating to the poles. It may have been just such evolutionary pressures that produced mankind and initiated his technological development. From this, one can argue that mankind will find it most difficult to detect solar neutrinos just at the time that he first starts looking.

Acknowledgments. I am indebted to Dr. D. O. Gough for some recent correspondence, as well as to most of the other actors mentioned in this drama for useful conversations.

This has been supported in part by grants from the National Science Foundation and the National Aeronautics and Space Administration.

REFERENCES

Bahcall, J. N., and R. L. Sears, Solar neutrinos, *Annu. Rev. Astron. Astrophys., 10,* 25, 1972.

Broecker, W. S., In defense of the astronomical theory of glaciation, *Meteorol. Monogr., 8,* 144, 1968.

Dilke, F. W. W., and D. O. Gough, The solar spoon, *Nature, 240,* 262, 1972.

Ezer, D., and A. G. W. Cameron, Effects of sudden mixing in the solar core on solar neutrinos and ice ages, *Nature, 240,* 180, 1972.

Fairbridge, R. W. (Ed.), *The Encyclopedia of Atmospheric Sciences and Astrogeology,* Reinhold, New York, 1967.

Fenyves, E., Y. Jani, G. Bozoki, K. Lande, and C. K. Lee, The solar neutrino puzzle and the temperature history of the earth, Proceedings of 6th Texas Symposium on Relativistic Astrophysics, N.Y. Acad. Sci., in press, 1973.

Fowler, W. A., What cooks with solar neutrinos?, *Nature, 238,* 24, 1972.

Parker, P. D., D. J. Pisano, M. E. Coburn, and G. H. Marks, Failure to discover the '11.5-MeV' state in ^{6}Be, *Bull. Amer. Phys. Soc., 17,* 929, 1972.

Rood, R. T., A mixed-up sun and solar neutrinos, *Nature, 240,* 178, 1972.

Sakurai, T., The evolution of the solar inner rotation by the Eddington-Sweet type circulation under the influence of the solar wind torque, *Publ. Astron. Soc. Jap., 24,* 153, 1972.

Schwarzbach, M., *Climates of the Past,* D. Van Nostrand, London, 1963.

Trimble, V., and F. Reines, The solar neutrino problem—A progress(?) report, *Rev. Mod. Phys., 45,* 1, 1973.

Part IV

STRATIGRAPHIC EPISODICITY

Editor's Comments on Papers 21 Through 27

21 **HALLAM**
Secular Changes in Marine Inundation of USSR and North America Through the Phanerozoic

22 **VAIL, MITCHUM, and THOMPSON**
Excerpt from *Seismic Stratigraphy and Global Changes of Sea Level, Part 4: Global Cycles of Relative Changes of Sea Level*

23 **SLOSS**
Areas and Volumes of Cratonic Sediments, Western North America and Eastern Europe

24 **EICHLER**
Excerpts from *Origin of the Precambrian Banded Iron-Formations*

25 **GARRELS and MACKENZIE**
Sedimentary Rock Types: Relative Proportions as a Function of Geological Time

26 **RONA**
Worldwide Unconformities in Marine Sediments Related to Eustatic Changes of Sea Level

27 **MOORE et al.**
Excerpts from *Cenozoic Hiatuses in Pelagic Sediments*

The seven papers included here represent three related fields of study that together testify to the importance of long-term episodicity and cyclicity in the stratigraphic record: changes of sea level (Papers 21 and 22), sedimentary cycles (Papers 23–25), and episodicity of oceanic sedimentation (Papers 26 and 27).

CHANGES OF SEA LEVEL

The historical development of concepts of global changes of sea level was examined by Fairbridge (1961). Although changes in the relative levels of land and sea have been recognized since earliest antiquity, the idea of relative changes of sea level on a global scale was first postulated by Eduard Suess (1904–1924). A. W. Grabau (1936a and b, 1940) was foremost among the early pioneers, but his revolutionary revision of the stratigraphic nomenclature was highly controversial. The far more influential approach of Umbgrove (1942, 1947) was founded more securely on physical geology. These and other works (e.g., Schuchert, 1910, 1916, 1932; Stille, 1924; Kuenen, 1939) prepared the ground for recent studies of episodic and cyclic relative changes of sea level. (Grabau's *The Rhythm of the Ages* [1940] is now available in facsimile edition (Huntington, N.Y.: Krieger, 1978].)

As noted by Hallam (Paper 21), Egyed (1956a and b) used world paleogeographic maps to produce curves of water-covered continental areas since Early Cambrian time. His curves show an apparent emergence of the continents during the Phanerozoic and superimposed long-term oscillations of sea level. Egyed (1956b, 1957) proposed that 8½ such oscillations of transgression and regression occurred during the past 400 m.y., and that their mean period of 47 m.y. reflected the periodic accumulation and release of global tectonic stresses. Wise (1974) drew attention to pitfalls in the use of paleogeographic maps to produce curves of relative sea level. He used Schuchert's North American atlas (1955) to provide curves that show relative rises and falls of sea level of overall period (~35–45 m.y.) similar to the period proposed by Egyed but with essentially constant "freeboard" (relative elevation of continents with respect to sea level) through time. Revised curves of relative sea level determined by Hallam (Paper 21) from the lastest paleogeographic and facies maps of the USSR and North America lend weight, however, to Egyed's notion of oscillations of sea level superimposed on secular withdrawal of the sea during the Phanerozoic.

Seismic stratigraphy provides a different approach to the determination of relative changes of sea level. This important technique is described by P. R. Vail and co-workers at the Exxon exploration laboratories, Houston, in a series of eleven papers in Payton (1977), an excerpt from which is included here as Paper 22. Hallam (1978), who acknowledged this work as "one of the most significant advances in stratigraphy for many years," summarized the technique thus:

> The essence of the technique adopted by Vail and his colleagues is to define so-called "depositional sequences," which are major unconformity-bounded stratal units, in a given region. Relative rise of sea level is indicated by coastal onlap of maritime deposits, stillstand by coastal toplap and relative fall by downward shift of coastal onlap. "Sea level curves" are constructed by making chronostratigraphic correlation charts of successive depositional sequences and plotting the apparent rise and fall of sea level through time (correction is made as far as possible for differential subsidence). If good correlation exists between three or more regions in different parts of the world, eustatic control is inferred.

Modal averages of correlative regional cycles provide a hierarchy of global cycles of relative changes of sea level, which have durations of 200–300, 10–80 and 1–10 m.y. The second- and third-order cycles are characterized by a gradual rise, a period of stillstand, and a rapid fall of sea level. Paper 22 provides support for the constant freeboard concept of Wise (1974).

Despite differing methods of derivation, the curves of relative sea level of Wise (1974), Hallam (Paper 21), and Vail et al. (Paper 22) have basic similarities: (1) in broad terms, overall high sea levels during the early to middle Paleozoic and the middle to late Mesozoic and relatively low sea level during the late Paleozoic to early Mesozoic, which suggest a cycle – 300 m.y. long; and (2) the superimposed more rapid oscillations of sea level—the "second-order" cycles of Vail et al.—which have durations of 10–80 m.y. and mean periods between 35 and 55 m.y. long. Moreover, each study demonstrates a *tendency* for Period boundaries to correspond to relatively low sea levels, a circumstance that may be fundamental to our understanding of revolutions in the history of life (see Paper 32 and Part VI).

The dominant cause of long-term changes of sea level may be a combination of continental thickening by orogeny and variation in the volume, cumulative length, and spreading rates of the oceanic ridge system (Paper 21; Rice and Fairbridge, 1975; Donovan and Jones, 1979). Vail et al. (Paper 22, p. 94, page not included here) concurred that "the cause for the first-order and some second-order cycles may be related to geotectonic mechanisms;" they noted that first-order cycles may correspond to patterns of sea-floor spreading rates (cf. Paper 11, Figure 8) and orogeny, and second-order highstands in general to orogenic movements and volcanism. Thus the changes are *tectono-eustatic,* not glacio-eustatic.

Turcotte and Burke (1978) concluded that highstands over the past 500 m.y., which they ascribed to episodes of volume increase of the oceanic ridges and rapid plate accretion, which in turn is related to plate consumption and orogeny, may "prove to be the best indicators of the episodicity of orogeny." No agreement exists, however, that high sea levels correlate with rapid sea-floor spreading and orogeny; several authors have correlated marine *regression* with orogeny (e.g., Umbgrove, 1942, 1947; Grasty, 1967; Damon, Paper 9) or rapid sea-floor spreading (Flemming and Roberts, 1973). A further problem is presented by the frequent rapid falls of sea level proposed by Vail et al. (Paper 22); Hallam (1978) pointed out that "changes of oceanic ridge volume appear to happen too slowly to account for the dramatic phenomena in question." The evidence of sea-floor spreading rates suggests, in contrast, rapid transgressions due to accelerations followed by slow crustal cooling and gradual regression (Rice and Fairbridge, 1975).

Eustatic control alone of the episodic emergence and submergence of continents was rejected by Sloss and Speed (1974) and Sloss (1972, 1973, and Paper 23), who proposed instead that the continents have moved up and down in concert in response to some underlying globally effective mechanism. Finally, we should add that regional changes of the geoid may occur also, creating anomalies in opposite hemispheres. Wegener (1924) and later Fairbridge (1961) pointed out that any rapid pole migration would lead to simultaneous geoidal shifts of sea level that would be in opposite senses in different quadrants of the earth's surface.

In conclusion, the determination of global oscillations of sea level appears central to an understanding of the episodicity of diastrophism and of revolutions in the history of life. Despite general agreement on the times of major highstands and lowstands, disagreement persists concerning the overall constancy of continental freeboard and proposed causes of sea-level oscillations.

SEDIMENTARY CYCLES

Over the past thirty years long-term episodicity and cyclicity in the stratigraphic record have been recognized by numerous workers, led by the Russians N. M. Strakhov and A. B. Ronov and the American L. L. Sloss. The three papers included in this section are representative of studies of (1) long-term sedimentary cycles related to relative changes of sea level; (2) the overall time dependence of certain sedimentary facies; and (3) the recycling of sedimentary rocks through geological time.

The Russian school of lithology has devoted much attention to the study of long-term stratigraphic episodicity and cyclicity (e.g., Strakhov, 1949, 1967, 1969, 1971; Vinogradov and Ronov, 1956a; Beloussov, 1962; Ronov et al., 1969); these lengthy works proved difficult to abridge and so are not included here. The concept of pervading cyclicity (sensu stricto) is present in many Russian works. Such periodicity usually is attributed to oscillatory tectonic movements (e.g., Beloussov, 1962, pp. 392–410; Odessiky and Aynemer, 1969; Ronov et al., 1969; Bgatov and Kazarinov, 1970; Krylov and Mal'tseva, 1977), although exogenetic, galactic control also has been invoked (Malinovskiy, 1977). The concept of strict cyclicity is however not without critical analysis by compatriots (e.g., Yanshin, 1974).

In a key work, Sloss (1963) reviewed the concept of major rock-stratigraphic units of interregional scope and discussed its applications; he recognized (Sloss, 1963, 1964) that the Phanerozoic sedimentary cover of the North American craton is divisible into six major rock-stratigraphic units, each separated by an interregional unconformity. In subsequent studies Sloss (1972 and Paper 23) established the synchronism of Phanerozoic depositional and erosional events of a few millions or tens of millions of years duration on the North American and east European cratons. Paper 23 (1976) is a revised and updated synopsis of stratigraphic data for roughly equal areas of the two continents. As discussed in the previous section, Sloss believes that the temporal patterns of sediment areas and volumes on large cratons are not explained by eustasy or by accompanying isostatic compensations; he has proposed instead that cratons have undergone vertical motions in response to some endogenetic, globally effective mechanism. Other comparable studies include that of Quilty (1977), which established four pulses of marine deposition of "sedimentation cycles" for the Cenozoic of Western Australia, and the study of Soares et al. (1978), which identified six cyclic successions of erosional and depositional events that are synchronous on the Brazilian craton and are correlative with events of other cratons.

With regard to the stratigraphic record in general most sedimentary facies appear to be time-dependent to some extent, although in certain cases this time-dependence may well reflect a latitude-dependence. The episodicity of glaciogenic facies, curiously enough, seems independent of paleolatitude (see Part III). Other facies that appear to be nonuniformly distributed in time include: continental molasse, which reflects the episodicity of orogenic activity (Willard, 1950; Fairbridge, 1958; Van Houten, 1969; Logvinenko, 1976, p. 240); red beds, whose episodicity Willard (1950) attributed to repetitive climatic conditions; Phanerozoic evaporites (Meyerhoff, 1970; Malinovskiy, 1977); carbonate rocks (Anhaeusser and Button, 1976; Veizer, 1976); phosphorites (Piper and Codispoti, 1975); coal and oil shale (Malinovskiy, 1977); and iron ores, which are discussed in more detail in the next paragraph. The occurrence of Precambrian paleosols in the USSR (Mats, 1972) and of Phanerozoic bauxites (Valeton, 1972, pp. 61–65; Malinovskiy, 1977; Paper 35) also appear to be time-dependent.

The most enigmatic and among the most important economically of sedimentary facies are the Precambrian banded iron-formations (BIF). J. Eichler (of Ferteco Mineracao S/A, Rio de Janeiro) discusses in Paper 24 the main features and temporal distribution of BIF. As reviewed by Eichler and also recognized by Young (1976a), Veizer (1976), Anhaeusser and Button (1976), and Williams (Paper 35), among others, the BIF are not uniformly distributed in time. The plot (Paper 24, Figure 2) of the relative abundance through time of iron-formations, together with the frequency histogram of igneous and metamorphic age determinations (Paper 2, Figure 1), suggests that the development of iron-formations took place during intervals of overall anorogenesis preceding and partly overlapping the initial stages of major magmatic and metamorphic events during the Precambrian (see Papers 1–7 and accompanying Editor's Comments). The peaks shown in iron-formation abundance also coincide quite closely with apparent "benchmarks" in Precambrian biological evolution (see Part VI). The model of Cloud (1972, 1974) brings together such diverse events, but it appears contradicated by the wide occurrence of BIF of late Proterozoic age (see Williams, 1975b) and by the occurrence of large volumes of sulfate evaporites in mid-Proterozoic times (Walker et al., 1977). An alternative hypothesis (Williams, Paper 35, and 1973) links long-term fluctuations in global paleoclimates and environments with geotectonic megacycles through secular changes in planetary dynamics.

Finally, we have the question of secular change in lithic budgets through geological time. In Paper 25 Garrels and Mackenzie proposed that the distribution of mass of sedimentary rocks and the proportions of sedimentary rock types in the geological column vary as a function of geological age. Their Figure 1, with peaks at ~2500, ~1800, ~1000, and ~400 m.y. ago, illustrates the important influence that geotectonic megacycles (Papers 1–7) probably have exerted on the mass distribution of sedimentary rocks with age. Paper 25 and other more comprehensive works (Garrels and Mackenzie, 1971, chapter 10; Garrels et al., 1972; Veizer, 1973) draw attention to another type of long-term cyclicity of the sedimentary record—that is, the continuous recycling of sediments through geological time. Despite the important influence episodic tectonism and climatic change must exert on recycling rates, long-term secular changes in sedimentation rates have not been convincingly demonstrated. Although Holmes (1959, p. 208) believed that Phanerozoic average maximum sedimentation rates reflected "the interplay of large-scale rhythmic or cyclic processes superimposed on a progressive increase," Hudson (1964) concluded that overall sedimentation rates were essentially constant during the Phanerozoic. Gregor (1968, 1970), in contrast, considered that an abrupt reduction in denudation rates took place 300–400 m.y. ago, perhaps in response to the advent of land plants in Early Devonian time.

OCEANIC SEDIMENTATION

The deep ocean floor is the ultimate repository for sediments, and until recently it was believed that sedimentation in the deep oceans is essentially continuous. As discussed by Moore et al. (Paper 27), however, the common occurrence of hiatuses in Deep Sea Drilling Project cores has frustrated, in part, attempts to recover an unbroken geological record from the world ocean. We now know that processes of sedimentation and erosion in the ocean deeps typically are episodic.

The Deep Sea Drilling Project, which commenced in 1968, has established that hiatuses up to tens of millions of years and of wide extent occur in the stratigraphic records of the principal ocean basins (e.g., Kennett et al., 1972, 1975; Edwards, 1973; Pimm, 1974; Sigal, 1974; Davies et al., 1975; van Andel et al., 1975). The worldwide occurrence of certain unconformities in oceanic sediments was first recognized in 1973 by Rona (Paper 26). Dr. Rona (personal communication, 1978) comments that "this concept diverged from that prevalent at the time the paper was written, which held that all unconformities were local features of the stratigraphic record; indeed the authors of the various Deep Sea Drilling Reports from which these data were drawn ascribed the Oligocene and Paleocene unconformities to local factors." In 1978 Moore et al. (Paper 27) presented a major synthesis of Cenozoic hiatuses in oceanic sediments. They recognized the occurrence, for all ocean basins, of maxima in hiatus abundance at the Cenozoic-Mesozoic boundary, in the late Eocene-early Oligocene, and in the middle to late Miocene, and of minima in hiatus abundance in the middle Eocene, early to middle Miocene, and Quaternary. This temporal distribution of hiatus abundance clearly demonstrates the episodicity of oceanic sedimentary processes.

The episodicity of sedimentary processes in ocean basins is demonstrated by other sedimentological data. Van Andel et al. (1977) considered that during the past 125 m.y. of depositional history of the South Atlantic Ocean, long-term fluctuations occurred in the calcite compensation depth and the carbonate accumulation rate and that the accumulation of certain facies (e.g., siliceous oozes) was episodic.

Moore and Heath (1977) discussed the sedimentary recycling implied by the hiatuses in oceanic deposits. A greater proportion of older marine sediments than that predicted by the simple linear model of Garrels and Mackenzie (1971; cf. the constant mass model outlined in Paper 25) is preserved, which suggests that burial does provide sediments some protection from destruction. Moore and Heath (1977, p. 76) pointed out, however, that their study "also implies that deep-sea deposits are never buried deeply enough to be totally immune from destruction, even though the short duration of most hiatuses (less than 10 m.y.) decreases the likelihood of destruction rapidly with time."

Numerous workers (e.g., Kennett et al., 1972, 1975; Davies et al., 1975; van Andel et al., 1975; Berggren and Hollister, 1977; Moore et al., Paper 27) have attributed erosion and unconformity in deep-sea sediments to paleocirculation changes resulting from continental dispersal and climatic deterioration. Rona (Paper 26) considered, however, that such an interpretation alone does not explain the worldwide distribution and timing of certain hiatuses. Dr. Rona (personal communication, 1978) comments that Paper 26 presented the concept "that the occurrence of worldwide unconformities is basically controlled by eustatic changes in sea level. This concept is developed in more detail in a companion paper [Rona, 1973b] which relates the eustatic changes in sea level primarily to reversible volume changes of ocean ridges." Rona's work therefore has focused attention on global cycles related to tectono-eustatic events.

21

Reprinted from *Nature* **269**:769–772 (1977)

Secular changes in marine inundation of USSR and North America through the Phanerozoic

A. Hallam

Department of Geological Sciences, University of Birmingham, P.O. Box 363, Birmingham, UK

AN apparent secular withdrawal of the sea from the continents during the course of the Phanerozoic, superimposed on shorter-term oscillations of sea level, has been inferred from world palaeogeographic maps[1,2] and held to support the notion of slow earth expansion of ~0.5 mm yr^{-1} during that time[3–6]. This interpretation has been challenged by Armstrong[7] and Hallam[8], who argue that the secular withdrawal can be adequately explained without invoking a change in the Earth's radius, and by Wise[9], who disputes the very fact of secular change. The problem is re-examined here using the analysis of palaeogeographic and facies maps of the USSR and North America, which leads to a discussion on the relative merits of the various hypotheses that can be put forward to account for Phanerozoic sea-level changes. Attention is devoted to the longer-term secular changes rather than the shorter-term transgressions and regressions.

Soviet Union

The series of palaeogeographic and facies maps of the Soviet Union published in the 1960s[10], as yet unmatched for any other extensive region, allow a detailed analysis of changes in the areal extent of sea through the whole Phanerozoic by a technique outlined elsewhere[11]. All palaeogeographic maps involve, of course, a degree of inference, but there can be little doubt from the abundant data provided that the Russian maps give a close approximation to the area of continent flooded at several intervals for each successive geological period, although areas of present continental shelf cannot be included because of lack of data.

As shown in Fig. 1, the sea spread slowly in the early part of the Palaeozoic to cover approximately 60% of the USSR by middle Ordovician times. Thereafter there was a faster withdrawal to reach a minimum value at the Silurian–Devonian boundary, followed by a relatively rapid restoration to middle Ordovician values in the late Devonian and early Carboniferous. Subsequently a progressive withdrawal culminated in a minimum at the Permian–Triassic boundary. During the Mesozoic, there were two transgressive peaks, in the late Jurassic and late Cretaceous, interrupted by an early Cretacous minimum and followed by a sharp Cainozoic decline interrupted by a small peak in the middle to late Eocene, and a further, more dubious, peak in the Pliocene. None of the Mesozoic and Cainozoic peaks attained the values of the Palaeozoic maxima and in fact barely extended beyond the Silurian–Devonian trough.

The close spacing of data points, back at least to the Devonian, rules out a systematic temporal bias, with the length of time intervals chosen increasing the probability of overestimating the extent of sea the earlier the period[9]. Indeed, the probability of losing the stratigraphic record through subsequent deep burial, metamorphism and erosion must increase with time, and so the maps for older periods are more likely to underestimate than overestimate the former extent of marine cover. Furthermore, cratonic areas such as the Russian Platform and margins of the Siberian Shield, in fact almost everywhere outside eugeosynclinal regions, contain much more substantial proportions of carbonates and evaporites to terrigenous clastics in the Palaeozoic than Mesozoic or Cainozoic, implying more areally restricted and topographically subdued sediment sources (mechanical denudation rates on the present continents show a tendency towards an exponential increase with increasing topographic elevation[12]). Hence the Russian data support strongly the notion of Phanerozoic secular withdrawal of sea.

North America

The only comprehensive series of palaeogeographic maps available for North America is that published in the atlas of Schuchert[13], which was used by Wise[9] to demonstrate a condition of essentially constant continental freeboard throughout the Phanerozoic, interrupted by short-term oscillations of sea level, 80% of which remained within about 60 m of a normal freeboard level ~20 m above the present level. These maps were drawn, however, several decades before the atlas was published, and therefore do not take

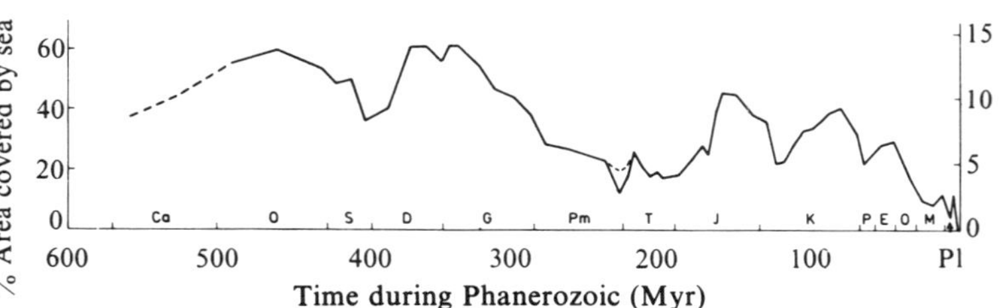

Fig. 1 Area of the Soviet Union covered by sea at different times during the Phanerozoic, expressed in both relative and absolute terms. Dating based on Geological Society of London Phanerozoic time scale. Ca, Cambrian; O, Ordovician; S, Silurian; D. Devonian; C, Carboniferous; Pm, Permian; T, Triassic; J, Jurassic; K, Cretaceous; P, Palaeocene; E, Eocene; O, Oligocene; M, Miocene; Pl, Plio–Pleistocene. Broken line signifies extent of non-marine Permo–Triassic.

into account extensive modern discoveries of strata, especially in the Arctic. Moreover, Schuchert left large areas of the cordilleran region blank for the Palaeozoic, but facies changes from areas further east clearly indicate a marine eugeosynclinal regime[14]. As with the USSR, Palaeozoic rocks of the North American central craton are much richer in carbonates than younger rocks, and testify to a correspondingly greater spread of sea, so that Schuchert's maps, based on erosional remnants, tend to be cautious and substantial underestimates for times earlier than the Mesozoic.

Data from a new atlas of facies distribution through time[15] are here used to propose a revised graphical interpretation of North American data. In the absence of up-to-date palaeogeographic maps for individual stages, an attempt was made to determine the approximate maximum extent of sea for each period, using references cited in the atlas and making fairly conservative inferences. This allows a general comparison with Schuchert's data and the world data of the Termiers and Strakhov, plotted in a similar way (Fig. 2).

Schuchert considerably underestimated the former extent of seas from the Cambrian to the Lower Carboniferous. Subsequently, the Schuchert and revised North American curves match quiet closely with only a relatively slight but systematic increase in inferred extent of flooding in the revised version. The revised curve matches the similarly plotted Russian curve much more closely than the Schuchert curve. The main differences are that the North American seas attained their maximum in the late Ordovician rather than the middle Ordovician and late Devonian–early Carboniferous, although the values are remarkably similar at around 60%, and the late Jurassic transgression was less extensive than the late Cretaceous. Because the two continents comprise approximately one-third of the Earth's land surface it might reasonably be inferred that the changes in question are primarily expressions of eustatic changes in sea level, with the relatively minor differences being attributable to regional differences in topography and rates of subsidence. This is confirmed by the close match of the two curves with those based on the Termier and Strakhov world palaeogeographic maps. A eustatic interpretation is both simpler and more plausible than the alternative proposed by Sloss and Speed[16], that the continents have moved up and down in concert. An additional point is that the Permo–Triassic minimum must be due to a genuine eustatic fall and not to the silting up of shallow sea during a phase of stillstand. This is because the extent of end Permian and beginning Triassic non-marine sediments is negligible in both the USSR (Fig. 1) and North America.

Discussion

Several different explanations can account for the long-term secular changes under consideration. Earth expansion: besides the fact that the notion of slow earth expansion poses serious physical problems[17] it has already been pointed out that the Phanerozoic overall regression can be explained more conservatively without considering a change in the earth's radius. In addition, the regression ought to have continued back into the Proterozoic, but the end of the Proterozoic was in fact a regressive phase. Thus Lower Cambrian deposits transgress widely on to much older rocks, often Archaean or early Proterozoic in age, a fact that has given rise to the term Lipalian interval. This regressive interval is probably comparable in magnitude with the end-Palaeozoic and late Cainozoic examples.

The alternative expansion model, which involves a much more rapid increase in the earth's radius since the early Mesozoic, leading to dispersal of the fragments of Pangaea[6,18], is even more clearly refuted by physical arguments[19–21] and is incompatible with the dominantly transgressive character of Mesozoic seas. Holmes[4] even interpreted the Quaternary secular fall in sea level, superimposed on glacially-controlled oscillations, as a consequence of rapid earth expansion.

Glacial control: the onset of the Cainozoic ice age might be considered to have been responsible for a substantial component of the contemporary drop in sea level. It has been estimated that sea level would rise 46 m if all polar ice was melted[7]. An Antarctic ice sheet first formed in the late Miocene, although ice apparently started to accumulate as early as the late Eocene[22]. Abstraction of ice from the ocean system can only account for a small proportion of the lowering of sea level since the late Cretaceous high stand, however, if we accept the figure of slightly over 500 m computed from estimated changes through time of oceanic ridge volume[23], and supported independently by an estimate of the overall continental margin sedimentary offlap through the Cainozoic (P. R. Vail, personal communication).

The glacial explanation is even less successful in accounting for the major late Permian–early Triassic regression, which took place after the melting of the extensive late Carboniferous–early Permian Gondwana ice cap and the concomitant establishment of a more equable world climate.

Orogeny: plate tectonics provides a ready mechanism for the thickening and hence uplift of elongate sectors of crust at continental margins overlying subduction zones, by frictional melting of the descending slab giving rise to ascending magmas[8], and perhaps more locally by the underthrusting of one continental edge beneath another. Thus closure of the Iapetus Ocean at the end of the Lower Palaeozoic in north-west Europe and eastern North America has converted a marginal basin–island arc–trench regime to a tectonically more stable zone of continent which has not been the site of subsequent orogeny but has persisted in most places as a positive area welded to the adjacent Precambrian shields and only locally and occasionally inundated by shallow sea.

With the detailed Russian data of Fig. 1, the clearest indication of orogeny correlating with regression concerns the substantial post-Valanginian eastward withdrawal of the sea in eastern Siberia, which is bound up with the younger Cimmerian Orogeny[24]. The subsequent Aptian to Campanian transgression never reached as far west as the late Jurassic seas, and the Cimmerian event is the principal reason for the Russian late Cretaceous extent of sea being less than the late Jurassic, hence contradicting the world picture.

Similarly, the late Carboniferous–Permian regression is associated with the multiphase Hercynian Orogeny, commencing in the middle Carboniferous and reaching a climax in the early Permian, and converting the old Uralian and Angara geosynclines into cratonic regimes. More tentatively, the regression commencing in the late Ordovician and reaching a maximum at the end of the Silurian seems to be connected with the Russian equivalent of the Caledonian Orogeny, as manifested around the western and southern margins of the Siberian Shield. In the far east of the country, a notable end-Cretaceous orogeny is reported locally[24], and a number of minor Tertiary orogenic phases in the same region might help to account for part of the Cainozoic overall regression.

The North American data of Fig. 2 are, of course, less precise, but it is apparent that a component of the post-middle Ordovician Palaeozoic regression may be attributed to the progressive conversion of the Appalachian and neighbouring regions by the successive Taconic, Acadian and Appalachian orogenies from a geosynclinal-island arc zone into a positive upland area welded to the central North American craton, which lay beyond the reach of subsequent Mesozoic and Cainozoic transgressions. Like-

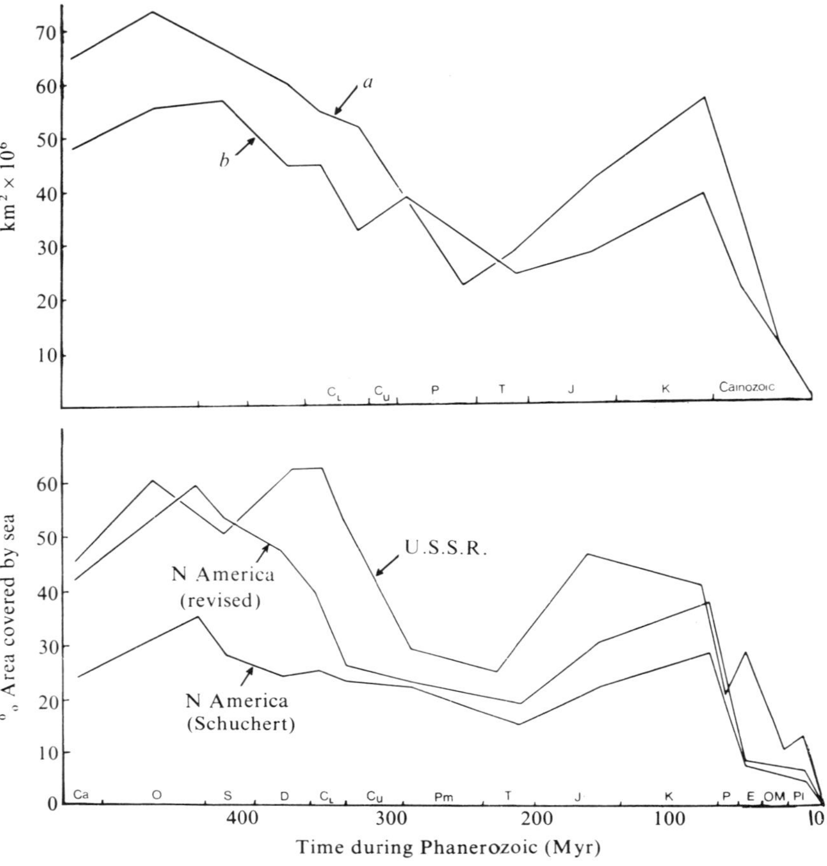

Fig. 2 Maximum degree of marine inundation for each geological period for USSR and North America (according to Schuchert and as revised in this paper, and comparison with the world data of Strakhov (*a*) and the Termiers (*b*), derived from ref. 8.

wise, the pronounced post-Cretaceous regression correlates with the Laramide Orogeny in the west of the continent.

That orogeny cannot account for all the regression is shown by examination of the history of extensive shield and platform areas of both continental masses, which were unaffected by Phanerozoic diastrophism. Thus the Russian Platform was extensively covered by shallow sea thoughout the great bulk of Phanerozoic time but was completely emergent by the end of the Permian. A post-Cretaceous emergence is also apparent after the extensive late Mesozoic transgressions. Similarly, the Siberian Shield was substantially transgressed in the Cambro-Ordovician but a major withdrawal of the sea commenced in the west towards the close of the Silurian, with a land area emerging between the Ural and east Siberian seas. From middle Jurassic times onwards the sea covered the huge West Siberian depression but withdrew definitively in the Oligocene.

In North America, shallow seas covered a large part of the Canadian Shield from the Ordovician to the Devonian, but withdrew definitively in the Lower Carboniferous, as they did in the American mid-west in the Upper Carboniferous and Permian, and in the USA and Canadian Western Interior after the Cretaceous.

More generally, only a small percentage of the present land surface is occupied by 'Alpine' orogenic zones, yet it has been estimated that continental relief has increased by a factor of about 2.5 in the late Cainozoic[25].

Continental underplating: to account for the apparent emergence through the Phanerozoic of such cratonic areas I formerly argued for a slight thickening by sialic underplating, using a term proposed for the African cratons by Shackleton[26]. Several lines of evidence suggest, however, that old cratons have remained tectonically inert after the early Proterozoic and have been subjected to relatively little uplift or erosion since their time of formation, an interpretation supported by low heat flow at present[27]. Persistence of ancient structural patterns and distinctive types of mineral assemblage is also incompatible with significant change at the base of the crust[28]. Underplating, in fact, invokes a petrological process not required by plate tectonics and not supported by independent evidence. It is probably only relevant, if at all, to the Archaean.

Variations in rate of seafloor spreading: Hays and Pitman[23] attribute the late Cretaceous major transgression to a rapid seafloor spreading in the Pacific and Atlantic from 110 to 85 Myr, with a correspondingly greater extent of 'hot', buoyant oceanic ridge displacing seawater on to the continents. This interpretation relies on the accuracy of correlation of magnetic anomalies and has been questioned by Berggren *et al.*[30], who proposed an amended time scale consistent with more-or-less uniform rate of spreading. Even disregarding this criticism, the Hays and Pitman interpretation does not account satisfactorily for the marked secular regression through the Cainozoic. According to the data presented by Larsen and Pitman[29], the spreading rate for the Atlantic remained more or less constant from Palaeocene time onwards, and matched that for the late Jurassic and early Cretaceous. The rate was similarly constant for the Phoenix–Pacific plate boundary, while the fuller data for the Farallon–Pacific plate boundary shows a fall at 85 Myr to a lower rate from the Coniacian to the Palaeocene. Thereafter there was a sharp but small reduction at the Palaeocene–Eocene boundary, a rise to a maximum in the Oligocene and subsequently a stepwise decline to the present. The Oligocene, however, is clearly regressive with respect to the Palaeocene and Eocene[31].

Taking Indian Ocean data into account only complicates the picture further, and in a way not obviously consistent with the Hays and Pitman interpretation[32]. India moved away from Antarctica at a very rapid rate in the early Tertiary, and was followed by a period when little or no spreading took place west of the 90°E Ridge. Spreading from the Carlsberg Ridge has proceeded at a more or less constant rate of 1.2–1.3 cm yr^{-1} over the past 30–35 Myr, back into the Oligocene, following a period of very slow spreading back to the late Palaeocene. In the central Indian Ocean the spreading rate has not been constant as in the Atlantic and Pacific, but changed from 5.7 cm yr^{-1} in the Campanian to $\sim$ 9 cm yr^{-1}, persisted at that rate until the middle Palaeocene and then decelerated.

Changes in length of ocean ridge system: the most plausible alternative to the spreading rate interpretation to account for the spectacular late Cretaceous transgression, which was obviously a very unusual geological event, is that a large volume of seawater was displaced by creation of new oceanic ridges in the South Atlantic and Indian oceans, since this was the time when effective dispersal of the fragments of Gondwana commenced. Such an interpretation would imply that the Cainozoic regression was due at least partly to the progressive loss of ridges elsewhere by subduction. That this is likely to have happened is evident from the analysis of Larsen and Pitman[29]. According to their interpretation, in the middle Cretaceous there were four major plates in the Pacific region, the Kula, Farallon, Phoenix and Pacific plates. In the past 110 Myr there has been progressive consumption through subduction beneath Asia and the Americas of ridges separating the Kula from the Farallon, and the Farallon from the Phoenix plate, together with the Pacific–Kula ridge.

On the other hand, the early Tertiary saw the creation of new ridges between Greenland and Scandinavia as the North Atlantic opened, and between Antarctica and Australia. This might account for the worldwide Eocene transgression[31] shown for the USSR in Fig. 1, which is against the secular Cainozoic trend.

Assessment of the role of ridges in pre-Cretaceous times must be highly speculative because the marine record is lost, but it seems possible that the more or less progressive Jurassic transgression relates in part to the creation of a ridge within the newly-opened central Atlantic, while the striking late Palaeozoic regression could to some extent be due to the consumption of spreading sea floor including ridges that had previously driven continents together to effect the Hercynian Orogeny.

The dominant cause of the longer-term Phanerozoic changes in sea level and overall regression is likely to have been a combination of continental thickening by orogeny consequent on subduction or collision, and variation in the cumulative length of the oceanic ridge system. The best prospect for testing these ideas quantitatively is in the Cainozoic, which has by far the fullest record of geological events per unit of time, both on the continents and ocean floor.

Received 28 June; accepted 2 September 1977.

1. Termier, H. & G. *Histoire Géologique de la Biosphère* (Masson, Paris, 1952).
2. Strakhov, N. M. *Outlines of Historical Geology* (Government Printing Office, Moscow, 1948).
3. Egyed, L. *Geofis. Pura Apl.* 33, 42 (1956).
4. Holmes, A. *Principles of Physical Geology* 2nd edn (Nelson, London, 1965).
5. Dearnley, R. *Physics and Chemistry of the Earth* (eds Ahrens, L. H. *et al.*) **7**, 1 (1966).
6. Carey, S. W. *Earth Sci. Rev.* **11**, 105 (1975).
7. Armstrong, R. L. *Nature* **221**, 1042 (1969).
8. Hallam, A. *Nature* **232**, 180 (1971).
9. Wise, D. U. *The Geology of Continental Margins* (eds Burk, C. A. & Drake, C. L.) 45 (Springer, New York, 1974).
10. Vinogradov, A. P. (ed.) *Atlas of the Lithological–Palaeogeographic Maps of the USSR* 4 vols (Ministry of Geol., Moscow, 1967–69).
11. Hallam, A. *Earth Sci. Rev.* **5**, 45 (1969).
12. Garrels, R. M. & Mackenzie, F. T. *Evolution of Sedimentary Rocks* (Norton, New York, 1971).
13. Schuchert, C., *Atlas of Paleogeographic Maps of North America* (Wiley, New York, 1955).
14. Johnson, J. G. *Bull. geol. Soc. Am.* **82**, 3263 (1971).
15. Cook, T. D. & Bally, A. W. *Stratigraphic Atlas of North and Central America* (Princeton University Press, Princeton, 1977).
16. Sloss, L. L. & Speed, R. C. *S.E.P.M.* Spec. Publ. 22, 89 (1974).
17. Birch, F. *Phys. Earth planet. Inter.* **1**, 141 (1968).
18. Owen, H. G. *Phil. Trans. R. Soc.* **A281**, 223 (1976).
19. Hospers, J. & Van Andel, S. I. *Tectonophysics* **5**, 5 (1967).
20. Le Pichon, X. *J. geophys. Res.* **73**, 3661 (1968).
21. Stewart, A. D. *J. geol. Soc.* **133**, 281 (1977).
22. Hayes, D. E. & Frakes, L. A. *Init. Rep. D.S.D.P.* **28**, 919 (1975).
23. Hays, J. D. & Pitman, W. C. *Nature* **246**, 18 (1973).
24. Nalivkin, D. V. *Geology of the U.S.S.R.* (Oliver and Boyd, Edinburgh, 1973).
25. Flint, R. F. *Glacial and Pleistocene geology* (Wiley, New York, 1957).
26. Shackleton, R. M. *Proc. geol. Ass. Lond.* **81**, 549 (1970).
27. Watson, J. V. *Phil. Trans. R. Soc.* **A280**, 629 (1976).
28. Watson, J. V. *Phils. Trans. R. Soc* **A273**, 443 (1973).
29. Larsen, R. L. & Pitman, W. C. *Bull. geol. Soc. Am.* **83**, 3645 (1972).
30. Berggren, W. A., Mckenzie, D. P., Sclater, J. G. & Van Hinte, J. E. *Bull. geol. Soc. Am.* **86**, 267 (1975).
31. Hallam, A. *Am. J. Sci.* **261**, 397 (1963).
32. McKenzie, D. P. & Sclater, J. G. *Geophys. J.* **25**, 437 (1971).

22

Reprinted from pages 83–85 of *Am. Assoc. Petroleum Geologists Mem. 26*, pp. 83–97 (1977)

Seismic Stratigraphy and Global Changes of Sea Level, Part 4: Global Cycles of Relative Changes of Sea Level. [1]

P. R. VAIL, R. M. MITCHUM, JR.,[2] and S. THOMPSON, III[3]

Abstract Cycles of relative change of sea level on a global scale are evident throughout Phanerozoic time. The evidence is based on the facts that many regional cycles determined on different continental margins are simultaneous, and that the relative magnitudes of the changes generally are similar. Because global cycles are records of geotectonic, glacial, and other large-scale processes, they reflect major events of Phanerozoic history.

A global cycle of relative change of sea level is an interval of geologic time during which a relative rise and fall of mean sea level takes place on a global scale. A global cycle may be determined from a modal average of correlative regional cycles derived from seismic stratigraphic studies.

On a global cycle curve for Phanerozoic time, three major orders of cycles are superimposed on the sea-level curve. Cycles of first, second, and third order have durations of 200 to 300 million, 10 to 80 million, and 1 to 10 million years, respectively. Two cycles of the first order, over 14 of the second order, and approximately 80 of the third order are present in the Phanerozoic, not counting late Paleozoic cyclothems. Third-order cycles for the pre-Jurassic and Cretaceous are not shown. Sea-level changes from Cambrian through Early Triassic are not as well documented globally as are those from Late Triassic through Holocene.

Relative changes of sea level from Late Triassic to the present are reasonably well documented with respect to the ages, durations, and relative amplitudes of the second- and third-order cycles, but the amplitudes of the eustatic changes of sea level are only approximations. Our best estimate is that sea level reached a high point near the end of the Campanian (Late Cretaceous) about 350 m above present sea level, and had low points during the Early Jurassic, middle Oligocene, and late Miocene about 150, 250, and 200 m, respectively, below present sea level.

Interregional unconformities are related to cycles of global highstands and lowstands of sea level, as are the facies and general patterns of distribution of many depositional sequences. Geotectonic and glacial phenomena are the most likely causes of the sea-level cycles.

Major applications of the global cycle chart include (1) improved stratigraphic and structural analyses within a basin, (2) estimation of the geologic age of strata prior to drilling, and (3) development of a global system of geochronology.

INTRODUCTION

Cycles of relative change of sea level on a global scale are evident throughout Phanerozoic time. The evidence is based on the fact that many regional cycles determined on different continental margins are simultaneous and that the relative magnitudes of the changes generally are similar. Concepts and methods of determination of relative changes of sea level and regional cycles were given previously (Part 3, Vail et al, this volume). In this paper are presented charts of global cycles, the methods for constructing the charts from a modal average of correlative regional cycles based on seismic stratigraphy, and our estimates of the actual magnitudes of the sea-level changes.

Because the global cycles are records of geotectonic, glacial, and other large-scale processes, they reflect major events of Phanerozoic history. The timing and relative importance of these events are indicated by charts of the cycles. Such a composite record offers a means of subdividing Phanerozoic time into significant geochronologic units based on a single criterion.

Fairbridge (1961) summarized the historical development of concepts of sea-level change on a global scale, including the classic works of Haug (1900), Suess (1906), Stille (1924), Grabau (1940), Umbgrove (1942), Kuenen (1940, 1954, 1955), Arkell (1956), and others. These pioneer investigations laid the foundation for later work including ours. However, some developments have confused "transgressions and regressions" of the shoreline with "rises and falls" of sea level. Grabau (1924) recognized this problem. The charts we present in this paper show relative and eustatic rises and falls of sea level on a global scale, and differ from charts that show transgressions and regressions of the shoreline.

GLOBAL CYCLES

Figures 1 through 3 are charts of relative changes of sea level on a global scale. The vertical axis of each chart is scaled in millions of years (after Van Hinte, 1976 a, b), with standard periods and epochs plotted alongside. The horizontal axis shows relative positions of sea level and is scaled from 1.0 to 0.0, with 1.0 being the maximum relative highstand (65 M.A.) and 0.0 being the minimum relative lowstand (30 M.A.). Relative rises of sea level are plotted toward the left, and relative falls toward the right.

[1]Manuscript received, January 6, 1977; accepted, June 13, 1977.

[2]Exxon Production Research Co., Houston, Texas 77001.

[3]New Mexico Bureau of Mines and Mineral Resources, Socorro, New Mexico 87801.

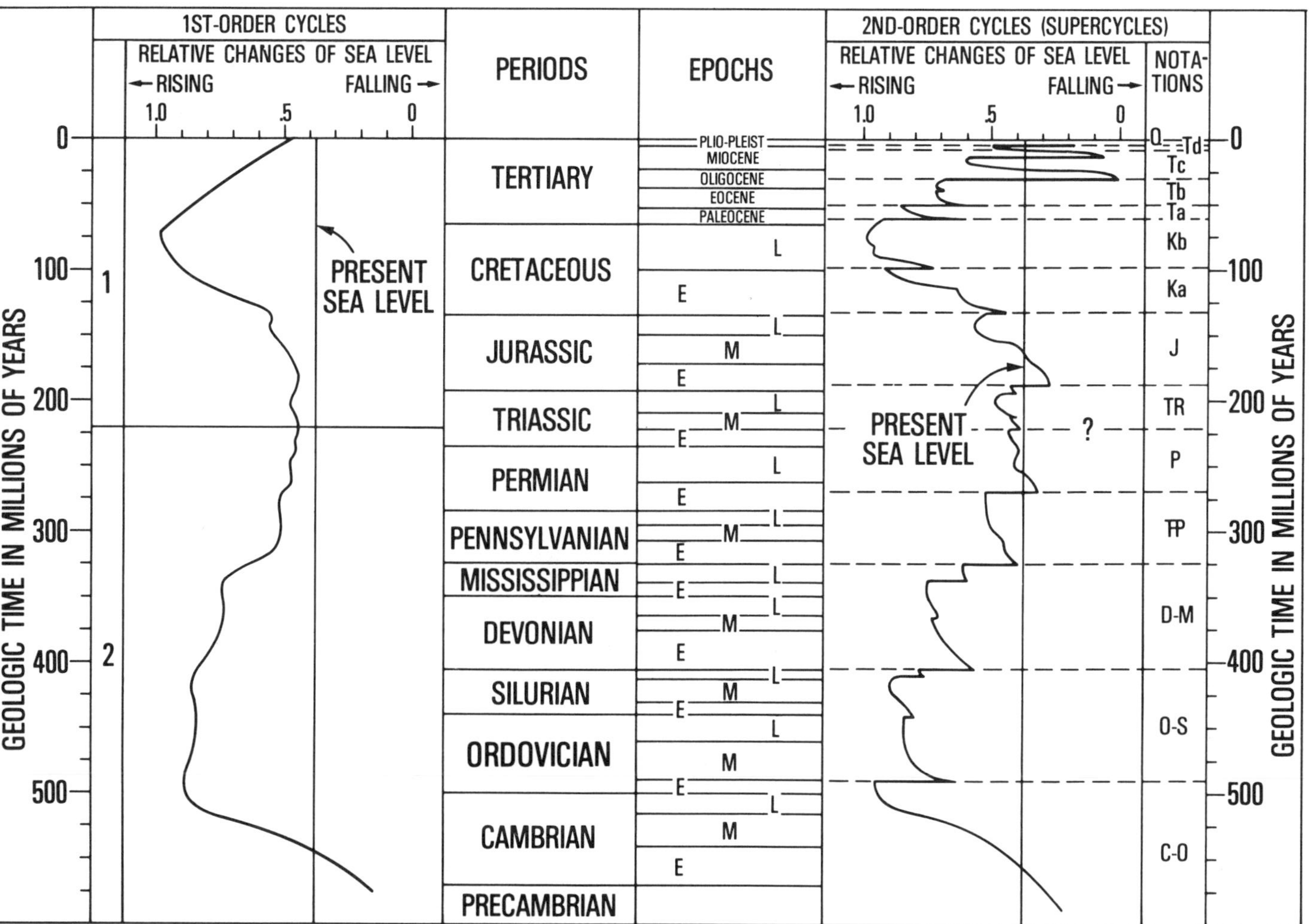

FIG. 1—First- and second-order global cycles of relative change of sea level during Phanerozoic time.

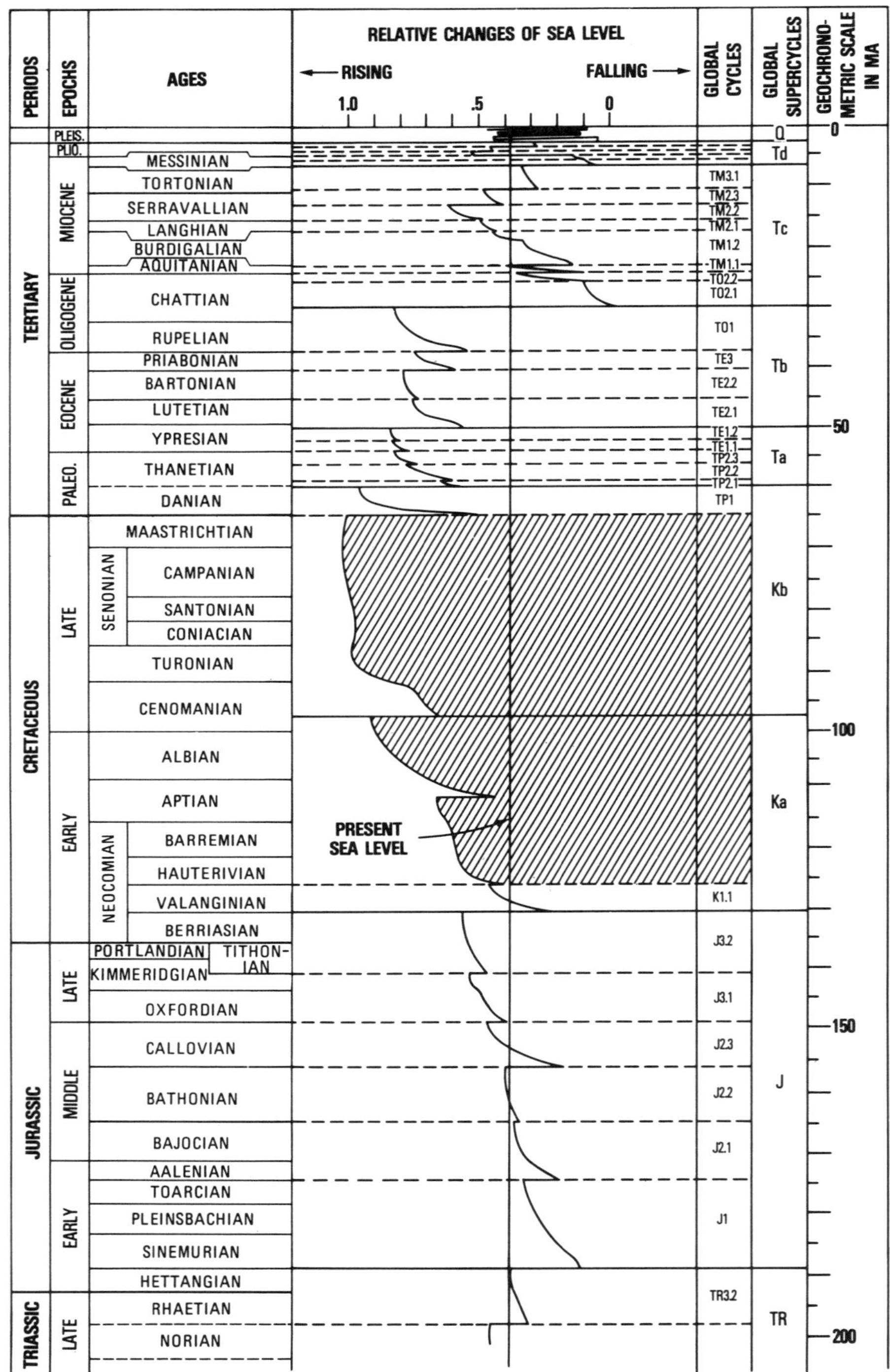

Jurassic-Cretaceous time scale after Van Hinte 1976 a, b

FIG. 2—Global cycles of relative change of sea level during Jurassic-Tertiary time. Cretaceous cycles (hatchured area) have not been released for publication.

[*Editor's Note:* Figure 3, which shows the sea-level curve for Cenozoic time, has been omitted; the curve is included in Figure 2.

In the remainder of this paper Vail et al. discuss global cycles of sea-level change, the construction of a global cycle chart, the estimation of eustatic changes of sea level, global highstands and lowstands and major interregional unconformities, and causes and applications of global cycles.]

REFERENCES

Arkell, W. J. 1956. *Jurassic Geology of the World*. London: Oliver & Boyd, 806 pp.

Fairbridge, R. W. 1961. Eustatic changes in sea level. *Phys. Chem. Earth* **4**:99–185.

Grabau, A. W. 1924. *Principles of Stratigraphy*. New York: Seiler, 1185 pp.

Grabau, A. W. 1940. *The Rhythm of the Ages*. Peking: Vetch, 561 pp.

Haug, E. 1900. Les geosynclinaux et les aires continentales. *Soc. Géol. France Bull.*, Ser. 3, **28**:617–711.

Kuenen, P. H. 1940. Causes of eustatic movements. *Pacific Sci. Congr., 6th, Proc.* **2**:833–837 (Univ. Calif. Press, Berkeley).

Kuenen, P. H. 1954. Eustatic changes of sea-level. *Geologie en Mijnbouw* **16**:148–155.

Kuenen, P. H. 1955. Sea level and crustal warping. *Geol. Soc. America Spec. Paper* **62**, pp. 193–204.

Stille, H. 1924. Grundfragen der vergleichenden Tektonik. Berlin: Borntraeger, 443 pp.

Suess, E. 1906. *The Face of the Earth*, vol. 2. Oxford: Clarendon Press, 556 pp.

Umbgrove, J. H. F. 1942. *The Pulse of the Earth*. The Hague: Nijhoff, 179 pp.

Van Hinte, J. E. 1976a. A Jurassic time scale. *Am. Assoc. Petroleum Geologists Bull.* **60**:489–497.

Van Hinte, J. E. 1976b. A Cretaceous time scale. *Am. Assoc. Petroleum Geologists Bull.* **60**:498–516.

23

Reprinted from *Geology* **4**:272–276 (1976)

AREAS AND VOLUMES OF CRATONIC SEDIMENTS, WESTERN NORTH AMERICA AND EASTERN EUROPE

L. L. Sloss, Department of Geological Sciences, Northwestern University, Evanston, Illinois 60201

ABSTRACT

Prevailing opinion is that alternating episodes of sedimentation and erosion on cratons are passive responses to eustatic variations in sea level. Comparative analysis of the preserved areas and volumes of Paleozoic and Mesozoic strata covering large regions of the North American and eastern European cratons reveals evidence of synchronous depositional and erosional events of a few million to tens of millions of years duration. Significantly, the study also reveals that the three-dimensional geometry of many sedimentary successions is not compatible with passive cratons subject only to eustatic controls. Rather, consideration must be given to some more complex global mechanism governing both continental freeboard and vertical motions within cratons.

INTRODUCTION

An earlier paper (Sloss, 1972) introduced comparative data on the areas covered by successive Paleozoic and Mesozoic stratigraphic units in western Canada and the Russian Platform; volumetric data were treated only in terms of their statistical distributions. The subsequent availability (Table 1) of updated syntheses of Russian stratigraphy and publication of a stratigraphic atlas of the Rocky Mountains–Great Plains region of the United States permit revision and expansion of the previous study to include roughly equal areas of two continents. Further, the development of absolute time scales with increased accuracy and precision makes it possible to calculate time-dependent volume measures with some degree of confidence. The major purpose of the present paper is to publish these data in the hope that they will be useful to the many workers concerned with the geotectonic implications of the sedimentation-erosion history of continents.

Data for western Canada and the western United States are separately displayed in Figure 1; combined North American data are compared with eastern European data in Figure 2. Details of the methodology employed appear in Appendix 1, but for many readers it is sufficient to know that the data were derived from digitization of isopach maps representing the preserved areal extent and thickness of successive time-stratigraphic units; no reconstruction of presumed original extent is involved, except where minor gaps in obvious original continuity were ignored by map compilers.

AREAS OF PRESERVATION

The preserved areal extent of each time-stratigraphic unit analyzed is plotted (Figs. 1, 2) versus the time at the end of deposition of the unit in question; successive points are then connected by straight lines that approximate a function of changing areal distributions through time. Zero values (as at ~420 and ~500 m.y. on all plots) represent major stratigraphic lacunas.

The limits of the North American stratigraphic sequences (for example, Sauk) are shown on the plots. These successions and their eastern European equivalents represent major cratonic cycles of subsidence and uplift. At the close of a cycle, the area occupied by a particular stratigraphic unit is determined by the area of original deposition less the area of postdepositional erosion of sufficient depth to remove the unit. Areas of preservation of units of a transgressive succession, depressed below base level by continuing subsidence during the deposition of superjacent units, are closer to original depositional area than are those of regressive units, which are exposed to erosion shortly after deposition.

Where the effects of several cratonic cycles are evident, areas of preserved sediment represent processes more complex than simple eustatic controls on deposition and erosion. Indeed, the areal preservation of individual sequences and their components is largely determined by the amplitude of vertical motions of cratons and of individual elements within cratons, including movements during and after—commonly long after—deposition of a particular sequence. Consider, for example, the marked differences in the extent of preservation of late Paleozoic strata in the United States and in Canadian study areas (Fig. 1; 250 to 350 m.y. B.P.). The lack of widespread Permian-Carboniferous sediments in the Canadian area is in part a response to the lack

TABLE 1. REGIONS CONSIDERED

	Area 10^6 km^2	Boundaries
Western Canada (McCrossan and Glaister, 1964)	~1	North to 60°N; south to 49°N; east to Canadian Shield; west to frontal zone ("disturbed belt") of Rocky Mountains
Western U.S. (Mallory, 1972)	~2	North to 49°N; south to 34°N; east to map limits (~100°W); west to 114°W or eastern limit Sevier thrust belt
Russian Platform (Vinogradov, 1968-1969)	~3½	North to Baltic Shield and Barents Sea; south to 44°N * (excluding Black and Caspian seas); east to west flank Ural Mountains; west to Baltic Sea and east flanks Ukrainian and Turano-Scythian arches

* *Southern boundary of study area of Paleozoic units is line connecting Volgograd-Uralsk-Orenburg-Aktyubinsk (limit of sub-Permian-salt data in Paracaspian Basin).*

of deeply subsiding basins such as existed in Colorado at this time; the more effective control, however, is the greater depth of sub-Zuni (chiefly pre–Middle Jurassic) erosion on the Sweetgrass arch in the Prairie provinces. Here, the preserved areal distribution is a poor measure of what may have been a broad extent.

PRESERVED VOLUMES

Interpretation of the preserved volume of a stratigraphic unit requires reference to the length of time during which deposition took place; therefore, volumes *per unit time* are shown by the bar diagrams in Figures 1 and 2. These are subject to error in estimation of individual time spans, but the major trends in rates of deposition of preserved volumes would not be seriously affected by errors of a factor of two. Isolated peaks, such as those shown for the Coniacian (85 to 86 m.y. B.P.) of Russia and the early Leonardian (277 to 281 m.y. B.P.) of the United States, are less easily defended.

It is the nature of the geometry of sedimentary basins (Sloss and Scherer, 1975) that at the close of a depositional episode, a high proportion of the *volume* of sediment lies in the interior of subsiding basins, whereas the largest *areas* of deposition are on basin margins and surrounding stable shelves. Further, basin interiors are least likely to suffer significant erosion, whereas basin margins and interbasin shelves are sites of recurring erosional events sufficient to strip thin sedimentary covers. Thus, minor uplift and erosion commonly destroy the stratigraphic record of vast regions and long periods of time, but the equivalent record remains almost intact within basin interiors. Therefore, except for the frailties introduced by uncertainties in the estimation of depositional time spans, preserved volumes carry significantly more information on the tectonic behavior of cratons than do areal distributions.

INTERPRETATION

As has been noted by a number of workers (for example, Hallam, 1963; Wise, 1972; Flemming and Roberts, 1973; Rona, 1973; Morgan, 1975), the volume of ocean basins, and thus the elevation of cratons, responds to the inflation of mid-oceanic ridges and to the thermal contraction of aging oceanic lithosphere. The power of this hypothesis is such as to lead prevailing opinion toward belief in largely eustatic control of sedimentation and erosion on cratons that passively submerge and emerge with shifting global sea levels. A recently reiterated (Sloss and Speed, 1974; Sloss and Scherer, 1975) opposing view, while not denying the significance of eustatic control, places greater emphasis on vertical motions of and within cratons and on global episodicity of modes of cratonal tectonism. The data presented here are subject to interpretation in support of this position.

On a rigid craton that is subject only to eustatic shifts in base level, the rates of change of preserved areas and of volumes per unit time would have the same sign and would approach proportionality; departures indicate syndepositional differential vertical motions within the craton. There is general conformity of the rates of change of areas and volumes in Cambrian to Silurian units of all study areas; this, in combination with low mean thicknesses ($V(TA)^{-1}$), suggests nondeforming cratonic areas. In contrast, the Kaskaskia sequence and equivalents (350 to 420 m.y. B.P.) on the Russian Platform show markedly higher accumulation rates and discordances in the rates of change of areas and volumes. Note that extreme area-volume peaks in the Frasnian (388 to 395 m.y. B.P.) are followed by sharp drops in volumes into Tournaisian time (370 to 378 m.y. B.P.) without proportional losses in area. These data reflect phases of rapid and slow subsidence of the Elk Point–Williston, Dneiper-Donetz, and Moscow basins. Other subsidence events (depophases of Johnson, 1971), marked by proportional increases in area and volume on both cratons, are observable for Early Permian and mid-Cretaceous time (for the latter, compare Canadian and Russian plots; the combined United States and Canadian data lack sufficient time resolution).

An earlier paper (Sloss and Speed, 1974) defined the time of deposition of the first three Phanerozoic sequences (roughly Cambrian–early Carboniferous as ***submergent episodes*** separated by *emergent episodes* (~420 and ~510 m.y. B.P.) marked by sequence-bounding unconformities. Analysis (Fig. 2) confirms the synchrony of these episodes on two widely separated cratons and shows substantial agreement of short (5-m.y.) events within the well-preserved Kaskaskia sequence. The same paper calls the time of the Absaroka sequence (roughly late Carboniferous–Triassic) an *oscillatory episode* characterized by widely varying tectonic styles within a single craton. The contrast between Absaroka preservation in study areas in the United States and Canada documents the latter point, but apparent synchrony between the area in the United States and the Russian Platform suggests intercontinental controls.

When hard data on both areas and volumes of sediments on large cratonic areas are considered, patterns emerge that cannot be explained by eustatic controls of marine transgression and regression or by accompanying isostatic compensations. Rather, the data suggest that the freeboard of cratons and the mode and amplitude of certain vertical motions within cratons are concomitant responses to some underlying globally effective mechanism.

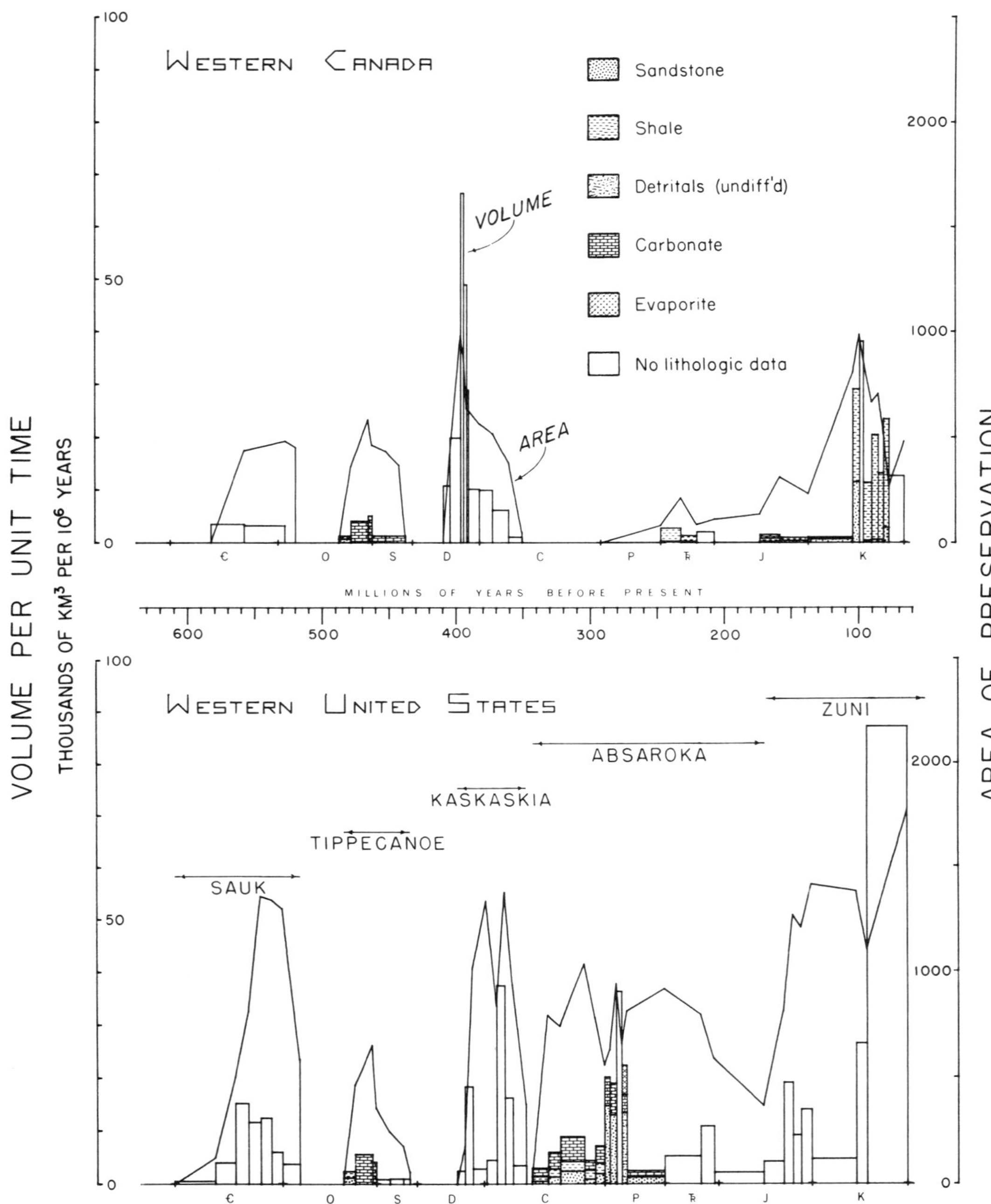

Figure 1. Volumes accumulated per unit time (bars) and areas (lines) of Paleozoic-Mesozoic strata preserved on North American craton in western Canada and western United States.

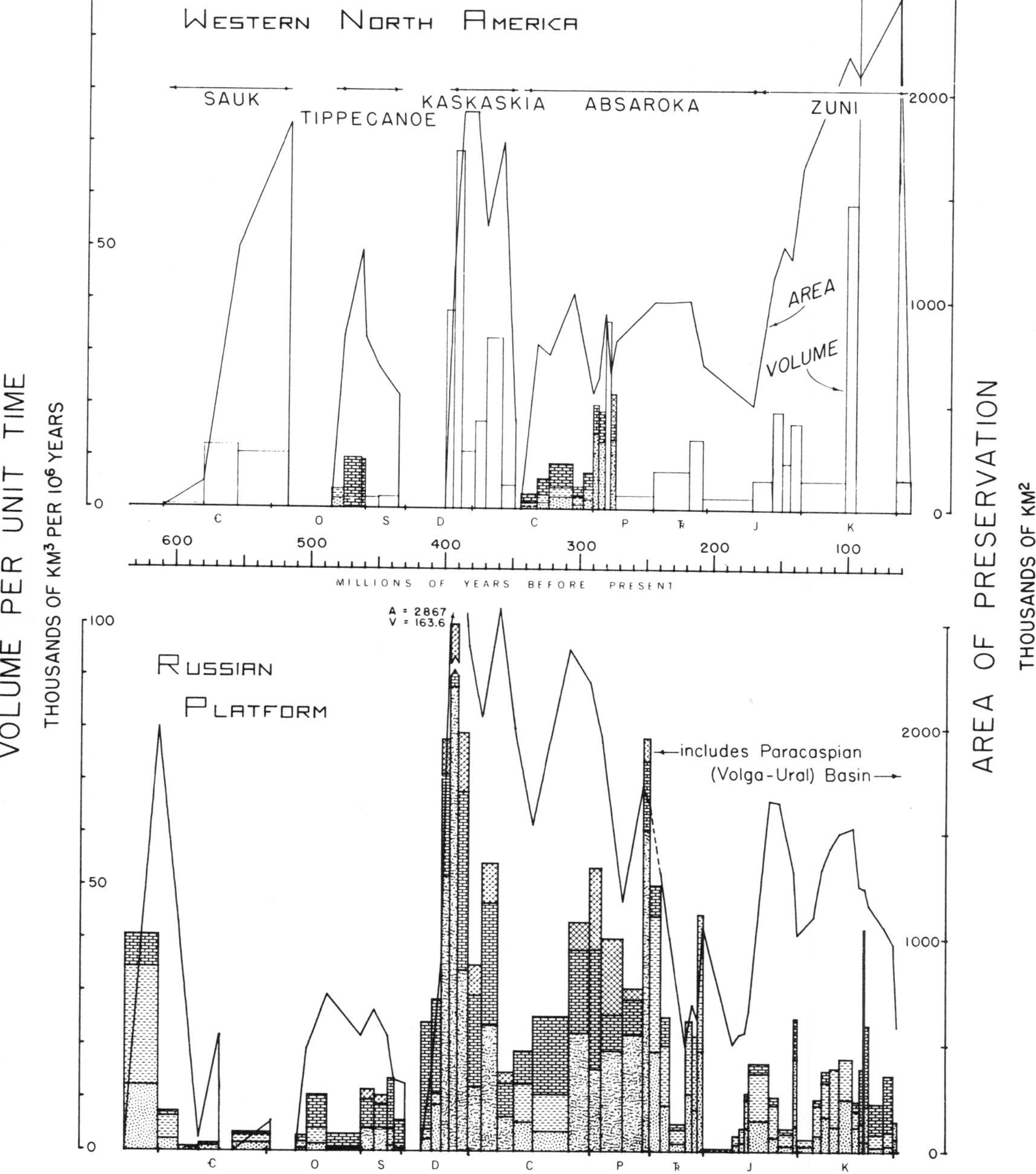

Figure 2. Volumes accumulated per unit time (bars) and areas (lines) of Paleozoic-Mesozoic strata preserved on western part of North American craton and on Russian Platform of the eastern European craton. Legend as in Figure 1.

APPENDIX 1. METHODS AND TIME SCALE

Methods. One hundred twenty-seven isopach maps have been digitized to give the thickness of successive stratigraphic units at equally spaced grid points (4,144 km^2 per datum point for North America; 9,073 km^2 per point for Russia). Where quantitative lithofacies information is presented, as on all Russian maps, lithologic ratios or percentages were entered for each point. The area occupied by each stratigraphic unit was calculated, together with its volume and, where possible, the volumes of lithologic components.

Calculations such as those presented here are subject to errors introduced by uncertainties of stratigraphic correlation and by insufficient density of subsurface data in certain regions. Less obviously, the figures are strongly influenced by the choice of regions to be excluded from study areas. The present analysis includes no extracratonic "eugeosynclinal" or ophiolitic terranes identified by petrologic suites and tectonic histories markedly different from those of cratons. Limiting consideration to cratonic rocks and regions has produced volumetric and compositional data significantly at variance with other, differently oriented, studies (for example, Ronov and others, 1969).

It is clearly inequitable to compare the distributions and volumes of units documented over large regions, by reason of abundant outcrop or economic importance, with those of other units that may be equally extensive but are mapped only in smaller areas of local drilling. Therefore, the study areas outlined in Table 1 are defined by the limits of documentation on the thickness (or absence) of the least known units. The western limit of the North American study area does not define the boundary of the Paleozoic and early Mesozoic craton; indeed, very large volumes of allochthonous cratonic sediment are ignored by this choice. However, neither the level of stratigraphic mapping in the thrust belt nor the degree of requisite palinspastic restoration is sufficient to support analysis.

The Canadian atlas treats short time units in the Devonian and Late Cretaceous, whereas Cambrian–Early Ordovician time is represented by two subdivisions. Conversely, the United States atlas presents maps of seven Sauk units and of eight Absaroka units not represented in the Canadian study area, but all of Late Cretaceous time is covered by a single isopach map. These inequalities, together with differences in the selection of stratigraphic limits of mapped units, reduce the time resolution of the combined analysis of western North America (Fig. 2) relative to the study of individual areas (Fig. 1).

In Table 1 the study area for Mesozoic strata of the Russian Platform (eastern European craton) is about 16 percent larger than that for Paleozoic strata. Thus, the post-Permian decline in preserved areas and volumes is more extreme than is suggested by the Russian Platform plot in Figure 2.

Time Scale. The Paleozoic time scale adopted is after Harland and Francis (1971), with subdivisions of the periods adjusted according to the "critical points" listed by Lambert (1971). Note that the age approximations used are based on a ^{87}Rb half-life of 50×10^9 yr. Pre-Late Cretaceous Mesozoic chronology is a subject of wide disagreement among authorities. I have accepted the upper limits suggested by Lambert (1971) and have juggled age-stage boundaries to harmonize with other data—within the Triassic to fit an apparently sound figure (R. W. Kistler, 1973, personal commun.) derived from early late Scythian volcanic rocks in western Nevada and in the Jurassic and Early Cretaceous according to the relative time spans of individual ages suggested by van Hinte (in Berggren and others, 1975). Late Cretaceous "dates," perhaps the most dependable of all, are the products of K-Ar analyses of biostratigraphically controlled bentonites (Obradovich and Cobban, 1975).

REFERENCES CITED

Berggren, W. A., McKenzie, D. P., Sclater, J. G., and van Hinte, J. E., 1975, World-wide correlation of Mesozoic magnetic anomalies and its implications: Discussion: Geol. Soc. America Bull., v. 86, p. 267–269.

Flemming, N. C., and Roberts, D. G., 1973, Tectono-eustatic changes in sea level and sea floor spreading: Nature, v. 243, p. 19–22.

Hallam, A., 1963, Major epeirogenic and eustatic changes since the Cretaceous and their possible relationship to crustal structure: Am. Jour. Sci., v. 261, p. 397–423.

Harland, W. B., and Francis, E. H., eds., 1971, The Phanerozoic time-scale, a supplement: Geol. Soc. London Spec. Pub. 5, 356 p.

Johnson, J. G., 1971, Timing and coordination of orogenic, epeirogenic, and eustatic events: Geol. Soc. America Bull., v. 82, p. 3263–3298.

Lambert, R.St.J., 1971, The pre-Pleistocene Phanerozoic time-scale—A review, *in* Harland, W. B., and Francis, E. H., eds., The Phanerozoic time-scale, a supplement: Geol. Soc. London Spec. Pub. 5, p. 9–33.

Mallory, W. W., ed., 1972, Geologic atlas of the Rocky Mountain region: Denver, Rocky Mtn. Assoc. Geologists, scale 1:5,000,000, 331 p.

McCrossan, R. G., and Glaister, R. P., eds., 1964, Geological history of western Canada: Calgary, Alberta Soc. Petroleum Geologists, scale 1:5,000,000, 232 p.

Morgan, W. J., 1975, Heat flow and vertical movements of the crust, *in* Fischer, A. G., and Judson, S., eds., Petroleum and global tectonics: Princeton, N.J., Princeton Univ. Press, p. 23–43.

Obradovich, J. D., and Cobban, W. A., 1975, A time scale for the Late Cretaceous of the western interior of North America, *in* Caldwell, W.G.E., ed., The Cretaceous system in the western interior of North America: Geol. Assoc. Canada Spec. Paper 13, p. 31–54.

Rona, P. A., 1973, Worldwide unconformities in marine sediments related to eustatic changes of sea level: Nature Phys. Sci., v. 244, p. 25–26.

Ronov, A. B., Migdisov, A. A., and Barskaya, N. V., 1969, Tectonic cycles and regularities in the development of sedimentary rocks and paleogeographic environments of sedimentation of the Russian Platform (an approach to a quantitative study): Sedimentology, v. 13, p. 179–212.

Sloss, L. L., 1972, Synchrony of Phanerozoic sedimentary-tectonic events of the North American craton and the Russian Platform: Internat. Geol. Cong., 24th, Montreal 1972, sec. 6, p. 24–32.

Sloss, L. L., and Scherer, W., 1975, Geometry of sedimentary basins: Applications to Devonian of North America and Europe, *in* Whitten, E.H.T., ed., Quantitative studies in the geological sciences: Geol. Soc. America Mem. 142, p. 71–88.

Sloss, L. L., and Speed, R. C., 1974, Relationships of cratonic and continental margin tectonic episodes, *in* Dickinson, W. R., ed., Tectonics and sedimentation: Soc. Econ. Paleontologists and Mineralogists Spec. Pub. 22, p. 98–119.

Vinogradov, A. P., ed., 1968–1969, Atlas of the lithological paleogeographical maps of the U.S.S.R.: Moscow, USSR Ministry of Geology and USSR Acad. Sciences, v. 1–3, scale 1:7,500,000.

Wise, D. U., 1972, Freeboard of continents through time, *in* Shagam, R., and others, eds., Studies in Earth and space sciences (Hess volume): Geol. Soc. America Mem. 132, p. 87–100.

ACKNOWLEDGMENTS

Reviewed by A. L. Howland and R. C. Speed.

Supported by National Science Foundation, Earth Sciences Section, Grants GA-22844 and DES74-22337.

Data compilation by Virginia DeGregorio.

MANUSCRIPT RECEIVED DECEMBER 5, 1975
MANUSCRIPT ACCEPTED FEBRUARY 24, 1976

24

Reprinted from pages 157–159 and 163–166 of *Handbook of Strata-Bound and Stratiform Ore Deposits*, vol. 7, K. H. Wolf, ed. Amsterdam: Elsevier, 1976, 656 pp.

ORIGIN OF THE PRECAMBRIAN BANDED IRON-FORMATIONS

JUERGEN EICHLER

INTRODUCTION

Banded iron-formations represent one of the most distinctive, though peculiar, types of rock which occur widespread in space and in time throughout the Precambrian on practically all shield areas of the world. Major iron-formations may form enormous blanket deposits with stratigraphic thicknesses of hundreds of metres and a lateral extent of several hundred to more than a thousand kilometres. The quantity of iron contained is immense – probably somewhere between 10^{14} and 10^{15} tons. Single iron-formations may have originally contained as much as 10^{14} tons, such as in the Hamersley basin of Western Australia (Trendall and Blockley, 1970) or, for example, $10^{12} - 10^{13}$ tons of iron in the Minas itabirites of Minas Gerais, Brazil (Dorr II, 1973b).

There is no doubt that iron-formations are by far the economically most important source of iron ore mined in the world and, because of their wide distribution, also the most prominent type of iron deposit.

However, despite countless detailed studies of single iron-formation deposits, the knowledge of the origin of this rock is still fragmentary. The problems start with the lack of a clear and generally acceptable definition of what is exactly meant by the term "Precambrian banded iron-formation"; thus, it is deemed necessary to start this paper with a discussion of terminology and definition of iron-formation.

The term "Precambrian banded iron-formation" includes coincidentally the key-words for the problems concerning the origin of these rocks. Why are banded iron-formations confined to the Precambrian? What is the explanation for the peculiar characteristic banding of interlayered chert and iron-rich layers? From where have the enormous iron concentrations been derived? What was the method of transportation and deposition of the iron and silica and of the separation of iron from other elements?

Many hypotheses as to the origin of iron-formations have been suggested for particular occurrences, but none of these have been able to offer a generally acceptable model to explain the genesis of all iron-formations. As it will be shown, iron-formations are diversified rocks and despite many commonly shared features, any hypothesis based on studies of a single deposit is most likely equivocal if applied on a universal scale.

The reason for the lack of a generally valid genetic model is probably that principles of actualistic geology are difficult to apply to Precambrian iron-formations which, with a few questionable exceptions, are strictly confined to the Precambrian. Some distinctive features, however, such as their depositional and paleogeographical environments, sedimentary rock assemblages and facies distribution in shallow-marine basins can be fittingly compared with younger iron deposits. There is strong evidence that iron-formations are chemical and possibly biochemical sediments, but the problem of from whence the iron and silica was derived and how the typical banding originated, is still unclear. The answer is probably connected with the indispensable assumption of fundamentally different environmental conditions in the Precambrian, first of all an atmosphere and a hydrosphere lacking oxygen during the deposition of iron-formations. If at all true, this is of course difficult to prove.

However, in further studies on a worldwide scale, iron-formations may turn out to be a helpful tool in the understanding of the sedimentary and tectonic development of the Precambrian crust. For example, Archean iron-formations, being most readily distinguishable and very widespread rock units, have already proved to be key-criteria in the recognition of Archean basins, as Goodwin (1973) has explicitly demonstrated for the Canadian Shield. They have also provided further evidence for the former connection of the Gondwana landmass and their regional distribution suggests that the bulk of the Proterozoic iron-formations was laid down in depositional environments which follow major fractures in a segmented ancient crust.

The present chapter is an outgrowth of a review of the comprehensive literature on iron-formations which tries to give a general view of the geology, the distribution and origin of this fascinating rock. The writer is aware of the fact that many important contributions on the subject may not have been adequately acknowledged in this paper, due to the vast amount of information to be digested. For more details the interested reader is referred to the recently published volumes of the Kiev symposium on the genesis of Precambrian iron deposits (*UNESCO, Earth Sciences*, Vol. 9, 1973) and to the topical iron-formation symposium in *Economic Geology*, Vol 68, 1973. The references herein cited comprise most of the more pertinent literature on iron-formation, including a very valuable inventory of the Russian literature.

TERMINOLOGY AND DEFINITION OF IRON-FORMATION

An internationally accepted nomenclature to describe the banded cherty iron-formation sediments in accordance with the principles of stratigraphic classification of sedimentary rocks has not yet been established. The usage of different terms in different countries for the same iron-rich sediments still complicates a common understanding as to what exactly is defined as "banded iron-formation", so much the more since identical terms are occasionally also used for different rocks.

A most valuable discussion of the problems of nomenclature for banded ferruginous-cherty sedimentary rocks has been given by Brandt et al. (1972), recommending the usage of the term "iron-formation" as a generic lithologic name, strictly on a parallel, for example, with "limestone". The words are in this case hyphenated but capitalized when using them in a stratigraphic sense as, for example, "Brockman Iron Formation" as suggested by Trendall and Blockley (1970).

Contracting the old term "iron-bearing formation" of Van Hise and Leith (1911) to "iron-formation", James (1954) supplied the first strict definition, now widely accepted by most geologists as "... a chemical sediment, typically thin-bedded or laminated, containing 15% or more iron of sedimentary origin, commonly but not necessarily containing layers of chert".

A similar definition is given by Gross (1965), but, extended also to layered iron-rich rock units of clastic sedimentary origin.

The typical features of iron-formation as defined above are physically a thin layering in the range of less than a millimetre to some centimetres, with lamination on a scale of sometimes less than 1 mm and chemically the composition of one or more iron-rich minerals interbanded with chert. Chert or jasper is a term loosely used for silica-rich material, regardless of degree of crystallinity. Thus, in metamorphosed iron-formations the original chert normally has been recrystallized into granular quartz.

The primary iron-rich minerals alternating with the silica-bands may consist of iron oxides, carbonates, silicates or sulfides, indicating the physico-chemical conditions of the environment of deposition of the iron-formation as shown by Krumbein and Garrels (1952) and Krauskopf (1967). James (1954) very usefully adapted these existing differences in the content of distinct iron minerals to divide the iron-formations into facies-types, such as oxide-facies iron-formation (hematite and/or magnetite), silicate-facies iron-formation, etc.

This concept of facies of iron-formations seems to be acceptable worldwide, as it offers by itself a simple, descriptive characterization of a certain kind of iron-formation, regardless of the degree of metamorphism. The latter factor should be excluded from the basic definition of this rock.

Gross (1965) suggests a classification of the Precambrian iron-formations of Canada into two major groups, according to general features and characteristics of their depositional environments and typical rock associations. The Algoma-type is characterized by relatively small lenticular bodies, closely associated with and genetically related to volcanic rocks and graywacke sediments in eugeosynclinal belts. The Superior-type iron-formation is associated with clastic sedimentary rocks of shelf or miogeosynclinal environments, such as quartzites, dolomites and pelitic rocks, without showing any direct relationships to volcanic associations. This type is extensively developed in Middle to Late Precambrian shield areas throughout the world and forms prominent iron ranges that can be followed over several hundred to more than a thousand kilometres, attaining thicknesses of up to several hundred metres.

[*Editor's Note*: In pages 160 to 163 Eichler first discusses other terms used for "iron-formation." He then outlines the regional distribution of banded iron-formations, concluding that there is no major cratonic area that lacks iron-formations as prominent members of its stratigraphic column.]

AGES OF PRECAMBRIAN BANDED IRON-FORMATIONS

Due to the lack of minerals within the iron-formations suitable for radiometric age determinations, a precise dating of these rocks has been possible only in rare cases. However, the progress in radiometric dating techniques and the great number of samples determined in recent years from associated metasedimentary and volcanic rocks makes it possible at least to mark the age limits of many prominent iron-formations of all parts of the world. Compilations of age determinations have been given by Lepp and Goldich (1964), Govett (1966), James (1966) and recently by Goldich (1973). The review in this chapter is essentially based on these papers.

Though banded iron-formations occur during a period of at least 3,000 m.y. throughout the Precambrian, their prevailing development is confined to a single epoch between 2,600 and 1,800 m.y. ago (Goldich, 1973). As Bayley and James (1973) have shown, this interval was clearly the principal time of deposition in North America, far outweighing by their extent and iron content the more limited older and younger iron-formations. The same is the case for many major iron-formations on other continents such as those of Russia (Krivoy Rog, Kursk), South Africa, Brazil and Western Australia, all of which seem to have been deposited approximately contemporaneously between 2,600–1,800 m.y. ago. In all these shield areas apparently, there is a progressive decrease in the abundance of iron-formations, in size as well as in quantity, in the post-1,800 m.y. period just as in the pre-2,600 m.y. epoch.

Following the framework classification given by Goldich (1973), the iron-formations should be grouped, according to their ages, in the categories: (1) older than 3,000 m.y.; (2) 3,000–2,600 m.y.; (3) 2,600–1,800 m.y.; and (4) younger than 1,800 m.y.

(1) With few but notable exceptions, the bulk of the Early Precambrian, Archean iron-formations, though not restricted to the pre-2,600 m.y. period, belong to the Algoma-type as defined by Gross (1965). They are closely related to greenstone belts and

form relatively small lenticular bodies. The oldest known iron-formations, close to or older than 3,000 m.y., are found in the Swaziland System of South Africa (3,000–3,400 m.y.), in the Pilbara and Yilgarn Blocks of Western Australia (2,670–3,000 m.y.), in the Imataca Complex of Venezuela (3,000–3,400 m.y.) and in the Ukrainian Shield in the Greater Krivoy Rog area (3,100–3,500 m.y.).

(2) In the range from 3,000–2,600 m.y. numerous occurrences are known on the Canadian Shield, most of them closely associated with greenstone belts. A great number of them was deposited in the remarkably brief interval from 2,750–2,700 m.y. ago (Sims and Morey, 1972). In South Africa iron-formations in the Sebakvian, Bulavayan and Shamvayan Systems, stratigraphically associated with volcanic rocks, were dated at more than 2,700 m.y. (see Beukes, 1973). The same ages of >2,700 m.y. are found in some small Algoma-type occurrences in the Rio das Velhas-Series of Brazil, in India in Mysore and in the "Iron Ore Series" of Bihar–Orissa and also in eastern Karelia on the Baltic Shield (for references see Goldich, 1973). Thus, it appears that the span of time between 3,000 and 2;600 m.y., especially around 2,700 m.y., marks a significant epoch, very favourable for the deposition of iron-formations.

(3) The most abundant development of iron-formations in the world is found in the interval between 2,600 and 1,800 m.y. On the Canadian shield the ages of the major iron-formations of Labrador and the Lake Superior region range between 2,200 and 1,900 m.y. The interval of the main deposition was probably much less, since for the iron-formations in Minnesota, thought to be largely contemporaneous with others in North America, a minimum age of 2,000 m.y. has been determined. In Australia the Brockman Iron Formation of the Hamersley Group is bracketed between 2,200 and 2,000 m.y. Similar ages are given for Krivoy Rog/Ukrainia (2,200–1,900 m.y.), for the Brazilian itabirites in Minas Gerais and in the Serra dos Carajás/Pará (about 2,000 m.y.) and also for iron-formations of the Svecofennian of the Baltic Shield (minimum age of 1,900 m.y.) and of the Aldan Shield on the southeastern margin of the Siberian platform (2,000–2,750 m.y.). In South Africa iron-formations in association with clastic rocks are predominantly developed during the interval from 2,300--3,000 m.y., whereas the shelf-carbonate and carbonate–manganese associations belong to the period prior to 2,300 m.y.

(4) The post-1,800 m.y. banded iron-formations known in the world are quantitatively much less important; compared with the overwhelming concentration in the Lower Proterozoic and Archean they should rather be considered as exceptions. If a clear definition of "iron-formation" should exist, it most probably would not fit the characteristics of "iron-formation", principally of the problematic younger iron-formations, described by O'Rourke (1961). Examples for these are occurrences on Nepal and in Mato Grosso/ Brazil (Dorr II, 1973a). These Late Proterozoic or Early Cambrian cherty iron-formations are also typically banded but, averaging about 56% Fe, they are much richer in iron than the older types. Their origin is still enigmatic.

In North America only a few minor occurrences of Late Precambrian age are known.

In the Rapitan Group of the Yukon and Northwest Territories the age of an iron-formation is correlated with that of the Windermere Group (800–600 m.y.). The iron-formation in the Yavapai Series of central Arizona was metamorphosed between 1,820 and 1,760 m.y. ago and may be coeval with those of the Lake Superior region. The iron-formation in the lower part of the Damara Supergroup of South West Africa has been deposited between 1,000 and 620 m.y. ago (see Beukes, 1973).

It is a most remarkable fact that major iron-formations all over the world can be fairly well grouped in several time intervals of the Precambrian, between 3,400 and 3,000 m.y., about 2,700 m.y. and outstandingly in one period between 2,600 and 1,800 m.y. ago with an accumulation at about 2,000 m.y. This last statement has been also made for the major iron-formations of the Soviet Union (Alexandrov, 1973). Many of them have metamorphic ages of about 2,000 m.y. Younger ages are thought to correspond to the age of the final orogeny in the evolution of the geosyncline, which went across the Russian platform so that they do not reflect the time of deposition. The iron-formations of the Algoma-type generally are limited in size, but occur over a greater span of time, from the earliest Precambrian through the Phanerozoic, whereas the Superior-type is strictly confined to the Precambrian, the bulk of it most probably to the period prior to about 2,000 m.y. ago.

It must be pointed out that the data for single iron-formations should be used with caution as many of them might represent not the true age of deposition but mixed ages from subsequent metamorphic or orogenic events which have overprinted the old craton areas, for instance, the 500 m.y. "Panafrican Thermal Event" advocated by Kennedy (1964). Nevertheless, these factors as well as uncertainties in the use of decay constants in the different laboratories should not affect the general statement that the deposition of many major iron-formations of the world was synchronous in certain epochs of the Precambrian.

Fig. 2 suggests that the development of iron-formations fell into relatively stable

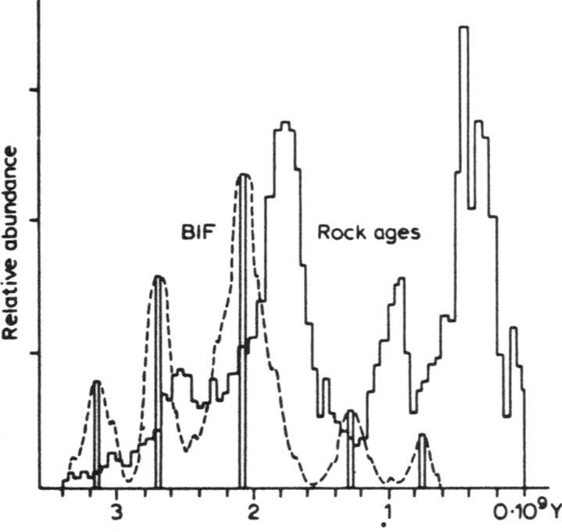

Fig. 2. Ages of major magmatic and metamorphic events in earth's history (mineral and rock ages compiled from Dearnley (1965); 3,400 age determinations) and relative abundance of iron-formations on the time scale.

epochs precursory to and partly overlapping with the initial stages of major magmatic and metamorphic cycles during the Precambrian.

Unequivocally, as it is also pointed out by Cloud (1973), the epoch between about 2,600 and 2,000 m.y. ago marks a profound change of conditions in the earth's evolution, terminating the vigorous growth of procaryotic organisms and giving rise to a development under increasingly oxidizing conditions.

[*Editor's Note*: In the remainder of this paper Eichler discusses in turn the geological setting of banded iron-formations with regard to depositional environments, sedimentary facies and sedimentary structures, diagenetic and metamorphic alterations of the mineral phases, organic remains, and chemical composition; the similarities and differences between banded and oolitic iron-formations; and hypotheses of origin of banded iron-formations with respect to environments of deposition, source of iron and silica, and transport, deposition and banding.]

REFERENCES

Alexandrov, E. A. 1973. The Precambrian banded iron ores of the Soviet Union. *Econ. Geology* **68**:1035–1062.

Bayley, R. W., and H. L. James. 1973. Precambrian iron-formations of the United States. *Econ. Geology* **68**:934–959.

Beukes, N. J. 1973. Precambrian iron-formations of southern Africa. *Econ. Geology* **68**:960–1004.

Brandt, R T., G. A. Gross, H. Gruss, N. P. Semenenko, and J. V. N. Dorr, II. 1972. Problems of nomenclature for banded ferruginous-cherty sedimentary rocks and their metamorphic equivalents. *Econ. Geology* **67**:682–684.

Cloud, P. 1973. Paleoecological significance of banded iron-formation. *Econ. Geology* **68**:1135–1143.

Dearnley, R. 1965. Orogenic fold-belts and continental drift. *Nature* **206**: 1083–1087; 1284–1290.

Dorr, J. V. N., II. 1973a. Iron-formation and associated manganese in Brazil. *Kiev. Symp., UNESCO, Earth Sci., Proc.* **9**:105–113.

Dorr, J. V. N., II. 1973b. Iron-formation in South America. *Econ. Geology* **68**:1005–1022.

Goldich, S. S. 1973. Ages of Precambrian banded iron-formations. *Econ. Geology* **68**:1126–1134.

Goodwin, A. M. 1973. Archean iron-formations and basins. *Econ. Geology* **68**:915–930.

Govett. G. S. 1966. Origin of banded iron-formations. *Geol. Soc. America Bull.* **77**:1191–1212.

Gross, G. A. 1965. Geology of iron deposits in Canada: General geology and evaluation of iron deposits. *Canada Geol. Surv. Econ. Geology Rept.* 22, 181 pp.

James, H. L. 1954. Sedimentary facies of iron-formations. *Econ. Geology* **49**:235–293.

James H. L. 1966. Chemistry of iron-rich sedimentary rocks. In *Data of Geochemistry, 6. U.S. Geol. Survey Prof. Paper 440(W),* 61pp.

Kennedy, W. Q. 1964. The structural differentiation of Africa in the Pan-African (±500 m.y.) tectonic episode. *Annual Rept. Sci. Results, 8th,* Leeds.

Krauskopf, K. B. 1967. *Introduction to Geochemistry*. New York: McGraw-Hill, 721 pp.

Krumbein, W. C., and R. H. Garrels. 1952. Origin and classification of chemical sediments in terms of pH and oxidation-reduction potentials. *Jour. Geology* **60**:1–33.

Lepp, H., and S. S. Goldich. 1964. Origin of Precambrian iron formations. *Econ. Geology* **59**:1025–1060.

O'Rourke, J. E. 1961. Paleozoic banded iron-formation. *Econ. Geology* **56**:331–361.

Sims, P. K., and G. B. Morey. 1972. Resumé of geology of Minnesota. In *Geology of Minnesota—A Centennial Volume,* P. K. Sims and G. B. Morey, eds. Minn. Geol. Surv., pp. 3–17.

Trendall, A. F., and J. G. Blockley. 1970. The iron formations of the Precambrian Hamersley Group, Western Australia. *Western Australia Geol. Surv. Bull.,* **119**, 366 pp.

Van Hise, C. R., and C. K. Leith. 1911. The geology of the Lake Superior region. *U.S. Geol. Survey Mon. 52,* 641 pp.

25

Reprinted from *Science* **163**:570–571 (1969)

Sedimentary Rock Types: Relative Proportions as a Function of Geological Time

R.M. Garrels and F.T. Mackenzie

The proportions of sedimentary rock types in the geologic column vary as a function of age; for example, evaporites amount to several percent of post-Precambrian sedimentary rocks whereas they are far less than 1 percent of Precambrian rocks of sedimentary origin. It is often assumed that the ratios of sedimentary rock types of a particular age represent the relative proportions of sediment types deposited at that time. We propose that age differences in ratios of rock types may be largely due to differential rates of overturn of sedimentary materials in which the various components of the sedimentary rock mass circulate at markedly different rates controlled by their erodibility.

Sediments have been continuously deposited and destroyed throughout geologic time. The rates of deposition and destruction certainly have not been constant, as evidenced today by the irregular distribution of sediment mass as a function of age; however, one can construct highly simplified models in an attempt to simulate the gross aspects of today's mass distribution.

Based on the assumptions of a total mass that is constant with time, constant and equal rates of deposition and destruction, and an equal probability of destruction of equal masses (independent of age) distributions of sedimentary mass as a function of age are predicted (Fig. 1). The 5x model is of the right order of magnitude to fit actuality; the 10x model predicts too small a mass of older sediments, and the 2x model far too much (Fig. 1). The choice of the 5x model, as opposed to either of the others, is compelling. This conclusion would not change even if there were large errors in estimates of the actual age distribution of existing sediments. The mass distribution required to fit the 10x and 2x models is contrary to the experience of many geologists.

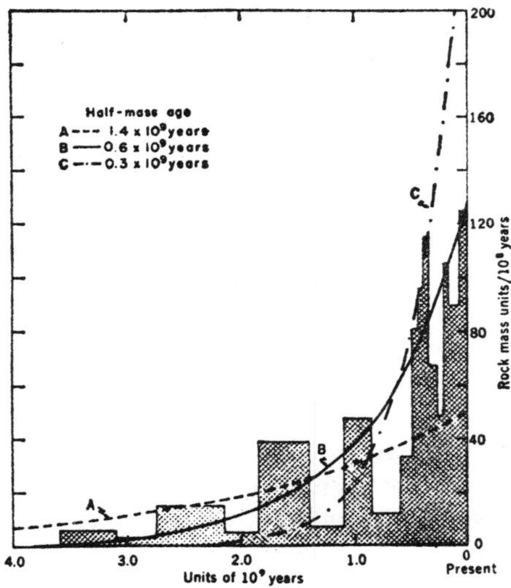

Fig. 1. Distribution of mass of sedimentary rocks as a function of age. Curves based on models that assume the total mass of sediments existing today has remained constant throughout geologic time. Curves A, B, and C represent total deposition of a mass of sediments equal to two, five, and ten times the existing mass, respectively. The corresponding half-mass ages are 1.4, 0.6, and 0.3 billion years. Hachured histogram is an estimate of the actual mass distribution based on observed occurrence (*1*) and our interpretation of the Precambrian distribution.

Approximately half the total mass of existing sediments is younger than 600 million years, whereas the rest is distributed irregularly over an interval of about 2500 to 3000 million years; that is, the half-mass age of all sedimentary rocks is about 600 million years. However, the half-mass ages of the various components of the sedimentary lithosphere appear to be different; that of carbonate rocks is about 300 to 400 million years and that of evaporites perhaps 200 to 300 million years (Fig. 1) (*2*).

The near absence of evaporite deposits in Precambrian rocks may then be largely attributable to a rapid turnover of these relatively soluble materials. The present mass–age relations indicate that a cycling rate two to three times that of the shales would be sufficient to account for present day distributions. Also, there has long been an apparent discrepancy between the 20 to 30 percent of carbonate rocks in the post-Precambrian record and the 5 to 10 percent predicted for all sedimentary rocks by geochemical balance calculations (*3*). If carbonate rocks cycle approximately 1.5 to 2 times faster than shale, they would be predicted to make up decreasing percentages of existing sedimentary rocks as a function of increasing age, even though their percentage of the total mass may always have been very nearly the same. The predicted present distribution of shale, carbonate, and evaporite as a function of age (Fig. 2) based on half-mass ages of 600, 300, and 200 million years, respectively, agrees favorably with the actual distribution (*2*). Finally, the possibility emerges that the very large percentages of cherty rocks, particularly of middle Precambrian age, may reflect the slow cycling of chert because of its high resistance to erosion and its characteristic protected position at the bottom of sedimentary basins.

From these qualitative relations we suggest that geochemical "uniformitarianism" should be strongly considered when interpreting the ratios of various rock types of any given age. In other words, the total mass of sediments of all ages existing at any given time in the geologic past may have had about the same ratios of rock types that we observe today. Differences in the ratios of rock types as a function of age in the rock mass existing today may depend on differential cycling rates of the components. We do not contend that all age differences in ratios of rock types existing in sedimentary rocks are caused by differential removal and deposition of various components of the sedimentary mass, but that geological conclusions based on use of today's ratios as indications of the ratios at the time of sedimentation should be tempered by consideration of the effect of differential cycling.

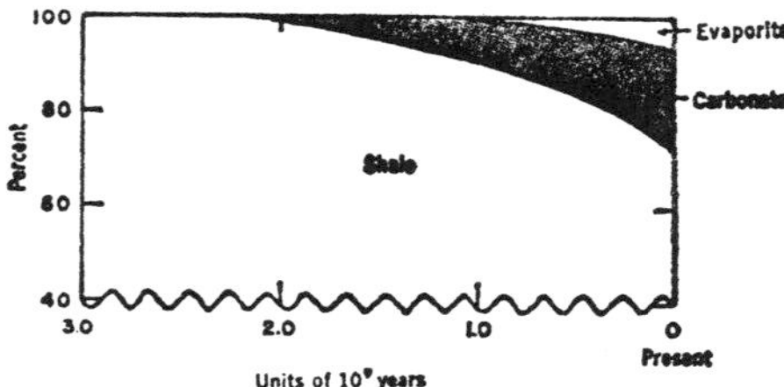

Fig. 2. Calculated present distribution of shale, carbonate, and evaporite as a function of time. The calculation is based on half-mass ages of shale, carbonate, and evaporite of 600, 300, and 200 million years, respectively.

References and Notes

1. M. Kay, in *Crust of the Earth*, A. Poldervaart, Ed. (Geologic Society of America, New York, 1955); A. G. Ronov, *Geochemistry* **5**, 493 (1959).
2. See Figure 1 in A. G. Ronov, *Geochemistry* **8**, 715 (1964). This figure illustrates the relative abundances and compositions of rock types as a function of geologic time. See also C. B. Gregor, *Kon. Ned. Akad. Wetensch. Proc.* **71**, 22 (1968), for a mathematical analysis of the mass distribution of sedimentary rocks.
3. F. J. Pettijohn, *Sedimentary Rocks* (Harper, New York, 1957); M. K. Horn and J. A. S. Adams, *Geochim. Cosmochim. Acta* **30**, 279 (1966).

4. We thank R. H. Leeper who aided in the calculations. Supported by the Petroleum Research Fund of the American Chemical Society and NSF grant GA-828.

26

Reprinted from *Nature Phys. Sci.* **244**:25–26 (1973)

WORLDWIDE UNCONFORMITIES IN MARINE SEDIMENTS RELATED TO EUSTATIC CHANGES OF SEA LEVEL

P. A. Rona

THE Deep Sea Drilling Project has discovered hiatuses of up to tens of millions of years in the stratigraphic record of every principal ocean basin. A hiatus is recognized by a discontinuity in palaeontologic ages and/or by an abrupt decrease in average sediment accumulation rates below that typical for the sediment type present within a cored interval. The physical expression of a hiatus may be a surface(s) of unconformity within the stratigraphic column resulting from an interval(s) of non-deposition and/or erosion.

A review of 171 sites of the Deep Sea Drilling Project reveals fifty-one sites in the Atlantic, Pacific, and Indian oceans and Caribbean Sea where significant hiatuses are recognized within the stratigraphic column penetrated (Fig. 1). The distribution of the hiatuses in time and in space forms patterns from which physical events of oceanwide and worldwide extent related to the development of the hiatuses can be inferred. In inferring the timing of a physical event associated with a hiatus, the change from non-deposition and/or erosion to accumulation of sediment at the end of the hiatus constitutes a datum for the termination of the associated physical event; the change from sediment accumulation to non-deposition and/or erosion at the onset of a hiatus may represent an erosional level and does not necessarily correspond to the time of initiation of the associated physical event.

Two hiatuses seem synchronous in all the principal ocean basins (Fig. 1). The first involves at least one half of Palaeocene time (65–53.5 m.y. BP) at twenty-four of thirty-eight sites where sediments Palaeocene or older in age were penetrated. At nine of the thirty-eight sites this hiatus ends in Palaeocene or early Eocene indicating that an associated physical event was active during the Palaeocene and terminated during the early Eocene. This hiatus accounts for 61% of the total Palaeocene time represented at thirty-three sites where the entire Palaeocene section was penetrated (Fig. 1). The second hiatus involves at least one half of Oligocene time (36–22.5 m.y. BP) at forty-four of the fifty-one sites where sediments Oligocene or older in age were penetrated. This hiatus ends by middle Miocene at thirty-three of the fifty-one sites indicating that an associated physical event of worldwide extent was active during the Oligocene and terminated during the early Miocene. This hiatus accounts for 73% of the total Oligocene time represented at fifty sites where the entire Oligocene section was penetrated (Fig. 1). The Oligocene hiatus extends through the Eocene at forty of forty-eight sites at which the entire Eocene was penetrated. The Palaeocene hiatus is less pronounced than the Oligocene.

The duration of the hiatuses is longest in the western North Atlantic extending from Middle Tertiary through Late Cretaceous at five of six sites where the entire Late Cretaceous is penetrated (Fig. 1). Next longest is the north-west Pacific where hiatuses extend from Middle through Early Tertiary at four of six sites where the entire Tertiary is penetrated. The hiatuses are relatively short in the eastern North Atlantic where only the Oligocene hiatus is present at five of seven sites

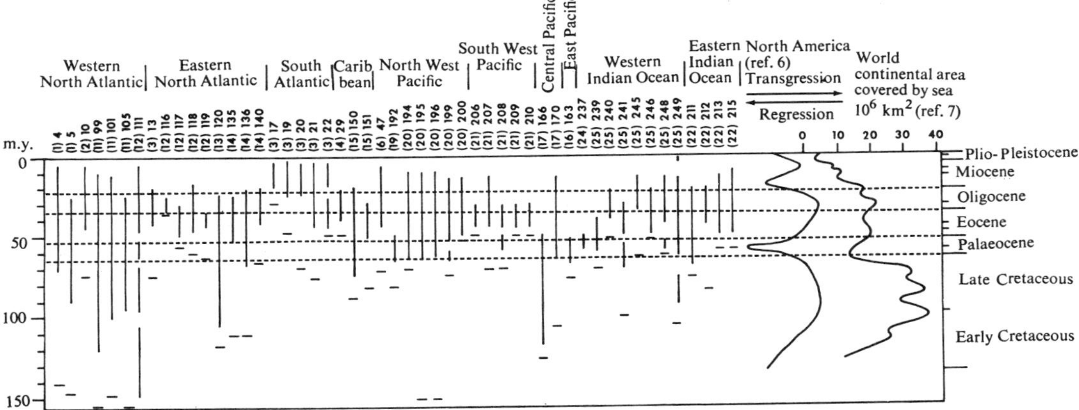

Fig. 1 Vertical lines indicate the duration of hiatuses within stratigraphic columns penetrated at sites of the Deep Sea Drilling Project identified by (leg) and site number. Horizontal bars indicate the oldest sediment penetrated. The two curves represent epicontinental marine transgressions and regressions of North America[6] and world continental area covered by sea[7]. Horizontal dashed lines indicate boundaries of the Palaeocene and Oligocene epochs. Time scales used are those of Berggren[15] for the Cainozoic and the Geological Society of London[16] for the Mesozoic.

which penetrate the entire Tertiary and in the eastern Pacific where a hiatus is recognized at only one site. Hiatuses are shorter in the southern than northern basins of the Atlantic and Pacific. Both Palaeocene and Oligocene hiatuses are recognized in the Indian Ocean where their pattern of expression is more complex than in the Atlantic and Pacific.

The longer duration of the hiatuses in the western than eastern basins of the Atlantic and Pacific indicates that erosion by bottom currents is a primary process in their development because deep circulation is predicted to be stronger along western than eastern boundaries of ocean basins[1,2]. The more pronounced expression of the Oligocene than the Palaeocene hiatus may reflect the development of thermohaline circulation following present circulation patterns, especially where source regions of cold water masses opened in the North Atlantic during the Eocene[3] and the south-west Pacific during the Oligocene[4] as a result of continental drift and climatic change. This interpretation has been advanced for the origin of the Oligocene hiatus in the North Atlantic[5] and south-west Pacific[4], but alone does not explain the worldwide distribution and timing of both the Palaeocene and Oligocene hiatuses.

The Palaeocene hiatus correlates with a time of reversal between epicontinental marine regression and transgression (Fig. 1), inferred from the distribution of marine strata on the North American continent and world continents. The Oligocene hiatus correlates with a time of worldwide epicontinental marine regression[6,7] (Fig. 1). Synchronous worldwide regressions other than glacio-eustatic are most likely controlled by eustatic lowering of sea level resulting from increase in the combined volumetric capacity of world ocean basins. A significant factor contributing to increase of ocean basin volume is volume decrease of the world mid-ocean ridge system[8,9] which may lower sea level up to hundreds of metres[10,11]. Eustatic lowering of sea level of this order would be expected to produce interacting physical and chemical effects on a global scale which may account for the worldwide hiatuses in marine sediments.

The existence of the hiatuses indicates that changes of ocean circulation resulting from lowered sea level and subsidence of the mid-ocean ridge system, in addition to the development of thermohaline circulation, favoured erosion by deep currents at the sites of the hiatuses. Climatic changes, expected to have accompanied the changes in ocean circulation, are evidenced by oxygen isotope measurements made on the tests of planktonic foraminifera which indicate climatic optima in the Pacific during the Palaeocene through early Eocene and the late Oligocene[12]. Expected changes in chemistry of the oceans, including downward shifts in depth dependent parameters such as the lysocline and increased input of elements eroded from emergent continents, may be evidenced by the relatively high average carbonate content of those Atlantic basin sediments preserved during the Palaeocene through early Eocene and of Oligocene sediments of both the Atlantic and Pacific[12]. Complications in the worldwide pattern of hiatuses, such as those observed in the Indian Ocean, imply the influence of more local oceanic and tectonic factors.

The eustatically lowered sea levels of the Palaeocene and Oligocene would have altered the hypsographic curve by emergence of continents resulting in rejuvenation of erosion on continents including continental margins. Average rates of sediment accumulation would be expected to have decreased and prominent surfaces of unconformity would have developed on continents, continental shelves, and continental slopes of the world. These effects have been documented on the North American continent[6], continental margins of the central North Atlantic[9,13,14], and require worldwide investigation. Consequently, input of sediment from continents to ocean basins is expected to have been relatively large during the Palaeocene and Oligocene. The net sedimentary budget of the ocean basins was accumulation in spite of the hiatuses in marine sediments that developed synchronously on continents, continental margins, and at certain sites in ocean basins.

1 Wüst, G., *Wiss. Erg. Deutsch Atl. Exp. "Meteor"* 1925–1927, **6** (2), 261 (1957).
2 Defant, A., *Physical Oceanography*, **1**, 729 (Macmillan, New York, 1961).
3 Rona, P. A., *J. Sedimentol. Petrol.*, **39**, 1132 (1969).
4 Kennett, J. P., Burns, R. E., Andrews, J. E., Churkin, M., Davies, T. A., Dumitrica, P., Edwards, A. R., Galehouse, J. S., Packham, G. H., and van der Lingen, G. J., *Nature phys. Sci.*, **239**, 51 (1972).
5 Hayes, D. E., *et al.*, *Initial Reports of the Deep Sea Drilling Project*, NSFSP-14, 975 (US Government Printing Office, Washington, 1972).
6 Sloss, L. L., *Bull. geol. Soc. Am.*, **74**, 93 (1963).
7 Holmes, A., *Principles of Physical Geology*, 1288 (Ronald Press, New York, 1965).
8 Hallam, A., *Am. J. Sci.*, **261**, 397 (1963).
9 Rona, P. A., *Bull. Am. Assoc. Petrol. Geol.*, **54**, 129 (1970).
10 Menard, H. W., *Marine Geology of the Pacific*, 271 (McGraw-Hill, New York, 1964).
11 Menard, H. W., and Smith, S. M., *J. geophys. Res.*, **71**, 4305 (1966).
12 Moore, jun., T. C., *Geotimes*, **17**, 27 (1972).
13 Rona, P. A., *Geol. Soc. Am., Abstr. With Programs*, **4** (7), 644 (1972).
14 Rona, P. A., *Bull. geol. Soc. Am.* (in the press).
15 Berggren, W. A., *Nature*, **224**, 1072 (1969).
16 *Q. Jl. geol. Soc. Lond.*, **120s**, 260 (1964).

27

Reprinted from pages 113–115, 116, 117, and 132–138 of *Micropaleontology* **24**:113–138 (1978)

CENOZOIC HIATUSES IN PELAGIC SEDIMENTS*

T. C. Moore, Jr., Tj. H. van Andel, C. Sancetta, and N. Pisias

ABSTRACT

Sediment accumulation, non-deposition or erosion in the deep sea depends on the balance between the rate of sediment supply and the rate of sediment removal. In most pelagic areas the rate of supply is related to the productivity of near-surface waters. The rate of removal depends on the advective velocity of bottom waters and their chemical nature (*i.e.*, their corrosiveness with respect to biogenic skeletal debris). Any change in the shape of the ocean basins or in the oceanic circulation patterns can have major regional and even global effects on the balance between sediment accumulation and removal. The temporal and spatial distribution of hiatuses can be determined for the Cenozoic section from the core material recovered by the Deep Sea Drilling Project. For all ocean basins, maxima in hiatus abundance occurred at the Cenozoic-Mesozoic boundary, in the Late Eocene (spanning the Eocene-Oligocene boundary in some areas) and in Middle to Late Miocene. Fewest hiatuses are found in the Middle Eocene, Early to Middle Miocene and Quaternary. There is some variation in the temporal and spatial distribution of hiatuses both within and between the major ocean basins.

It is suggested that the beginning of global cooling during the Cenozoic started with the rapid opening of the Atlantic Ocean and Norwegian Sea and the gradual closing of the Tethys Sea. The opening of the Australia-Antarctic seaway, the development of the circum-Antartic current, the opening of the Norwegian Sea to the Arctic, the lowering of the southern sill to the Norwegian Sea, and the opening of the Tasman passage to deep water flow, all may have contributed to the increased formation of hiatuses in the Late Eocene-Oligocene. The initiation of North Atlantic Deep Water caused by the further lowering of the sill to the Norwegian Sea and the subsequent formation of the Antarctic icecap and Antarctic Bottom Water led to the Miocene maxima. The onset of glaciation in the northern hemisphere is reflected by a small peak in hiatus abundance in all the major ocean basins.

INTRODUCTION

The common occurrence of hiatuses in Deep Sea Drilling cores has in part frustrated attempts to recover an unbroken geologic record from the world ocean. The presence of hiatuses is a paradox in that the deep ocean basins are the ultimate sediment catchments, yet a discontinuous stratigraphic record is preserved within them. Everything that is washed from the continents is carried finally to the ocean; the skeletal material of the organisms that live in the sea must fall to the bottom unless it is dissolved en route. The fact that hiatuses exist in the record belies the nineteenth century concept of a still and stagnant abyss.

The purpose of the present study is to present the spatial and temporal patterns of occurrence for Cenozoic hiatuses sampled by the Deep Sea Drilling Project and thus to shed light on the interrelationships between breaks in the stratigraphic record and the history of the oceans. Several regional studies suggest that hiatuses in the record of marine sediments do have distinct patterns in space and time (*e.g.*, Berger, 1972; Kennett *et al.*, 1972, 1974; Johnson, 1972*a*; Edwards, 1973; Rona, 1973; Sigal, 1974; Pimm, 1974; Davies *et al.*, 1975). In this study the world ocean is considered as a whole, and changes in the distribution of hiatuses are related to changes in the boundary conditions of the ocean basins.

Bottom waters and the occurrence of hiatuses

Erosion, non-deposition, or sediment accumulation on the sea floor is determined by the dynamic balance between the rate at which sediments are supplied to the bottom and the rate at which they are removed (text-figure 1). In pelagic regions, the supply rate is largely controlled by the productivity of the near-surface waters; therefore the areas most susceptible to hiatuses in the sedimentary record are those that lie beneath the relatively unproductive centers of the oceanic gyres. Removal depends on processes which either physically transport sediments to other regions of the sea floor or chemically dissolve them.

Recent investigations have demonstrated that the advective flow of bottom waters gives rise to well-defined, deep currents, and that these currents can cause non-deposition, erosion, the transportation of sedimentary material, and the construction of large depositional features on the sea floor (*e.g.*, Heezen, Hollister and Ruddiman, 1966; Schneider and Heezen, 1966; Jones *et al.*, 1970; Johnson and Johnson, 1970; Amos, Gordon and Schneider, 1971; LePichon, Ewing and Marek, 1971; Ruddiman, 1972; Johnson, 1972*a*, 1972*b*; Hollister, Johnson and Lonsdale, 1974). Most Pacific Ocean studies (*e.g.*, Wüst, 1929; Knauss, 1962; Panfilova, 1967; Burkov, 1969)

*Paper presented at the 3rd Planktonic Conference, Kiel, West Germany, September 1, 1974.

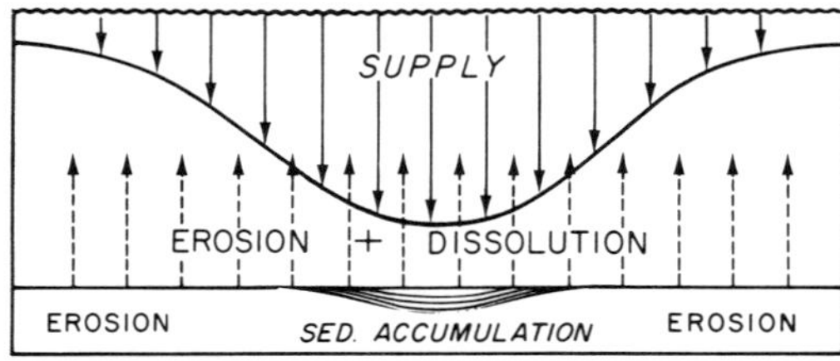

TEXT-FIGURE 1
Diagram of balance of sediment supply and removal rates. Sediment accumulation occurs only when rate of supply exceeds rates of erosion and dissolution.

agree that Antarctic Bottom Water moves northward along the western margin of the South Pacific, following the bathymetric contours into the northern hemisphere, then northwestward between the Marshall Islands and Mid-Pacific Mountains and eastward through gaps in the Line Island chain.

In the Indian Ocean, bottom waters flow northward in a western boundary current (Warren, 1974), and appear to circulate in a clockwise fashion, turning southward into the Central Indian Basin (Wyrtki, 1971). Bottom waters also cross the Southeast Indian Ridge and pass south of Australia and northward into the Wharton Basin. In the Atlantic, Antarctic Bottom Water flows northward in the western basins but does not cross into the eastern basins until it reaches the Romanche Fracture Zone near the equator (Wüst, 1933; Stommel and Arons, 1960).

The North Atlantic Ocean is presently the only northern hemisphere source for deep water. Waters in the Norwegian Sea mix with the salty waters that are carried into the North Atlantic by the Gulf Stream, and form North Atlantic Deep Water. This cold, salty water sinks and flows southward (Worthington and Volkman, 1965; Swallow and Worthington, 1969; Ried and Lynn, 1971).

The pattern of occurrence of hiatuses that result from erosion and non-deposition may be associated with the flow-path of these bottom waters. Furthermore, erosional hiatuses are most likely to occur where the flow is intensified, such as in the areas swept by western boundary currents or near prominent topographic highs and constrictions.

The rate of removal is determined not only by the rate of bottom flow, but also by the corrosiveness of the bottom waters with respect to biogenic detritus, particularly calcite and opaline silica. In general "young" bottom water (that water which has only recently formed at or near the sea surface) is largely stripped of dissolved silica and tends to be more corrosive with respect to the tests of diatoms, silicoflagellates, and Radiolaria than "old" bottom water which has gradually become more enriched in silica through the dissolution of these tests (Heath, 1974 and references therein). The effect is opposite for calcite. The highly oxygenated "young" bottom waters are not as corrosive to the tests of foraminifera and coccoliths as "old" bottom waters that are rich in CO_2 (Berger, 1970, 1973).

Because the chemical properties of ocean waters may change with depth it might be expected that the distribution of hiatuses would also be dependent on depth. This has been demonstrated for South Atlantic sites (Berger, 1972), but was not seen in sites from the Indian and southwestern Pacific Oceans (Kennett *et al.*, 1972, Davies *et al.*, 1975). In this presentation, spatial patterns of hiatus occurrence are interpreted solely in terms of the factors discussed above. Although depth is not considered specifically as one of these factors, the interpretation of the paths of bottom flow does require some qualitative estimate.

Everywhere in today's ocean, bottom waters appear to be under-saturated with respect to both calcite and opaline silica (Ingle *et al.*, 1973; Jones and Pytkowicz, 1973). Under these conditions, non-deposition and thus continued exposure to bottom waters might lead to the dissolution of biogenic sediments. This removal of sediment by corrosion is considered a special case of erosion. The fine residuum of insoluble wind-blown debris, authigenic materials, and trace fossils that are left after the dissolution of the carbonate and siliceous material, accumulates very slowly (on the order of 1 m. per million years). Because of the lack of biostratigraphic markers, unconformities in such a dissolution facies are nearly impossible to detect.

If in the past the rate of supply of these shell-building materials from the land to the ocean had been greatly reduced, it would have resulted in the increased dissolution of tests on the deep-sea floor. Because the residence time of these materials in the ocean is very short—for silica on the order of a few thousand years—the response to decreased input to the ocean would be geologically instantaneous (Heath, 1974).

Any change in the productivity of surface waters, the rate of bottom water flow or the chemical nature of the bottom waters, can affect the balance between accumulation and erosion and, therefore, the distribution of hiatuses. Tectonic movements associated with plate motions can drastically alter the path of flow or even the source of bottom water and thus change the pattern of occurrence of hiatuses. Finally, the migration of crustal plates can displace a site from an area of accumulation to one of erosion. Therefore, in spite of the paradoxical aspect of hiatuses in the marine record,

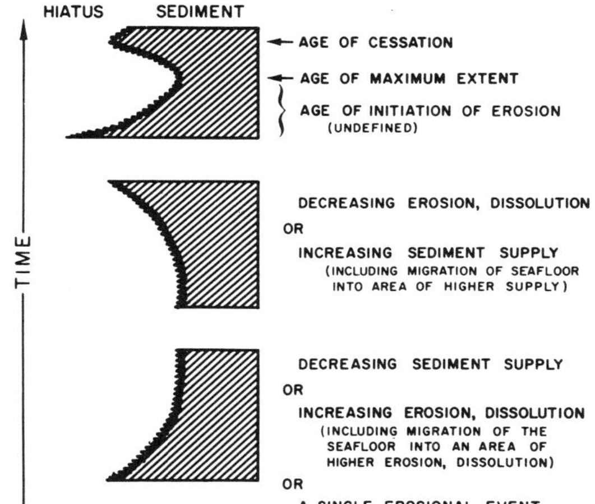

TEXT-FIGURE 2
Diagrammatic cross section of hiatuses with possible causes. Age of maximum extent and age of cessation can be defined. Age of initiation cannot.

the problem does not lie in finding plausible causes for their existence, but rather in finding the probable cause for each occurrence and for changes in spatial and temporal patterns of occurrence.

Definition of hiatuses

The multiplicity of explanations for various cross-sectional patterns of hiatus extent are illustrated in text-figure 2. If each hiatus is viewed as resulting from a single erosional event, then the time of maximum extent can be defined. The time of cessation can be defined if the overlying material has not been subsequently eroded by a second erosional event. Even partial erosion of such material may increase the apparent age of cessation. Thus, the observed age of cessation is a maximum value. The time of initiation of an erosional event cannot be precisely defined; however, regional synchroneity in the age of an unconformity does suggest that the age of initiation is very near the age of the underlying sediments (Berger, 1972).

All Deep Sea Drilling sites up to and including those of Leg 31 were considered in this compilation (text-figure 3a, b). Stratigraphies and therefore ages of the hiatuses, are primarily those given by Berggren (1972). We used the shore laboratory reports of Bukry (included in most volumes of the Initial Reports of the Deep Sea Drilling Project) to resolve disagreements and as supplementary confirmation of age assignments.

There are parts of the stratigraphic record in which the present zonation of the various microfossil groups does not provide sufficient resolution to actually define a hiatus of a few million years; however, the range of certain species may give a clear indication at some sites that a part of the record has been removed or intensively reworked. This is particularly true of the Upper Eocene, an interval that biostratigraphers have tended to lump into a single zone (*e.g.*, Dinkelman, 1973; Bukry, 1973). Based on the first appearance of such radiolarian species as *Lophocystis jacchia* (Ehrenberg) and *Theocyrtis tuberosa* Riedel (Eocene variety) and the last appearance of *Podocyrtis goetheana* (Haeckel), a Late Eocene hiatus may be suspected in certain drill sites; but they are not included as such in this compilation.

In most cases the existence of a hiatus is clearly defined by the absence of one or more biostratigraphic zones. Only these breaks in the stratigraphic record are accepted as unconformities and used in the compilation. In a few cases where the unrepresented period of time is short relative to the average sedimentation rate, it is possible that the section is present, but not sampled. Even in continuously cored sections in which the recovered core barrels were full, an average of 35% of the section may have been missed (Moore, 1972). This estimate is based on the abnormally frequent occurrence of biostratigraphic boundaries between successive cores in the same hole. Thus, of the 18 m. drilled and sampled by two successive cores, 6.3 m. (0.35 × 18 m.) of the actual section could be missed in the drilling and recovery process. If the average sedimentation rate at a site were 2 m./m.y. (million years), a hiatus of over 3 m.y. duration could be created by the vagaries of the sampling technique.

In the data set used here, hiatuses of less than 6 m.y. duration were carefully examined. If such a hiatus occurs within a single core, it is accepted as real. If it occurs between cores and is of a duration less than 6.3 m. divided by the average sedimentation rate (m./m.y.), above and below the gap, then the existence of the hiatus is questioned and it is not included as a reliable hiatus in the compilation.

For sites which were not continuously sampled, hiatuses are considered to exist only when the average sedimentation rate is significantly less than 1 m./m.y. Where average sedimentation rates are low (1 to 2 m./m.y.) or the preservation of microfossils does not permit accurate stratigraphic definition, hiatuses may be suspected, but are not included as such in this compilation.

In addition to the problem of how well unconformities are sampled in the cores of the Deep Sea Drilling Project there is the additional question as to whether a single drill site is representative of the surrounding region. Where more than one hole was drilled at a site the recovered sections generally show good agreement. However, in the central tropical Pacific, where Tertiary sediments are commonly found cropping out, a suprisingly large range of Tertiary ages may be sampled in a small area of sea floor. Two small areas have been surveyed and sampled in this region, one of about 190 km². at 8°20′N., 153°W. (Moore, 1970) and one of about 300 km.² at 7°30′N., 134°W. (Johnson, 1972*a*; Johnson and Parker, 1972). In both areas Tertiary sediments were recovered underlying a few centimeters of Quaternary material. The average age of these sediments was 23 (s.d. ±6) Ma (million years before present) in the eastern area and 20 (s.d. ±3) Ma in the western area. Based on these age distributions, it appears that the probability of a DSDP site sampling an unconformity within 3 m.y. of the average age of the outcropping Tertiary material would be between about p = 0.4 and p = 0.7. Site 69, the DSDP site closest to either of these two areas (290 km. south of the western area), sampled an unconformity with a basal age of 20 Ma, identical to the average age of the Tertiary material in the western area.

It should be pointed out that the two small areas studied by Moore and Johnson were sampled with the purpose of determining the maximum range of age represented by the outcropping sediments. The sampling program, therefore, was not strictly random. Likewise, DSDP sites are not selected in a random manner. In general, sites have been located where acoustic reflection profiles indicated an easily interpretable record. In a few areas, obviously eroded sections were drilled in order to obtain older sediments; but at most sites, the sections drilled have tended to be of normal or greater than normal thickness for the region surrounding the site. Either because of, or in spite of this careful planning, the distribution patterns of the hiatuses (studied previously and presented herein) are surprisingly coherent.

[*Editor's Note:* Text-figure 3, which shows sites of the DSDP, has been omitted.

In pages 117 to 132 Moore et al. discuss the temporal and spatial distributions of hiatuses and changes in the characteristics of the bottom waters.]

DISCUSSION AND SUMMARY

It is interesting to note that the stratigraphic boundaries of the Paleocene and Eocene tend to be marked by maxima in hiatus abundance, while those of the Miocene tend to be marked by minima. Many of the minor fluctuations in hiatus abundance that coincide with stratigraphic boundaries may be artifacts of the biostratigraphies used in dating the sections. However, the more prevalent and well-defined maxima (Maastrichtian to Paleocene, Paleocene to Early Eocene, Early to Middle Eocene, Middle to Late Miocene and Early to Late Pliocene) probably represent major changes in oceanic configuration which affected not only bottom waters and their circulation, but also the surface circulation. These changes could have given rise to a large scale alteration of the marine environment, to the disruption of previously established gene pools, and ultimately to significant evolutionary changes in the flora and fauna. Thus, the near-perfect synchroneity of the major hiatuses in different ocean basins may not be an artifact of the biostratigraphies, but may be caused by the changes in the shape of the ocean basins that lead to changes in circulation resulting in distinct changes in sedimentary facies and clearly marked biostratigraphic boundaries and hiatuses.

In the oceans, those biostratigraphic boundaries which coincide with maxima in hiatus abundance are usually clearly defined (*e.g.*, the Eocene-Oligocene boundary). There tends to be more confusion regarding those boundaries which match the minima in hiatus abundance (*e.g.*, the Oligocene-Miocene and Miocene-Pliocene boundaries). This happens even though the faunal and floral changes seen across those boundaries are distinct. It may be that those boundaries represent the biotic response to oceanographic changes that are more gradual in nature (such as the closing of Tethys; Frakes and Kemp, 1972) and are not necessarily accompanied by the rapid lowering of sills and the opening of new seaways to bottom flow.

The following summary attempts to tie the changes in the distribution of hiatuses to both the gradual and rapid changes in the configuration of the ocean basins.

The Cretaceous-Cenozoic Boundary ($\simeq$ 65 Ma)

Hiatuses: Maxima in all ocean basins. Spanning the equatorial region of the Pacific. Extent gradually decreasing with time.

Tectonics: Continued rapid widening of the Atlantic and Labrador Sea and closing of Tethys. Complete closure of a possible Arctic-Pacific seaway through Siberia (Herron, Dewey and Pitman, 1974).

58–59 Ma
Hiatuses: Minima in the Southern Ocean and associated basins. Area ceases to spread in Tasman basin (Hayes and Ringis, 1973).

Tectonics: Seafloor spreading starts between Norway and Greenland (Pitman and Talwani, 1972).

53–55 Ma
Hiatuses: Maxima in all ocean basins (preceded by minimum at 55–56 in the Atlantic).

Tectonics: Seafloor spreading starts between Australia and Antarctica (Weissel and Hayes, 1972). Spreading ceases on Carlsberg Ridge (McKenzie and Sclater, 1971). Spreading continues both sides of Greenland; widening of Atlantic.

52 Ma
Hiatuses: Pronounced minima in most ocean basins. Westward shift in area of hiatus occurrence between Africa and India.

Tectonics: Seafloor spreading continues between Australia and Antarctica. Sinking of Greenland-Faeroe sill begins (Laughton, Berggren *et al.*, 1972).

45–48 Ma
Hiatuses: Secondary minima in all basins except west Indian-South Atlantic, which is equivalent to basal age of hiatuses in Atlantic and west Indian basins; age of hiatus cessation in Pacific.

Tectonics: Seafloor spreading ceases in Labrador Sea. Spreading direction and rate changes in North Atlantic (45 Ma; Avery, Vogt and Higgs, 1969). Separation of Greenland and Spitsbergen (Pitman and Talwani, 1972).

42–43 Ma
Hiatuses: Pronounced maxima in all basins except Indian where maximum occurs somewhat earlier (44 Ma). Maximum extent in West Indian Basin. First appearance of hiatus northeast of Venezuela. Drift deposits forming in North Atlantic.

Tectonics: Continued lowering of Greenland-Faeroe sill. Possible opening of Drake Passage.

37–39 Ma
Hiatuses: Pronounced maxima in East Indian and Southwest Pacific. Lesser maximum in North Atlantic. Areal extent greatly increased in Southwest Pacific, decreased in West Indian, East Pacific.

Tectonics: Earliest possible opening of Tasman passage to deepwater flow. Continued restriction of Tethyan seaway. Closure of northern arm of Tethys. Subsidence of Greenland-Faeroe sill ceases.

30–21 Ma
Hiatuses: Gradually increasing in East Pacific and decreasing in other basins. Marked drop in abundance in Southwest Pacific. Sedimentation pulse in drift deposits.

Tectonics: Seafloor spreading starts on Carlsberg Ridge (33 Ma). Possible closure (earliest) of African-Asian seaway (Tethys; Dewey *et al.*, 1973). Full opening of Tasman passage to deep water flow (Kennett *et al.*, 1974).

21–22 Ma
Hiatuses: Pronounced minima in hiatuses in all ocean basins except Atlantic-Caribbean. Last hiatuses in eastern North Atlantic. Minimum in East Pacific reworking.

Tectonics: Possible closure (latest) of African-Asian seaway (Tethys). Possible closure of western Mediterranean to deep water flow.

17–18 Ma
Hiatuses: Maximum in abundance in all oceans except Caribbean-Gulf of Mexico. Pronounced in Southwest Pacific and Indian Ocean (the latter at 15 Ma). Major pulse of sedimentation in drift deposits of North Atlantic, maximum in reworking in East Pacific. Silica deposition in high latitudes of Southern Ocean and North Pacific starts. Silica deposition in Caribbean ceases.

Tectonics: Increased seafloor spreading rate in North Atlantic. Lowering of Greenland-Faeroe sill renewed. Possible closure of Indo-Pacific passage to bottom water flow.

10–12 Ma
Hiatuses: Maxima in all ocean basins except Indian. Pronounced in East Pacific. Extent reduced off North America and in Tasman Basin, increased off Venezuela and in North Pacific. Continued major pulse in sediment accumulation in North Atlantic, maximum in reworking in East Pacific (12–15 Ma).

Tectonics: Spreading direction in North Atlantic as well as other area changes. Continued restriction of Indo-Pacific passage. Mediterranean restricted and finally closed (7 Ma). Straits of Panama restricted.

4–5 Ma
Hiatuses: Minima in all ocean basins except West Pacific. Extent restricted in all ocean basins.

Tectonics: Straits of Gibraltar opened. Straits of Panama closed.

3–4 Ma
Hiatuses: Maxima in all ocean basins except West Pacific.

There are many small coincidences and possible triggering mechanisms that come to mind when reviewing the above outline. The major features will be discussed in the following scenario of plate movement and bottom water formation.

Hiatuses are most abundant at the Cretaceous-Tertiary boundary, a time of rapid opening of the Atlantic, particularly the North Atlantic, and possibly the closure of free circulation between the Arctic and Pacific. The details of this episode are unknown. The one basin that might hold a stratigraphic record containing some of these relevant details, the Arctic, is *terra incognita* to the Deep Sea Drilling Project and is likely to remain so. It is known that during this interval the great tropical seaway of Tethys was growing smaller as the longitudinal seaway of the Atlantic-Arctic grew larger.

At about 53–55 Ma Australia began to separate from Antarctica, and to judge from the oxygen isotope record (Douglas and Savin, 1971, 1973; Shackleton and Kennett, 1974*a*, 1974*b*) this coincided with the major cooling trend of the ocean waters that has continued almost unbroken since that time. The Eocene sinking of the Greenland-Faeroe sill allowed some passage from the Norwegian Sea to the North Atlantic. The separation of Spitsbergen and Greenland (at about 47 Ma) eventually allowed a deep passage between the Norwegian Sea and the Arctic Ocean. This influx of cold bottom waters probably brought an end to the Early Eocene minimum in hiatus abundance and marked the beginning of the silica-rich deposits of the Middle Eocene.

It is not known precisely when the South America-Antarctica passage formed and the circumpolar current became fully established. It is also uncertain exactly what gave rise to each hiatus maximum in the closely spaced series occurring in the Middle and Late Eocene (38–50 Ma). However, the creation of the Arctic-Atlantic Passage, the widening of the Australia-Antarctic seaway and the opening of the Drake Passage were all associated with the Eocene-Oligocene peak in hiatus abundance.

The events in the Northern hemisphere allowed free communication between the Arctic and North Atlantic basins, but only at shallow to intermediate depths (500–1000 m.; Laughton, Berggren *et al.*, 1972). This was apparently sufficient to provide bottom water to the North Atlantic, and the contour currents thus produced led to both drift deposits and hiatuses, depending on the location. The tectonic events of the southern hemisphere allowed surface waters to remain in high southern latitudes for a significantly longer period of time than was previously possible. They would thus lose more of their stored heat, contribute to a general cooling of the global climate, and create masses of cold bottom water. In this way a more vigorous bottom circulation came about, and with the shift in the locus of bottom water formation to the southern hemisphere the bottom waters took on a more "youthful" character.

A deep water connection may well have existed in the Early Tertiary between the North Atlantic and Indian Ocean and Tethys. In the West Indian basin, changes in the pattern of hiatus occurrence appear to reflect the closure of such a passage in the early Tertiary. With the gradual closure of this seway, another vast low latitude ocean basin was lost, and the time allowed a given parcel of surface water to gain thermal energy was diminished (Frakes and Kemp, 1972). In the Early Miocene, concurrent with this restriction of Tethys and its remnant, the Mediterranean Sea, the region of eastern and northern Europe rose above sea level (cf. text-figures 24, 25). This forced at least part of the Gulf Stream flow into the Norwegian Sea and provided relatively warm, salty water to this area. Upon mixing with cold waters of the northernmost Atlantic, an early version of North Atlantic Deep Water was formed. Aided by the further lowering of the Greenland-Faeroe sill (Laughton, Berggren, *et al.*, 1972; Vogt, 1972) the formation of this cold, salty deep water led to the major pulse of sedimentation in the North Atlantic drift deposits and to hiatuses in most ocean basins. It also may have led to the complex vertical structure of bottom, deep, and intermediate waters that we have today, the commencement of siliceous deposits in high latitudes, and the initial formation of a major icecap on East Antarctica (see Shackleton and Kennett, 1974*a*, 1974*b*; summary in Kennett *et al.*, 1974).

The 10–12 Ma maximum in hiatus abundance and extent is difficult to relate to a specific tectonic event, although continued constriction between North and South America, Eurasia and Africa, and Australia and Asia may have further reduced the residence time of surface waters in low latitudes and contributed to further cooling of the climate. The striking pattern of hiatus occurrence in the west Indian Ocean (text-figures 10, 18) suggests the importance of the formation of the Antarctic cap and its effect on the formation of Antarctic Bottom Water (*e.g.*, Gordon, 1971, 1973). The peaks in hiatus abundance (text-figure 10) suggest the possibility of three episodes of erosion by bottom flow: 1) at 18 Ma, associated with the formation of North Atlantic Deep Water; 2) at 15 Ma associated with the development of the East Antarctic icecap; and 3) at 10–12 Ma when the Antarctic ice sheet reached the sea and through the formation of very cold, salty water the modern version of Antarctic Bottom Water was formed.

The final peak in hiatus abundance appears to be related to the start of continental glaciation in the northern hemisphere (Laughton, Berggren *et al.*, 1972; Shackleton and Kennett, 1974*b*) and is more readily related to the general trend of climatic change than to a specific change in the boundaries of the ocean.

It is not known whether all aspects of the above scenario are valid; however, we feel that they are consistent with the present knowledge of the history of the ocean basins. To test these ideas, the exact timing and sequence of events should be more finely resolved. Once the conceptual model can be stated precisely, budget calculation and mathematical models of ocean circulation can be used to compare with observational data.

CONCLUSIONS

During the Cenozoic there have been eight times during which hiatuses reached peak abundance in the sedimentary record of most ocean basins. Seven of these can be grouped into three major episodes: the Maastrichtian-Paleocene, Eocene-Oligocene, and Miocene. The oldest episode is represented by hiatuses in about 80% of the sites sampling this interval and appears to be related to the rapid opening of the Atlantic and Arctic Oceans. The Eocene-Oligocene events apparently are related to the opening of the Australia-Antarctic seaway, the opening of the Norwegian Sea-Arctic passage, the sinking of the Greenland-Faeroe sill, the creation of the Drake Passage, and the opening of the Tasman Passage to deep water flow.

It is suggested that the Miocene events are related to the formation of North Atlantic Deep Water and the further lowering of the Greenland-Faeroe sill and to the subsequent formation of the East Antarctic icecap and Antarctic Bottom Water.

The final peak in hiatuses appears in the Pliocene. This break in the record may still be in the process of forming. It is apparently related to increased bottom water production and continental glaciation in the Northern hemisphere.

Patterns of hiatus occurrence, changes in the distribution of siliceous and carbonate debris, variations in the lateral reworking of sediments and the building of sedimentary ridges and banks in the deep sea all point to sediment erosion and corrosion by bottom waters as the prime cause for these widely distributed hiatuses. Periods of time when hiatuses were particularly abundant appear to be related to changes in the boundaries of ocean basins, and particularly to those changes that affected the high latitude regions of bottom water formation. The coherence of the patterns of change in both space and time agree with the basic tenets of oceanography: the ocean is a system, and a change in any part of that system may affect the system as a whole.

ACKNOWLEDGMENTS

We thank the scientists and staff of the Deep Sea Drilling Project, in particular Dr. Tom Davies, for their aid in compiling these data. Special appreciation is given to Dr. James D. Hays and Dr. Eugen Seibold for their encouragement at the inception of this project and to Dr. G. Ross Heath for his continued interest and helpful discussions. Drs. Jim Kennett, Tom Davies, and Bill Ruddiman provided preprints of their own work and valuable criticism of our interpretation. Dr. Fran Shabica was of great help in the treatment of the data and Mrs. Linda Haley worked hard on the illustrations. Finally, a special thanks is given to N.S.F. for grant no. DES 74–22295.

[*Editor's Note:* Text-figures 10, 18, 24, and 25 mentioned on page 134 of this paper have been omitted. They show, respectively, the change in proportion of sampled sections represented by hiatuses as a function of time for the East Indian Ocean, and the distribution of hiatuses in DSDP sites for age intervals 50–55, 20–25, and 15–20 m.y.]

APPENDIX I

Maximum number of sites represented by either recovered sediments or a hiatus for each of thirteen 5-million-year age intervals (0–65 Ma)

SITE GROUPS

Age Interval (Ma)	Total Ocean	Southern Ocean	West Pacific	East Pacific	South-west Pacific	Indian	East Indian	West Indian	Atlantic	North Atlantic	Caribbean
0–5	264	141	52	55	26	48	50	37	47	57	23
5–10	214	126	45	44	24	43	45	34	35	38	15
10–15	188	114	43	39	23	37	44	26	31	33	12
15–20	167	106	42	33	19	32	37	25	31	30	10
20–25	155	96	39	31	17	32	34	23	27	28	9
25–30	140	86	34	27	18	30	34	20	24	25	7
30–35	131	80	30	25	17	28	32	19	23	25	8
35–40	119	75	29	22	15	28	28	19	18	21	7
40–45	117	70	28	19	14	29	28	18	21	24	6
45–50	99	54	23	12	12	26	26	17	20	21	6
50–55	89	46	24	9	9	23	20	14	15	22	9
55–60	76	42	22	6	7	21	20	10	11	18	9
60–65	65	36	22	6	6	16	18	6	8	13	7

REFERENCES

AMOS, A. F., GORDON, A. L., and SCHNIEDER, E. D.
1971 *Water mass and circulation patterns in the region of the Blake-Bahamas Outer Ridge.* Deep-Sea Res., vol. 18, pp. 145–165.

AVERY, O. E., VOGT, P. R., and HIGGS, R. H.
1969 *Morphology, magnetic anomalies and evolution of the northeast Atlantic and Labrador Sea, p. II. Magnetic anomalies.* Amer. Geophys. Union, Trans., vol. 50, p. 184.

BENSON, W. E., GERARD, R. D., and HAY, W. W.
1970 *Cruise Leg synthesis, summary and conclusions.* In: Bader, R. G. *et al., Initial reports of the Deep Sea Drilling Project, Volume IV.* Washington, D.C.: U.S. Govt. Printing Office, pp. 659–676.

BERGER, W. H.
1970 *Biogenous deep-sea sediments; fractionation by deep-sea circulation.* Geol. Soc. Amer., Bull., vol. 81, pp. 1385–1402.

1972 *Deep sea carbonates: dissolution facies and age depth constancy.* Nature, vol. 236, pp. 392–395.

1973 *Deep-sea carbonates: evidence for a coccolith lysocline.* Deep-Sea Res., vol. 20, pp. 917–921.

BERGGREN, W. A.
1972 *A Cenozoic time scale: some implications for geology and paleobiogeography.* Lethaia, vol. 5, pp. 195–215.

BLANT, G.
1966 *Sedimentary basins of the African coasts. Vol. 1.* Assn. African Geol. Surv., Paris, pp. 1–304.

1973 *Sedimentary basins of the African coasts. Vol. 2.* Assn. African Geol. Surv., Paris, pp. 1–233.

BRAMLETTE, M. N.
1965 *Massive extinctions in biota at the end of Mesozoic time.* Science, vol. 148, pp. 1696–1699.

BRINKMANN, R.
1960 *Geologic evolution of Europe.* New York: Hafner Publishing Co., pp. 1–161.

BUKRY, D.
1973 *Coccolith stratigraphy, eastern equatorial Pacific.* In: van Andel, Tj. H., and Heath, G. R. *et al., Initial reports of the Deep Sea Drilling Project, Volume XVI.* Washington, D.C.: U. S. Govt. Printing Office, pp. 653–711.

BULLARD, E., EVERETT, J. E., and SMITH, A. G.
1965 *The fit of the continents around the Atlantic.* In: *A symposium on continental drift.* Royal Soc. London Philos., Trans., vol. 258, pp. 41–51.

BURKOV, V. A.
1969 *The bottom circulation of the Pacific Ocean.* Oceanology, vol. 9, pp. 179–188.

DAVIES, T. A., LUYENDYK, B. P., KIDD, R. B., and WESER, O. E
1975 *Unconformities in the sediments of the Indian Ocean.* Nature, vol. 253, no. 5486, pp. 15–19.

DEWEY, J. F., PITMAN, W. C., III, RYAN, W. B., and BONNIN, J.
1973 *Plate tectonics and the evolution of the Alpine systems.* Geol. Soc. Amer., Bull., vol. 84, pp. 3137–3180.

DINKLEMAN, M. G.
1973 *Radiolarian stratigraphy.* In: van Andel, Tj. H., and Heath, G. R., et al., *Initial reports of the Deep Sea Drilling Project, Volume XVI.* Washington, D. C.: U. S. Govt. Printing Office, pp. 747–813.

DOUGLAS, R. G., and SAVIN, S. M.
1971 *Isotopic analyses of planktonic foraminifera from the Cenozoic of the northwest Pacific.* In: Fischer, A. G., and Heezen, B. C., *et al., Initial reports of the Deep Sea Drilling Project, Volume VI.* Washington, D.C.: U. S. Govt. Printing Office, pp. 1123–1127.

1973 *Oxygen and carbon isotope analyses of Cretaceous and Tertiary foraminifera from the central North Pacific.* In: Winterer, E. L., Ewing, J. I., *et al., Initial reports of the Deep Sea Drilling Project, Volume XVII.* Washington, D.C.: U. S. Govt. Printing Office, pp. 591–606.

EDWARDS, A. R.
1973 *Southwest Pacific regional unconformities encountered during Leg 21.* In: Burns, R. E., Andrews, J. E. *et al., Initial reports of the Deep Sea Drilling Project, Volume XXI.* Washington, D.C.: U. S. Govt. Printing Office, pp. 641–692.

FRAKES, L. A., and KEMP, E. M.
1972 *Influence of continental positions on Early Tertiary climates.* Nature, vol. 240, pp. 97–100.

GORDON, A. L.
1971 *Oceanography of Antarctic waters.* In: Reid, J. L., Ed., *Antarctic oceanography I.* Antarctic Res. Ser., vol. 15, pp. 169–204.

1973 *Physical oceanography.* Anarctic Jour. U. S., vol. 8, no. 3, pp. 61–69.

HARRINGTON, H. J.
1962 *Paleogeographic development of South America.* Amer. Assoc. Petrol. Geol. Bull., vol. 46, no. 10, pp. 1773–1813.

HAYES, D. E., and RINGIS, J.
1973 *Seafloor spreading in the Tasman Sea.* Nature, vol. 243, pp. 454–458.

HEATH, G. R.
1969 *Carbonate sedimentation in the abyssal equatorial Pacific during the past 50 million years.* Geol. Soc. Amer., Bull., vol. 80, pp. 689–694.

1974 *Dissolved silica and deep-sea sediments.* In: Hay, W. W., Ed., *Studies in paleo-oceanography.* Tulsa: Soc. Econ. Pal. Min., Spec. Publ. no. 20, pp. 77–93.

HEEZEN, B. C., HOLLISTER, C.D., and RUDDIMAN, W. F.
1966 *Shaping of the continental rise by deep geostrophic contour currents.* Science, vol. 152, pp. 502–508.

HEIRTZLER, J. R., DICKSON, G. O., HERRON, E. M., PITMAN, W. C., III, and LEPICHON, X.
1968 *Marine magnetic anomalies, geomagnetic field reversals and motions of the ocean floor and continents.* Jour. Geophys. Res., vol. 73, pp. 2119–2136.

HERRON, E. M., DEWEY, J. F., and PITMAN, W. C., III
1974 *Plate tectonics model for the evolution of the Arctic.* Geology, vol. 2, pp. 377–380.

HOLLISTER, C. D., JOHNSON, D. A., and LONSDALE, P. F.
1974 *Current-controlled abyssal sedimentation, Samoa Passage, equatorial West Pacific.* Jour. Geol., vol. 82, no. 3 pp. 275–300.

INGLE, S. E., CULBERSON, C. H., HAWLEY, J. E., and PYTKOWICZ, R. M.
1973 *The solubility of calcite in seawater at atmospheric pressure and 35‰ salinity.* Mar. Chem., vol. 1, no. 4, pp. 295–308.

JOHNSON, D. A.
1972*a* *Ocean-floor erosion in the equatorial Pacific.* Geol. Soc. Amer., Bull., vol. 83, pp. 3121–3144.

1972*b* *Eastward flowing bottom currents along the Clipperton Fracture Zone.* Deep-Sea Res., vol. 19, pp. 253–257.

JOHNSON, D. A., and JOHNSON, T. C.
1970 *Sediment redistribution by bottom currents in the central Pacific.* Deep-Sea Res., vol. 17, pp. 157–170.

JOHNSON, D. A., and PARKER, F. L.
1972 *Tertiary Radiolaria and foraminifera from the equatorial Pacific.* Micropaleontology, vol. 18, no. 2, pp. 129–143.

JONES, E. J. W., EWING, M., EWING, J. I., and EITTREIM, S. L.
1970 *Influences of Norwegian Sea overflow water on sedimentation in the northern North Atlantic and Labrador Sea.* Jour. Geophys. Res., vol. 75, pp. 1655–1680.

JONES, M. M., and PYTKOWICZ, R. M.
1973 *Solubility of silica in seawater at high pressures.* Soc. Roy. Sci. Liege, vol. 42, pp. 125–127.

KENNETT, J. P., BURNS, R. E., ANDREWS, J. E., CHURKIN, M., Jr., DAVIES, T. A., DUMITRICA, P., EDWARDS, A. R., GALEHOUSE, J. S., PACKHAM, G. H., and VAN DER LINGERN, G. J.
1972 *Australian-Antarctic continental drift, paleo-circulation changes and Oligocene deep-sea erosion.* Nature Phys., Sci., vol. 239, pp. 51–55.

KENNETT, J. P., HOUTZ, R. E., ANDREWS, P. V., EDWARDS, A. R., GOSTIN, V. A., HAJOS, N., HAMPTON, M., JENKINS, D. G., MARGOLIS, S. V., OVERSHINE, A. T., and PERCH-NIELSEN, K.
1974 *Cenozoic paleo-oceanography in the southwest Pacific Ocean and the development of the circum-Antarctic current.* In: Kennett, J. P., Houtz, R. E. *et al., Initial reports of the Deep Sea Drilling Project, Volume XXIX.* Washington, D.C.: U. S. Govt. Printing Office, pp. 1155–1169.

KLING, S. A.
1973 *Radiolaria from the eastern North Pacific Deep Sea Drilling Project, Leg 18.* In: Kulm, L. D., von Huene, R. *et al., Initial reports of the Deep Sea Drilling Project, Volume XXVIII.* Washington, D.C.: U. S. Govt. Printing Office, pp. 617–672.

KNAUSS, J. L.
1962 *On some aspects of the deep circulation of the Pacific.* Jour. Geophys. Res., vol. 67, pp. 3943–3954.

KUMMEL, B.
1961 *History of the Earth.* San Francisco: W. H. Freeman and Co., pp. 1–610.

LAUGHTON, A. S., BERGGREN, W. A., *et al.*
1972 *Sites 116 and 117.* In: Laughton, A. S., and Berggren, W. A. *et al., Initial reports of the Deep Sea Drilling Project, Volume XII.* Washington, D.C.: U. S. Govt. Printing Office, pp. 395–672.

LEPICHON, X., EWING, M., and TRUCHAN, M.
1971 *Sediment transport and distribution in the Argentine Basin 2: Antarctic bottom current passage into the Brazil Basin.* Phys. Chem. Earth, vol. 8, pp. 49–77.

LING, H. Y.
1973 *Radiolaria, Leg 19, of the Deep Sea Drilling Project.* In: Creager, J. S., Scholl, D. W. *et al., Initial reports of the Deep Sea Drilling Project, Volume XIX.* Washington, D.C.: U. S. Govt. Printing Office, pp. 777–798.

MCELHINNY, M. W.
1973 *Paleomagnetism and plate tectonics.* Cambridge: Cambridge University Press, pp. 1–358.

MCKENZIE, D. P., and SCLATER, J. G.
1971 *Evolution of the Indian Ocean since Late Cretaceous.* Roy. Astron. Soc. Geophys. Jour., vol. 25, pp. 437–528.

MINSTER, J. B., JORDAN, T. H., MOLNAR, P., and HAINES, E.
1974 *Numerical modeling of instantaneous plate tectonics.* Roy. Astron. Soc., Geophys. Jour., vol. 36, no. 3, pp. 541–576.

MOORE, T. C., Jr.
1970 *Abyssal hills in the central equatorial Pacific: sedimentation and stratigraphy.* Deep-Sea Res., vol. 17, pp. 573–593.

1971 *Radiolaria.* In: Tracey, J. I., and Sutton, G. B. *et al., Initial reports of the Deep Sea Drilling Project, Volume VIII.* Washington, D.C.: U. S. Govt. Printing Office, pp. 727–775.

1972 *DSDP: successes, failures, proposals.* Geotimes, vol. 17, July, pp. 27–31.

1973 *Radiolaria from Leg 17 of the Deep Sea Drilling Project.* In: Winterer, E. L., Ewing, J. I. *et al., Initial reports of the Deep Sea Drilling Project, Volume XVII.* Washington, D.C.: U. S. Govt. Printing Office, pp. 797–870.

MURRAY, G. E.
1961 *Geology of the Atlantic and Gulf Coastal Province of North America.* New York: Harper Bros., pp. 1–692, American maps.

PANFILOVA, S. C.
1967 *Bottom temperature and salinity in the Pacific Ocean.* Oceanology, vol. 7, pp. 690–694.

PIMM, A. G.
1974 *Sedimentology and history of the northeastern Indian Ocean from late Cretaceous to Recent.* In: von der Broch, C. C., and Sclater, J. G. *et al., Initial reports of the Deep Sea Drilling Project, Volume XXII.* Washington, D.C.: U. S. Govt. Printing Office, pp. 771–773.

PITMAN, W. C., III, and TALWANI, M.
1972 *Sea floor spreading in the North Atlantic.* Geol. Soc. Amer., Bull., vol. 83, pp. 619–646.

REID, J. L., and LYNN, R. J.
1971 *On the influence of the Norwegian-Greenland and Weddell Sea upon the bottom waters of the Indian and Pacific Ocean.* Deep-Sea Res., vol. 18, pp. 1063–1088.

RIEDEL, W. R., and SANFILIPPO, A.
1973 *Cenozoic Radiolaria from the Caribbean Deep Sea Drilling Project, Leg 15.* In: Edgar, N. T., Saunders, J. B. *et al., Initial reports of the Deep Sea Drilling Project, Volume XV.* Washington, D.C.: U. S. Govt. Printing Office, pp. 705–752.

RONA, P. A.
1973 *Worldwide unconformities in marine sediments related to eustatic changes of sea level.* Nature Phys. Sci., vol. 244, pp. 25–26.

RUDDIMAN, W. F.
1972 *Sediment redistribution on the Reykjanes Ridge: seismic evidence.* Geol. Soc. Amer., Bull., vol. 83, pp. 2039–2062.

SCHNEIDER, E. D., and HEEZEN, B. C.
1966 *Sediments of the Caicos outer ridge.* Geol. Soc. Amer., Bull., vol. 77, pp. 1381–1398.

SCLATER, J. G., and FISHER, R. L.
1974 *Evolution of the East-Central Indian Ocean with emphasis on the tectonic settling of the Ninetyeast Ridge.* Geol. Soc. Amer., Bull., vol. 85, no. 5, pp. 683–702.

SHACKLETON, N. J., and KENNETT, J. P.
1974*a* *Paleotemperature history of the Cenozoic and the initiation of Antarctic glaciations; oxygen and carbon isotope analyses at DSDP sites 77, 279, and 281.* In: Kennett, J. P., and Houtz, R. E., *et al., Initial reports of the Deep Sea Drilling Project, Volume XXIX.* Washington, D.C.: U. S. Govt. Printing Office, pp. 743–755.

1974*b* *Late Cenozoic oxygen and carbon isotopic changes at DSDP site 284; implication for glacial history of the northern hemisphere and Antarctica.* In: Kennett, J. P., and Houtz, R. E., *et al., Initial reports of the Deep Sea Drilling Project, Volume XXIX.* Washington, D.C.: U. S. Govt. Printing Office, pp. 801–807.

SIGAL, J.
1974 *Comments on Leg 25 sites in relation to Cretaceous and Paleogene stratigraphy in the eastern and southeastern Africa coast and Madagascar regional setting.* In: Simpson, E. S. W., and Schlich, R. *et al., Initial reports of the Deep Sea Drilling Project, Volume XXV.* Washington, D.C.: U. S. Govt. Printing Office, pp. 687–723.

STOMMEL, H., and ARONS, A. B.
1960 *On the abyssal circulation of the World Ocean II, an idealized model of circulation pattern in oceanic basins.* Deep-Sea Res., vol. 6, pp. 217–233.

SWALLOW, J., and WORTHINGTON, L.
1969 *Deep currents in the Labrador Sea.* Deep-Sea Res., vol. 16, pp. 77–84.

VAN ANDEL, TJ. H.
1974 *Cenozoic migration of the Pacific plate shift of the axis of deposition, and paleobathymetry of the central equatorial Pacific.* Geology, vol. 2, pp. 507–510.

VAN ANDEL, TJ. H., HEATH, G. R., MOORE, T. C., JR.
1975 *Cenozoic history and paleoceanography of the central equatorial Pacific Ocean.* Geol. Soc. Amer., Mem. 143, pp. 1–134.

VAN ANDEL, TJ. H., and MOORE, T. C. JR.
1974 *Cenozoic calcium carbonate distribution and calcite compensation depth in the central equatorial Pacific.* Geology, vol. 2, pp. 87–92.

VOGT, P. R.
1972 *The Faeroe-Iceland-Greenland aseismic ridge and the western boundary undercurrent.* Nature, vol. 239, pp. 79–81.

WARREN, B. A.
1974 *Deep flow in the Madagascar and Mascerene basins.* Deep-Sea Res., vol. 21, pp. 1–21.

WATKINS, N. D., and KENNETT J. P.
1971 *Antarctic bottom water: major change in velocity during the late Cenozoic between Australia and Antarctica.* Science, vol. 173, pp. 813–817.

1972 *Regional sedimentary disconformities and upper Cenozoic changes in bottom water velocities between Australia and Antarctica.* In: Hayes, D. E., Ed., *Antarctic oceanology II: the Australian-New Zealand sector.* Amer. Geophys. Union, Antarctic Res. Ser., vol. 19, pp. 273–294.

WEISSEL, J. K., and HAYES, D. E.
1972 *Magnetic anomalies in the southeast Indian Ocean.* In: Hayes, D. E., Ed., *Antarctic oceanography II, the Australian-New Zealand sector.* Amer. Geophys. Union, Antarctic Res. Ser., vol. 19, pp. 165–196.

WOODRING, W. P.
1954 *Caribbean land and sea through the ages.* Geol. Soc. Amer., Bull., vol. 65, pp. 719–732.

WORTHINGTON, L. V., and VOLKMANN, G. H.
1965 *The volume transport of the Norwegian Sea overflow water in the North Atlantic.* Deep-Sea Res., vol. 12, pp. 667–676.

WÜST, G.
1929 *Schichtung und Tiefenzirkulation des pazifischen Ozeans.* Veroff, Inst. Meersek. Univ., Berl., N. F. (A. Geogr-Naturwiss.), vol. 20, pp. 1–64.

1933 *Schichtung und Zirkulation des atlantischen Ozeans-Lieferung. Das Bodewasser und die Gliederung des atlantischen Tiefsee.* Wiss. Eegebn. Atlant. Exped. Meteor, vol. 6, bd. 1, no. 1, pp. 1–106.

WYRTKI, K.
1971 *Oceanographic atlas of the International Indian Ocean Expedition.* Washington, D.C.: Nat. Sci. Found., pp. 1–531.

Part V

GEOCHEMICAL CYCLES

Editor's Comments on Papers 28 Through 31

28 WEBER
Palaeoclimatic Significance of δ-Oxygen-18-Time Trends Observed by Oxygen Isotopic Analysis of Freshwater Limestones

29 WEBER
Possible Changes in the Isotopic Composition of the Oceanic and Atmospheric Carbon Reservoir over Geologic Time

30 PETERMAN, HEDGE, and TOURTELOT
Excerpts from *Isotopic Composition of Strontium in Sea Water Throughout Phanerozoic Time*

31 HOLSER
Catastrophic Chemical Events in the History of the Ocean

The important subject of long-term variations in the earth's chemistry is divisible into the two broad fields of elemental and stable isotope geochemistry. The four papers included here deal with the stable isotope geochemistry of oxygen, carbon, strontium, and sulfur. Most papers on elemental geochemical cycles are too long to be included, but key works are discussed below.

ELEMENTS

The study of long-term variations in the elemental composition of the crust, hydrosphere and atmosphere follows on logically from that of sedimentary cyclicity (Part IV), since the former study is based largely on the analysis of sedimentary rocks. The lead in this field has been given by A. B. Ronov, the late A. P. Vinogradov, and colleagues at the V. I. Vernadskiy Institute of Geochemistry and Analytical Chemistry in Moscow.

Vinogradov and Ronov (1956a) demonstrated that the Phanerozoic carbonate rocks of the Russian Platform exhibit major

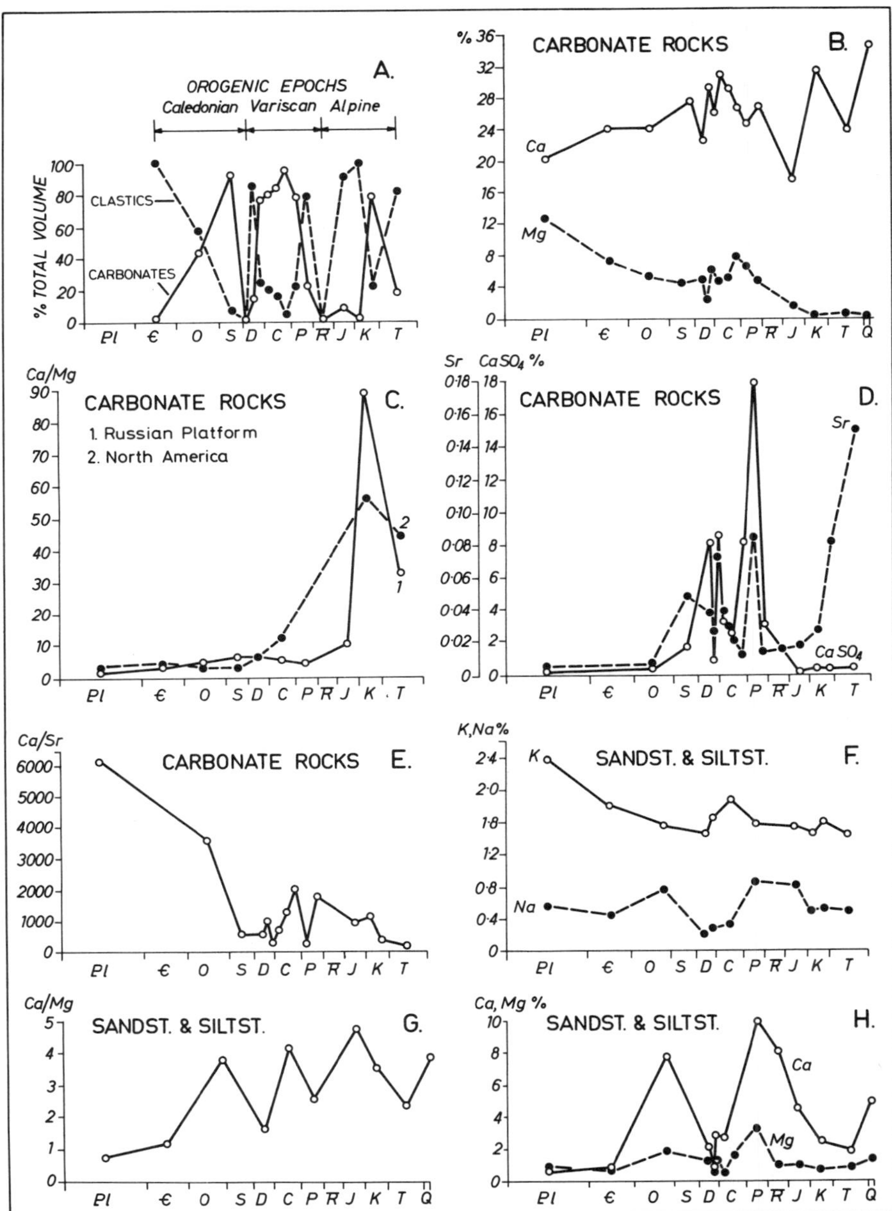

Figure 7 Composition of the sedimentary rocks of the Russian Platform. Adapted from Vinogradov and Ronov (1956a, Figures 2–6, 9–11).

variations with time in the average content of Ca, Mg, $CaSO_4$, Sr, Ba, and Th and in Ca/Mg and Ca/Sr, and sandstones and siltstones major variations in the average content of K, Na, Ca, and Mg and in Ca/Mg (Figure 7). They concluded (p. 552):

> These changes are most clearly shown by those rocks in which a given element is essential (for example, Mg in the carbonate rocks, K in clays, etc.). But they are exhibited clearly enough by the rocks in which a given element plays a secondary role. This undoubtedly indicates that the initial material of the sedimentary rocks had a common source and that their evolution has been a directed historical process. Against this general background there occurred regular variations in the abundance of the elements under study, coinciding with the periodic development of oscillatory tectonic movements in the ancient source areas and sedimentation basins.

Ronov et al. (1965) determined temporal variations in the chemical and mineralogical compositions of late Precambrian through Quaternary sandy rocks of the Russian Platform; they concluded that such variations reflect the cyclicity of tectonic movements in the source areas and sedimentary basins (which influence the prevailing composition of the source rocks), the climatic conditions governing their weathering, and the depositional and diagenetic environments. In other works, Ronov (1959) discussed changes through time in the average content of organic carbon and in Ca/Mg of Phanerozoic rocks of the Russian Platform and of North America, Ronov et al. (1970) described variations in Li, MgO, and insoluble residue for Riphean through Quaternary rocks of the Russian Platform, and Vinogradov and Ronov (1956b) discussed changes in the chemical composition of clays of the Russian Platform.

Fewer such studies have been conducted outside the USSR. Chilingar (1956), for example, in a study of the relationship between Ca/Mg and geological age for Phanerozoic carbonate rocks, determined an overall increase in that ratio through time (cf. Figure 7C) with superimposed periodic fluctuations. Chilingar (p. 2256) considered that the observed cyclic occurrence of dolomite and calcitic limestones "suggests that the formation of dolomites was occurring in cycles with a gradual accumulation of magnesium in sea water until favorable conditions were established for the formation of dolomites." Weaver (1967) considered that episodic changes occurred in the composition of the clay mineral suite of North American shales from Precambrian through Tertiary time, perhaps in response to relative changes in the amounts of Na^+, Mg^{2+}, and K^+ released to the world ocean from the terrestrial environment.

More recent studies have provided evidence of long-term geochemical variation and episodicity from Archean through Cenozoic time. Veizer (1977) illustrated major temporal variations in Sr/Ca of carbonate rocks; Engel et al. (1974, p. 843) considered that variations in the composition of common rock assemblages and their weighted average K_2O/Na_2O "suggest profound episodicity in crustal evolution and global tectonics;" Jackson and Moore (1976) showed a complex pattern, punctuated by several maxima and minima, for the total phosphorus content of pre-extracted sedimentary rocks; and Jackson (1973) illustrated systematic variations in the molecular properties of humic matter isolated from sedimentary rocks, which he interpreted as representing major events in the evolutionary history of pre-Paleozoic algae. In Paper 33 (see Part VI) Fairbridge proposed five great revolutions since early Precambrian time that reflect principally the changing biogeochemical role of carbon. Such geochemical episodicity in general is superimposed on *secular* trends in the chemical composition of the crust, hydrosphere, and atmosphere (e.g., Ronov, 1968, 1972; Ronov and Migdisov, 1971; Schwab, 1978; Veizer, 1978; Veizer and Garrett, 1978).

STABLE ISOTOPES

Oxygen

An early indication of long-term variations in stable isotope geochemistry was provided by paleothermometry. As reviewed by Emiliani (1966, p. 851), oxygen-isotope analysis of belemnite rostra between 1951 and 1964 "revealed two successive major temperature cycles having an amplitude of several degrees Celsius and extending across about 60 million years of Mesozoic time." Temporal variations in oxygen isotope paleotemperatures have been determined also for Tertiary marine fossils from New Zealand (Devereux, 1967) and from the North Sea area (Buchardt, 1978). Savin (1977) reviewed the history of the earth's surface temperature during the past 100 m.y.

Expansion of oxygen isotope analyses to include pre-Mesozoic sedimentary carbonates revealed additional long-term variation. In 1964 the late Jon N. Weber (Paper 28) showed long-term cyclic trends in $\delta^{18}O$ values for North American freshwater limestones, which he attributed to temperature changes in response to the Permo-Carboniferous and Quaternary glaciations. Keith and

Weber (1964), however, in presenting the same curve, drew attention to the few samples (four) in the Devonian age group and the relatively large within-group variability for post-Cretaceous samples. Dontsova et al. (1973) concluded that variations in the oxygen isotope composition of Phanerozoic carbonate rocks mostly from the Russian Platform reflect episodic changes in the intensity of volcanism and in the aridity of paleoclimates. In an important study, Veizer and Hoefs (1976) determined $\delta^{18}O$ values for limestones and dolomites of Archean through Tertiary age. Their histograms show higher values in the Ordovician and Permian than at other times in the Paleozoic; this pattern could reflect the Ordovician and Permo-Carboniferous ice ages, but Veizer and Hoefs were unsure whether or not the fluctuations are real features.

Evidence exists that the $\delta^{18}O$ composition of ancient evaporite sulfates shows temporal changes (Sakai, 1972; Hoefs, 1973), although Hoefs considered that more data are needed before an "age effect" may be established. Variations in oxygen isotope composition of Phanerozoic cherts have been ascribed to past climatic temperature fluctuations (Knauth and Epstein, 1976).

The temporal variations in oxygen isotope compositions identified in ancient carbonate rocks and cherts are superimposed on secular trends of increasing $\delta^{18}O$ values with decreasing age (e.g., Dontsova et al., 1973; Knauth and Epstein, 1976; Veizer and Hoefs, 1976; Knauth and Lowe, 1978; Perry et al., 1978), although geochemists disagree as to the meaning of such trends. An overall decrease in global surface temperatures through geological time interpreted from this isotopic trend in cherts (Knauth and Epstein, 1976) would, if confirmed, militate against Öpik's solar model for terrestrial climatic change (Paper 19).

Carbon

In an early proposal of long-term fluctuation in isotopic abundance, Jeffery et al. (1955) identified apparent cyclicity in the carbon isotope composition of Phanerozoic limestones and coals, which they attributed to changes in the isotopic composition of the hydrosphere and atmosphere that resulted from periodic diastrophism. Keith and Weber (1964) pointed out deficiencies in the sampling of Jeffery et al. (1955), and from isotopic analyses of five hundred samples of Phanerozoic marine and freshwater limestones and fossils, concluded (p. 1797) that "there is no evidence of a pronounced age effect." They implied, however, that Weber was more receptive to the idea of cyclic isotopic changes than was Keith.

In 1967 Weber (Paper 29) proposed cyclic changes in $\delta^{13}C$ composition of marine and freshwater limestones, which he ascribed (p. 2343) to "small but possibly real variations in the carbon isotopic composition of the active exchange carbon reservoir throughout geologic time." Weber suggested that such variations were a response to paleogeographic and paleoclimatic changes; his curves of $\delta^{13}C$ composition of freshwater limestones (Paper 29, Figure 2) and of $\delta^{18}O$ composition of similar rocks (Paper 28) do reflect quite closely the rhythm of late Paleozoic and Cenozoic glaciation. Weber's curve for marine limestones (Paper 29, Figure 1) has a total variation of less than 3‰, however, and Garrels and Perry (1974, p. 322) pointed out that the data therefore should be interpreted with caution.

No definite age trend is evident in carbon isotope data for either limestones or dolomites (e.g., Keith and Weber, 1964; Veizer and Hoefs, 1976). According to Oehler et al. (1972) and Oehler and Smith (1977) the reduced carbon in very old sediments seems enriched in ^{13}C compared with that in most younger, sedimentary kerogens and may reflect a major event in biological evolution ~3500–3300 m.y. ago. Brooks et al. (1973) suggested, however, that this isotopic discontinuity could be due to volcanogenic heating of the older rocks.

Strontium

In 1970 Peterman et al. (Paper 30) reviewed studies on the past isotopic composition of strontium in seawater, and from further isotopic analyses of strontium in primary fossil carbonate they presented a curve of variations in $^{87}Sr/^{86}Sr$ of seawater during the Phanerozoic. They suggested that certain features of the curve are related to episodes of widespread volcanism and to the episodic emergence and submergence of continents. In an alternative explanation of the curve of Peterman et al., Armstrong (1971) noted the correspondence of times of highest $^{87}Sr/^{86}Sr$ with the late Paleozoic and late Cenozoic glaciations; he commented (p. 132) that "erosion of Pre-Cambrian shields by continental glaciers would be an effective means of adding radiogenic ^{87}Sr to the dissolved load input to the oceans."

Veizer and Compston (1974), in an isotopic study of carbonate

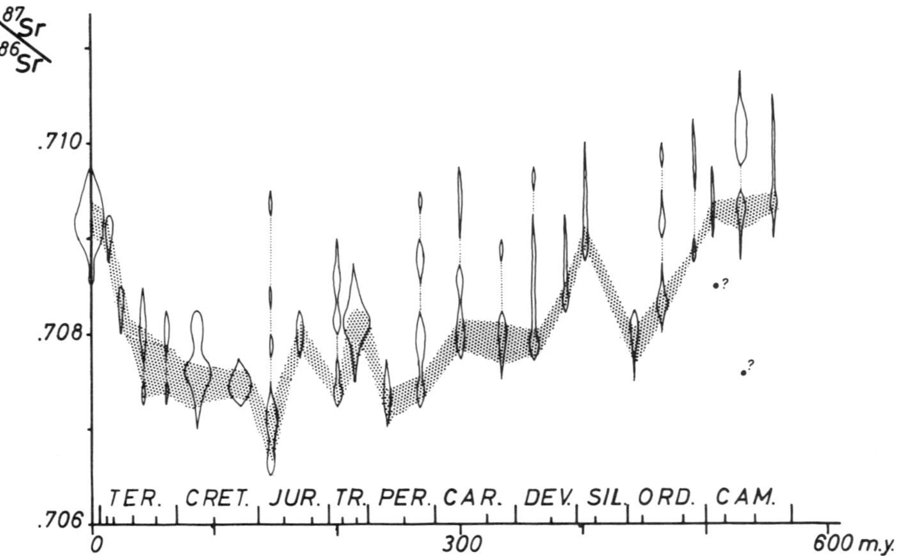

Figure 8 Variations of $^{87}Sr/^{86}Sr$ in seawater during the Phanerozoic, according to Veizer and Compston (1974, Figure 5). The points with question marks represent metamorphosed limestones.

rocks aimed at checking the observations of Peterman et al. (Paper 30) for the Phanerozoic and adding to them for the Paleozoic, presented an updated curve of strontium isotope variations during the Phanerozoic (Figure 8). Some correspondence with the variations of Peterman et al. was found; however, Veizer and Compston considered that the Triassic peak of Peterman et al. requires additional confirmation, and the Pennsylvanian peak was not confirmed. Veizer and Compston (p. 1475) concluded:

> The general trend of the curve [Figure 8] seems to show a rough correlation with the denudation data for the Phanerozoic (cf. Gregor, 1970 [see Part IV]) but the uncertainties in the validity of the Early Palaeozoic isotopic ratios are considerable. If this general trend is valid, it may support the conclusion that the Sr isotopic ratio of seawater is determined by the rates of Sr input from the crust and mantle, but again more definite data (not only for Sr isotopes) are required to decipher the exact causes of the observed fluctuations.

Veizer and Compston (1976) discussed variations in $^{87}Sr/^{86}Sr$ of "seawater" since early Precambrian time (see Figure 3). They correlated three sharp changes with successive tectonic transitions from the ensimatic "greenstone belt" regime during the Archean, to the ensialic "mobile belt" regime of the Proterozoic, to the mostly ensimatic "plate tectonics" regime during the Phanerozoic.

Secular trends in $^{87}Sr/^{86}Sr$ of seawater were further discussed by Veizer (1976).

Sulfur

As reviewed by Holser (Paper 31), general agreement exists that the sulfur isotope ratio in the surface of the world ocean has undergone major excursions during the Proterozoic-Phanerozoic that could be documented from evaporite sulfate studies. The updated sulfur isotope age curve presented by Holser exhibits three large, abrupt rises in $\delta^{34}S$ that he considered represent catastrophic "chemical events." Holser attributed such events to episodic mixing of stored brines heavy in $\delta^{34}S$ with the surface ocean, initiated by episodic tectonic destruction of the storage basins.

Other explanations of major variations in the $\delta^{34}S$ content of the ocean with time propose recycling of ancient evaporites containing sulfur isotopically different from that of ocean water, the episodic addition of "magmatic" sulfur to the oceans, inflow of freshwater sulfate released by weathering, and bacterial reduction of sulfate to sulfide (Thode and Monster, 1965); the large-scale transfer of sulfur to or from the shale reservoir (Nielsen, 1965; Holser and Kaplan, 1966); and the transfer of sulfur among seawater and the shale and evaporite reservoirs (Rees, 1970; Nielsen, 1972).

28

Reprinted from *Nature* **203**:969–970 (1964)

PALAEOCLIMATIC SIGNIFICANCE OF δ-OXYGEN-18-TIME TRENDS OBSERVED BY OXYGEN ISOTOPIC ANALYSIS OF FRESHWATER LIMESTONES

J. N. Weber

AFTER McCrea[1] and Epstein *et al.*[2] established the carbonate–water palæotemperature technique whereby the temperature of calcium carbonate precipitation can be determined by measuring the oxygen-18 content of the calcite, this tool has been widely utilized by a large number of workers for marine carbonates ranging in age from early Mesozoic to the present. Since the fractionation factor, that is, the isotopic composition of calcite divided by that of sea-water, is temperature dependent, deposition temperatures can be determined by: (1) measuring the $^{18}O/^{16}O$ ratio of ancient carbonate; (2) assuming a value for the isotopic composition of sea-water at that time.

So far, this technique has not been applied to fresh-water carbonates, presumably because of large present-day variations in the isotopic composition of fresh-water, shown, for example, by the data of Dansgaard[3], and Epstein and Mayeda[4]. If the $^{18}O/^{16}O$ ratio of fresh-water varied to this extent in large ancient lakes, temperatures determined by this method would be meaningless.

This need not be the case, however, at times when ice caps were non-existent, presumably over much of geological time. The isotopic composition of meteoritic water formed by the evaporation of sea-water is approximately -7 per mil. relative to the mean ocean water standard, and Epstein and Mayeda claim that isotopic fractionation which changes the oxygen-18 content of meteoritic water from about -7 per mil. to about -20 per mil. occurs in areas close to and possibly in the permanently glaciated regions. These authors further suggest that precipitation over the larger part of the Earth has an oxygen-18 content similar to that resulting from the initial evaporation of sea-water from the surface of the ocean, that is, -7 per mil.

Thus, in times of non-glaciation, the isotopic composition of large bodies of fresh water, especially in latitudes where calcium carbonate precipitation is possible, should not be a large variable. If enough samples of fresh-water limestones are used to suppress random errors or variations, it should be possible to determine palæotemperatures for these carbonates. Moreover, since these bodies of water are much smaller than the oceans, the response of water temperature to changes in climate are more rapid. Hence, fresh-water carbonate sequences may provide very detailed climatological trends and serve as sensitive palæoclimatic indicators.

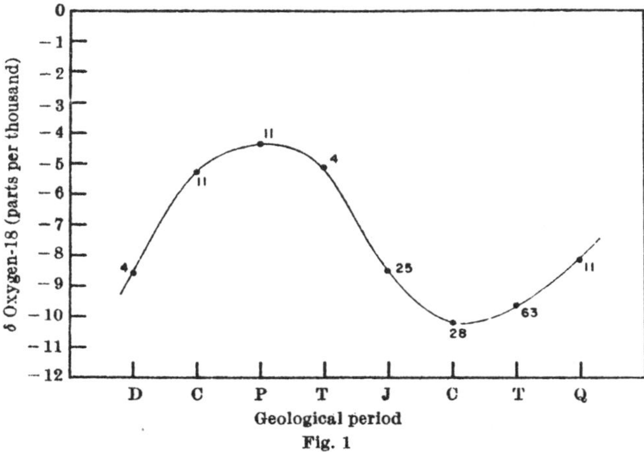

Fig. 1

In this light, a select collection of 157 samples of fresh-water limestones assembled by Keith and Weber[5] might be examined. Almost all these specimens are from North American localities, and range in age from Devonian to the present. Each sample was identified on the basis of fresh-water fossils and was analysed for $^{18}O/^{16}O$ ratio in duplicate using a 6 in., 60° sector mass spectrometer with isotope ratio recording apparatus. Grouped according to age, mean δ-oxygen-18 values relative to the Chicago P.D.B. standard, with the standard deviation given in brackets, are: Devonian −8·57 (1·3), Pennsylvanian −5·25 (1·6), Permian −4·36 (3·5), Triassic −5·12 (0·91), Jurassic −8·52 (3·0), Cretaceous −10·22 (3·5), Tertiary −9·65 (4·5), and Quaternary −8·15 (2·1). The results illustrated in Fig. 1 appear to generate a sinusoidal curve with $^{18}O/^{16}O$ maxima in the periods Permo-Carboniferous and Quaternary, and a minimum in the Cretaceous. The striking fit of the mean δ-oxygen-18 values to the curve shown in Fig. 1, the magnitude of the difference (about 6 per mil.) between the maximum and the minimum, and the occurrence of relatively high oxygen-18 contents in the carbonates precipitated during geological periods (Permo-Carboniferous and Quaternary) when glaciation was known to have occurred, strongly suggest that the data based on 314 analyses of 157 samples reflect temperature differences in large lakes located in central North America.

There are a number of variables on which the isotopic composition of fresh-water carbonate depends; these are discussed by Keith and Weber[5]. If the effects of some of these variables can be eliminated, for example, by proper sampling, random variation should cancel if enough samples are taken.

[1] McCrea, J. M., *J. Chem. Phys.*, **18**, 849 (1950).
[2] Epstein, S., Buchsbaum, R., Lowenstam, H. A., and Urey, H. C., *Geol. Soc. Amer. Bull.*, **62**, 417 (1951).
[3] Dansgaard, W., *Geochim. et Cosmochim. Acta*, **3**, 253 (1953).
[4] Epstein, S., and Mayeda, T., *Geochim. et Cosmochim. Acta*, **4**, 213 (1953).
[5] Keith, M. L., and Weber, J. N., *Geochim. et Cosmochim. Acta* (in the press).

29

Reprinted from *Geochim. et Cosmochim. Acta* **31**:2343–2351 (1967)

Possible changes in the isotopic composition of the oceanic and atmospheric carbon reservoir over geologic time

JON N. WEBER
Materials Research Laboratory, Pennsylvania State University
University Park, Pennsylvania

Abstract—Carbon isotope ratio analyses of 326 limestones and fossils ranging in age from Precambrian to Quaternary suggest small but possibly real variations in the carbon isotopic composition of the active exchange carbon reservoir throughout geologic time. Variation of $C^{13}:C^{12}$ ratios in freshwater limestones with time may be correlated with the geologic evidence concerning the development and distribution of land plants.

INTRODUCTION

THE possibility that the $C^{13}:C^{12}$ ratio of the active exchange reservoir of the carbon cycle may not have been constant throughout the period of Paleozoic to Cenozoic time has interested earth historians for some time. Since changes in the isotopic composition of the carbon of the cycle may provide information pertaining to orogenic episodes, the gradual development of life, and the differentiation between a biogenic and non-biogenic origin for certain Precambrian graphite deposits (RANKAMA, 1954; CRAIG, 1954; HOERING, 1961), some effort has been expended in exploratory surveys to determine, first of all, if variations in carbon isotopic composition, as reflected in the carbon fixed or removed from the cycle by the deposition of carbon-bearing sediments, exist, and secondly, to uncover any systematic time–compositional trends should they exist. Data bearing on this problem have originated from two sources: (1) Carbon isotopic analyses of materials designed to illustrate certain features of the carbon isotope distribution in nature (e.g. WICKMAN, 1953; CRAIG, 1953), some analyses of which might be interpreted as illustrating an "age effect", and (2) carbon isotopic analyses of carefully selected marine limestones and coals which are specifically assembled to investigate the possibility of $C^{13}:C^{12}$ variation with time (e.g. COMPSTON, 1960; JEFFERY *et al.*, 1955).

Analyses of the former group often suffer somewhat in the selection of material adequate for a proper age interpretation of the compositional variations, since numerous conditions of the ambient environment of deposition of the carbon-bearing sediments may to some degree effect considerable variations in the $C^{13}:C^{12}$ ratio. Both types of investigation are limited in the number of samples examined. It is unlikely, therefore, that investigations specifically designed to uncover time–compositional trends have adequately sampled the materials which may reflect $C^{13}:C^{12}$ changes in the active carbon reservoir.

A large number of freshwater and marine limestones, carefully selected to represent the original $C^{13}:C^{12}$ ratios, was isotopically analyzed in duplicate by WEBER and KEITH (1962) (KEITH and WEBER, 1964) to evaluate the technique of differentiating ancient marine and freshwater environments on the basis of carbon and oxygen isotopic composition. Although the samples were not selected specifically to investigate a possible δC^{13} variation with time, and hence this study would rank in type (1) above, the analyses might be applied to the "age effect" problem. This is especially appropriate because of the large number of specimens studied of Cambrian to Recent age (1076 analyses of 538 samples), and because of the close scrutiny given these samples as far as the preservation of isotopic composition and the marine* nature of the depositional environment are concerned. While certain periods of time are better represented than others, an interesting pattern of marine limestone composition emerges and correlates to a surprising degree with the compositional–time trend of freshwater limestones. The large number of samples, the relatively small variance of analyses within each age period, and the cyclic nature of the pattern suggests that the relationships shown here may not be wholly fortuitous.

Nature of the age effect

The mechanisms involved in a possible age effect have been discussed by SILVERMAN (1961, 1962) who draws attention to the fact that carbonate deposition and photosynthesis are responsible for the fixation of carbon from carbon dioxide of the reservoir. These modes of fixation have formed nearly all of the carbon compounds of the bio-, hydro-, and sedimentary portion of the litho-sphere. Because photosynthetic activity preferentially extracts the light isotope, C^{12} (NIER, 1950; WICKMAN, 1952; CRAIG, 1953, 1954; PARK and EPSTEIN, 1960), while carbonate precipitation preferentially removes C^{13} from the reservoir† changes in the relative rates of formation and of decomposition of limestones and coals will cause changes in the average isotopic composition of the oceanic and atmospheric reservoir. The development of land plants, and the variation in their abundance from time to time, and the variation in formation and decomposition rates of carbonates with time imply that atmospheric carbon dioxide should not be isotopically constant throughout geologic time, and SILVERMAN predicts that corresponding changes should be recognized in carbon-bearing sediments of different ages.

At the present rate of carbon fixation in sediments, BORCHERT (1951) showed that in about 250 million yr no "free" carbon would remain in the cycle. As this catastrophe has obviously not occurred, JEFFERY *et al.* (1955) propose a recycling mechanism by which old or fixed carbon is recycled by organic activity so that the oceanic and atmospheric reservoirs are replenished. As the cycle is not closed, shallow, warm water limestones deposited in epeiric seas constituting a "leak" in the system, a net accumulation of fixed carbon implies that the entire reservoir of free carbon will slowly alter and become lighter in isotopic composition. Coals, relatively concentrating the lighter isotope, have an opposite though undoubtedly much smaller effect. In summary, JEFFERY *et al.* predict gradually lighter limestones from the start of each period to the end when relatively heavy carbon is reinstated in the cycle. Because of the net accumulation of great volumes of limestone of all ages, the authors suggest an overall isotopic lightening of the carbon reservoir with time.

---

* Versus brackish, estuarine-deltaic, and other complex or "mixed" environments.

† Shown theoretically by calculations of UREY and GREIFF (1935), experimentally by McCREA (1950) and empirically by CRAIG (1953) and others.

Selection and analysis of the sample material

A collection of 326 limestones and fossils of known marine origin and 183 limestones and fossils of known freshwater genesis was assembled to serve as a model for the isotopic differentiation of these two environments (WEBER and KEITH, 1962). Environmental classification was based solely on the presence of a marine fauna in the marine limestones, and on both organic fossil content and paleogeographic evidence in the case of freshwater carbonate. Specimens which appeared altered in any way which might have changed the original isotopic composition were rejected, as were all samples which may have been deposited under brackish water or estuarine conditions. Addition of secondary carbonate to the original limestone also served to disqualify a specimen from the preferred collection. In summary, the analyses presented here represent the isotopic composition of apparently unaltered, undolomitized, unmetamorphosed marine and freshwater carbonate.

After completion of the analyses, certain specimens appeared to have abnormal isotopic composition as evidenced not only by the failure to fall within or near the range of other carbonates of the same age (which were fairly closely grouped in composition), but in some cases by the appearance of marine limestone isotope ratios in the well-defined range of isotopic composition shown by freshwater carbonates. Further study disclosed that the samples providing anomalous isotopic composition could be grouped into several categories, and that the divergent $C^{13}:C^{12}$ ratios could be explained by certain mechanisms which act to decrease the $C^{13}:C^{12}$ ratio of the carbonate. The divergent samples include corals of a group known to precipitate $CaCO_3$ out of equilibrium with their external environment (LOWENSTAM and EPSTEIN, 1957). Twelve samples of reef and atoll limestone, largely from Eniwetok and Kita Daito Jima, exhibited very low δC^{13} values which are attributed to either the Ghyben–Herzberg freshwater lens derived from downward percolation of meteoric water, or in some cases are attributed to a Pleistocene stand above sea level (GROSS, 1964). While enclosing remnants of a marine fauna, some samples contained a large percentage of sand or silt, and probably represent sediments deposited in the near-shore environment where contribution of terrestrial HCO_3^- may account for the exceptionally high abundance of the lighter carbon isotope. Black, organic-rich limestones exhibiting pelagic nektonic organisms such as ammonites as the major organic fossil, form the fourth category, comprising 5 specimens. Precipitating in a reducing environment with stagnant or very poor circulation of the surrounding water, it is highly probable that the latter group of carbonates incorporated carbon from the decay and decomposition of the isotopically light carbon-bearing organic matter, or carbon oxidized by the anaerobic sulphate-reducing bacteria. Similar effects were observed by LANDERGREN (1954) on comparing the composition of the Limbata (*L. Ordovician*) limestones deposited in well-mixed and well-aerated, highly oxidizing waters with those of the same group laid down under conditions of poor circulation and aeration in a reducing environment. Several samples of limestone with divergent composition were derived from lenses and concretions in shale. Whether these specimens represent fine-grained secondary carbonate or have obtained isotopically light carbon from decomposing organic materials in the surrounding shale is uncertain. Twelve additional samples, whose anomalous composition could not be ascribed to any apparent mechanism, complete the group of divergent marine samples. In all, the divergent samples account for 49 of the 326 marine carbonates analyzed.

Mass spectrometric analytical techniques have been described in detail by WEBER and RAUP (1966), and the isotope data are expressed in delta notation:

$$\delta C^{13} = 1000 \left(\frac{C^{13}:C^{12}_{sample}}{C^{13}:C^{12}_{std}} - 1 \right) \text{permil},$$

where 'std' refers to the Chicago PDB standard CO_2.

Table 1. Isotopic composition of marine limestones*

	No. of specimens	Mean δC^{13}†	Standard deviation†
Precambrian (marbles)‡	24	−0·26	3·09
Lower Cambrian	1	−0·46	—
Middle Cambrian	8	−0·54	0·906
Upper Cambrian	16	+0·18	1·04
All Cambrian	25	−0·07	1·02
Lower Ordovician	1	−0·51	—
Middle Ordovician	25	+0·47	1·41
Upper Ordovician	1	−0·31	—
All Ordovician	30	+0·37	1·32
Middle Silurian	3	+2·02	2·71
Upper Silurian	3	+1·00	0·859
All Silurian	12	+1·41	1·83
Lower Devonian	4	+0·27	1·06
Middle Devonian	20	+0·33	1·00
Upper Devonian	18	+0·94	2·12
All Devonian	42	+0·59	1·58
Mississippian	5	+1·72	1·54
Pennsylvanian	9	−0·94	0·828
Permian	12	+1·09	1·92
Triassic	13	+0·12	1·63
Jurassic	21	+0·38	1·69
Jurasso-Cretaceous	5	+2·62	1·26
Cretaceous	36	+1·05	1·43
Tertiary	45	−0·05	1·05
Quaternary	22	+1·96	1·55

* Sample identification, stratigraphic position and localities can be found in KEITH and WEBER (1964).

† Relative to the Chicago PDB standard CO_2, in permil.

‡ Included for comparison only.

Discussion of the results

The carbon isotopic composition of the marine and freshwater carbonates is presented in Tables 1 and 2 respectively. Graphic representation is made in Figs. 1 and 2.

Table 2. Isotopic composition of freshwater limestones*

	No. of specimens	Mean δC^{13}†	Standard deviation†
Middle Devonian	3	−1·86	1·10
Upper Devonian	1	−3·38	—
All Devonian	4	−2·24	1·18
Mississippian	1	−4·63	—
Pennsylvanian	11	−6·83	2·40
Permian	11	−6·21	2·85
Triassic	4	−4·20	3·06
Jurassic	25	−3·76	1·53
Cretaceous	28	−4·81	2·54
Tertiary	63	−5·27	3·15
Quaternary	11	−4·06	1·71

* Sample identification, stratigraphic position and localities can be found in KEITH and WEBER (1964).

† Relative to the Chicago PDB standard carbon dioxide, in permil.

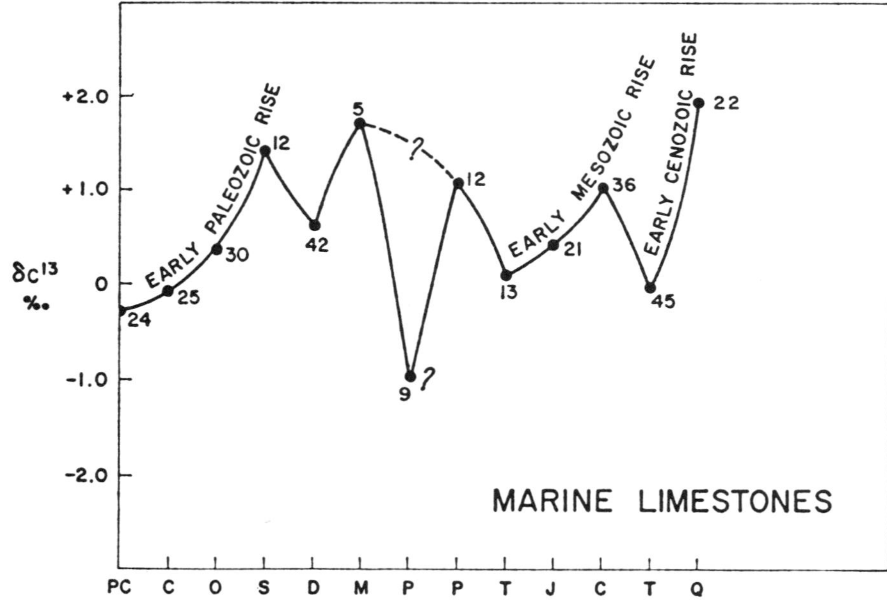

Fig. 1. Carbon isotopic composition of marine limestones vs. age. The number of specimens analyzed in each age group from Precambrian to Quaternary is shown.

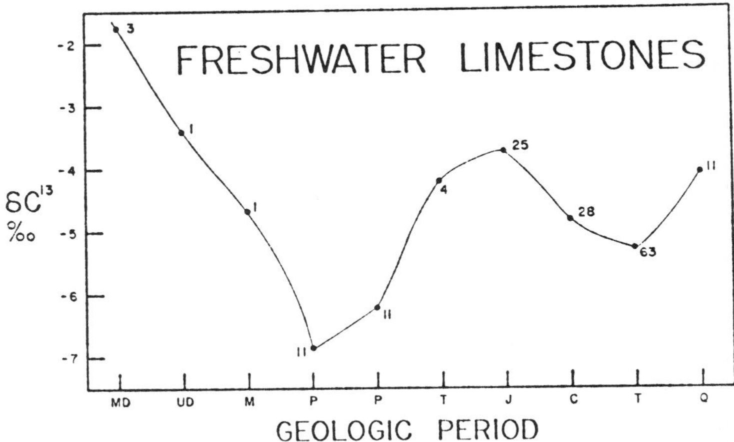

Fig. 2. Carbon isotopic composition of freshwater limestones from Middle Devonian to Quaternary. The number of specimens in each age group is shown.

(1) Considering first the entire span of time shown in Fig. 1, there appears to be a general but very poorly defined trend toward increasing δC^{13} values with decreasing age. The trend, if real, is much complicated by differences in the mean composition of the carbonates of certain periods but the correlation coefficient for the variables isotopic composition and time is positive, supporting the visual observation.

(2) With the exception of the Pennsylvanian marine samples, the mean values of the various age groups appear to form a fairly regular, more or less sinuous curve with peaks around the Silurian, Mississippian, Cretaceous and Recent times. It is interesting to note that the four major peaks appear at more or less regular intervals of 100 to 150 million yr and may correspond to significant carbon 13 additions to the reservoir in the BORCHERT cycle cited above. The very low mean δC^{13} value for the Pennsylvanian sediments may or may not be real. It is possible that the true isotopic composition of Pennsylvanian limestones may be somewhere close to the dashed line shown on Fig. 1, since Pennsylvanian sediments are poorly represented, only four of the ten samples having been obtained from areas remote from the great Pennsylvanian coal swamps and paludal environments of the central north-eastern United States. These four specimens display much more positive δC^{13} values.

(3) The time trends suggest a gradual and consistent increase in δC^{13} from Lower Cambrian to Silurian followed by an abrupt decrease in the Lower Devonian (Early Paleozoic Rise). The upward trend in the lower Paleozoic is repeated from Lower Devonian to Mississippian and is again followed by a decline from Pennsylvanian (?) to Permian. A third, Early Mesozoic cycle begins with the Triassic, with δC^{13} values becoming consistently more positive until after Cretaceous time. Finally, an early Cenozoic rise is suggested by the change from negative δC^{13} values in the Tertiary to positive values in the Quaternary.

(4) Within a range of 2 to 3 permil, the average composition of the active exchange carbon reservoir appears unchanged from the Cambrian to Recent.

(5) The carbon isotopic composition of the freshwater samples shown in Fig. 2

also presents a rather distinct pattern, with a sharp decrease in δC^{13} values from Devonian to Pennsylvanian, an abrupt but distinct increase in $C^{13}:C^{12}$ ratios from the Pennsylvanian to Jurassic, and a less abrupt but similar cycle from the Jurassic to the present. While it is true that few samples represent the Devonian to Triassic periods owing to the notable scarcity of freshwater limestones earlier than Jurassic (a result of gradational processes), it appears unlikely that the relationship is fortuitous. Because of the probable variation in the rates of C^{13}-enriched carbon from the erosion of marine limestones, and of C^{12}-enriched carbon from organic sources, a pattern of δC^{13} values from freshwater limestones might not be expected. The opposite is true of marine limestones where the dominant influence on their composition is the carbon isotope ratio of the active exchange reservoir.

(6) Ignoring the Devonian and Mississippian freshwater limestones, a poorly defined trend of increasing δC^{13} with decreasing age is also suggested for the freshwater limestones.

Interpretation of the results

The cyclic nature of the δC^{13} values shown on Figs. 1 and 2 may arise from unknown mechanisms which periodically and gradually return the carbon fixed by sedimentary processes to the active exchange reservoir. If the present rate of carbon fixation in sediments is more or less typical of the past, BORCHERT's (1951) calculations show a life span for the carbon cycle of some 250 million yr, and the continued presence of life from Cambrian to the present attests to the fact that this cycle has never been seriously depleted in free carbon.

JEFFERY *et al.* (1955) suggest that orogenic activity may be responsible for the recycling of fixed carbon by the decomposition of sedimentary rock entrapped in the cores of tectogenes. Should this be the case, they predict gradually lighter isotopic carbon in limestones deposited from the beginning to the end of each geologic period of orogenic quiesence. Of the data presented here, two periods are well enough represented to test this prediction, the Cambrian and the Devonian.

Twenty-five samples of normal marine limestone were obtained, representing the major areas of Cambrian deposition in the United States and Canada. Since land plants had not yet evolved, the isotopic ratios of the limestones are not complicated by concurrent C^{12}-enrichment in continental deposits. The trend is apparently towards heavier, rather than lighter limestones in the Upper Cambrian.

The Devonian period is equally well covered with 42 samples from North America and Europe, those of Europe representing the major limestone deposits, both areally and stratigraphically. The trend is identical with that of the Cambrian, with more positive δC^{13} values in the latter portion of the period.

The well-defined inverse relationship of $C^{13}:C^{12}$ ratios in the freshwater and marine limestones from Devonian to Lower Carboniferous is also of interest and raises the question of whether the trend, if real, can be related to the development of land plants. If abundant carbon 12 were removed from the exchange reservoir and fixed on the continents, perhaps for the first time, freshwater limestones would be enriched in this isotope through biological processes. The increasing content of the heavy carbon isotope in the reservoir would be reflected in the deposition of marine carbonates.

Since the enrichment of C^{12} in freshwater limestones results either directly or indirectly from the incorporation of carbon fractionated by plants in the photosynthesis process, it is likely that the size of the plant component of the biosphere will significantly affect the carbon isotopic composition of lacustrine and fluvial limestones. It is probable, that freshwater Silurian and older limestones were similar in isotopic composition to the marine limestones, with δC^{13} values about 0 to -1 permil. The evolution of land plants from Lower Devonian to Lower Carboniferous and their transgression of the continents is suggested by increasing C^{12} content of the freshwater limestones which exhibit the lowest $C^{13}:C^{12}$ ratios in the period of the great Pennsylvanian coal swamps. With the onset of widespread arid conditions* of the Permian, Triassic and Jurassic, the restriction of luxuriant continental plant activity resulted in a decreased contribution of C^{12} to freshwater limestones. The restoration of a humid environment over much of North America during the Cretaceous and Tertiary permitted land plants to spread widely over the continent, and a correlative trend of decreasing $C^{13}:C^{12}$ ratios is evident. Finally with the elevation of North America and the spread of continental ice sheets in the Quaternary, plant life was once again restricted, and presumably contributed less light carbon to the freshwater carbonates.

As in the case of the data provided by COMPSTON (1960), and JEFFERY *et al.* (1955), considerable overlap of δC^{13} values occurs in some periods, and the magnitude of the variation is indicated by the standard deviations of Tables 1 and 2. The consistency of the results, however, and the relationship of the isotopic data presented here to paleogeographic and paleoclimatic evidence found in the geologic column suggest random variation around mean values which appear to be different from period to period.

One further point is worthy of note. In the study of JEFFERY *et al.*, only three periods are well represented, and the authors suggest that the distinctly different δC^{13} values imply an age effect. They conclude that the Devonian seas were heavier than both the Cambrian and Tertiary. This relationship is evident in the data presented here.

CONCLUSION

The data presented here constitute a further contribution to the line of investigation which may ultimately yield an answer to the intriguing question of a possible age effect and its interpretation in the carbon cycle. Since the samples were analyzed (in duplicate) only after careful scrutiny to eliminate the complicating effects of brackish water environments and possible alteration of the original isotope ratios, and because of the large number of specimens studied, these results may have some significance. Their value, however, can only be assessed in the light of future investigations of this nature. Reliable conclusions on the possible existence of a carbon isotope age effect will require a thorough study of all major carbonate deposits to which properly weighted factors must be applied to take into account the estimated volume of each specific deposit. In any case, changes in the isotopic composition of the active exchange reservoir with time appear to be small, and if linear extrapolation of the Phanerozoic data into Precambrian time is valid, one of the basic assumptions

* In the areas sampled.

in RANKAMA's (1954) method of assigning a biogenic or non-biogenic origin to ancient graphite deposits may be strengthened.

Acknowledgments—The author is indebted to Prof. M. L. KEITH, E. G. WILLIAMS and G. M. ANDERSON who kindly reviewed the manuscript and participated in many discussions. The author, however, alone bears the responsibility for the interpetations presented. Support from the National Science Foundation (currently GA-290) is gratefully acknowledged.

REFERENCES

BORCHERT H. (1951) On the geochemistry of carbon. *Geochim. Cosmochim. Acta* **2,** 62–75.

COMPSTON W. (1960) The carbon isotopic composition of certain marine invertebrates and corals from the Australian Permian. *Geochim. Cosmochim. Acta* **18,** 1–22.

CRAIG H. (1953) The geochemistry of the stable carbon isotopes. *Geochim. Cosmochim. Acta* **3,** 53–92.

CRAIG H. (1954) Geochemical implications of the isotopic composition of carbon in ancient rocks. *Geochim. Cosmochim. Acta* **6,** 186–196.

GROSS M. G. (1964) Variations in the O^{18}/O^{16} and C^{13}/C^{12} ratios of diagenetically altered limestones in the Bermuda Islands. *J. Geol.* **72,** 170–194.

HOERING T. C. (1961) Carbon isotope ratios in carbonates and reduced carbons from ancient sediments (Abstr.). *Geol. Soc. Amer. Spec. Papers* **68,** 199.

JEFFERY P. M., COMPSTON W., GREENHALGH D. and DE LAETER J. (1955) On the carbon 13 abundance of limestones and coals. *Geochim. Cosmochim. Acta* **7,** 255–286.

KEITH M. L. and WEBER J. N. (1964) Carbon and oxygen isotopic composition of selected limestones and fossils. *Geochim. Cosmochim. Acta* **28,** 1787–1816.

LANDERGREN S. (1954) On the relative abundance of the stable carbon isotopes in marine sediments. *Deep-Sea Res.* **1,** 98–119.

LOWENSTAM H. A. and EPSTEIN S. (1957) On the origin of sedimentary aragonite needles of the Great Bahama Bank. *J. Geol.* **65,** 364–375.

McCREA J. M. (1950) On the isotope chemistry of carbonates and a paleotemperature scale. *J. Chem. Phys.* **18,** 849–857.

NIER A. O. (1950) A redetermination of the relative abundances of carbon, nitrogen, oxygen, argon and potassium. *Phys. Rev.* **77,** 789–793.

PARK R. and EPSTEIN S. (1960) Carbon isotope fractionation during photosynthesis. *Geochim. Cosmochim. Acta* **21,** 110–126.

RANKAMA K. (1954) The isotopic constitution of carbon in ancient rocks as an indicator of its biogenic or non-biogenic origin. *Geochim. Cosmochim. Acta* **5,** 142–152.

SILVERMAN S. R. (1961) Evidence for an age effect in the carbon isotopic composition of natural organic materials (Abstr.). *Geol. Soc. Amer. Spec. Papers* **68,** 272.

SILVERMAN S. R. (1962) Effect of evolution of land plants on C^{13}/C^{12} ratios of natural organic materials (Abstr.). *J. Geophys. Res.* **67,** 1657.

UREY H. C. and GREIFF L. J. (1935) Isotope exchange equilibria. *J. Amer. Chem. Soc.* **57,** 321–327.

WEBER J. N. and KEITH M. L. (1962) Isotopic composition and environmental classification of selected limestones and fossils (Abstr.). *Geol. Soc. Amer. Spec. Papers* **73,** 259–260.

WEBER J. N. and RAUP D. M. (1966) Fractionation of the stable isotopes of carbon and oxygen in marine calcareous organisms—the Echinoidea. *Geochim. Cosmochim. Acta* **30,** 681–736.

WICKMAN F. E. (1952) Variations in the relative abundance of the carbon isotopes in plants. *Geochim. Cosmochim. Acta* **2,** 243–254.

WICKMAN F. E. (1953) Wird das Häufigkeitsverhältnis der Kohlenstoffisotopen bei der Inkohlung verändert? *Geochim. Cosmochim. Acta* **3,** 244–252.

30

Reprinted from pages 105–108 and 111–118 of *Geochim. et Cosmochim. Acta* **34**: 105–120 (1970)

Isotopic composition of strontium in sea water throughout Phanerozoic time*

ZELL E. PETERMAN, CARL E. HEDGE and HARRY A. TOURTELOT
U.S. Geological Survey, Denver, Colorado 80225

Abstract—Isotopic analyses of strontium in primary fossil carbonate reveal significant variations in Sr^{87}/Sr^{86} of sea water during the Phanerozoic. The strontium isotopic composition may have been uniform from the Ordovician through the Mississippian, with an average Sr^{87}/Sr^{86} of 0·7078. A subsequent decrease in this value into the Mesozoic is interrupted by two provisionally documented positive pulses in Sr^{87}/Sr^{86}—one in the Early Pennsylvanian and one in the Early Triassic. The lowest observed value (0·7068) occurred in Late Jurassic time, and this was followed by a gradual increase to 0·7075 in the Late Cretaceous and a more rapid increase through the Tertiary to 0·7090 for modern sea water. These variations are thought to be the result of a complex interplay of periods of intense volcanism and epeirogenic movements of the continents on a worldwide scale.

INTRODUCTION

Statement of problem

The isotopic composition of strontium in sea water has long been of interest to geochemists. WICKMAN (1948) predicted that it might be possible to determine ages of marine carbonates on the basis of their Sr^{87}/Sr^{86} ratios. His calculations were based on poorly known estimates for the crustal abundances of Rb and Sr, and GAST (1955) subsequently showed that any strontium isotopic variations were exceedingly small—too small to be accurately measured at that time. HEDGE and WALTHALL (1963) reported strontium isotopic compositions for modern sea water and for two Phanerozoic and two Precambrian limestones; the total variation observed in Sr^{87}/Sr^{86} ratios was about 1 per cent. HURLEY and colleagues (MASSACHUSETTS INSTITUTE of TECHNOLOGY, 1965) determined initial Sr^{87}/Sr^{86} for some early Precambrian graywackes and interpreted it as representing the strontium isotopic composition of sea water. Combining this with other data for limestones of known age and for modern sea water, they plotted a nearly linear increase in Sr^{87}/Sr^{86} for sea water from 2700 m.y. ago to today. Although the variations in the strontium isotopic compositions of sea water are not as great as predicted by WICKMAN (1948), there is evidence for increasing Sr^{87}/Sr^{86} at least since early Precambrian time.

The present study is directed towards determining, as precisely as possible, the strontium isotopic composition of sea water during Phanerozoic time. Since the predominant source of strontium in sea water at any given time is the total exposed continental areas, small nonlinear variations in the Sr isotopic composition could probably occur. This would seem to be reasonable in view of the complex geologic history of continental areas throughout Phanerozoic time as evidenced by

* Publication authorized by the Director, U.S. Geological Survey.

repeated emergence and submergence, periodicity of volcanic and plutonic activity, possible changes in the character of weathering, and any other factors which could produce a change in the mean isotopic composition of strontium entering the ocean basins.

Method of approach

The Sr isotopic composition of sea water in the geologic past is directly determinable by analyzing the shell carbonate of mollusks of different ages (appendix). The agreement in isotopic composition of modern shells and of the water in which the animals lived has been shown by HEDGE and WALTHALL (1963) and FAURE *et al.* (1965) and the present study. Additional evidence is provided by the uniform composition of concretions and the fossils of Late Cretaceous which they contain (TOURTELOT and PETERMAN, 1967). The concretions were formed from the sea water in which the animals lived. Although different biologic groups of animals may utilize different proportions of the Sr available in their environment (LOWENSTAM, 1964), no evidence for natural fractionation of Sr isotopes has yet been found.

Shell material of macrofossils also provides an independent control on the quality of the sample. Recrystallization, replacement, or physical deformation, any of which could be accompanied by isotopic exchange with the adjacent environment, is easily detectable in fossils because of the destruction of original shell structure. Such changes may not be evident in limestone. In addition, shell carbonate is virtually pure calcium carbonate, either aragonite or calcite, so the problem of having an insoluble residue which might contribute radiogenic strontium is eliminated.

The possibility of isotopic exchange with the components of enclosing rocks without destruction of original shell structure is largely eliminated by the uniformity of the data for fossils from the same age group. Some fossils come from limestone and some from shale, and these rock types undoubtedly have different Sr isotopic compositions. For most of the age groups, however, the intragroup variation in isotopic composition is compatible with the analytical uncertainty.

It is difficult to prove that the water which determined the isotopic composition of the fossils was equivalent isotopically to the open ocean waters of the world. Most of the fossils analyzed come from rocks deposited in epicontinental seas rather than in deep ocean basins. The analyses themselves, however, suggest that world oceans and epicontinental seas were well mixed with respect to strontium isotopes at most times in the past.

ANALYTICAL METHODS

Selection and preparation of samples

Only fossils with primary shell material were used. The shells were carefully cleaned of any matrix material by scraping or ultrasonic treatment in distilled water. Where the shells were filled with secondary material, a sample was taken with the aid of a small diamond saw or steel dentist's tool to ensure that only the shell material was obtained.

Samples of approximately 0·1–0·2 g were then dissolved in 2·5 N HCl and passed through an ion-exchange column to obtain a purified strontium. For most samples this procedure involved approximately 100–400 μg of strontium.

Table 1. Isotopic data for strontium standards and correction factors for adjustment of data. (s is standard deviation; $2\sigma_{\bar{x}}$ is the 95 per cent confidence level on the mean)

Instrument	No. of analyses	Time interval	Mean $(Sr^{87}/Sr^{86})_n$	s	$2\sigma_{\bar{x}}$
		U.S.G.S. strontium standard			
(1) 6-in.	14	3-1-66 to 9-18-67	0·71006	0·000147	0·00008
(2) 6-in.	9	12-5-67 to 9-12-68	0·71041	0·000090	0·00005
(3) 12-in.	14	1963	0·70926	0·000893	0·00051
		Eimer and Amend strontium standard			
(4) 6-in.	5	1966	0·70791	0·000116	0·00014
		Correction factors to adjust data to (1)			
Difference of (2) — (1)			0·00035	0·00005	
Difference of (1) — (3)			0·00080	0·00024	

Table 2. Sr^{87}/Sr^{86} values for standard strontium (United Mining and Chemical Corp. Lot B-857)

Date	Sr^{86}/Sr^{88}	$(Sr^{87}/Sr^{86})_n$	Date	Sr^{86}/Sr^{88}	$(Sr^{87}/Sr^{86})_n$
3-1-66	0·1197	0·7099	12-5-67	0·1196	0·7103
4-19-66	0·1198	0·7101	12-6-67	0·1198	0·7105
6-16-66	0·1197	0·7101	1-9-68	0·1197	0·7104
10-7-66	0·1197	0·7101	1-16-68	0·1202	0·7103
10-10-66	0·1197	0·7099	1-17-68	0·1197	0·7103
1-13-67	0·1198	0·7098	1-19-68	0·1197	0·7106
2-9-67	0·1197	0·7102	1-22-68	0·1198	0·7104
2-10-67	0·1198	0·7103	2-21-68	0·1196	0·7104
2-17-67	0·1197	0·7101	3-12-68	0·1201	0·7105
4-5-67	0·1198	0·7103	6-3-68	0·1203	0·7104
5-22-67	0·1198	0·7100	8-29-68	0·1200	0·7104
6-1-67	0·1197	0·7100	9-2-68	0·1200	0·7105
9-6-67	0·1198	0·7103	9-11-68	0·1199	0·7104
9-18-67	0·1199	0·7099	9-12-68	0·1197	0·7103
			9-13-68	0·1200	0·7105
Mean $(Sr^{87}/Sr^{86})_n$		0·71006			0·71041
Standard deviation		±0·00015			±0·00009
95 per cent confidence limits on the mean		±0·00008			±0·00005

Isotopic measurements

Strontium isotopic measurements were made on a solid source, 6-in., 60-degree mass spectrometer employing a triple-filament mode of ionization. Approximately 10–30 μg of Sr was loaded on each Re filament as $SrCl_2$, taken to dryness, and the filaments were heated to a dull red glow for 1 min. After loading the sample into the spectrometer and attaining a tube pressure less than 2×10^{-7} mm Hg, stable Sr signals were usually obtained within 10 to 15 min. Data was taken by chart recorder using an expanded scale modification and the same scales were used for all of the samples and standards. Data were accumulated by magnet current switching in a pattern of repetitive peak hopping of the three consecutive masses, 86 to 87 to 88 to 86 etc., with initial and final monitoring for Rb^{85}. Five to six complete scans of the spectrum were obtained and Sr^{87}/Sr^{86} and Sr^{86}/Sr^{88} ratios were calculated at each peak position by extrapolation of the adjacent peaks. For uniformity, attempts were made to obtain a Sr^{86} signal between 150 and 300 mV and thus most of the data have Sr^{86} peaks on an expanded scale factor of 5–9.

In order to reduce any instrumental drift, the vibrating reed electrometer and the recording system were periodically calibrated for scale factors and linearity and a strontium standard was periodically analyzed. Results for this standard are summarized in Table 1, and individual analyses on the 6-in. mass spectrometer are shown in Table 2. During 1966 and most of 1967 there was no discernible drift or discontinuities in these results. This conclusion is reinforced by comparing results for actual groups of samples of the same age that were analyzed over considerable time periods (Fig. 1).

Just prior to the completion of this study the Sr isotopic values Sr^{87}/Sr^{86}, for the standard shifted slightly (Tables 1 and 2); however, data collected after this shift can be reduced to the 1966–67 scale without introducing significant additional error. For purposes of comparison with data obtained by other laboratories the Eimer and Amend strontium standard was analyzed five times during 1967 (Table 1).

Treatment of data

Analytical results are given in Table 3 and Fig. 2. Normalization to Sr^{86}/Sr^{88} of 0·1194, statistical calculations, and adjustments where necessary were made on these data prior to rounding off the results to four significant figures. During the investigation 21 samples were analyzed in duplicate and the analytical precision was determined from these measurements.

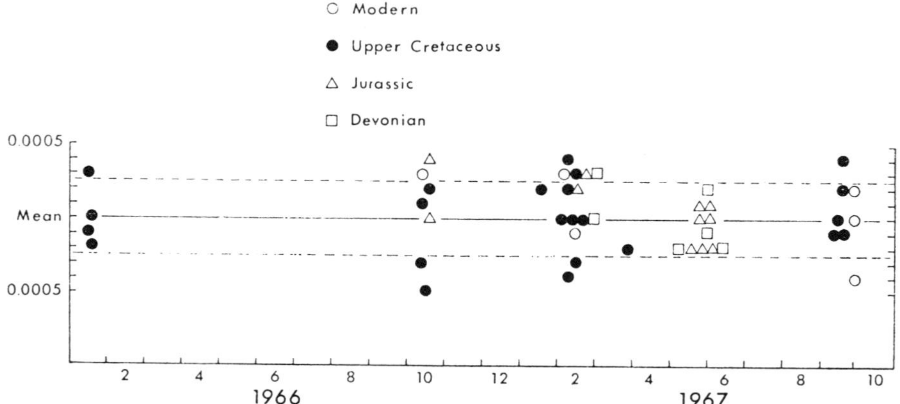

Fig. 1. Sr^{87}/Sr^{86} values for replicate determinations of samples plotted against date of analysis. Because the groups represent different absolute values, the points are plotted as the difference between the observed value for a single analysis and the mean for the group ($Sr^{87}/Sr^{86}{}_{\text{single analysis}}$-$Sr^{87}/Sr^{86}{}_{\text{mean}}$). The dashed lines represent $\pm$ one standard deviation for a single analysis and 71 per cent of the points fall within this interval.

Based on these data the standard deviation for a single Sr^{87}/Sr^{86} measurement is $\pm 0{\cdot}00025$.

Two groups of data are shown in Table 3 for which it was necessary to make an adjustment to the observed Sr^{87}/Sr^{86} values for purposes of comparison. This first group of data, modern shells and sea water, were obtained on a 12-in. mass spectrometer in 1963 prior to the present study, and the Sr^{87}/Sr^{86} ratios are significantly lower than subsequently obtained on the 6-in. mass spectrometer (see data for standard, Table 1). These 12-in. data are correspondingly increased by 0·00080 and the variance of the adjusted data is increased accordingly.

Sample SS-5 (Table 3) was analyzed in duplicate on the 12-in. mass spectrometer and subsequently analyzed six times on the 6-in. mass spectrometer. The mean Sr^{87}/Sr^{86} value for the six analyses is 0·70903 compared to a mean of 0·70891 for the adjusted 12-in. data. These are analytically identical values thus substantiating the adjustment of the 12-in. data as based on the mean values for the strontium standard.

The second set of data that requires an adjustment is that which was obtained on the 6-in. mass spectrometer after November 1967. As shown in Table 2, Sr^{87}/Sr^{86} for the standard strontium increased significantly at about this time. The values for the standard are precisely known for these intervals and the difference of 0·00035 in Sr^{87}/Sr^{86} has an extremely small uncertainty (standard deviation of 0·00005); thus, the increased variance to the adjusted ratios is negligible.

[*Editor's Note:* Table 3 on pages 109–111, which lists isotopic data for individual samples, has been omitted. The individual analyses are summarized as means and errors in Table 4.]

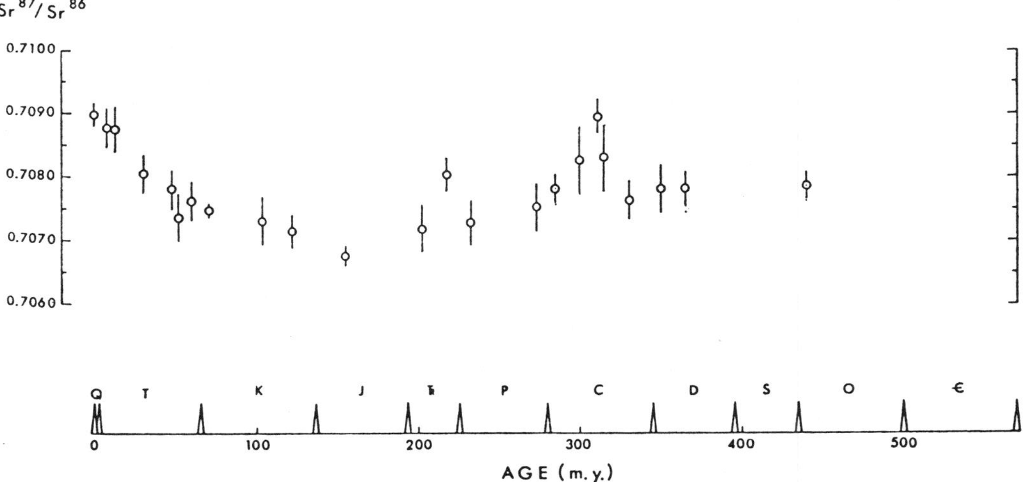

Fig. 2. Variation in Sr^{87}/Sr^{86} values for sea water during most of Phanerozoic time. The vertical bars represent the 95 per cent confidence level about the mean points.

Table 4. Mean Sr^{87}/Sr^{86} values according to age

Age	Number*	Sr^{87}/Sr^{86}	95 per cent confidence level†	
			Calculated	Predicted
Holocene (6-in. data)	7	0·70903	0·00024	0·00020
(12-in. data)	12	0·70891	0·00026	—
(combined)	19	0·70898	0·00018	—
Pliocene (data from Obradovich)	5_1	0·70877	0·00031	—
Late Miocene	2_1	0·70875	—	0·00037
Middle Oligocene	3	0·70804	0·00029	0·00031
Middle Eocene	3	0·70780	0·00041	0·00031
Early Eocene	2_1	0·70735	—	0·00037
Paleocene	3	0·70761	0·00087	0·00031
Late Cretaceous	16_8, 9	0·70746	0·00011	0·00011
Early Cretaceous (Albian)	2_1	0·70730	—	0·00037
Early Cretaceous (Hauterivian)	4_2	0·70713	0·00008	0·00027
Late Jurassic	2_1, 9	0·70675	0·00014	0·00016
Late Triassic	2	0·70719	—	0·00037
Early Triassic	4	0·70804	0·00041	0·00027
Late Permian	2	0·70726	—	0·00037
Early Permian	2	0·70751	—	0·00037
Late Pennsylvanian	4_2	0,70783	0·00045	0·00027
Middle Pennsylvanian	1	0·7083	—	0·00053
Early Pennsylvanian				
Late Morrow	2_1, 2	0·70897	0·00053	0·00027
Middle Morrow	1	0·7083	—	0·00053
Late Mississippian	3	0·70763	0·00038	0·00031
Late Devonian	2_1	0·70780	—	0·00037
Middle Devonian	2_1, 2	0·70780	0·00047	0·00027
Total Devonian	4_2, 2	0·70780	0·00022	0·00022
Late Ordovician	6_3	0·70786	0·00031	0·00022

* Number of samples and analyses. A number without a subscript refers to single analyses of that many samples. A number with a subscript indicates replicate analyses, the subscript being the number of samples. Except for the Holocene and Pliocene data the multiple analyses are all duplicates. Thus 16_8, 9 refers to 8 samples analyzed in duplicate (16 analyses) and 9 samples analyzed singly.

† Uncertainty on the mean determined at the 95 per cent confidence level. The predicted value is that calculated from the expected analytical uncertainty. The calculated value is determined from the individual values within a single group and increased according to Student's *t* factor for a limited number of samples, this being determined only when the number of analyses is three or more.

In Table 3 the $(Sr^{87}/Sr^{86})_c$ column is used only when it was necessary to adjust the data. Otherwise, the values given in the $(Sr^{87}/Sr^{86})_n$ column are those determined during 1966 and part of 1967.

The individual analyses (Table 3) are summarized as means and errors in Table 4. For many of the age groups two uncertainties are shown. The predicted uncertainty is calculated from analytical uncertainty as determined from the 21 samples analyzed in duplicate ($2\sigma_{\bar{x}} = 0{\cdot}00053/\sqrt{n}$, where n is the number of analyses within an age group). Where there are three or more individual analyses within a group, an internally calculated uncertainty is determined for the mean. When the number of samples becomes large such as for the Upper Cretaceous and Jurassic the two uncertainties (predicted and internal) are identical.

For the groups of data where both uncertainties are given the two variances are F-tested according to the method given by YOUDEN (1951). The variances of none of the groups exceed the predicted analytical variance at the 95 per cent level. For many of the groups the degrees of freedom are small and hence the F-test is not too definitive. As an additional test, the within group variance was estimated using the single analyses excluding duplicates. These data, 53 samples and 39 degrees of freedom, gave a pooled standard deviation of 0·00023 (see YOUDEN, 1951, p. 13) which is identical to that calculated from the duplicate analyses. The F-test and pooling of the single analyses are strongly suggestive that all of the within-group variations are analytical. On this basis the predicted uncertainties for the means are considered as being the most realistic uncertainties and these will be used in the subsequent interpretation of the data.

Results

Modern sea water

Measurements on samples of modern sea water from the Atlantic, Pacific and Indian oceans indicate that the isotopic composition of strontium is uniform within experimental error of the analyses (Faure *et al.*, 1965; Hamilton, 1966; Murthy and Beiser, 1968). The long residence time of strontium and the constancy of the composition of dissolved salts in the oceans were cited as supporting arguments for a uniform strontium isotopic composition (Murthy, 1964, and Faure *et al.*, 1965, respectively). The present data (Table 3) obtained on samples from the Atlantic, Pacific, and Arctic Oceans adds substance to the conclusion that open oceans are uniformly mixed with regard to strontium.

Tertiary

Strontium isotopic data for samples representing the Tertiary are given in Tables 3 and 4. Included here (Table 4) is the mean of five determinations on a sample of Pliocene foraminifera for which Obradovich (1968) reported a mean Sr^{87}/Sr^{86} of 0·7093 $\pm$ 0·00026 (95 per cent confidence level). His analyses were completed on our 6-in. mass spectrometer in 1965 prior to the present study during which time the mean Sr^{87}/Sr^{86} ratio for our standard strontium was $0{\cdot}7105_9$ (15 analyses), and an adjustment to his value is made (Table 4).

Data for the Tertiary define a nearly linear increase from a Sr^{87}/Sr^{86} of 0·7076 for the Paleocene to 0·7088 for the Upper Miocene and Lower Pliocene. Except for the Pliocene, the samples are restricted to the Gulf Coast, Mississippi, Alabama, and Florida, and therefore the geographic spread is not as great as would be desirable. However, the remarkably systematic increase in Sr^{87}/Sr^{86} and the agreement of the end-member time periods with values for adjacent periods suggest that this trend is very likely the actual variation in strontium isotopic composition of sea water during this time interval.

Cretaceous

The Upper Cretaceous, or more precisely the Campanian and Maestrichtian Stages, is represented by the largest number of samples and analyses (Tables 3 and 4). The initial phase of this study was begun on fossils from the Pierre Shale of the Great Plains region (Peterman and Tourtelot, 1968) and it was thought that the unusually low Sr^{87}/Sr^{86} ratios obtained for the fossils may have indicated incomplete mixing of the Pierre sea with open oceans at least with regard to strontium. Subsequently, additional fossils of equivalent age from the Gulf Coast, Atlantic Coast, Pacific Coast and from near the Arctic Ocean were analyzed and these had strontium isotopic compositions identical to those from the Pierre Shale (Tourtelot and Peterman, 1967). The locations for these samples and the areas of sea and exposed land masses during Pierre time are shown on Fig. 3. The uniformity of the strontium isotopic compositions of these samples now makes the early interpretation of incomplete mixing (Peterman and Tourtelot, 1968) untenable and we conclude that Sr^{87}/Sr^{86} for sea water at this time was 0·7075 $\pm$ 0·0001 and that the strontium in the epicontinental portion of this sea was mixed isotopically with that of the open oceans.

8

Duplicate analyses have been made on three samples of Early Cretaceous age (Tables 3 and 4). The sample of Albian Age (Early Cretaceous) has Sr^{87}/Sr^{86} indistinguishable from the mean for the Late Cretaceous. The two samples of Hauterivian Age (Early Cretaceous) from Alaska gave a Sr^{87}/Sr^{86} ($0{\cdot}7071_3$) that is statistically distinct from the mean value for Late Cretaceous at the 95 per cent

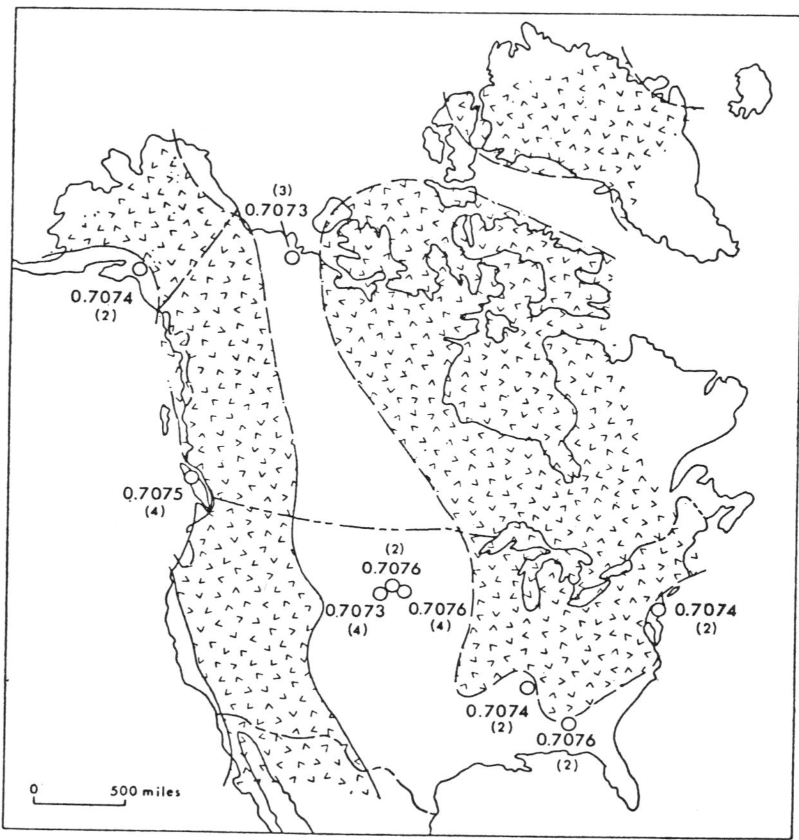

Fig. 3. Paleogeologic map for the Late Cretaceous showing mean Sr^{87}/Sr^{86} values. Numbers in brackets represent the number of analyses. Pattern areas indicate emergent land areas and the unpatterned areas are the Late Cretaceous seas.

but not the 99 per cent confidence level. Little more can be said about these except they do add to the conclusion that the Sr^{87}/Sr^{86} values for the Cretaceous were unusually low and they fall in an orderly progression between two much more precisely documented time points—the Late Cretaceous and Late Jurassic.

Upper Jurassic

Ten samples of Late Jurassic belemnites represent the Oxfordian, the time of greatest marine transgression in the Jurassic. Samples from the United States are from the Sundance Formation or its equivalents in Wyoming, Montana,

Idaho, Utah, and South Dakota. The host rocks represent a variety of sedimentational environments ranging from very near shore to geosynclinal. Strontium isotopic compositions of the belemnites are uniform with a mean Sr^{87}/Sr^{86} of $0{\cdot}7067_5$ (Tables 3 and 4) which is the lowest ratio obtained for any of the age groups studied. The internal agreement for the fossils suggests that the Sundance sea had a uniform Sr isotopic composition, and the agreement between these and the duplicate value obtained for the Alaskan sample, the location of which was much closer to open oceans, suggests that this value was probably that for oceanic strontium in Oxfordian time.

Triassic

We were not able to obtain suitable Triassic sample material from North America. Two samples representing the Upper Triassic were obtained from the Taimyr Peninsula in northwestern Siberia. The Lower Triassic samples are from the Olenek River area in the northern part of the Yakutsk Autonomous Soviet Socialists Republic. Isotopically, these two sets of data are distinct, with the Lower Triassic mean value of $0{\cdot}7080_4$ being higher than both the Upper Triassic, $0{\cdot}7071_9$, and the Upper Permian, $0{\cdot}7072_6$. If it is assumed that these values represent the strontium isotopic composition of sea water during this interval, it would appear that Early Triassic time was marked by a rather sudden increase in the influx of strontium more radiogenic than that indicated by the Permian samples.

Permian

The Upper Permian is represented by two samples both of *Stenocisma* sp., one from Alaska and one from Pakistan. These have identical strontium isotopic compositions with a mean Sr^{87}/Sr^{86} ratio of $0{\cdot}7072_6$. The two samples from the Lower Permian are from Texas and Kansas.

Pennsylvanian

Eight samples are available for the Pennsylvanian (Tables 3 and 4). Three of these are of Late Pennsylvanian age (Virgil) from Kansas; one of Middle Pennsylvanian (Des Moines) from Oklahoma; and four are of Lower Pennsylvanian (Morrow) from Arkansas and Kentucky. One of the Lower Pennsylvanian samples, (2849a, Table 3) is of middle Morrow age and the others (22512Br, 22512Be and Coral) are of late Morrow age. Strontium isotopic data on these samples indicate that there were some rather rapid changes during this time period. The samples of late Morrow age have significantly higher Sr^{87}/Sr^{86} than those samples of earlier Paleozoic age. This high value is followed by a systematic decrease in Sr^{87}/Sr^{86} through the Middle and Upper Pennsylvanian which apparently continued through most of the Permian.

Mississippian, Devonian and Ordovician

Samples from the remaining Paleozoic all have strontium isotopic compositions that are analytically identical (Tables 3 and 4). If all of the data are considered together rather than by age groups a mean Sr^{87}/Sr^{86} of $0{\cdot}7077_9$ can be calculated and a standard deviation for a single analysis of $\pm 0{\cdot}00024$ which is identical to that predicted from analytical uncertainties alone.

Geologic Implications

In extrapolating Sr^{87}/Sr^{86} determined on fossils (Fig. 2) to variations in sea water, different degrees of reliability must be attached to some of the time intervals studied. The best documented intervals, present oceans, Cretaceous, Jurassic, and Permian, are those for which samples of wide geographic spread have been analyzed. The systematic variation in Sr^{87}/Sr^{86} through the Tertiary is highly suggestive that this trend is also real. Two time periods, the Pennsylvanian and Triassic, which show rapid apparent changes in Sr^{87}/Sr^{86} are not documented on a regional basis; hence, added uncertainty must be considered in the interpretation of these data.

The strontium isotopic variations are probably related to a complex interplay of geologic events which results in varying proportions of different isotopic compositions of strontium being delivered to the ocean basins (e.g. Faure *et al.*, 1965). The response of Sr^{87}/Sr^{86} in the oceans to changing input strontium would be a function of both strontium residence time and of rapidity of changing input Sr^{87}/Sr^{86}. Lowering of Sr^{87}/Sr^{86} in the oceans could be accomplished by an increased contribution of strontium from young volcanic rocks whereas times of increasing Sr^{87}/Sr^{86} may be related to times of continental emergence resulting in a greater contribution of more radiogenic strontium derived from older crystalline rocks. Since the strontium isotopic composition of sea water is some form of integrated record of input strontium on a world scale, it is tempting to try to correlate changes with specific geologic events. However, such attempts are hampered either by lack of adequate documentation or of adequate preservation of the geologic record.

Jurassic and Cretaceous seawater contained the most unradiogenic strontium of any of the time intervals. Intense igneous activity and worldwide continental submergence may have been effective in lowering the Sr^{87}/Sr^{86} of sea water during these periods. This speculation is supported by Ronov's (1959, Fig. 2) suggestion that periods of unusually intense volcanism are coincident with periods of continental submergence and higher rates of carbonate deposition. That the latter part of the Mesozoic was characterized by continental submergence is shown by the transgression curves of Damon and Mauger (1966) for North America and of Egyed (1956) for the world. There is also little doubt that the Mesozoic was a time of unusually intense igneous activity as shown by the extent of Mesozoic plutons in the circum-Pacific province (Bateman and Eaton, 1967, Fig. 2) and by the comparison of the extent of Mesozoic batholiths in North America with those of other time intervals (Knopf, 1955). Since the emergent areas during this time period were probably those that were orogenically active in contrast to the stable platforms and Precambrian shield areas which probably tended towards submergence, the increased contribution of strontium from young igneous rocks may well have been accompanied by an attendant decrease in strontium derived from older marine carbonates and crystalline rocks.

The rapid increase in Sr^{87}/Sr^{86} during the Tertiary and beginning about 50–70 m.y. ago may have resulted from worldwide continental uplift during this time (see DAMON and MAUGER, 1966; EGYED, 1956). As a result of this uplift, a greater contribution of strontium from crystalline rocks and marine carbonates might be expected. The near-linear increase of Sr^{87}/Sr^{86} during the Tertiary could be explained either by an increased residence time of strontium or, more likely, by a parallel change in the Sr^{87}/Sr^{86} of the input strontium.

The apparent pulses in Sr^{87}/Sr^{86} during the Early Triassic and Early Pennsylvanian and the subsequent systematic decreases are more difficult to relate to known geologic events. Although these correspond to periods of emergence as shown by transgression curves (EGYED, 1956; RONOV, 1959) and possibly to periods of increased volcanism (RONOV, 1959), the correlations are not well defined. The lack of correlation between Sr^{87}/Sr^{86} in sea water and other fluctuations on the transgression and volcanic intensity curves (RONOV, 1959) suggests that there may be other significant factors which are also important in controlling variations in the strontium isotopic composition of sea water.

Acknowledgments—We are especially grateful to the following people for supplying us with suitable sample material: W. A. COBBAN, J. T. DUTRO, JR., M. GORDON, JR., R. W. IMLAY, D. L. JONES, R. B. NEUMAN, N. F. SOHL, E. L. YOCHELSON, and R. E. ZARTMAN, all of the U.S. Geological Survey; A. J. ROWELL of the University of Kansas; N. J. SILBERLING of Stanford University; and P. N. VARFOLOMEEV of the Ministry of Geology of the U.S.S.R. Leningrad. The constructive criticisms of Drs. WILLIAM COMPSTON, P. W. GAST, N. F. SOHL, M. TATSUMOTO and K. TUREKIAN are appreciated. We are also indebted to numerous colleagues in the U. S. Geological Survey with whom we have had many valuable discussions of the problem.

REFERENCES

BAIN H. F. (1895) Central Iowa section of the Mississippian series. *Amer. Geol.* **15,** 317–325.

BATEMAN P. C. and EATON J. P. (1967) Sierra Nevada batholith. *Science* **158,** 1407–1417.

DAMON P. E. and MAUGER R. L. (1966) Epeirogeny-orogeny viewed from the Basin and Range province. *Soc. Min. Eng. Trans.* **235,** 99–112.

EGYED L. (1956) The change of the earth's dimensions determined from paleogeographical data. *Geofis. Pura Appl.* **33,** 42–48.

FAURE G., HURLEY P. M. and POWELL J. L. (1965) The isotopic composition of strontium in surface water from the north Atlantic Ocean. *Geochim. Cosmochim. Acta* **29,** 209–220.

GAST P. W. (1955) Abundance of Sr^{87} during geologic time. *Geol. Soc. Amer. Bull.* **66,** 1449–1454.

HAMILTON E. I. (1966) The isotopic composition of strontium in Atlantic Ocean water. *Earth Planet. Sci. Lett.* **1,** 435–436.

HEDGE C. E. and WALTHALL F. G. (1963) Radiogenic strontium-87 as an index to geologic processes. *Science* **140,** 1214–1217.

KNOPF A. (1955) Bathyliths in time. In *Crust of the Earth—A Symposium* (editor A. Poldervaart). *Geol. Soc. Amer. Spec. Paper* **62,** 685–702.

LOWENSTAM H. A. (1964) Sr/Ca ratio of skeletal aragonites from the recent marine biota at Palau and from fossil gastropods. In *Isotopic and Cosmic Chemistry*, Chap. 10, pp. 114–132. North-Holland.

MASSACHUSETTS INSTITUTE OF TECHNOLOGY (1965) Evidence from Western Ontario of the isotopic composition of strontium in Archean seas. *Massachusetts Inst. Technology, Dept. Geology and Geophysics, 13th Ann. Prog. Rept.*, 1965, pp. 145–147.

MURTHY V. R. (1964) The significance of Sr^{87} in ocean water [abs.]. *Trans. Amer. Geophys. Union* **45,** 113.

MURTHY V. R. and BEISER E. (1968) Strontium isotopes in ocean water and marine sediments. *Geochim. Cosmochim. Acta* **32,** 1121–1126.

OBRADOVICH J. D. (1968) The potential use of glauconite for late-Cenozoic geochronology. *Proc. Int. Assoc. Quaternary Res. 7th Cong.*, pp. 267–279.

PETERMAN Z. E. and TOURTELOT H. A. (1968) Sr^{87}/Sr^{86} values in the Upper Cretaceous Pierre Shale. In *Abstracts for 1966. Geol. Soc. Amer. Spec. Paper* **101,** 161.

RONOV A. B. (1959) On the post-Precambrian geochemical history of the atmosphere and hydrosphere. *Geochemistry* **5,** 493–506.

STAINBROOK M. A. (1944) The devonian system in Iowa. In Symposium on Devonian stratigraphy. *Illinois Geol. Surv. Bull.* **68,** 182–188.

TOURTELOT H. A. and PETERMAN Z. E. (1967) Strontium isotopic composition of fossils and variation in composition of sea water in time. *7th Int. Sedimentol. Congr., Program.*

WICKMAN F. W. (1948) Isotope ratios—a clue to the age of certain marine sediments. *J. Geol.* **56,** 61–66.

YOUDEN W. J. (1951) *Statistical Methods for Chemists,* 126 pp. John Wiley.

[*Editor's Note:* The appendix of sample localities on pages 118–120 has been omitted.]

31

Reprinted from *Nature* **267**:403–408 (1977)

Catastrophic chemical events in the history of the ocean

William T. Holser

Max-Planck-Institut für Chemie, D-6500 Mainz, Germany (BRD)

Department of Geology, University of Oregon, Eugene, Oregon 97403

Catastrophic chemical events are characterised by sharp rises in $\delta^{34}S$ in the surface of the whole world ocean, and by greater overshoots locally. Three events are recognised and named for the formation in which they are most sharply displayed. The sharpness of the rise in $\delta^{34}S$ suggests that the sulphide deposition necessary to explain it must have been accumulating residual high-$\delta^{34}S$ seawater for some tens of millions of years out of contact with the surface ocean. A modified geological model is presented: brine generated by evaporite deposition is stored in deeps of a mediterranean basin; underneath the brine, pyrite precipitation builds a store of brine heavy in $\delta^{34}S_{SO_4}$, whose corresponding buildup of $\delta^{18}O_{SO_4}$ may balance the decrease of $\delta^{18}O_{SO_4}$ from evaporite deposition. Catastrophic mixing of the brine and the surface ocean, initiated by destruction of the storing basin, is the source of the sharp rise in the sulphur isotope age curve detected world-wide in evaporites. These events have important implications not only for modelling of the chemical history of the ocean, atmosphere, and sediments, but also for the explanation of faunal crises, and the many aspects of geology that depend on the composition and circulation of the oceans and their peripheral basins.

It has been known for some time that the sulphur isotope ratio in the world ocean surface has undergone major excursions during the period (Proterozoic–Phanerozoic) that could be documented from evaporite sulphate samples[1–3]. The resulting sulphur isotope age curve has recently been updated with many new analyses of sulphur isotopes[4,5], supplemented by measurements of oxygen isotopes from some of the same sulphates[5,6]. The interpretations of these isotope age curves have focused mainly on explaining the trends of $\delta^{34}S$ as a result of widespread variations in biological reduction of sulphur into shales; on subsequent oxidative erosion into the sea; and with possible coupling of the carbon cycle to maintain atmospheric oxygen (refs 3, 5, 7–12). My purpose here is to discuss some more dramatic but less well-documented features of the isotope age curves, and to discuss whether they record catastophic chemical events that are specific in time and place.

Upward jumps in $\delta^{34}S$

Fig. 1 shows as a solid line the mean best estimate for $\delta^{34}S$ in sulphate evaporite minerals in equilibrium with the surface of the world ocean[4,5]. In constructing this curve greater weight was attributed to values whose world-wide distribution bore out their origin in the ocean surface as a whole, and less weight to results that were found in a narrow range of time and place. But it is worth noting that those of the latter deviations that are not so widespread may have considerable significance.

Marks on Fig. 1 highlight three rises in $\delta^{34}S$; and a solid line in Fig. 2 repeats these parts of the curve on an enlarged scale. These rises are so large and take place over such a short time, that I have labelled them chemical events by analogy with the nomenclature of magnetic reversals. Table 1 lists some characteristics of these three events. There are differences that may be significant, but some common features are: (1) The mean curve established by world-wide evaporite sampling rises sharply.

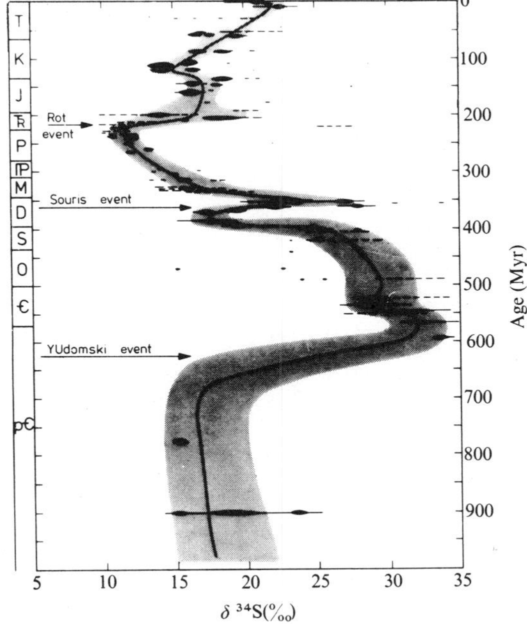

Fig. 1 Summary sulphur isotope age curve for marine sulphate, from Claypool, *et al.*[4,5], who give detailed plots and lists of analyses. Data are shown here by areas or lines that indicate qualitatively the number of analyses, plotted at their most probable age. The heavy line is the best estimate for $\delta^{34}S$ in equilibrium with the world surface ocean of that time; the shaded area is an estimate of the uncertainty of that determination. The named catastrophic chemical events are sharp rises in the best-estimate curve, accompanied by very high values ('overshoots') found only locally.

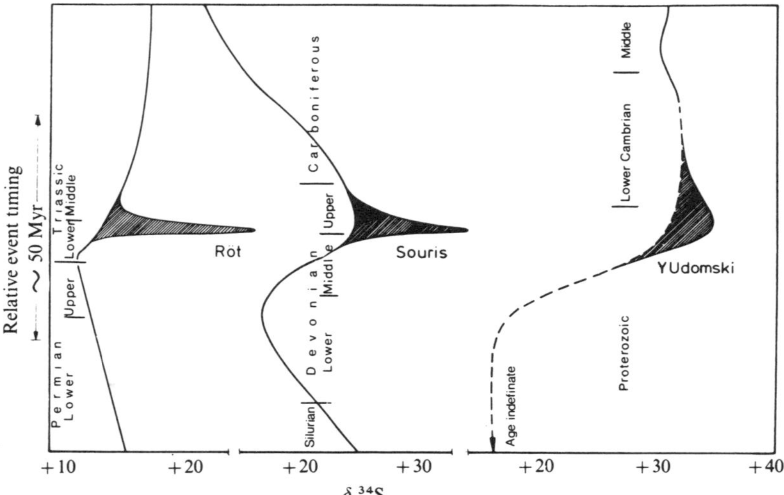

Fig. 2 The chemical events labelled in Fig. 1 are shown here schematically on greatly enlarged scale. Solid line: best estimate for $\delta^{34}S$ of sulphate mineral in equilibrium with world surface ocean; shaded area: local high overshoot. The timing is based approximately on stratigraphic stages, so that the absolute time scale is very approximate.

(2) The rise is highly accentuated by a positive overshoot, in one formation in one basin, which gives its name to the event. The overshoot peak is a step function within the limits of stratigraphic control. (3) It is difficult to fix the absolute time interval within which the event is bracketed. Each seems to have taken place within one stratigraphic stage, probably much less than 5 Myr. (The times given in Table 1 are only scaled by interpolation from radioactive dates that are for the most part separated by a whole geological period[13].) (4) Each of the events is followed by a long period of mean oceanic $\delta^{34}S$ that is higher than the pre-event level, and of course lower than the overshoot peak. This post-event level is either constant (YUdomski and Röt Events) or relatively slowly decreasing (Souris Event) over a substantial period of time. (5) The mean curve for $\delta^{18}O_{SO_4}$ in marine sulphates[5] at equivalent times displays either no discernible rise (YUdomski and Souris Events) or a relatively small one (Röt Event).

Nature of the chemical event

Models for interpretation of the sulphur isotope age curve explain general rises as indicative of marine biological precipitation of sulphide predominant over weathering of sulphide, and falls as oxidative sulphide weathering (refs 4, 5, 8, 11, 12), although the models differ widely in assumed constants, residence times, and couplings. Processes of a similar nature must also explain the sharp events.

Detailed consideration of the isotope geochemistry of oxygen has led Claypool *et al.*[4,5] to conclude that when $\delta^{34}S$ rises a corresponding rise in $\delta^{18}O_{SO_4}$ can only be prevented if the sulphide deposition is accompanied by extensive evaporite deposition. This is possible because with oxygen, much more than with sulphur, the isotope effect of evaporite deposition tends to offset the effect of sulphide deposition. Consequently, the chemical events seem to have involved major and contemporaneous deposition of sulphide into anaerobic muds and of sulphate into evaporities.

But these rises in the sulphur curve, and especially the local overshoots, are so precipitous that an extremely high rate of sulphide precipitation would be required to change the whole world ocean so much in so short a time. One can first of all see qualitatively that if the residence time of sulphur in the ocean (with cyclic sea salt subtracted) is about 20 Myr, as at the present time[3], then it will be very difficult to generate changes of 5–15‰ in $\delta^{34}S$ within 5 Myr, and still more so in 1 Myr. Nielsen[2] calculated that the rate of change of $\delta^{34}S$ probably would not exceed 0.3‰ per Myr. In making this calculation, he estimated the present rate of sulphur erosion ($\simeq$ deposition) of 15 Mt yr^{-1}, but the maximum rate would have meant the complete suspension of erosion to get the calculated rate of rise from sulphur deposition. This seems unlikely, especially as present rates of erosion may be higher than usual[14]. Nielsen had also to use the current value of the sulphur content of the ocean. It is true, however, that if the ocean were smaller, the same rate of sulphide deposition would raise $\delta^{34}S$ correspondingly faster. Holland[11] also assumed the present sulphur content of the ocean, as well as the present rate of sulphur input by rivers (uncorrected for cyclic sea salt and anthropogenic sulphur). His values of rises in available oxygen for the intervals that span my chemical events can be converted to amounts (of sulphur deposited) that are correspondingly large, and when compressed into an actual rise time even as long as 5 Myr, the rate of net sulphur deposition figures out to about 150 Mt yr^{-1}. Using the simultaneous solution of mass balances for both sulphur and oxygen isotopes I have also calculated similarly extreme rates of sulphide deposition. These calculated rates all seem unlikely, and suggest that the deposition of sulphide that was responsible for the rise in the age curve was accumulated over a long period of time, but was read into the evaporite record in a very short time.

The first evidence of the event is also the strongest peak, and in at least the YUdomski and Souris Events is found in only one basin. This suggests a chemical overshoot, possibly in the very area where the event was generated. The post-event level is seen on the whole ocean's surface, and the eventual decline corresponds to a more uniform decay through the weathering process.

A geological model

To explain the 'Permo-Triassic faunal crisis' (which is coincident with my Röt Event) Fischer[15] suggested that brine was generated on shelves from evaporite or near-evaporite conditions, and then slid down as a density current to accumulate in the ocean deeps. This mechanism generates a stratified ocean that is at least temporarily stable, with brine below and a brackish surface layer. The characteristics of the isotopic events suggest the following modifications of Fischer's proposal.

Fischer required brine that was to some undefined degree dense enough to stably resist mixing; he allowed but did not

Table 1 Catastrophic chemical events

Suggested name	YUdomski Event	Souris Event	Röt Event
Stratigraphic stage	Pre-Aldan (Latest Proterozoic)	Frasnian (Lower Upper Devonian)	Scythian (Upper Lower Triassic)
Approximate age (Myr)	575	355	215
Pre-event level $\delta^{34}S$			
Stratigraphic stage	Late Proterozoic	Eifelian (Lower Middle Devonian)	Mittlerer Buntsandstein (M. L. Triassic)
Approximate age (Myr)	635–850	365	220
Mean $\delta^{34}S$, ‰	+15	+18	+13
Earliest peak in $\delta^{34}S$			
Formation	YUdomski Horizon	Souris River and others	Röt
Location	Eastern Siberian Platform	Saskatchewan	Netherlands to DDR
Other localities	Saline Series, Pakistan Hormuz Series, Iran	—	Moenkopi of Utah
Maximum $\delta^{34}S$	+35	+34	+27
Maximum excursion in $\delta^{34}S$	+20	+16	+14
Post-event level $\delta^{34}S$			
Stratigraphic stage	Lower to Middle Cambrian	Famenian (Upper Upper Devonian)	Muschelkalk (Middle Triassic)
Mean level	+30	+23	+19
Rise of mean $\delta^{34}S$	+15	+5	+6
Rise of mean $\delta^{18}O$	0?	+0.5	+2.5

require that evaporation proceed as far as evaporite precipitation. The oxygen isotope data indicate that evaporites of at least the calcium sulphate facies were generated in large amounts. The resulting brine, of density $1.10–1.20 \times 10^3$ kg m^{-3} has therefore all the more likely to be hydraulically stable in the bottom of a basin. Other proposals[16,17] have been based on evaporite deposition in the halite facies, and that is not precluded here.

Fischer suggests a deep-sea depression as the storage site for brine. That is not necessary, and it may not be correct. The bottom of any basin will do for brine storage, so long as it is large enough to hold the amount of brine that must be sequestered to account for the change of $\delta^{34}S$. A rift graben or other mediterranean has the advantage that it is more susceptible to to the quick destruction suggested by the fast rise of the $\delta^{34}S$ curve, which represents chemical mixing of the deep brine and the surface layer.

The suggestion of a large basin in which evaporite deposition (as well as sulphide deposition) is important will probably remind the reader of the desiccating sub-sealevel basin[18,19], which enjoys inflow from the ocean without reflux. One problem with that arrangement is that in conditions approaching desiccation, the required mixing of the brine is inhibited. Valiashko[20] has pointed out that even at the beginning of the potash evaporite facies, the volume of remaining brine may be accommodated in the porosity of the previously deposited salts. But the basin could be sub-sealevel, and non-refluxing, so long as it was not dessicating.

It therefore seems most likely that the site for brine accumulation is neither a deep-sea basin, nor a sub-sealevel desiccating basin, but rather a large mediterranean isolated from the ocean, especially with regard to its deeps.

A particular requirement for the chemical event is that evaporite deposition be accompanied by sulphide deposition, within the same basin complex if not actually the same basin. Although many evaporite sequences are grey and evidently associated with reducing conditions, the amount of sulphide is usually small. So it may be that the major part of the sulphide deposition occurs under brine accumulated nearby. Some present saline deeps support bacterial reduction of the large proportion of sulphate that they still retain, while others are sterile. The differences are not understood[21–23].

Sulphide deposition requires not only the generation of sulphide by sulphur-reducing bacteria, but also a sufficient supply of iron (ferrous or ferric) to fix the sulphide in the sediment. Not much sulphide can be retained in solution in the brine, and even this would be ineffective for the chemical event if it were simply remixed with the ocean. Although sulphide deposition in ordinary bottom muds is usually supposed to get its iron from the sediments themselves (probably from hematite or goethite coating sand grains)[24], the amount required for building the large reservoir of fixed sulphide suggests that supplies could well be augmented from a more plentiful source such as submarine volcanism[25,26].

According to Fischer[15] the fossil record suggests that during the Triassic the surface ocean returned to normal salinities. The

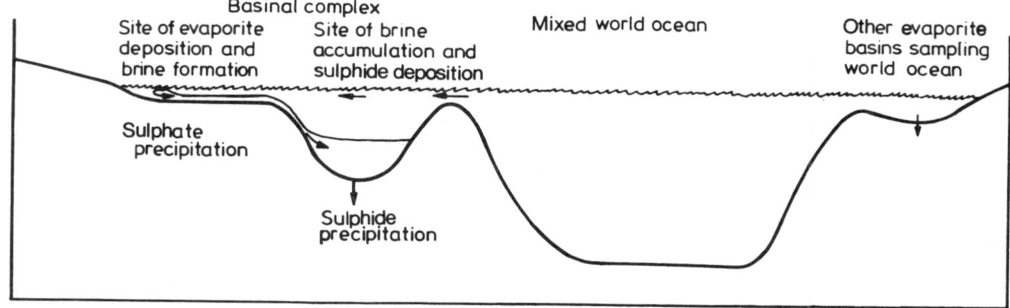

Fig. 3 Geological model to explain the chemical events, somewhat modified from Fischer[15]. A basin or basin complex precipitates sulphate, and generates and stores brine, under which reduced sulphide is precipitated. The evaporites of the world (both left and right) sample only the surface ocean, out of contact with sulphate residues of the sulphide reduction, until its accumulated effect is catastrophically mixed with the surface waters at the time the basin complex is destroyed.

sulphur isotope record also suggests this—but the sharp onset indicates a more catastrophic mixing of brine and surface ocean. At that time, in addition to the restoration of normal salinity described by Fischer, we may have a catastrophic inflow in various places of waters rich in salts, organic matter, sulphide ions, nutrients, oxygen demand, and $\delta^{34}S$. The local nature of the overshoot in $\delta^{34}S$ is further evidence for the catastrophic nature of the mixing. The location of such an overshoot suggests but does not necessarily prove that the brine was stored in a basin nearby.

In accordance with current custom much of the above can be stated in terms of the theory of plate tectonics. The rifting Atlantic is prominently associated with evaporites (refs 19, 27 and 28); and Kinsman (refs 17, 29 and 30) has generalised and explained this association in the following way: rifting leads not only to initial rift valleys like the Red Sea, but also continuously generates subsiding continental margins, as well as transverse tensional grabens. These are all potential sites of restricted circulation and evaporite deposition along the trailing continental edge. Isolation may be aided by island-forming volcanism in the rift. Kinsman[17] also pointed out the fast timescale of evaporite generation on specific sections of a continental margin

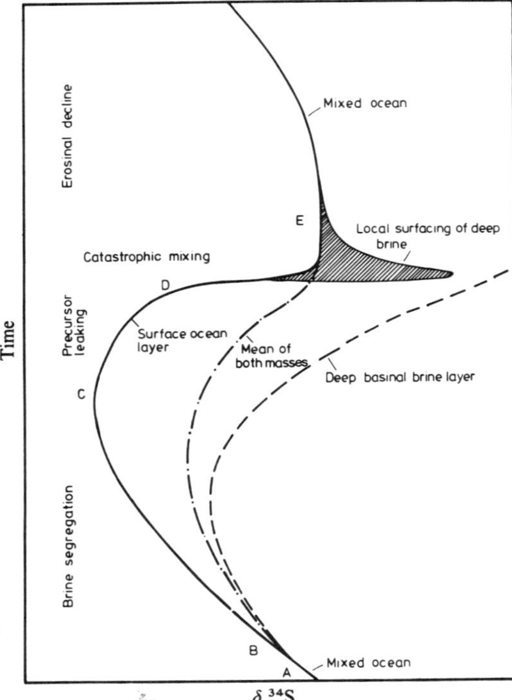

Fig. 4 Schematic representation of sulphur isotope effects during the course of a chemical event. At the earliest time, *A*, $\delta^{34}S$ is slowly decreasing as erosion of sulphide exceeds deposition of sulphide. At *B* brine starts to accumulate in a basin complex as a result of evaporite deposition (Fig. 3), and biological reduction under the brine begins to slow and then reverses the decline of $\delta^{34}S$ in the residual sulphate in the brine, but out of contact with the surface ocean which continues to decline in $\delta^{34}S$ (as sampled by worldwide evaporites). At *C* the brine begins to mix somewhat with the surface ocean, and at *D* mixing becomes catastrophic, with local overshoot due to surfacing of deep brine. At *E* a level for the now mixed ocean is a weighted mean of the previous two reservoirs of sulphate; the course of this fictive mean isotope composition during the event is shown by the middle line.

—his time limit of 10 Myr is compatible with the chemical event reported here. I need only add to this description that brine sliding down and accumulating in the rift basin will help reduction of sulphide, and volcanic activity or associated thermal waters may supply iron and other heavy metals to fix the sulphide in giant deposits[25,26]. Eventual full separation of the continents mixes brine and surface ocean. This scenario is a possible one, but other tectonic situations may also generate the required structures and geography.

History of a chemical catastrophe

The story line is indicated in Fig. 4. At time *A* we find the sulphur isotope age curve slowly declining as a result of erosion of predominantly sulphide material into a mixed ocean. This erosion exceeds the sulphide deposition. At time *B*, dense residual brines from evaporite deposition begin to accumulate in the deep of a large nearby basin. Reduction to sulphide continues and $\delta^{34}S$ in the brine gradually begins to reverse and then increase, while at the same time its mass continues to grow as a larger and larger proportion of the ocean's sulphate is processed through the sulphate–sulphide production system. Meanwhile, the surface layer becomes more brackish, and its $\delta^{34}S$ continues to fall under the influence of the continuing erosion of older sedimentary sulphide. At point *C* the accumulated brine with high $\delta^{34}S$ begins to leak or mix slightly with the surface body of the ocean, slowing and even reversing the downward trend of $\delta^{34}S$ recorded in evaporites everywhere. At point *D* the climax of the event is reached as the brine is released, perhaps through tectonic destruction of the basin. Local appearance of the high $\delta^{34}S$ may be sampled by evaporites as an isolated record of $\delta^{34}S$ overshoot. Mixing of brine and surface water leads to an intermediate level of $\delta^{34}S$ in the subsequent evaporite sampling of the now mixed ocean. As erosion assumes greater importance again, at point *E*, $\delta^{34}S$ begins to decrease. This may continue until the required combination of evaporite production and basin geometry again induces brine accumulation.

Point *A* does not leave a record that I know how to recognise, so I don't know how long before the event brine segregation and accumulation began. It is conceivable that all the depressed sections of the $\delta^{34}S$ curve represent segregation and that therefore erosion never really takes over, but only becomes more evident when its depositional counterpart is hidden in a deep.

I realise that the explanation proposed for the chemical events is only a skeleton, and I exhibit it here so that others may join me in either giving it flesh or burying it. Whether it is viable or not the events are real and demand another explanation if not this one.

Geological model

Each of the events displays some individual character in the isotope age curves—these will now be described. And to follow up on the conjecture that the location of brine storage is flagged by the overshoot, the palaeogeography will be examined for conditions favourable to brine accumulation and sulphide reduction.

YUdomski Event. This event occurs just below the boundary of the Proterozoic and the Cambrian, between the Shaler Group of the Arctic Islands that can only be bracketed at 630–850 Myr (ref. 31), and the YUdomski Horizon of the Siberian Platform that is said to be dated at 600 ± 10 Myr (ref. 32). So the abruptness of the rise shown in Figs 1 and 2 is drawn by analogy to the later two events. Within the vague limits of the available data, $\delta^{18}O$ is unchanged throughout the whole event[5]. The level $\delta^{34}S \simeq 17‰$ in the Proterozoic pre-event stage is puzzling. It could be an artefact of an imperfect record. Perhaps brine was accumulating all that time, leading to the climax of the chemical event at the Cambrian boundary. But in that case one would have expected the surface ocean as sampled by the evaporites to have shown a declining curve due to erosional input.

The earliest record of this event is an overshoot to 33‰ in our new analyses of the YUdomski or Motskaya Suite, which crops out around the southern edge of the Irkutsk Amphitheatre in Siberia. This horizon lies confortably below the Cambrian[33], which especially in its middle stage developed into one of the world's largest evaporite basins[34]. There are earlier evaporites, as yet unanalysed for isotopes, in Siberia that precede the YUdomski[32]. All of these Siberian basins opened in the direction of the present Arctic, and that could have been the locale for brine accumulation, but its palaeogeography and tectonics are unknown to me.

A possible alternative site for the main event is suggested by one analysis of high $\delta^{34}S$ in the Cambrian of the Salt Range of Pakistan[6]; this horizon evidently overlies the extensive Saline Series of possible Late Precambrian age[35], which may also be correlative with the very extensive Hormuz Series evaporites that extend across southern Iran[36,37]. This has been confirmed by unpublished high values of $\delta^{34}S$ obtained by Heimo Nielson (personal communication) from several samples of the Hormuz Series.

Souris Event. A special character of this event is a well-documented gradual rise in worldwide $\delta^{34}S$ during the Middle Devonian, that preceeds the sharper rise to the peak in the Famenian Stage[5,7]. In terms of the model, this represents a more gradual leak or mixing from the stored brine into the ocean.

A strong overshoot is displayed in the Souris River Formation and related beds near the top of a giant sequence of evaporites that accumulated in the Elk Point Basin of Western Canada, beginning in the Middle Devonian. The basin was recharged from the north, across the strongly developed Presquile Reef in the North-west Territories. Beyond the reef black marine shales were extensively deposited[38], and this area may have been the place of brine accumulation and sulphide reduction. It lies at the edge of the Eurasian basin, whose palaeogeography and tectonics are still controversial[39,40].

Röt Event. This event occurs sometime between well-established low levels of $\delta^{34}S = 11‰$ in the latest Permian, and the moderate levels of the Middle Triassic and later Mesozoic stages. Although in many areas stratigraphic subdivision of the Lower Triassic has not been successful, analyses from the well divided Buntsandstein of East Germany[41] and West Germany (unpublished data by Dietrich Rambow and Heimo Nielsen) show that low values of the Permian continue through the Lower and Middle part of the Lower Triassic.

The Buntsandstein exhibits a most dramatic overshoot, up to $\delta^{34}S = +27‰$, in the Röt (Upper Lower Triassic), whose evaporite facies extend from Poland across Germany (East and West) into England[42,43] and touch Greenland[44]. The main development of Röt evaporites coincides areally with much of the thickest Zechstein evaporites. If, as seems plausible[42,43], the Röt and Zechstein had a common entrance to the north ('Skandik Sea'), this might furnish a suitable site for brine accumulation; although an alternative opinion is that the Röt was open to the Tethys Sea to the southeast[45].

A few analyses[5] suggest a possible alternative site for this event, in the Lower Triassic areas of the Western United States.

Note that all three events have possible connections with brine accumulations in the region of the Palaeozoic geography that lies in the present Arctic.

Comparison with other theories

Thode and Monster[1] proposed that certain basins would have high values of $\delta^{34}S$, due to greater amounts of sulphur reduction, and they therefore assumed as a value for the world ocean the lowest $\delta^{34}S$ in any basin at a given time. In the present instance reduction in a basin is also proposed, on an even larger scale, but its residues remain out of contact with the world ocean surface until the catastrophic event of mixing.

Pilot *et al.*[41] dealt specifically with the Röt event: They proposed that an accumulation of sulphide reduction in the Tethys Sea had been going on since the Carboniferous, and that the Röt evaporites were drawn from that sea, in contrast with the Zechstein which was drawn from the northern sea. For my model it is not sufficient to isolate Tethys and deposit sulphide there so that the Röt can tap it with a high $\delta^{34}S$ evaporite. The sulphide-depositing basin also has to derive a large amount of sulphate from the world ocean, while at the same time holding back its high $\delta^{34}S$ until a critical moment. I accomplish this with the evaporite-generated brine layer. This could have happened in Tethys, but for the reasons described above I think the northern sea is more likely.

Other speculations concerning the history of the ocean, while not specifically directed towards the interpretation of isotope age curves, are related to major chemical events. Buerlen[16] proposed that the faunal crisis at the Permian–Triassic boundary might be connected with a decreasing salinity of seawater due to the extensive precipitation of evaporites during the Permian. As indicated above, Fischer[15] had a similar explanation for this palaeontological problem, but made the surface ocean brackish by the accumulation of deep brines rather than necessarily by deposition of salt. His paper was the real ancestor of this one, but its significance was not appreciated until the isotope data had been accumulated in sufficient detail. Kinsman[17] showed how the kind of events proposed by Beurlen (although he did not refer to either Fischer or Beurlen) may have been very strong during the early phases of a continental rifting. None of these proposals connected the generation of brackish surface waters with sulphide reduction, which is the main point of the new proposal. More recently, Degens and Stoffers[46] proposed that density stratification, of unspecified origin, may have been a common phenomenon in earlier seas, on the model of the present Black Sea. Reducing conditions in the brine and the sediment, and accumulation of organic material, are important aspects of the model of Degens and Stoffers. What I am adding to their model is an emphasis on sulphide deposition in the sub-brine sediment, a suddenness of re-mixing with the surface ocean, and a method of specifically detecting such situations in the geological record by the interpretation of the isotope age curves.

Catastrophic chemical events—implications for future work

(1) Modelling of ocean history and the calculations of oxygen demand (positive or negative) associated with sulphide–sulphate transfers will be more complicated than it has been[4,11,12] under the assumption of a mixed ocean. In principle it will be possible to make new calculations with the sea sulphate divided between two reservoirs, surface and brine. Even without such calculations, one can say qualitatively that the overall decrease of $\delta^{34}S$ in the pre-event stage will be less than that for the surface ocean alone, so that the maximum oxygen demand associated with a $\delta^{34}S$ minimum will be less than would have been calculated assuming a mixed ocean.

(2) The dominant sulphide reduction during the pre-event stage will be under the brine reservoir. Therefore during the deep minimum in the $\delta^{34}S$ evaporite curve, a corresponding $\delta^{34}S$ curve for these particular sulphides need not show such a deep minimum or, indeed, any minimum at all. Whether sulphides of the briny deep are mainly in black shales, as seems most likely, or also in large stratiform sulphide deposits, remains to be determined. The limited sulphur isotope data presently available for both styles of deposition have been reviewed by Schwarcz and Burnie[48], but these represent only a handful of occurrences.

(3) The combination of evaporite deposition and brine formation decreases surface salinity (in all its components) of the ocean during the pre-event phase. The surface ocean should also show a corresponding increase of δD and $\delta^{18}O_{H_2O}$, which should be looked for in fossils, limestones, and cherts, but on a finer time scale than previously tried[15,48].

(4) The decrease of sulphate in the surface ocean may allow the buildup, from erosional input, of elements such as barium and strontium; they may be precipitated during the subsequent mixing. Some sedimentary barite deposits[49,50] coincide exactly with the Souris event; and peaks of strontium in shales are found

near the YUdomski and Röt Events[51]. The age curve for $^{87}Sr/^{86}Sr$ shows a number of sharp rises[52-54], and each of the chemical events postulated here has one of these jumps in $^{87}Sr/^{86}Sr$ associated with it: YUdomski Event—from 0.7075 in the Late Proterozoic (570 to 850 Myr?) to 0.7095 in the Lower Cambrian; Souris Event—from 0.7080 in the Givetian and Frasnian to 0.7093 in the Famenian; and the Röt Event—0.7073 in the Kazanian and Tartarian to 0.7083 in the Early Triassic. Veizer and Compston[53] noted that the fine structure of the strontium isotope curve might "... indicate that the fluctuations in Sr isotopic composition of seawater could develop within several million years." But the difficulty here is similar to sulphur—a residence time in a mixed ocean that is much longer than that.

(5) The possible effects of these various chemical changes on biological processes, ecology, and evolution may be far-reaching. Some of these have already been discussed in earlier models: pre-event increase in oxygen demand[11] (but less than previously assumed[12]), and decrease of surface salinity[15-17]. The new model would generate the added stress of a quick return to more or less normal salinity. The remixing of anoxic brines into the ocean may have profound effects, particularly in local regions of overshoot. Nutrients may be removed from effective participation in the biological cycle—temporarily with all seawater constituents, or by sedimentation under the brine. In the post-event stage re-mixing restores to the productive surface waters those nutrients still in solution, with consequent effects on organic productivity and possibly increased deposition of organic carbon and decrease of atmospheric p_{CO_2}. A post-event increase in phosphate in surface waters may account for the proliferation of apatite-secreting metazoa in the Cambrian[55].

I thank my colleagues G. Claypool, I. Kaplan, H. Sakai, and I. Zak for many discussions in our quest to find the reasons for the isotope age curves. I am grateful for an appointment as Senior Scientist under the programme of the Alexander von Humboldt Stiftung, and for the hospitality of C. Junge and M. Schidlowski at the Max-Planck-Institut für Chemie, Mainz, Germany, where this paper was written. I thank H. Nielsen for stimulating discussion and access to unpublished data.

Received 3 February; accepted 29 March 1977.

1 Thode, H. G. & Monster, J. *Am. Ass. petrol. Geol. Mem.* **4**, 367–377 (1965).
2 Nielsen, H. *Geol. Rundsch.* **55**, 160–172 (1965).
3 Holser, W. T. & I. R. Kaplan *Chem. Geol.* **1**, 93–135 (1966).
4 Claypool, G. E., Holser, W. T., Kaplan, I. R., Sakai, H. & Zak, I. *Geol. Soc. Am. Programs Abstr.*, **4**, 473 (1972).
5 Claypool, G. E., Holser, W. T., Kaplan, I. R., Sakai, H. & Zak, I. (Unpublished).
6 Sakai, H. *Earth planet. Sci. Lett.* **15**, 201–205 (1972).
7 Holser, W. T. & Kaplan, I. R. *Geol. Soc. Am. Spec. Pap.* **101**, 97 (1968).
8 Rees, C. E. *Earth planet. Sci. Lett.* **7**, 366–370 (1970).
9 Nielsen, H. *UNESCO Earth Sci.* **7**, 91–102 (1972).
10 Nielsen, H. in *Mezhdunar. Geokhim. Kong.* (*Dokl.*), *1st, 1971* (ed. Vinogradov, A. P.) **4**, 1, 127–140 (1973).
11 Holland H. D. *Geochim. Cosmochim. Acta* **37**, 2605–2616 (1973).
12 Schidlowski, M., Junge, C. E. & Pietrek, H. *J. geophys. Res.* (in press).
13 Harland, W. B., Smith, A. G. & Wilcock B. (eds) *Q. Jl geol. Soc. Lond.* **120S** (1964).
14 Garrels, R. M. & Mackenzie, F. T. *Evolution of Sedimentary Rocks* (Norton, New York, 1971).
15 Fischer, A. G. in *Problems in Palaeoclimatology* (ed. Nairn, A. E. M.) 566–574 (Discussion: p. 575–577) (Interscience, London, 1964).
16 Buerlen, K. *Z. Deutsch. Geol. Gesell.* **108**, 88–99 (1956).
17 Kinsman, D. J. J. *Nature* **255**, 375–378 (1975).
18 Hsu, K. H. *Earth Sci. Rev.* **8**, 371–396 (1973).
19 Burke, K. *Geology* **3**, 613–616 (1975).
20 Valiashko, M. G. *Freiberger Forschungsh.* **A123**, 197–233 (1958).
21 Trüper, H. G. in *Recent Heavy Metal Deposits in the Red Sea* (eds Degens, E. T. & Ross, D. A.), 263–271 (Springer–Verlag, Berlin, 1969).
22 Watson, S. W. & Waterbury, J. B. in *Recent Heavy Metal Deposits in the Red Sea* (eds Degens, E. T. & Ross, D. A.), 272–281 (Springer–Verlag, Berlin, 1969).
23 Nissenbaum, A. & Kaplan, I. R. *Limnol. Oceanog.* **17**, 570–582.
24 Berner, R. A. *Am. J. Sci.* **268**, 1–23 (1970).
25 Corliss, J. B. *J. geophys. Res.* **76**, 8128–8138.
26 Dymond, J., Corliss, J. B., Heath, G. R., Field, C. W., Dasch, E. J. & Veeh, H. H. *Geol. Soc. Am. Bull.* **34**, 3355–3372 (1973).
27 Olson, W. S. & Leyden, R. J. *Can. Soc. Petrol. Geol. Mem.* **2**, 720–732.
28 Bonatti, E., Ball, M. & Schubert, C. *Naturwissenschaften* **57**, 107–108 (1970).
29 Kinsman, D. J. J. in *Petroleum and Global Tectonics* (eds Fischer, A. G. & Judson, S.), 84–126 (Princeton University Press, Princeton, N. J., 1973).
30 Kinsman, D. J. J. in *4th Symposium on Salt* (ed. Coogan, A. H.), **2**, 255–259 (Northern Ohio Geological Society, Cleveland, Ohio, 1974).
31 Young, G. M. *Precambrian Res.* **1**, 13–41 (1974).
32 Kulibakina, I. B., Rabotnov, V. T., Goncharov, E. S. & Ahobot, M. R. *Dokl. Akad. Nauk. S.S.S.R.* **201**, 672–674 (Transl. *Dokl. Earth Sci. Ser.* **201**, 86–88, 1971).
33 Pisarchik, YA. K., Minaeva, M. A. & Rusetskaya, G. A. *Trudy Vses. Issled. Geol. Inst.* **215**, (1975).
34 Zharkov, M. A. *Dokl. Akad. Nauk. S.S.S.R.* **184**, 913–914 (1969) (transl. *Dokl. Earth Sci. Ser.* **184**, 72–74, 1969).
35 Schindewolf, O. H. & Seilacher, A. *Akad. Wiss. Literat. Mainz Abhandl. Math.-Naturwiss. Kl.*, **1955**, 260–446 (1955).
36 Stöcklin, J. *Geol. Soc. Am. Spec. Publ.* **88**, 159–179 (1968).
37 Ala, M. A. *Am. Ass. petrol. Geol. Bull.* **58**, 1758–1770 (1974).
38 Grayston, L. D., Sherwin, D. F. & Allan, J. F. in *Geological History of Western Canada* (eds McCrossan, R. G. & Glaister, R. P.), 49–59 (Alberta Society of Petroleum Geologists, Calgary, 1964).
39 Meyerhoff, A. A. *J. Geol.* **78**, 406–444 (1970).
40 Churkin, M., Jr *Am. Ass. petrol. Geol. Mem.* **19**, 485–489 (1973).
41 Pilot, J., Rössler, H. V. & Müller, P. *Neue Bergbautech.* **2**, 161–168 (1972).
42 Warrington, G. *Qu. J. Geol. Soc. Lond.* **126**, 183–223 (1970).
43 Wills, L. J. *Qu. J. Geol. Soc. Lond.* **126**, 225–283 (1970).
44 Birkelund, T. & Perch-Nielsen, K. in *Geology of Greenland* (ed. Escher, A. & Watt, W. S.), 305–339 (Grønlands Geologiske Undersogelse, Copenhagen, 1976).
45 Richter–Bernberg, G. *UNESCO Earth Sci. Ser.* **7**, 275–287 (1972).
46 Degens, E. T. & Stoffers, P. *Nature* **263**, 22–27 (1976).
47 Schwarcz, H. P. & Burnie, S. W. *Mineral. Deposita* **8**, 264–277 (1973).
48 Knauth, L. P. & Epstein, S. *Geochim. Cosmochim. Acta* **40**, 1095–1108 (1976).
49 Buschendorf, F. & Puchelt, H. *Geol. Jahrb.* **82**, 499–582.
50 Hoffman, T. *Mineral Deposita* **4**, 260–274 (1969).
51 Reimer, T. *Neues Jahrb. Mineral. Abhandl.* **116**, 167–195 (1972).
52 Veizer, J. & Compston, W. *Geochim. Cosmochim. Acta* **38**, 1461–1484.
53 Veizer, J. & Compston, W. *Geochim. Cosmochim. Acta* **40**, 905–914 (1976).
54 Tremba, E. L., Faure, G., Katsikatsos, G. C. & Summerson, C. H. *Chem. Geol.* **16**. 109–120 (1976).
55 Muller, K. J., *Naturwissenschaften* **51**, 36 (1964).

Part VI

REVOLUTIONS IN THE HISTORY OF LIFE

Editor's Comments on Papers 32 and 33

32 NEWELL
Excerpts from *Revolutions in the History of Life*

33 FAIRBRIDGE
Excerpts from *Carbonate Rocks and Paleoclimatology in the Biogeochemical History of the Planet*

As pointed out in 1967 by Newell (Paper 32) in the eloquent essay from which this Part takes its title, at the time of appearance of Darwin's *Origin of Species* in 1859 virtually all paleontologists believed that the history of life on earth had been marked by a succession of catastrophic extinctions. Although Darwin and Lyell rejected catastrophism and the concept of world revolutions, today abundant evidence exists that the history of life has not been a steady evolution but rather a succession of crises of extinction followed by marked adaptive radiations. The episodicity of evolution has been recognized this century by numerous paleontologists, among whom the names of Lull (1918, 1929), Schindewolf (1950, 1962), G. G. Simpson (1953), Newell (Paper 32), and Valentine (1973, 1977) stand out.

In Paper 32 Newell clearly demonstrated the marked episodicity of extinction and radiation in the animal kingdom during the Phanerozoic. Newell (p. 76) concluded that the most critical times of extinction for all major faunal groups occurred near the close of Periods. This conclusion is supported by the numerical analysis of the fossil record by Cutbill and Funnell (1967); Figure 9 shows their computation of "lasts" (extinctions) and "firsts" for all groupings (plants, invertebrates, vertebrates). The study by Cutbill and Funnell also indicates a reduction in the total number of taxa at the end of the Paleozoic, due to a high proportion of extinctions rather than to any fall-off in "firsts," and the rather random incidence of "firsts" with respect to Period boundaries. McAlester (1973) and Valentine (1973) likewise demonstrated that times of episodic net extinction generally coincided with Period

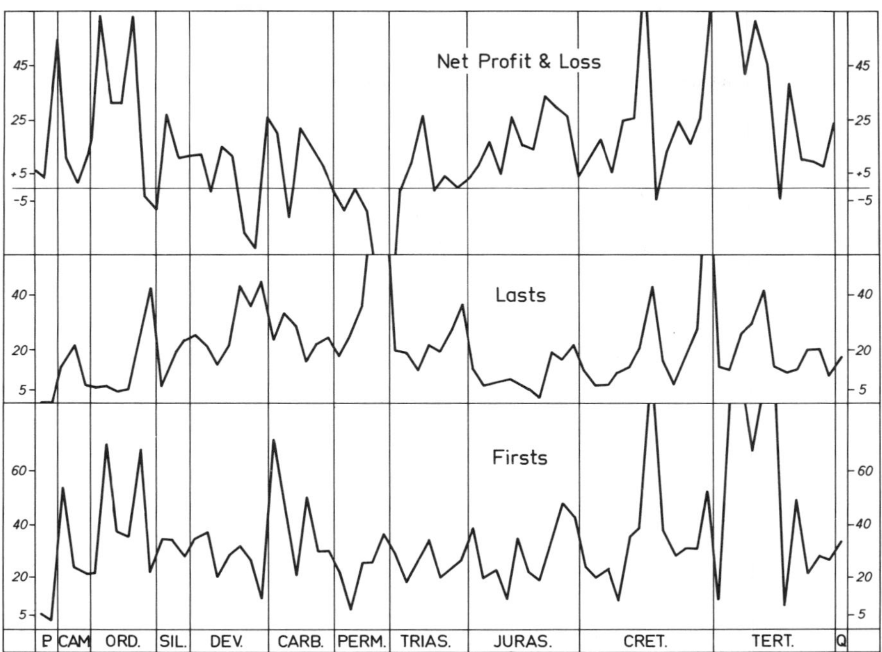

Figure 9 Number of firsts, last (extinctions), and net profit and loss of higher taxa of all biological groups. The durations of Periods are not drawn to scale. Adapted from Cutbill and Funnell (1967, Figure 22).

boundaries. McAlester (p. 11) further concluded that "Phanerozoic intervals of diversification and extinction have tended to occur simultaneously in all fossilized groups—marine and terrestrial, animal and plant." Newell (Paper 32) nonetheless rejected the term "neocatastrophism" (see Schindewolf, 1962) because of its "emotional connotation that implies calamity and destruction." The concept of neocatastrophism in the stratigraphic and fossil records has however been championed more recently by Ager (1973, 1976).

In 1967 Fairbridge (Paper 33) provided a broader perspective to the study of evolutionary patterns by proposing five great revolutions since early Precambrian time, based principally on the changing biogeochemical role of carbon. His "law of biological continuity" indicates that no revolution in the history of life transgressed globally the metabolic limits of organisms. A complementary subdivision of earth and biological history, which explores the role of oxygen in the evolution of the crust, hydrosphere and atmosphere, was presented by Cloud (1972, 1974). (See K. A. Kvenvolden, ed., *Geochemistry and the Origin of Life*, Benchmark Papers in Geology, vol. 13, 1975, which includes Cloud [1972] and other important papers on geochemical and organic evolution during earth history.) As discussed in Part IV, however, the important occurrences of banded iron-formations of late Proterozoic age and of sulfate evaporites in the middle Proterozoic indicate that revision of Cloud's scheme is required.

Paleontological research during the 1970s has provided further evidence of important "events" in Precambrian biological evolution, although no detailed synthesis of such evidence is at present available. As noted in Part V, carbon isotope analyses of very old sediments may indicate a major event in biological evolution ~3500–3300 m.y. ago. The dates of 2000^{+300}_{-100}, 1350±150, and 800±150 m.y. include all known major events in the evolution of Proterozoic stromatolitic microbiotas (see Schopf, 1977, Figure 12). These three broad dates appear to be of particular importance to Precambrian evolution in general. Awramik and Barghoorn (1977, p. 121), for example, considered that the microbiota in the ~2000 m.y. old Gunflint Iron Formation "is sufficiently diverse as to represent a 'benchmark' in the level of evolution at that time only somewhat less than intermediate between the origin of the earth and the present." Moreover, an abrupt expansion or "explosion" of stromatolites, including both carbonate and siliceous forms, commenced ~2300–2000 m.y. ago (Button, 1973; Awramik et al., 1976). The time of appearance of the first eukaryotic (nucleated) organisms is controversial; the date has been placed tentatively at ~1900 m.y. ago (Tappan, 1976) and at 1400±100 m.y. ago (Schopf and Oehler, 1976). The increase in taxonomic diversity of prokaryotic (bacterial and blue-green algal) microorganisms was most marked ~1400 m.y. ago (Schopf, 1977). The date of ~800 m.y. marks the abrupt decline in diversity of columnar stromatolites (Awramik, 1971), which may have been a response to a very important evolutionary event—the advent of the metazoa. Indeed, Francis et al. (1978) referred to the appearance of the Ediacaran metazoan faunas ~680 m.y. ago as an evolutionary "benchmark," and suggested that this date may represent the earliest appearance of eukaryotic organisms. The coincidence of the apparent evolutionary "benchmarks" at 2000^{+300}_{-100}, 1350±150 and 800±150 m.y. ago with the three successive *anorogenic* intervals of the Proterozoic (see Papers 2, 3, and 7) and with peaks in iron-formation abundance (Paper 24) may be significant.

The causes of biological revolutions and mass extinctions have been the subject of much speculation; reviews of hypotheses are provided by Newell (1962; Paper 36, pp. 81–88 [pages not included here]), Flessa and Imbrie (1973), Meyerhoff (1973), Valentine (1973, pp. 395–408), and Ager (1976). As pointed out by Newell (1962, p. 603), "the extinction of cosmopolitan groups surely requires severe world-wide environmental changes that proceed more rapidly than the organisms can evolve." In Paper 32 (p. 84) Newell advanced an hypothesis of widespread, approximately synchronous environmental disturbances and greatly increased selection pressure stemming from three different sources: (1) animal migrations; (2) severe climatic change; and (3) great and relatively rapid changes in distributions of land and sea. Newell stressed the importance of relative changes of sea level, concluding (p. 88) that "rapid emergence of the continents would result in catastrophic changes in both terrestrial and marine habitats and such changes might well trigger mass extinctions among the most fragile species." Dr. Newell (personal communication, 12 April 1978) has provided the following appraisal of his conclusions of 1967:

> Since 1967 the eustatic model of mass extinctions, then favored, has been strengthened by an extensive literature linking major extinction events with overriding plate movement control of ocean-basin volumes, numbers of paleogeographic provinces, and oscillations between mild and stressful climates especially disruptive of reproductive patterns. Within any given habitat, diversity would have been area-dependent and many marine and terrestrial habitats would have shrunk with increasing continentality. Diversity, however, would also be strongly influenced by the number of provinces and by accessibility and spread of fossiliferous deposits. These qualifications and the enigmatic common association of paraconformities with major paleontological discontinuities are outstanding problems that require more field data.

These ideas are further discussed by Newell (1977).

The "eustatic model" was advanced also by T. J. M. Schopf (1974) in explanation of Permo-Triassic extinctions and shares common elements with the proposal that Phanerozoic plate tectonics has controlled faunal diversity through changing land-sea patterns (Valentine and Moores, 1970, 1972; Flessa and Imbrie, 1973). A corollary of the eustatic model of mass extinctions is that times of widespread marine regression should approximate the close of Phanerozoic Periods, which were the most critical times of extinction for all major faunal and floral groups. A tendency for correspondence between Period boundaries and lowstands of sea level is indeed apparent (see Part IV). Other recent hypotheses propose that Phanerozoic evolutionary events and mass extinctions were related to geomagnetic reversals (e.g., Simpson, 1966;

Crain, 1971; Hays, 1971; see also Part II), to past variations in the concentration of atmospheric oxygen (McAlester, 1970), to variations in the uranium content of the sea (Neruchev, 1976), to the episodic injection of hypersaline waters into the world ocean (Thierstein and Berger, 1978), to influxes of high-energy cosmic radiation (Schindewolf, 1962), and to exogenetic galactic events (e.g., Steiner, 1967; Hatfield and Camp, 1970; Meyerhoff, 1973; Firsov, 1977).

Fewer workers have attempted to explain why Proterozoic evolutionary history appears divisible by "benchmarks" into intervals with an average length of ~600 m.y. Williams (Paper 35) termed such prolonged evolutionary intervals *life megacycles* and proposed control through long-term changes of climate in response to secular changes in planetary dynamics. Possible explanations of the seemingly common meter of long-term events evident in the earth's biological, climatic, and tectonic records are further discussed in Parts IV and VII.

32

Reprinted from pages 63–66, 69–81 and 88–91 of *Geol. Soc. America Spec. Paper 89*, pp. 63–91 (1967)

Revolutions in the History of Life

NORMAN D. NEWELL
American Museum of Natural History
New York, New York

Paleontology's ". . . involvement in geology and inclusion in that science as well as in biology are primarily due to the fact that the history of organisms runs parallel with, is environmentally contained in, and continuously interacts with, the physical history of the earth. It is of less philosophical interest but of major operational importance that paleontology, when applicable, has the highest resolving power of any method yet discovered for determining the sequence of strictly geological events." (Simpson, 1963, p. 26).

Catastrophe Versus Uniformity

At the time of publication of Darwin's *Origin of Species*, in 1859, practically all paleontologists shared the belief of Georges Cuvier (1769–1832) and Alcide d'Orbigny (1802–1857) that the history of life on earth had been marked by a succession of catastrophic extinctions. The stratigraphic column, characterized by discontinuities and contrasting changes in lithologic type and fossil content, by its very heterogeneity, seemed to give support to nineteenth-century catastrophism. The rocks had been neatly classified in eras, systems, and stages outlining an arrangement still in general use today. Previous experience had convinced paleontologists that earth history fell naturally into books and chapters, each characterized by its organic remains and delimited by widespread interruptions in the sequence of fossils.

The usefulness of this method of classification has not been a matter of serious dispute, but many geologists, arguing that time and organic evolution are both continuous and uniform, have had reservations with respect to the periodicity inherent in the standard scale of geologic time. The purpose of this essay is to demonstrate that the history of life on which our geologic time scale is based, has been episodic rather than uniform, and to show that modern paleontology must incorporate certain aspects of both catastrophism and uniformitarianism while rejecting others.

Today, the concept of change in geologic history is almost universally accepted. Organic evolution, the transmutation of chemical elements, endless changes in geography, atmosphere, climate, and other aspects of physical and biological environments are taken for granted; but many geologists, still influenced by Lyell's views on uniformitarianism, think of most of these changes as smooth and gradual rather than episodic, uniform and predictable rather than variable and stochastic.

The stratigraphic record supplies abundant evidence that physical and biological processes have fluctuated greatly in extent and rate in the past, at times certainly exceeding limits observed during the course of human history. Environments have changed unceasingly and biologic reactions to those environments likewise have varied. Present conditions have unique aspects of relief, climate, and biota, and therefore cannot be regarded as an average sample of geologic history. Hence the present resembles the past only in the special sense of the timelessness of many geologic and biologic processes. But conditions in space and time do change, and the same processes operating in a context of changing environments will, of course, produce varied results (Simpson, 1963; Kitts, 1963; Gould, 1965). The search for analogies between past and present must take these changes into account. On the other hand, the intuitive conclusion that many of the *basic principles* of biology, physics, and chemistry are unaffected by the passage of time gains some support from repeated direct observations and from experimentation.

Geologists generally agree that the present provides many keys to the past. But this concept should not be equated with the more limiting principle of strict uniformity. Geological uniformity, as was pointed out long ago (*e.g.*, by Kelvin, 1864; Huxley, 1869; Sollas, 1900), implied to many constancy instead of evolution and minimized the great secular and episodic changes—the revolutions—that highlight earth history. Lyell and Darwin did not believe in world revolutions.

In the development of geology, the concepts of catastrophe and uniformity have both been useful. Formerly regarded as diametrically opposed, they have gradually become modified and blended together so that there is no longer a clearly defined distinction between them. If we overlook mystical and religious overtones, catastrophism rightly emphasized the episodic character of geologic history, the rapidity of some changes, and the difficulty of drawing exact analogies between past and present. Aside from the objectionable implications of constancy of rate, uniformitarianism (*actualism* of European writers) stressed the application of scientific principles to geologic history. Many modern geologists unconsciously embrace the best features of both points of view (Bülow, 1960). Uniformitarianism and catastrophism, however, no longer exist in their original form as separate doctrines. They have been replaced by the concept of an evolving universe in which erratic changes, conditioned by pre-existing states, take place at greatly fluctuating rates.

The cataclysmic working of certain natural processes was an aspect of, but not peculiar to, Cuvier's philosophy of catastrophic changes. For example, the pioneers of uniformitarianism, Mikhail Lomonosov (1711–1765), James Hutton (1726–1797), C. E. A. von Hoff (1771–1837), and Charles Lyell (1797–1875), all regarded floods, volcanic eruptions, and earthquakes as normal geologic phenomena. Herein lies the semantic confusion between the two philosophies. The most significant element in Cuvier's catastrophism was, however, not suddenness of change, but that observed conditions were considered to be inadequate to explain the history of the earth (Cuvier, 1812). This view was completely consistent with the belief generally held in his day that the earth was only a few thousand years old. He believed that sufficient time simply was not available to explain the earth by known processes. However, the supernaturalism and multiple creations envisioned by many catastrophists did not stem from Cuvier, who regarded new faunas as migrants from remote regions. He used uniformitarian methods of analogy when he drew detailed and accurate comparisons of anatomy in fossil and living animals.

I think that the evidence requires the conclusion that many significant episodes in geologic history took place during comparatively brief intervals of time and that some of these probably involved unusual conditions for which there are no modern close parallels. Growing acceptance of this view has been regarded by the Soviet geologists N. S. Shotski and D. L. Stepanov (Stepanov, 1959) as a trend toward *neo-catastrophism*. This, however, like catastrophism, is a term with an emotional connotation that implies calamity and destruction, and as such it is not appropriate in any scientific context.

On a local or even regional scale, many sedimentary processes operate intermittently and at high rates: for example, the emplacement of deposits of turbidity currents, landslides, and stream floods. Estimates of the average rates of accumulation of any sequence of sediments undoubtedly fall far short of maximum rates.

Zangerl and Richardson (1963), in a monumental study of the paleoecology of certain Pennsylvanian rocks in Illinois, have shown that great fresh-water swamps were flooded by the sea within a few hours or days without a gradual transition in salinity (Fig. 1). This was deduced from known rates of aerobic decomposition of animal tissues from which the rate of burial of fish carcasses was computed at about 1 mm in 5 days. The character of the associated fossil communities also shows that the increase in salinity was abrupt. It is salutary to keep in mind the fact that many geologic processes today are highly uneven in their operation; there is no *a priori* reason to think that rates known today are representative for all of geologic history.

[*Editor's Note:* Figure 1, which shows an idealized Mid-Continent Pennsylvanian cyclothem, has been omitted.

In pages 66 to 69 Newell briefly discusses the historical development and principles of biostratigraphy.]

Mass Extinctions

Rejecting catastrophism, Darwin (1859, p. 281) thought that the suddenness with which some major groups, such as the ammonites, disappeared was more apparent than real and attributable to deficiencies in the record. Unconformities and barren sequences do enhance the sharpness of widespread faunal interruptions, but, the magnitude of a paleontologic discontinuity does not bear a simple relationship to the time represented by a barren interval or an unconformity.

Physical gaps in a sedimentary sequence range from insignificant diastems to breaks of enormous temporal value. In the more stable regions of the cratons, they commonly are bounded by parallel sequences of strata, sometimes loosely termed conformable because of the lack of conspicuous angularities. Physical gaps in the record result from geographic and climatic changes which may be sufficiently severe to alter greatly the character of organic communities through changing environments and migrations. Hence an unconformity of small temporal value may be marked by a great paleontologic change not directly attributable to evolution. Evolutionary changes are likely to be conspicuous only after an appreciable lapse of time—hundreds of thousands of years for rapidly evolving species and millions of years for rapidly evolving genera.

Can stratigraphic breaks of short duration be widely distributed? We lack the means to measure an hiatus accurately, but circumstantial evidence suggests that obscure paraconformities and compressed sequences marked by striking paleontologic discontinuities are abundant in the fossil record and may, indeed, be widespread and thus significant in terms of earth history. The meaning of these discontinuities constitutes a major problem for which many observational data are needed.

The boundaries at the base and top of the Mesozoic Era are in many places paraconformities where the reworking of sediments over very flat surfaces produced contacts that are now obscure (Newell, 1967).

Paleontologists are sometimes charged with inventing invisible unconformities of convenience, but in many circumstances the paleontologic and paleogeographic evidence taken together suggest at least short intervals of nondeposition or erosion without conspicuous channeling, leaching, or other unequivocal indications of exposure to subaerial agencies. Studies of the physical relationships at these contacts usually provide no basis for judging the significance of the stratigraphic interruption. Stratigraphic breaks, in any case, can be evaluated only by reference to intermediate strata and transitional fossils preserved in contiguous areas.

Continued field studies show, however, that many persistent paleontological discontinuities are not being bridged by new discoveries. Consequently these breaks in the record are important both historically and stratigraphically. As in the days of Lyell and Darwin,

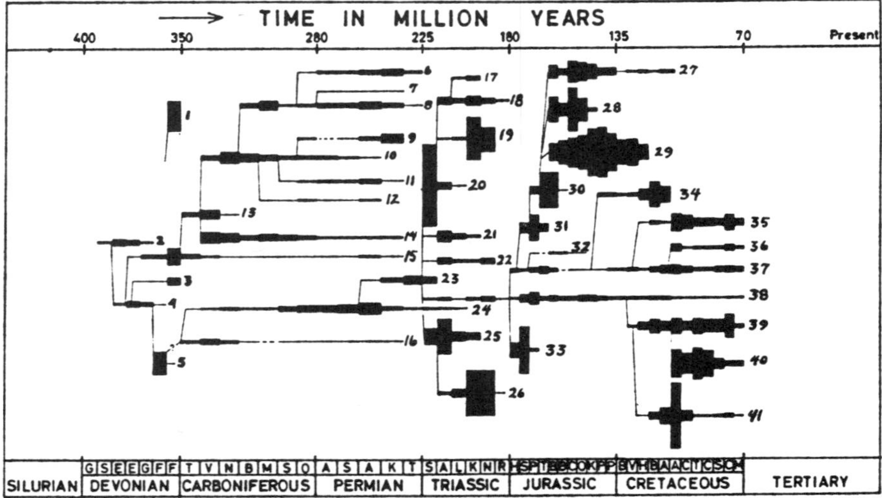

Figure 2. History of the 41 superfamilies of ammonoids showing mass extinctions at the close of the Devonian, Permian, Triassic, and Cretaceous periods. Height of the blocks is proportional to the number of genera known from each stage. Number 26 represents 48 genera (*after* House, 1963). Letters in boxes above the names of the periods are the initial letters of the standard stages.

they constitute a major problem in the history of life. Cuvier thought that mass extinctions of entire faunas took place quickly. Judging from the evidence, many of them did. In other cases they were simply the culminating phases of long declining trends, and we should focus attention on the trends as well as the final disappearance.

The history of the ammonoids is well known and may be taken as an illustration of the uneven course of a major group of animals (Fig. 2). They arose out of the nautiloids in the late Silurian, diversified slowly to a maximum of 10 superfamilies by mid-Permian time, then quickly dropped to two superfamilies by the close of that period. One of these repopulated the Triassic seas and soon underwent explosive radiation, increasing in diversity until near the close of the period when the order suddenly was reduced again, this time to a single surviving superfamily. The early Jurassic was the scene of a third major evolutionary burst, followed by a recession in the early Cretaceous and a minor flowering again in the late Cretaceous. Finally, after a premonitory

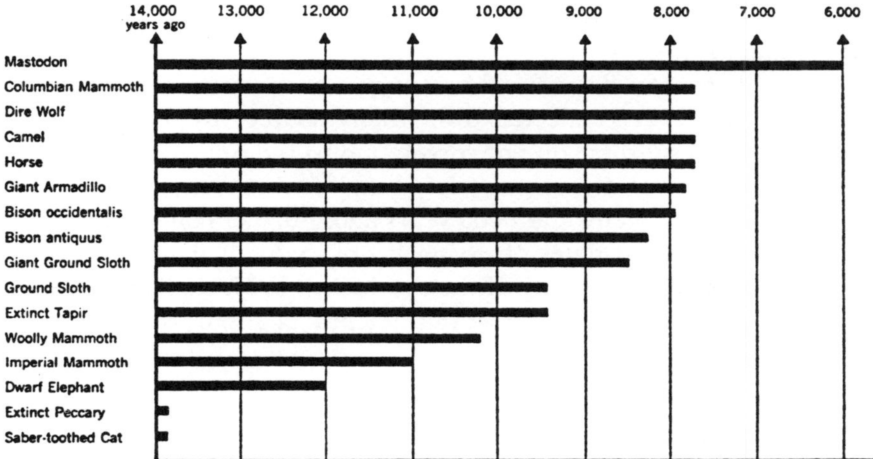

Figure 3. Late Quaternary extinctions of North American mammals based on radiocarbon dates (*from* MacGowan and Hester, 1962, courtesy of Doubleday and Company, Inc.)

dwindling of numbers of genera in the latest Cretaceous, the seven remaining superfamilies dropped out together, along with representatives of a host of other animals. As might be expected, the periods of mass extinction were relatively brief as compared to the expanding phases. It appears that these events took place within the time span of a single stage—say, 5 or 6 million years at most. This establishes the maximum interval within which the extinctions probably took place, but the time may have been much shorter.

A fascinating illustration of mass extinction was the disappearance outside of Africa of very many of the larger grazing and browsing mammals around the end of the Pleistocene epoch. In North America, for example, roughly three quarters of the larger herbivorous mammals died out within a very short time, geologically speaking (Fig. 3).

This event apparently was compressed mainly between the Wisconsin glacial maximum, some 11,000 years ago, and the close of the hypsithermal epoch, some 3000–4000 years ago. Much evidence remains to be collected and interpreted about this event, but the general facts now are clear (Martin, 1958). Many large mammals of quite diverse habitats were affected but there were very few extinctions among other animals or plants. The cause of this very recent event is unknown; doubtlessly it is significant that it corresponds in time to a marked drying of climate over much of the world and rapid advances in human technology.

The radiocarbon dates for the youngest known occurrences of the extinct species were collected from many sources by James Hester (1960). Obviously these do not correspond to the time of death of the last representatives of each species, but it is interesting that many dates fall between 7000 and 8000 years ago. Further collecting should reduce the apparent dates of extinction of many of the species. The data clearly show that many of the extinctions in North America took place within an interval so short that for practical purposes it was instantaneous.

The Fossil Record of Animal Revolutions

Notwithstanding the imperfections of the fossil record and the limitations of paleontologists, innumerable inductive conclusions have been confirmed so often that the chances of future contradictions seem small. That the probability of contradiction is, however, never zero, is shown by the recent discoveries of living coelacanths, monoplacophorans, and other surviving organisms long believed to have been extinct. It is, of course, possible, but not probable, that some ammonites and dinosaurs lived beyond the end of the Cretaceous Period. As in the physical sciences, conclusions in paleontology are statements of probability, not certainty. Everything is conditioned by the chance factors of evolution, circumstances of preservation, and the fortunes of discovery. But even if future discoveries show that a few stragglers persisted beyond the Cretaceous-Paleocene boundary, the image of a late Cretaceous crisis in the animal world would not be much affected. Experience suggests that such stragglers probably would show evolutionary differences from typical Cretaceous forms. The age of the enclosing rock would not be judged solely from the

TABLE 4. FIRST AND LAST KNOWN OCCURRENCES OF MAIN ANIMAL GROUPS

			First Occurrence		Last Occurrence	
Series	Duration (million years)	Total Families	Families *in* per cent	Families per million years	Families *in* per cent	Families per million years
Neogene	32.5	856	17	.52	10	.30
Paleogene	32.5	834	44	1.35	15	.46
Upper Cretaceous	35.0	629	26	.74	26	.74
Lower Cretaceous	35.0	520	35	1.00	11	.31
Upper Jurassic	15.0	377	15	1.00	11	.73
Middle Jurassic	15.0	346	14	.93	7	.47
Lower Jurassic	15.0	328	50	3.33	10	.66
Upper Triassic	16.6	278	19	1.14	35	2.11
Middle Triassic	16.6	254	25	1.50	10	.60
Lower Triassic	16.6	235	36	2.17	18	1.08
Upper Permian	25.0	303	21	.84	50	2.00
Lower Permian	25.0	330	15	.60	27	1.08
Upper Pennsylvanian	10.0	315	11	1.10	11	1.10
Middle Pennsylvanian	10.0	318	5	.50	5	.50
Lower Pennsylvanian	10.0	308	6	.60	3	.30
Upper Mississippian	17.5	318	8	.46	9	.51
Lower Mississippian	17.5	315	28	1.60	7	.40
Upper Devonian	20.0	327	14	.70	30	1.50
Middle Devonian	20.0	326	15	.75	13	.65
Lower Devonian	20.0	298	21	1.05	7	.35
Upper Silurian	6.6	272	8	1.21	13	1.96
Middle Silurian	6.6	265	13	1.96	5	.75
Lower Silurian	6.6	243	28	4.24	5	.75
Upper Ordovician	25.0	228	10	.40	24	.96
Middle Ordovician	25.0	221	33	1.32	7	.28
Lower Ordovician	25.0	192	76	3.04	23	.92
Upper Cambrian	33.3	97	34	1.02	52	1.56
Middle Cambrian	33.3	82	50	1.50	22	.66
Lower Cambrian	33.3	55	100	3.00	25	.75

presence or absence of ammonites and dinosaurs, but from the character of the entire biota.

It is consistent with the uniformitarian doctrine to interpret abrupt first and last appearances of fossil faunas in local sequences as the effects of migrations, hiatus, or contrasting depositional environments. As Schindewolf has pointed out however, where these abrupt changes occur in relatively complete sequences over a large part of the earth, they indicate episodes of greatly increased rate of extinction and evolution (Schindewolf, 1955).

Table 4 and the graphs of Figures 4–7 show fluctuating extinction rates in the main groups of fossil animals as inferred from available information on stratigraphic ranges. The plotted data represent not total extinctions but percentages of the families that drop out in each stratigraphic series.

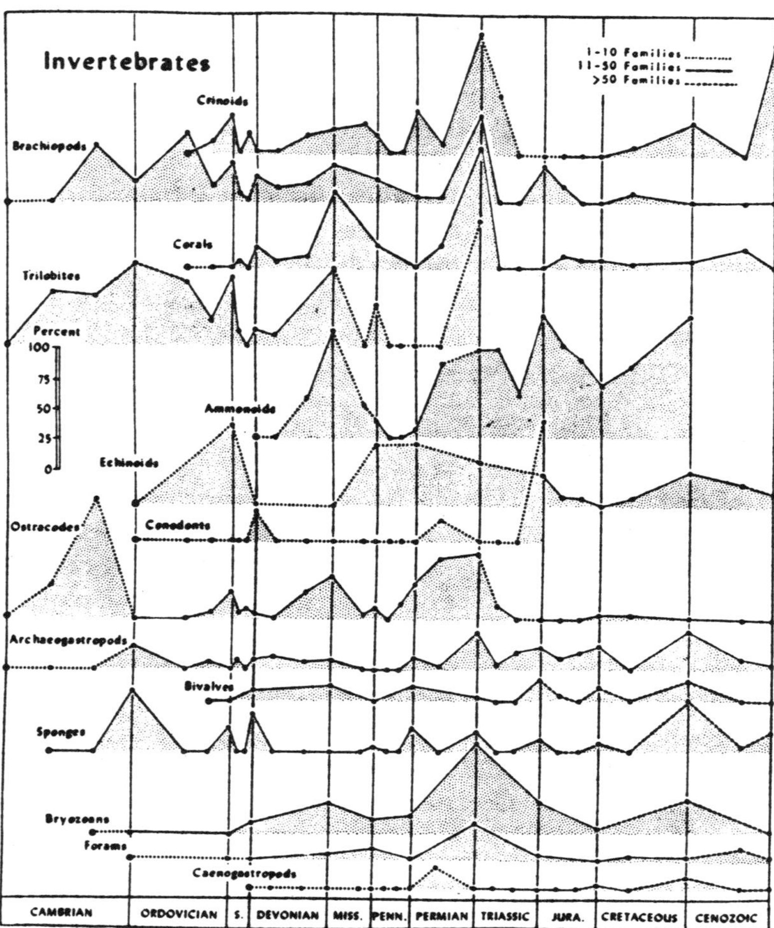

Figure 4. Percentage extinction of families of the chief groups of fossil invertebrates. There is strong correlation of extinction peaks near the end of the Devonian, Permian, Triassic, and Cretaceous periods.

There are, of course, two kinds of extinctions. Even though a family ceases to exist, it still may have contributed to the heredity of a descendant family. But a different kind of extinction is involved when a family dies without descendents. In paleontologic studies it frequently is difficult or impossible to distinguish between these two and no attempt has been made to do so in the present study.

Estimates of rates of change in fossils based on taxonomic groups are subject, of course, to many limitations and errors. Fossils provide only an incomplete and biased record of past life, limited mainly to the forms of shallow waters and lowlands. The reported stratigraphic ranges probably generally are less than the true ranges of the original organisms. Apparent ranges are affected by the intensity and extent of collecting, by environmental facies, accidents of preservation and discovery, and by taxonomic treatment, which may vary with different authors. For example, splitting tends to shorten ranges and lumping may extend them. Furthermore, the degree of stratigraphic precision in reporting ranges has much bearing on the result. These sources of error are recognized here, but certain over-all regularities within the data suggest that they are meaningful and allow some general conclusions.

In the first place, the data agree closely with the common knowledge that the era and certain other stratigraphic boundaries are sharply defined by mass extinctions. Secondly, there is an intriguing coincidence of peaks of extinction in quite unrelated groups. For example, marine invertebrates (Fig. 4) show accelerated extinction near the close of the Cambrian (trilobites, archaeogastropods, and sponges), late Ordovician (crinoids, brachiopods, trilobites, and echinoids), late Devonian (brachiopods, corals, trilobites, ammonoids, ostracodes, archaeogastropods, bryozoans, and foraminifers), late Triassic (brachiopods and ammonoids), and late Cretaceous (crinoids, ammonoids, echinoids, archaeogastropods, bivalves, sponges, and bryozoans). The lower vertebrates (Fig. 5) also show coincident peaks of extinction in the late Devonian, late Permian, late Triassic and late Cretaceous. There is an intriguing parallelism in extinction patterns of ammonites and reptiles (Fig. 6).

These relationships are even clearer if the data on all animal groups are synthesized. In Figure 7, the episodes of extinction are shown (solid line) for all the families (a total of some 2250) of the groups represented in the preceding figures. Taking all major fossil groups of animals into consideration, the most critical times of extinction occurred near the close of the periods, especially the late Cambrian (52 per cent extinction); late Devonian (30 per cent extinction), late Permian (50 per cent extinction), late Triassic (35 per cent extinction), and late Cretaceous (26 per cent extinction). During relatively short spans, one fourth to a half of all the families of animals of the world disappeared. The average extinction rate for all the geologic series was about 17 per cent, and the minimum about 3 per cent. As might be anticipated from the involved history of classification of the geologic systems, the Silurian-Devonian boundary, the two boundaries of the Pennsylvanian, and the upper boundary of the Jurassic are not as well characterized paleontologically as are some of the others.

Times of accelerated extinction were followed by episodes of ex-

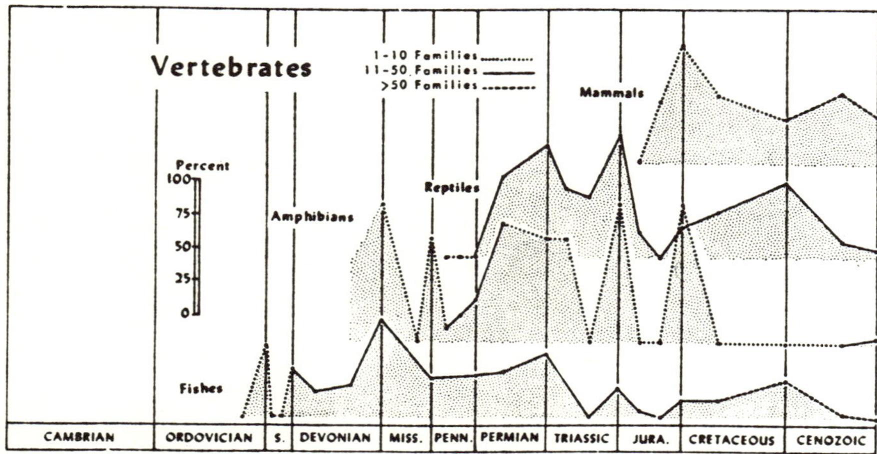

Figure 5. Percentage extinction of families of the chief groups of fossil vertebrates. There is strong correlation of extinction peaks near the end of the Devonian, Permian, Triassic, and Cretaceous periods.

ceptional radiation into vacated and new niches. It often has been noted that the pace of evolutionary diversification increases during and immediately after an episode of mass extinction (Fig. 7, broken line). This would appear to be an effect of the freeing of ecologic niches for reoccupancy by new forms added to the steady multiplication of new microhabitats consequent on the spread of life itself. Thus there appears to be a tendency for oscillations between times of accelerated extinction, which in many cases have been taken as the closing phases of chapters of geologic history, and episodes of radiation at the beginning of the subsequent intervals. This is true of many of the more distinctive series and stages, which provide subdivisions of the systems.

These tabulations were based on families of animals at the scale of stratigraphic series. More detailed tabulations of the fossils of separate stages or zones certainly would also show sporadic replacement of faunas at the taxonomic level of species or genera. Seventy-five per cent or more replacement of species in successive faunas is not exceptional in the fossil record. This may occur at a bedding plane or above an interval of barren strata. The abrupt replacement of old by

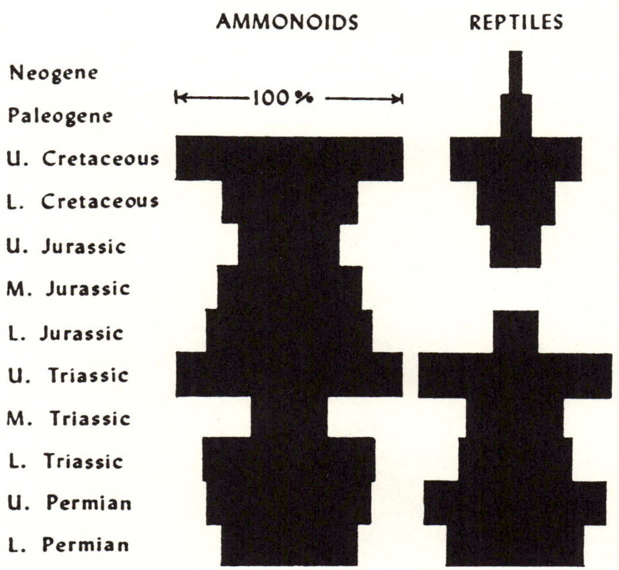

Figure 6. Parallelism in extinction of families of ammonoids and reptiles

new faunas is so common in the fossil record as to suggest general and recurring causes capable of simultaneously affecting diverse groups of animals.

The losses by extinction were quickly replaced and over-all diversity increased by continued evolution of surviving stocks (Fig. 8). There is some suggestion that their average increase was exponential at times since the early Cambrian. It seems probable that new organisms, by providing new microenvironments and new opportunities for further radiation, have stimulated evolution in a sort of ecological feed-back. A long pause occurred in this expansion during the Triassic period, as shown in Figure 8, and another may be taking place at the present as a result of the late Quaternary and modern extinctions. Very probably there always has been at least brief reduction in diversity during an episode of mass extinction, but the stratigraphic column is not so narrowly zoned, nor our correlations so precise, that we can measure exactly the duration of episodes of decreasing or increasing diversity. The calculated trends represent a greatly generalized average.

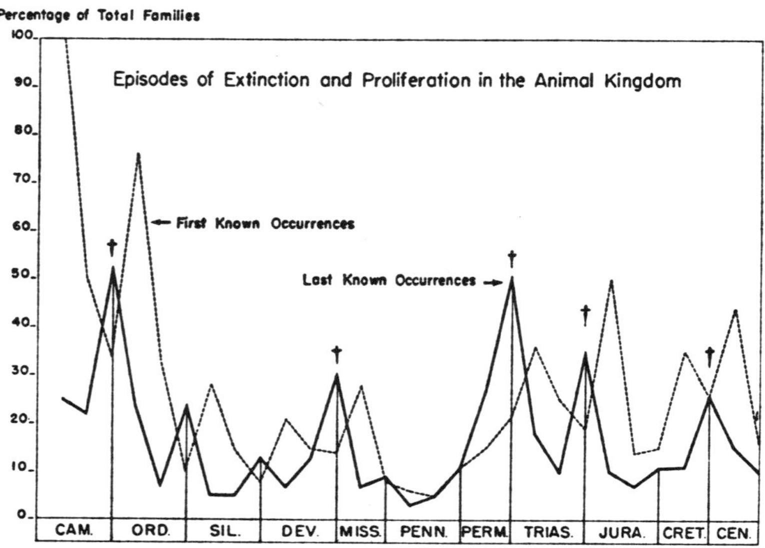

Figure 7. Percentage of first and last appearances of animal families through geologic time. Main episodes of extinction were near the close of the Cambrian, Devonian, Permian, Triassic, and Cretaceous periods.

The record of fossil plants is not nearly as well known as that of animals (Fig. 9). No detailed modern survey of the stratigraphic distribution of fossil plants is available; hence, we are unable to attempt correlations between the histories of animals and plants. Since animals are dependent on plants for food and shelter, there must have been many parallel developments in the two kingdoms. The spread of prairies and coincident adoption of the grazing habit by several groups of mammals in mid-Tertiary times may be cited as one example; the simultaneous deployment of pollinating insects and flowering plants during the Cretaceous period is another.

It seems that the greatest events in the plant kingdom were con-

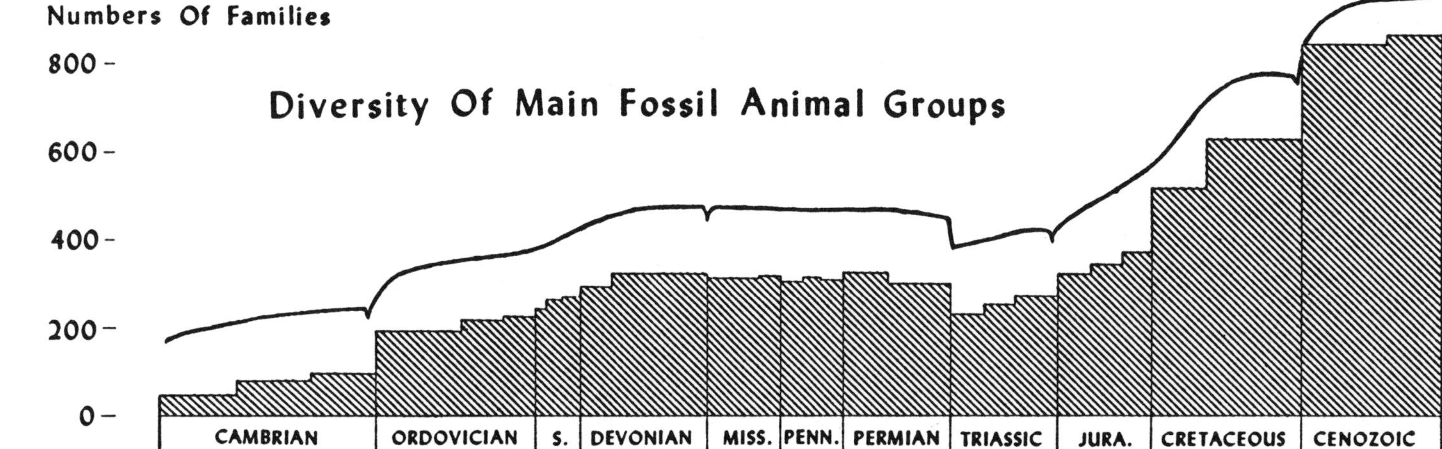

Figure 8. Expansion of the animal kingdom during geologic time as expressed in numbers of known families of the main fossil groups. Heavy line indicates inferred fluctuations during most critical times of faunal revolution.

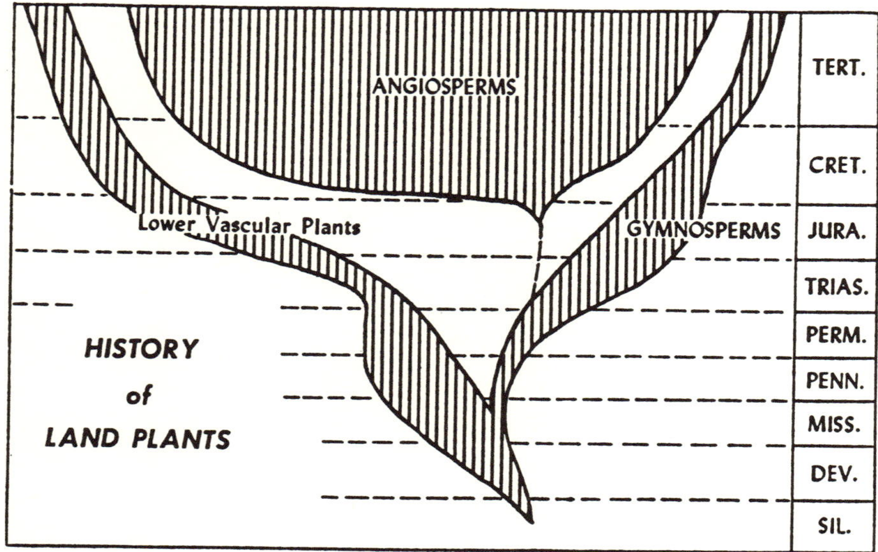

Figure 9. Trends in plant evolution (*after* Dorf, 1955). Apparently, none of the great events in plant history coincides with major revolutions in the animal kingdoms.

cerned with adaptive radiation and not with mass extinction, *e.g.*, the colonization of the lands by primitive vascular plants in the early and middle Paleozoic, the rise and spread of the gymnosperms in the Carboniferous, and the explosive radiation of the angiosperms in the early Cretaceous. It appears that the plants as a whole have been more conservative and ecologically more resilient than animals.

[*Editor's Note:* In pages 81 to 88 Newell discusses possible causes of animal revolutions. See Editor's Comments for Part VI.]

Conclusions

The fossil record shows that animal history has been marked by brief episodes of exceptionally high rates of extinction commonly accompanied and followed by somewhat more gradual evolutionary recovery. These revolutions were largely independent of the much more regular trends in plant history. The differences in the histories of animals and plants suggest that the latter were on the average more conservative and tolerant of environmental change than were animals.

All evidence indicates that episodes of mass extinction may occur within a few hundreds or thousands of years, while the development of a new replacing fauna probably requires hundreds of thousands or millions of years. There is nothing in the record to give support to catastrophism, as Cuvier understood it, nor to the literal uniformity of Lyell which emphasized slow and uniform instead of episodic changes. Yet the record of past revolutions in the animal kingdom is understandable by application of basic principles of modern science. In this sense, the present is the key to the past.

References Cited

BEURLEN, K., 1956, Der Faunenschnitt an der Perm-Trias grenze: Zeitsch. deutsch. Geol. Ges., v. 108, p. 88–99

BRAMLETTE, M. N., 1965, Problem of the massive extinctions in the biota at the end of Mesozoic time: Science, v. 148, p. 1696–1699

BULOW, K. VON, 1960, Der Weg des Aktualismus in England, Frankreich und Deutschland: Geol. Ges. Deutschl. Dem. Rep., Ber., v. 5, p. 160–174

CHORLEY, R. J., 1963, Diastrophic background to twentieth-century geomorphological thought: Geol. Soc. America Bull., v. 74, p. 953–970

CLOUD, P. E., JR., 1959, Paleoecology—Retrospect and prospect: Jour. Paleontology, v. 33, p. 926–962

CUVIER, L. C. F. D. G., 1812, Recherches sur les ossemens fossiles de quadrupèdes, discours préliminaire, tome 1: Paris, Deterville

DARWIN, C., 1859, On the origin of species by natural selection: London, John Murray, 490 p.

D'ORBIGNY, A. D., 1841 (1840), Paléontologie francaise, terrains crétacés, tome 1: Paris, 662 p.

DORF, E., 1955, Plants and the geologic time scale, p. 575–592 *in* Poldervaart, Arie, *Editor*, Crust of the Earth: Geol. Soc. America Special Paper 62, 762 p.

DUNBAR, C. O., and RODGERS, J., 1957, Principles of stratigraphy: New York, John Wiley & Sons, 356 p.

ELTON, C. S., 1958, The ecology of invasions by animals and plants: London, Methuen & Co., Ltd., 181 p.

FARRAND, W. R., 1961, Frozen mammoths and modern geology: Science, v. 133, p. 729–735

FISCHER, A. G., 1965, Brackish oceans as the cause of the Permo-Triassic marine faunal crisis, p. 566–575 *in* Nairn, A. E. M., Problems in palaeoclimatology: London, Interscience Publishers, 705 p.

GILLULY, J., 1949, Distribution of mountain building in geologic time: Geol. Soc. America Bull., v. 60, p. 561–590

GOULD, S. J., 1965, Is uniformitarianism necessary?: Am. Jour. Sci., v. 263, p. 223–228

HAPGOOD, C. H., 1958, Earth's shifting crust: New York, Pantheon Books, Inc., 438 p.

HENBEST, L. G., *Editor*, 1952, Significance of evolutionary explosions in geologic time (A symposium): Jour. Paleontology, v. 26, p. 298–394

HESTER, J. J., 1960, Late Pleistocene extinction and radiocarbon dating: Am. Antiquity, v. 26, p. 58–77

HOUSE, M. R., 1963, Bursts in evolution: Adv. Sci., v. 19, p. 499–507

HUXLEY, T. H., 1869, The anniversary address of the president: Geol. Soc. London, Quart. Jour., v. 18, p. xxviii–liii

KELVIN, W. T., 1864, On the secular cooling of the earth: Roy. Soc. Edinburgh, Trans., v. 23, pt. 1, p. 157–169

KITTS, DAVID B., 1963, The theory of geology, p. 49–68 *in* Albritton, Claude C., Jr., *Editor*, The fabric of geology: Stanford, Calif., Freeman, Cooper, & Co., 372 p.

LARRABEE, E., 1963, Scientists in collision: Was Velikovsky right?: Harper's Mag., v. 227, p. 48–55

MACGOWAN, K., and HESTER, J. A., JR., 1962, Early man in the New World: New York, The Natural History Library Anchor Books, Doubleday & Co., 333 p.

MARTIN, P. S., 1958, Pleistocene ecology and biogeography of North America, p. 375–420 *in* Zoogeography: Am. Assoc. Adv. Sci., Pub. 51

MAYR, E., 1965, Avifauna: turnover on islands: Science, v. 150, p. 1587–1588

NEWELL, N. D., 1962, Paleontological gaps and geochronology: Jour. Paleontology, v. 36, p. 592–610

—— 1963, Crises in the history of life: Sci. American, v. 208, p. 77–92

—— 1965, Mass extinctions at the end of the Cretaceous period: Science, v. 149, p. 922–924

—— 1966 Problems of geochronology: Philadelphia, Proc. Acad. Nat. Sci., v. 118, p. 63–89

—— 1967, Paraconformities, p. 349–367 *in* Teichert, C., and Yochelson, E. L., *Editors*, Essays in paleontology and stratigraphy, R. C. Moore Commemorative Volume: Lawrence, Univ. Kansas Press

PHILLIPS, J., 1841, Figures and descriptions of the Palaeozoic fossils of Cornwall, Devon, and west Somerset: London, Longman, Brown, Green, and Longman, 231 p.

SANDERSON, I. T., 1960, Riddle of the frozen giants: Saturday Evening Post, v. 232 (Jan. 16), p. 39

SCHATZ, A., 1957, Some biochemical and physiological considerations regarding the extinction of the dinosaurs: Pa. Acad. Sci. Proc., v. 31, p. 26–36

SCHINDEWOLF, O. H., 1955, Über die möglichen Ursachen der grossen erdgeschichtlichen Faunenschnitte: Neues Jahrb. Geol. und Paläont., Jahrg. 1954, p. 457–465

SIMPSON, G. G., 1940, Review of the mammal-bearing Tertiary of South America: Am. Phil. Soc. Proc., v. 83, p. 649–709

—— 1963, Historical science, p. 24–48 *in* Albritton, Claude C., Jr., *Editor*, The fabric of geology: Stanford, Calif., Freeman, Cooper & Co., 372 p.

SOLLAS, W. J., 1900, Evolutional geology: Science, v. 12, p. 745–756, 787–796

STEPANOV, D. L., 1959, Neocatastrophism in paleontology of these days (in Russian): Paleont. Zhur., Acad. Nauk, SSSR, No. 4, p. 11–16

UFFEN, R. J., 1963, Influence of the earth's core on the origin and evolution of life: Nature, v. 198, p. 143–144

VELIKOVSKY, I., 1955, Earth in upheaval: New York, Doubleday, 301 p.

WACE, N. M., 1966, Last of the virgin islands: Discovery, v. 27, p. 36–42

WELLER, J. M., 1960 Stratigraphic principles and practice: New York, Harper & Brothers, 725 p.

WILSER, J. L., 1931, Lichtreaktionen in der fossilen Tierwelt: Berlin, Gebrüder Borntraeger, 192 p.

ZANGERL, R., and RICHARDSON, E. S., JR., 1963, The paleoecological history of two Pennsylvanian black shales: Fieldiana—Geol. Mem., v. 4, 352 p.

33

Reprinted from pages 399 and 415–420 of *Carbonate Rocks. Origin, Occurrence and Classification*, G. V. Chilingar, H. J. Bissell, and R. W. Fairbridge, eds., Developments in Sedimentology 9A. Amsterdam: Elsevier, 1967, 471 pp.

CARBONATE ROCKS AND PALEOCLIMATOLOGY IN THE BIOGEOCHEMICAL HISTORY OF THE PLANET

RHODES W. FAIRBRIDGE

Columbia University, New York, N.Y. (U.S.A.)

SUMMARY

Physicochemical characteristics of most carbonate minerals are matched by the ecologic distribution of carbonate-secreting organisms to make the carbonate sediments of the present day strongly temperature dependent, and their concentration thus inversely related to degrees of latitude. It is believed that this correlation holds generally for the geological past; refinements added include determination of Ca/Mg, Ca/Sr, Ca/Fe/Ti ratios, the taxonomic habit of the organisms, and so on.

Going back in time, however, numbers of difficult problems arise: the size of the earth, the volume of ocean water, and its salinity, alkalinity, and pH. None of these can be uniquely tested at present. The patterns of carbonate sedimentation have changed in very important ways through time, so that a strictly uniformitarian status quo cannot be assumed for the past.

Available evidence suggests five great biogeochemical events, each corresponding to a certain threshhold peak: *Revolution I* (ca. $3.8 \cdot 10^9$ years), *First Life* at the atmosphere (CH_4, NH_3, H_2O)–water interface; *Revolution II* (ca. $2.9 \cdot 10^9$ years), *First Photosynthesis* with evolution of O_2, followed at an undetermined stage within the next billion years by the first primitive animals; *Revolution III* (ca. $6 \cdot 10^8$ years), *First Carbonate Shells*, appearing on organisms up to highest invertebrate level; *Revolution IV* (ca. $3 \cdot 10^8$ years), *Great Coal Age*, following long Paleozoic history of carbonate removal as limestone and the carbon as coal, marked by secular drop in p_{CO_2} and rise in p_{O_2}; and *Revolution V* (ca. $1 \cdot 10^8$ years), *First Carbonate Pelagic Foraminifera* (and coccoliths), shifting major site of carbonate sedimentation from neritic to abyssal realm and initiating major withdrawal of carbonates from the geologic cycle.

[*Editor's Note:* In pages 399 to 415 Fairbridge discusses paleoclimatic indicators (paleontological, stratigraphic, isotopic), the present latitudinal distribution of marine carbonates, the present range of continental carbonates, certain geochemical attributes of modern marine carbonates (Ca/Mg, Ca/Sr and Ca/Fe/Ti ratios), and geochemical variables in the past.]

BIOLOGICAL PROBLEMS OF THE PAST AND THE FIRST CARBONATES

There is a broad controlling rule in paleoclimatology that must not be broken. This is the "Law of biological continuity", which essentially states that inasmuch as life has continued throughout most of geological time, at no stage, not even for an instant, could the metabolic limits of any of the succeeding organisms be transgressed on a world-wide basis. Thus are ruled out many of the wilder flights of imagination, proposed as ad hoc hypotheses. Examples of these involve an earth totally frozen over, or involved in catastrophic planetary collisions or tidal moon-births; or even eustatic oscillations of such large dimensions that the salinity of the ocean would be seriously affected. "Biologic continuity" does not, however, imply evolutionary stagnation of the geochemical components. Far from it; there is increasing evidence that the natural environments, the earth's atmosphere and ocean with the present balance of atmospheric gases and ocean salts are in part the *result* of organic activity.

Revolution I

There is good reason for believing that the first organisms evolved in and from an atmosphere of methane (CH_4), ammonia (NH_3), and H_2O vapor. The appearance of this first life is *Revolution I* in earth history (say, about $3.8 \pm 0.3 \cdot 10^9$ years ago). Hydrogen originally present would gradually drift off into the upper atmosphere and away, and this would lead to the instability of methane and ammonia. The latter, in turn, will give rise to free nitrogen, which is the principal gas of our present atmosphere. Volcanic activity gradually added Cl^-, SO_2 and CO_2. As there was no soil in this early stage of earth history, rain-water solution would be slight (due to rapid runoff) and the supply of Ca^{2+} and Na^+ to the ocean rather meager. Ocean water would be slightly acid and salinity (NaCl) very low. Limestone formation was improbable at first. Any O_2 liberated by photo-dissociation of H_2O would immediately be withdrawn by mineral oxidation. Bacterial attack on SO_3^{2-} led to production of H_2S. Reducing conditions permitted easily oxidized minerals (FeS_2, UO_2, etc.) to exist as detrital sediments (RAMDOHR, 1957).

Revolution II

At a certain stage, it is evident that primitive chlorophyllic organisms (probably bacterial autotrophs) began to use CO_2 and liberate oxygen as a by-product of sugar synthesis. The beginning of this *Revolution II* ("Eparchean") probably took place shortly before the time of the oldest radio–isotope–dated stromatolitic limestones, say about $2.9 \pm 0.2 \cdot 10^9$ years ago. This removal of CO_2 from solution in sea water would raise the pH, especially if it took place in shallow coastal lagoons, and $CaCO_3$ would be precipitated. The structures in the early carbonates, however, do not suggest evaporites or primary precipitates; they are probably organogenic rocks. The first carbonate fossil traces are stromatolitic algal structures, and are found all over the world in Precambrian formations, the oldest of which are the Bulawayan dolomites that are over $2.7 \cdot 10^9$ years old by absolute dating methods. Measurements of the $^{12}C/^{13}C$ ratios in these rocks may be helpful (WICKMAN, 1956); organic carbonates tend to be enriched in the ^{13}C isotope. The stromatolites (*Collenia*, *Cryptozoon*, etc.) are not skeletal material secreted by multicellular plants, but are wrinkled mats of $CaCO_3$ precipitated against the outer surface of unicellular green Algae colonies, just as observed today in Florida (GINSBURG, 1960) and in Shark Bay, W. Australia (LOGAN, 1961). The growth of this mat in shallow lagoons may well have sheltered the early cyanophytes from excess UV radiation. This revolution gradually led to the build-up of an oxygen-rich atmosphere, though initially all the O_2 would have been reabsorbed by mineral oxidation. RUTTEN (1965) suggested that about $1.6 \cdot 10^9$ years ago the O_2 level reached 0.01 P.A.L. (present atmospheric level). The so-called "Pasteur Level" (p_{O_2} at 1% of the present atmospheric level) was probably achieved only at the end of the Precambrian (HOLLAND, 1962; BERKNER and MARSHALL, 1964). Nevertheless, it permitted the evolution of the ancestors of all modern animals. Rutten suggested that the first primitive fauna began to evolve about $1.0 \cdot 10^9$ years ago.

The end of the Precambrian era is sometimes called the "Lipalian interval" (WALCOTT, 1910) to cover an imaginary epoch, when modern shell-bearing invertebrates were supposed to have evolved. This evolution must have been fantastically rapid and complex, for the Lower Cambrian discloses already the trilobites, representatives of the highest phylum of the invertebrates together with many other forms. Trilobites are, crudely speaking, ancestral to the modern horse-shoe crab, and such organisms are provided with a highly developed nervous system, brain, eyes, prehensile organs, digestive system, articulated skeletal system, complex musculature and a sophisticated bisexual reproductive system. It has been postulated that such organisms burst out from single-celled primitives in a brief evolutionary explosion at the very close of the Precambrian time to fill a newly created ecologic niche (CLOUD and ABELSON, 1961). Modern genetics offer no support for such a revolutionary event. In any case, the very existence of a world-wide stratigraphic gap corresponding to the Lipalian was disproven with the discovery of many fine, unmetamorphosed sedimentary sequences that spanned the whole interval, e.g., the Adelaide System of South Australia (DAVID and BROWNE, 1950; GLAESSNER, 1966).

One is forced, therefore, to conclude that the oxygen-breathing invertebrates evolved during an extended time-span of the middle and late Precambrian, although traces of these organisms are extremely sparse and restricted to some impressions, tracks and worm casts. The geochemical environmental conditions appear to have been reasonably acceptable for organisms similar to those living today with but one exception: the sea-water composition was such that they could not secrete carbonate skeletons (SCHINDEWOLF, 1956; CHILINGAR and BISSELL, 1963b; FAIRBRIDGE, 1964a). Examples of the fauna are known, and worm tracks are relatively common. The Ediacara fauna of South Australia, with its hydrozoa, etc., appreciably antedates the late Precambrian Sturtian Tillite (GLAESSNER, 1962; GLAESSNER and DAILY, 1959). The primitive segmented fauna, presumably arthropods, of the Belt Series of Montana, formerly considered of very late Precambrian or even Lower Cambrian age, is now considered to be a "Middle" Precambrian stage (GILLULY, 1963; PFLUG, 1965a). Also about one billion years old are the probable Foraminifera reported by PFLUG (1965b).

It would appear, therefore, that the middle and late Precambrian sea must have been acceptable for modern invertebrate life in every way, but for this peculiar feature about the carbonate shell secretion. Only the intertidal or lagoonal Algae seemed to have been able to lead to such precipitation. In an acid environment this process would be applicable. In lakes and rivers, modern fresh-water arthropods secrete chitinous shells (with only small clots of calcite) by raising the pH of their own body fluids, regardless of the low pH (5–7) of the environment. It seems possible, therefore, that up till the end of the Precambrian, the oceanic pH did not exceed 7 or a little over. As a matter of interest, all lakes in acid igneous rock areas today have a pH below 7 (HUTCHINSON, 1957); and HOUGH (1958) has demonstrated that the great silica-iron deposits that are known all over the world in the Middle and Late Precambrian, *but at no other time*, could be readily explained by the geochemistry of a lacustrine regime. A predictable low p_{O_2} would also favor this deposition (LEPP and GOLDICH, 1964) (Fig.13).

There seems to be good justification for the idea of BERKNER and MARSHALL

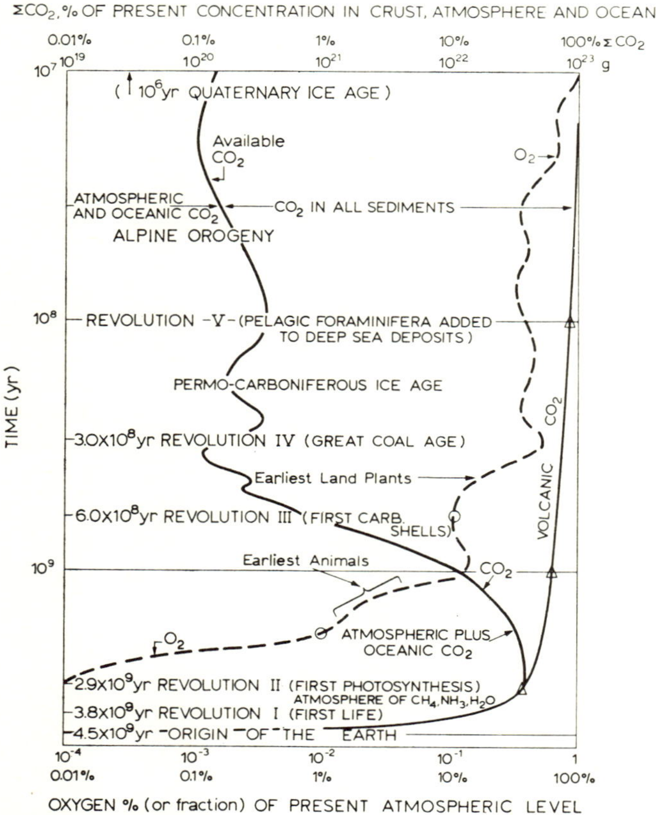

Fig.13. Great biological revolutions of the earth's history. Tentative suggestions of the variation in partial pressures of atmospheric oxygen (largely by photosynthesis; partly by photodissociation of H_2O) and CO_2 (from volcanic emanation).

(1964) that with the extremely thin O_2 and H_2O vapor atmospheric blanket in the early Precambrian, UV-synthesized ozone would be formed directly on the land surface. Inasmuch as ozone, rather like hydrogen peroxide, is a very strong oxidizing agent, the rate of chemical erosion would be very high and continue until such time as the atmospheric blanket is thickened. The production of the enormous silica-iron deposits would be greatly facilitated by such erosion. It must not be forgotten, however, that very important mechanical, unweathered deposits also occur during this period. Inasmuch as the iron deposits are not continuous, a series of oscillatory ozone build-ups may have alternated with less reactive epochs. Their world-wide distribution merely suggests a rather brackish, acid ocean. Alkalinity, however, may have been moderately high owing to the presence of borates and sulphates, as well as limited amounts of the principal sea salt (NaCl) which slowly increased through time. Late Precambrian gypsum and magnesite are known but there is no trace of halite evaporites.

Revolution III

Revolution III ("Eocambrian") occurred about $6 \pm 0.3 \cdot 10^8$ years ago. The first evaporites (halite) appeared in the Cambrian of Jordan, Iran, and India. At the beginning of the Paleozoic, CO_2 was being withdrawn in ever-increasing amounts (raising the pH), adequate concentrations of Ca^{2+} had now accumulated in sea water salts, and unlimited formation of calcite and aragonite shells became practicable for the invertebrates. It is possible that such shells were the inevitable consequences of a rising pH and rising Ca^{2+} concentration in sea waters, a natural sequence of events that could be predicted as the weathering and solution of silicate rocks on the continents gradually proceeded. Erosion during the Precambrian was dominantly mechanical (cf. graywackes and arkoses), but not exclusively so; during the Paleozoic time chemical erosion (soil development) became more important as the vegetation cover of the land surfaces grew richer and more varied.

During Paleozoic times, limestones became very widespread in a belt extending from North America, through western Europe and the U.S.S.R., to Australia. This seems to be a former equatorial belt that persisted (through most of that era), and it is quite unnecessary to propose the existence of climates on a world-wide basis that were warmer than, say, those of the mid-Tertiary. Indeed the mean temperature may have been somewhat cooler; and certainly in the early Paleozoic climates may have been somewhat more extreme, because there was no major vegetation to conserve and transpire moisture. Previously, no doubt, through much of the Precambrian time, there were probably soil bacteria, cyanophytic Algae, fungi, etc. (FISCHER, 1965), but they were probably limited to swampy places.

Land vegetation (major plants), according to the fossil record available so far, first evolved in Australia in Silurian times, and was widespread already in the Devonian. By Carboniferous times these vast swamp floras were providing for the greatest coal deposits of geologic history. In the Permian the locus of the great coal swamps turned to the Southern Hemisphere, but after the end of the Paleozoic, coal development in general became more restricted. The total coal preserved today is about $6 \cdot 10^{12}$ metric tons, but much has been lost by erosion.

Geochemically one may observe that during the Paleozoic, certain animals (e.g., corals) and plants (Algae, and coal swamp flora) were responsible for withdrawing very large quantities of carbon from the atmosphere–hydrosphere, and additional supplies of O_2 were liberated. Both the C (coal and oil) and $CaCO_3$ (limestones) by sedimentological processes became buried on a semi-permanent basis. Thus a new, biologically activated pattern was set up, on a scale that was greater than that which existed during the Precambrian time. There is no reason to imagine that volcanic supply of CO_2 rose to meet the demand; indeed, it may have been reduced in comparison to the Precambrian rate. On the other hand, the spreading of vegetation onto the land permitted the development of soils and their associated bacteria, which provided for a much more continuous contact of H_2CO_3

with rock surfaces than had ever been possible under Precambrian conditions. Thus the soil chemistry was changing to accelerate the supply of the ions of the principal silicate minerals to the sea: Ca^{2+}, Na^{+}, Mg^{2+}, preferentially over K^{+}, Fe^{2+}, etc.

Revolution IV

The time of *Revolution IV* ("Permo-Carboniferous") was $2.5 \pm 0.3 \cdot 10^8$ years ago. In the ocean it was marked by a rising alkalinity from the supply of alkali metals and, due to the removal of atmospheric and oceanic carbon, there may have been a lowering of the general p_{CO_2}. Thus there was a rise of oceanic pH, perhaps from 7 to about 8. During this transition certain marine invertebrate groups became extinct. The end of the Permian was not a time of wholesale slaughter of Paleozoic life, but rather marked a threshhold or end-point in a long series of geochemical changes that proved too much for certain taxonomic classes.

Revolution V

It seems that a "modern" atmospheric and oceanic geochemistry had been established early in the Mesozoic. Primitive land mammals, which require oxygen and a low atmospheric p_{CO_2} were able to evolve. Nevertheless, in the late Mesozoic a new revolution was developing in the ocean. This concerned principally the evolution of small floating organisms which secrete carbonate tests. This represents, then, the last great biogeochemical event, *Revolution V* ("Cretaceous"), dated about $1.0 \pm 0.2 \cdot 10^8$ years ago. For some reason, not well understood, small, pelagic carbonate organisms do not appear to have existed in the Paleozoic. They only gradually emerged during the Mesozoic, and exploded in an immense "flowering" all over the world oceans in the Cretaceous time. These organisms, mainly the Coccolithophoridae and pelagic Foraminifera, are the principal components of the famous Cretaceous chalk formations. They are found from the cliffs of Dover to Australia and from Texas to Antarctica. The high rate of $CaCO_3$ withdrawal is reflected by the mean Ca/Mg ratio of the Cretaceous carbonates of about 56 (weight ratio) compared with 16 in the late Paleozoic. Since Cretaceous time it has dropped to about 40. That such an event must have fundamentally changed the earth's carbonate economy was recognized by KUENEN (1950). The earlier limestones were neritic and epicontinental, and therefore, very liable to be recycled, whereas the pelagic carbonates are oceanic and almost permanently withdrawn from surface circulation (see also CHILINGAR, 1956). It should be remembered, however, that below the "calcium carbonate compensation depth" (now about 4,500 m on the average), all carbonate is redissolved. This level is probably set by a self-adjusting mechanism, related to Ca^{2+} and HCO_3^- fluxes.

[*Editor's Note:* In the remainder of this paper Fairbridge discusses principally some biological problems in paleoecology, and possible causes of fluctuations of epicontinental shore-lines.]

REFERENCES

Berkner, L. V., and L. C. Marshall. 1964. The history of growth of oxygen in the earth's atmoshere. In *The Origin and Evolution of Atmosphere and Oceans,* B. J. Brancazio and A. G. W. Cameron, eds. New York: Wiley, pp. 102–126.

Chilingar, G. V. 1956. Relationship between Ca/Mg ratio and geologic age. *Am. Assoc. Petroleum Geologists Bull.* **40**:2256–2266.

Chilingar, G. V., and H. J. Bissell. 1963b. Note on possible reason for scarcity of calcareous skeletons of invertebrates in Precambrian formations. *Jour. Paleontology* **37**:942–943.

Cloud, P. E., and P. H. Abelson. 1961. Woodring conference on major biologic innovations and the geologic record. *Natl. Acad. Sci. Proc.* **47**:1705–1712.

David, T. W. E., and W. R. Browne. 1950. *The Geology of the Commonwealth of Australia,* vol. 1, 747 pp; vol. 2, 618 pp; vol. 3, maps. London: Arnold.

Fairbridge, R. W. 1964a. The importance of limestone and its Ca/Mg content to paleoclimatology. In *Problems in Palaeoclimatology,* A. E. M. Nairn, ed. New York: Wiley, pp. 431–530.

Fischer, A. G. 1965. Fossils, early life, and atmospheric history. *Natl. Acad. Sci. Proc.* **53**:1205–1215.

Gilluly, J. 1963. The tectonic evolution of the western United States. *Geol. Soc. London Quart. Jour.* **119**:133–174.

Ginsburg, R. N. 1960. Ancient analogues of Recent stromatolites. *Internat. Geol. Congr. 21st, Rep.* **22**:26.

Glaessner, M. F. 1962. Precambrian fossils. *Biol. Rev.* **37**:467–494.

Glaessner, M. F. 1966. Precambrian palaeontology. *Earth-Sci. Rev.* **1**:29–50.

Glaessner, M. F. and B. Daily. 1959. The geology and late Precambrian fauna of the Ediacara fossil reserve. *South Australian Mus. Rec.* **13**: 363–401.

Holland, H. 1962. A model for the evolution of the earth's atmosphere. In *Petrologic Studies: A Volume in Honor of A. F. Buddington,* A. E. J. Engle, H. L. James and B. L. Leonard, eds. New York: Geol. Soc. America, pp. 447–477.

Hough, J. L. 1958. Fresh-water environment of deposition of Precambrian banded iron formations. *Jour. Sed. Petrology* **28**:414–430.

Hutchinson, G. E. 1957. *A Treatise on Limnology,* vol. 1. New York: Wiley, 1015 pp.

Kuenen, P. H. 1950. *Marine Geology*. New York: Wiley, 551 pp.

Lepp, H., and S. S. Goldich. 1964. Origin of Precambrian iron formations. *Econ. Geology* **59**:1025–1060.

Logan, B. 1961. *Cryptozoon* and associated stromatolites from the Recent, Shark Bay, Western Australia. *Jour. Geology* **69**:517–533.

Pflug, H. D. 1965a. Organische Reste aus der Belt-Serie (Algonkium) von Nordamerika. *Pälaont. Zeitschr.* **39**:10–25.

Pflug, H. D. 1965b. Foraminiferen und ähnliche Fossilreste aus dem Kambrium und Algonkium. *Palaeontographica* **125(A)**:46–60.

Ramdohr, P. 1957. Die Uran- und Goldlagerstätten Witwatersrand, Blind River District, Dominion Reef, Serra de Jacobina: Erzmikroskopische Untersuchungen und ein geologischer Vergleich. *Deutsch. Akad. Wiss. Berlin Abh., Kl. Chem., Geol. Biol.* **3**, 35 pp.

Rutten, M. G. 1965. Geologic data on atmospheric history. *Palaeogeography, Palaeoclimatology, Palaeoecology* **2**:47–57.

Schindewolf, O. H. 1956. Über präkambrische Fossilien. In *Geotektonisches Symposium zu Ehren von Hans Stille*, F. Lotze, ed. Stuttgart: Ferd. Enke, pp. 455–480.

Walcott, C. D. 1910. Abrupt appearance of the Cambrian fauna on the North American continent. *Smithsonian Misc. Colln.*, **57**:1–16.

Wickman, F. E. 1956. The cycle of carbon and the stable carbon isotopes. *Geochim. et Cosmochim. Acta* **9**:136–153.

Part VII

EPISODICITY IN PLANETOLOGY AND COSMOLOGY

Editor's Comments on Papers 34 Through 40

The American and Russian space missions since the early 1960s have revealed strong indications of episodicity and successive stages in planetary evolution and also have prompted scientists to look anew at the earth as a planet. The evidence for long-term episodicity in this new discipline of planetology occurs in four fields represented by the seven papers included here: planetary dynamics (Papers 34 and 35), lunar history (Papers 36 and 37), comparative planetology (Papers 38 and 39), and paleogravity and cosmology (Paper 40).

PLANETARY DYNAMICS

Short-term ($<10^7$ years) variables in the earth's planetary motions that have been established by astronomical *observation* include the obliquity of the ecliptic, the precession of the equinoxes, the eccentricity of the orbit, and the rate of rotation; the first three variables provide the basis for the Milankovitch astronomical theory of short-term climatic change. The search for any long-term changes in planetary dynamics cannot, however, rely on direct observation but must instead attempt to *deduce* such changes from the rock record. Long-term variability so proposed includes changes in the rate of the earth's rotation (Paper 34) and in the obliquity of the ecliptic (Paper 35).

As reviewed by Whyte (Paper 34), the cooperative efforts between paleontologists and geophysicists since the early 1960s have confirmed in general an overall secular deceleration of the earth's rotation during the Phanerozoic. The paleontological data of Pannella et al. (1968) and Pannella (1972) suggested that this deceleration has not been uniform, a conclusion supported by the independent data of Berry and Barker (1975). Creer (1975) correlated suggested discontinuities in the rate of change of the earth's spin with changes in the earth's geomagnetic polarity bias and reversal frequency. In 1977 Whyte (Paper 34) broadened the correlations made by Creer to incorporate the refined freeboard curve of Wise (1974), the number of "lasts" (extinctions) of all biological groups (see Part VI, Figure 9), and times of Phanerozoic glaciation. Such proposed correlations may be premature, however, since the present consensus among paleontologists cautions against the too-literal interpretation of previously published "paleontological-clock" data (see Rosenberg and Runcorn, 1975); Pannella (1975, p. 253) himself stated that "an analysis of published data in the light of the most recent advances suggested that there are many potential sources of error in the [growth pattern] counts and that a complete restudy of all the material is needed." Nonetheless, discontinuities in the rate of change of the earth's rotation may be important to global tectonism (Peyve, 1961; Williams and Austin, 1973; Kaz'min, 1975; Austin and Williams, 1978) and to climatic change (Orlova, 1963). Synthesis among different geological variables along the lines of that in Paper 34 therefore should be continued as further "clock" data become available.

The concept of *major* change in the obliquity of the ecliptic

(ϵ)—that is, former variation well beyond the calculated late Cenozoic limits of $\sim$22–24.5°—was a subject of lively debate between geologists and astronomers last century. Modern geological and geophysical research has revived the concept, and a variety of arguments now exists that at certain times during the geological past the earth's obliquity departed significantly from its present value of 23.5°:

1. Allard (1948) presented botanical evidence based on floral photoperiodism for equal ("12-hour") days and nights over wide latitudes, and hence near-zero obliquity, during the Cretaceous.
2. Wolfe (1969, 1977, 1978) reasoned that the existence of broad-leaved evergreen forest at high paleolatitudes during the early Cenozoic requires annual equability of *light,* and hence reduced obliquity, at that time.
3. Williams (Paper 35, and 1975b) and Kröner (1977) argued that the enigmatic features of late Precambrian glaciation, viz. predominantly low-to-middle-paleolatitude glaciation (confirmed by McWilliams, 1978, and Frakes, 1979), apparent marked seasonality extending into low paleolatitudes, and the characteristic association of tillites with rocks normally indicative of warm conditions (dolostones, evaporites, banded iron-formations), may be resolved if the earth's obliquity were substantially increased ($\epsilon > 54°$) in late Precambrian time ($\sim 750 \pm 200$ m.y. ago).
4. Young (1976a and b) considered that the puzzling association of Huronian glaciogenic rocks ($\sim$2200 m.y. old) with dolostones, iron-formations, and severely chemically-weathered sandstones may be explained by the proposal (Paper 35) of an increased obliquity also during the middle Precambrian ($\sim 2000 \pm 200$ m.y. ago).
5. The above proposal finds support also in Trendall's (1972) implication that an episode of marked seasonality and increased obliquity may be recorded by the enigmatic middle Precambrian ($\sim 2000 \pm 200$ m.y. ago) banded (*varved*) iron-formations.

All these proposed departures of obliquity from today's value can be integrated in a cyclic scheme of *secular* obliquity-change with a basic period of $\sim$2500 m.y. (Paper 35). The required mean rate of change of obliquity ($\lesssim$0.05 seconds of arc per century) is three orders of magnitude less than the *observed* rate of obliquity oscillation ($\sim 47''$ cy^{-1}) and is below the level of detectability by current astronomical observations. This scheme provides an explanation for the known paleolatitudinal distribution and facies associations of ancient glacial deposits and incorporates impor-

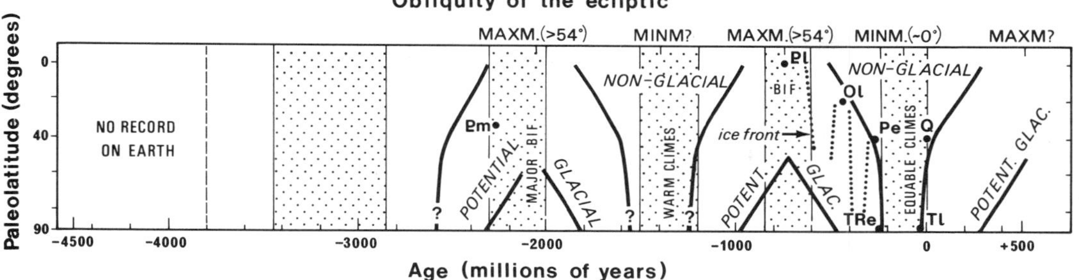

Figure 10 Scheme of obliquity maxima and minima based mainly on the distribution of continental glaciation in time and space, ages of major banded (*varved*) iron-formations (BIF), and the occurrence of high-paleolatitude evergreen floras and globally equable climates during the Mesozoic-early Cenozoic. Equatorward extent of ice sheets plotted for the late Precambrian (Pl), Late Ordovician (Ol), Early Permian (Pe), Early Triassic (Tre), late Tertiary (Tl), and Quaternary (Q). Paleolatitude of middle Precambrian (Pm) glaciogenic rocks also plotted (see text). Shaded areas represent times of overall anorogenesis as proposed by Fitch and Miller (Paper 7).

tant sedimentary, pedological, and biological events in earth history. Episodic glaciation (Papers 16 and 17) and a proposed 9.5 m.y. cycle of obliquity fluctuation (Wolfe, 1978) are superimposed on the much longer rhythm of obliquity change. An updated diagram of this change (Figure 10) takes account of the main peak in BIF ages at 2100 m.y. (Paper 24), the determination of a paleolatitude of deposition of 34° (Roy and Lapointe, 1976) for the Firstbrook Member of the glaciogenic Huronian Gowganda Formation, and the discovery of important evaporite deposits of Middle Proterozoic age (Walker et al., 1977; McClay and Carlile, 1978). Times of minimal obliquity (and overall minimal albedo) correspond to the warm, equable climates of the Mesozoic–early Cenozoic and to the warm climate of the Middle Proterozoic (see Figure 3). Moreover, Figure 10 shows that times of postulated minimal and maximal obliquity correspond to times of overall anorogenesis (with taphrogenesis) as proposed by Fitch and Miller (Paper 7). An hypothesis linking very long-term periodicity in climatic change and in geotectonism through obliquity variation is advanced by Williams (1973), and some related astronomical "coincidences" are pointed out in the Introduction.

With regard to other planets, Ward (1979) calculated that the obliquity of Mars varies between ~10.8° and ~38.0° with a period of ~10^5–10^6 years, and Murray et al. (1973) determined that Mars' eccentricity varies between 0.141 and 0.004 with a period of 2 m.y. Ward (1973) and Sagan et al. (1973) considered that such obliquity fluctuation could have caused episodic climatic changes on Mars.

Moreover, prior to the formation of the Tharsis volcanic dome on Mars' equator, the obliquity of Mars intermittently may have been as low as ~9° and as high as 46° or more (Ward et al., 1979). Dr. W. K. Hartmann (personal communication, 1978; see section on comparative planetology) considers that such major changes of obliquity may have influenced the climatic histories of both Mars and the earth.

LUNAR HISTORY

Lunar chronology and evolution are outlined by Wasserburg et al. (Paper 36). The chronology of known major events, illustrated by the cartoon in Paper 36, provides evidence of long-term rhythm during the first 1.5 b.y. of lunar history. A "megarhythm" of ~600–800 m.y. is defined by the proposed major igneous and metamorphic events at ~4.5 and ~3.9 b.y. ago and by the commencement of mare volcanism at ~3.9 b.y. and its apparent cessation at ~3.1 b.y. (Paper 36; Geiss et al., 1977). Although the proposal of a "terminal lunar cataclysm" at ~3.9 b.y. has been disputed (Baldwin, 1974; Hartmann, 1975), this date nonetheless has significance in recording either an abrupt peak in the cratering rate or a termination of accretion (Turner, 1977). As Wasserburg et al. (Paper 36) pointed out, the occurrence of major impacts 600 m.y. after the formation of the solar system is of considerable importance and presents the problem of a storage mechanism for the bombarding objects. Widely spaced significant events at ~4.45, ~4.0–3.9, and ~2.2 b.y. are suggested also by U-Th-Pb systematics of lunar samples (Nunes et al., 1973; Tera and Wasserburg, 1974).

Episodicity of several hundred million years appears to be superimposed on this overall "megarhythm." From U-Th-Pb systematics, Nunes et al. (1974) postulated semidistinct events at 4.5–4.4, ~4.2, and ~4.0 b.y. that were perhaps related to planetesimal bombardment. Turner (1977) proposed a minimum of three outgassing events, at 4.25, 4.0, and ~3.9 b.y., from the observed distribution of highland ages. Furthermore, the mare basalts are divisible into two broad groups: an older, high-titanium group with ages of ~3.9–3.5 b.y. and a younger, low-titanium group with ages of ~3.4–3.1 b.y. (Papike et al., 1976; Geiss et al., 1977).

Study of the lunar regolith may reveal past variations in solar and cosmic radiation and micrometeoroid flux. In Paper 37, Lindsay and Srnka recognized cyclic variations in the slope of the log-flux versus log-mass plot of meteoroids larger than ~10^{-7} grams, with depth in the lunar soil (time). Although the depth-age relation is uncertain, they speculated that the cycles may reflect an increase in the small-particle flux due to passage of the solar system through dust lanes bordering galactic spiral arms with a period of ~10^8 years (see Paper 18). Moreover, variation of $^{15}N/^{14}N$ with cosmic ray exposure ages and with depth in drill core of the lunar regolith may indicate greatly enhanced solar flare activity in the past (Crozaz et al., 1977). Studies of the lunar regolith are further discussed in *The Ancient Sun: Fossil Record in the Earth, Moon and Meteorites* (Lunar and Planetary Institute, Houston, 1979).

An apparent similarity in meter of the lunar "megarhythm" described above and subsequent geotectonic megacycles on the earth (Papers 1–7) was first noted by Williams (1973). Future study of the tectonic chronologies of the other terrestrial planets may determine whether this meter reflects a basic pulse in the development of the solar system.

COMPARATIVE PLANETOLOGY

Scientists have long recognized that an understanding of the evolutionary development of the other terrestrial planets would help resolve whether the long-term episodicity and cyclicity that appear to characterize earth history are reflections of endogenetic or exogenetic processes. Shapley (1949) was aware of this problem when he wrote (p. 37):

> Before this thought-provoking rough relation between galactic rotation and cyclic climates is written off as purely coincidental, it would be of great interest (I remark hopelessly) to obtain a record of the paleoclimatology of Mars, for on that planet the climatic effects from the cosmic dust clouds are probably not so seriously confused by orogenic movements.

Shapley's forlorn hopes may soon be fulfilled, for space exploration has indeed provided evidence of long-term episodicity in the paleoclimatic record of Mars and of multiple stages in the dynamic evolution of Mars and Mercury.

Two important sedimentological features of Mars—major channels apparently of fluvial origin and polar layered deposits—have been interpreted as evidence that the past climate of Mars was subject to episodic and perhaps periodic changes (e.g., Sagan et al., 1973; Hartmann, Paper 38, and 1978; Arvidson et al., 1978). In an important synthesis, Hartmann (Paper 38) brought together evidence of episodic climatic change on earth (see Papers 16 and 17 for more recent data on times of past terrestrial glaciations), Öpik's concept of a slow-flickering sun (Paper 19), the suggestion that the sun's luminosity at present is anomalously low (Paper 20), and evidence for episodic fluctuations of Martian climate, by postulating that climatic changes on earth and Mars have occurred simultaneously in response to episodic changes in solar luminosity on a time-scale of a few hundred million years. Future confirmation of coincident changes of planetary climates would indeed establish the importance of exogenetic processes in paleoclimatology but would not necessarily distinguish among the various hypotheses invoking such processes (e.g., Papers 16–20).

Alternatively, climatic changes on Mars may reflect large-scale variations of Mars' obliquity (e.g., Ward, 1973; Sagan et al., 1973). Recent calculations by dynamicists have emphasized the potential importance of obliquity changes to Mars' climatology, as recognized by Dr. W. K. Hartmann (personal communication, 11 April 1978) in the following addendum to Paper 38:

> In addition to any climatic changes caused by solar variations [as discussed in Paper 38], a further factor has been added to the Martian case. In recent studies by Burns et al. (1977), dynamical calculations show that the obliquity of Mars may have been changed by the build-up of the massive Tharsis volcanic dome. This dome is capped by several volcanic shield volcanoes, including the Olympus Mons structure, largest volcano known in the solar system. The calculations suggest that while today's Martian obliquity never exceeds 37°, the pre-Tharsis obliquity could have reached values as high as 45°. Values that high would have allowed water ice to melt entirely at the summer pole. This, in turn, would have allowed water transport from possible fluvial channel-forming activity. Since the pre-Tharsis obliquity, like the present obliquity, oscillated, this opportunity for fluvial activity would have occurred only episodically during the pre-Tharsis portion of Martian history. According to age estimates based on crater counts, this may have been the first few billion years, and the last channels may have formed on the order of a billion years ago. Thus, Mars' climatology, and perhaps the earth's as well may have been simultaneously modulated by both internal factors such as solar luminosity "flickers" and planetary perturbative effects on obliquities and orbits. I have discussed these ideas further in Hartmann (1977, 1978).

Planetary evidence for solar variations on several time scales is reviewed in *The Ancient Sun: Fossil Record in the Earth, Moon and Meteorites* (Lunar and Planetary Institute, Houston, 1979). Another possible explanation for past fluvial activity and floods on Mars is that such events were triggered by episodic volcanism

(Masursky et al., 1977). V. Gornitz (ed., *Geology of the Planet Mars*, Benchmark Papers in Geology, vol. 48, 1979) treats these problems in more detail.

Following receipt of the Mariner 10 pictures of Mercury in 1974, numerous workers have compared and contrasted the apparent evolutionary histories of the terrestrial planets (e.g., Murray et al., Paper 39; Kaula, 1975; Janle, 1977; Simonds et al., 1978). In the first significant such study Murray et al. (Paper 39) in 1975 inferred a sequence of events broadly similar for Mercury and the moon, viz.: (1) accretion and differentiation; (2) terminal heavy bombardment at ~4 b.y.; (3) basin filling and formation of smooth plains by volcanic activity; and (4) general quiescence and light cratering from ~3 b.y. They suggested similar cratering histories for Mars and possibly also Venus and concluded (p. 2513) that the Mercury picture data demonstrate that "planetary history and geological history are almost convergent nearly 4 b.y. ago." Wetherill (1975) agreed that many similarities exist amongst the cratering records of the terrestrial planets and proposed a ~3.9-b.y. "marker horizon" in those records throughout the inner solar system. Chapman (1976), however, was severely critical of the conclusions reached by Murray et al. (Paper 39) and Wetherill (1975), stating (p. 523) that "it is premature to establish a chronology for Mars and Mercury, relative to the known lunar chronology, to better than an order of magnitude." Subsequent studies (e.g., Neukum and Wise, 1976; Janle, 1977; Simonds et al., 1978) have supported Murray et al. and Wetherill's correlation of planetary cratering episodes and lent weight to the proposal of an intense bombardment of the terrestrial planets ~3.9–4.0 b.y. ago.

As pointed out by Papanastassiou et al. (1974) and Turner (1977), one method of establishing times of intense bombardment throughout the inner solar system is through determining shock-, metamorphic-, and exposure-ages of meteorites. Papanastassiou et al. noted that several meteorites have ages of 3.8–3.89 b.y., which accords with evidence for bombardment at about that time throughout the inner solar system. They concluded (p. 256) that "the question as to whether the cratering time scale determined for the moon can be applied to other planets (e.g. Mars) depends on the extent to which it can be demonstrated (from the study of meteorites) to be a solar system wide phenomenon." Evidence exists that metamorphic- and exposure-ages of meteorites are nonuniformly distributed throughout the ~4.5 b.y. of planetary history (e.g., Anders, 1964; Podosek and Huneke, 1973; Bogard et al., 1976), and future synthesis may identify episodicities in such data that can be related to other planetary events.

PALEOGRAVITY AND COSMOLOGY

Numerous workers have postulated gravitational variation as a cause of long-term geological episodicity. Lavrov (1962) attempted correlation of the gravity field with the galactic position of the solar system, the earth's mass supposedly increasing and orogeny occurring at times of perigal. Kropotkin and Trapeznikov (1965) proposed a long-term cyclic fluctuation of the gravitational "constant," which causes alternating "compression and dilation" in the earth. In 1967 Machado (Paper 40) postulated that a pulsating gravitational "constant" causes cyclic earth expansion and contraction, recorded by episodic orogeny and variations of sea level and climate. Other papers advocating variation of the gravitational constant as a cause of long-term episodicity and periodicity in the geological record include Steiner (1967; see Part III) and Kropotkin (1970). Proposals for long-term pulsation in gravitation must, however, be regarded as speculative; studies of quantitative limits to *g* (the acceleration due to gravity at the earth's surface) (Stewart, 1977, 1978) have yet to discern any such cyclicity (A.D. Stewart, personal communication, 16 November 1978).

Conversely, the study of long-term cyclicity might help resolve whether the gravitational constant *G* has indeed remained essentially constant or has undergone monotonic secular decrease during the lifetime of the solar system. This determination might be achieved by establishing whether the rhythms of systems dependent on *g* or *G*, as indicated by long-term episodicity and periodicity in the rock record, have varied little or have undergone systematic change over long intervals of time. For example, Stewart (1970, p. 423) noted that "if mantle convection is discontinuous the frequency of overturn should depend upon *g*. This might be reflected not only in frequency of orogeny but also in maximum rates of sedimentation." Stewart also suggested that "rates of continental drift might be expected to reflect changes in the gravity field." Systematic changes in the frequency of orogeny since early Precambrian time are not convincingly displayed by available data (see Part I), however, and rates of sedimentation show little secular change during the Phanerozoic (see Part IV). Furthermore, Phanerozoic rates of apparent polar wander have fluctuated periodically rather than changed monotonically (Paper 14). With regard to exogenetic systems, the nature of any convective processes in the sun would be dependent on g_{sun}; Öpik's (Paper 19) model of periodic convective overturn of material within the sun therefore may carry the implication of invariant gravitation. Moreover, since the orbital radius and period of a satellite about its primary are dependent on G (Vinti, 1974), a corollary of hypotheses advocating galactic control of *periodic* planetary events (e.g., Papers 16–18; Crain and Crain, 1970; Hatfield and Camp, 1970; Macintyre, 1971), therefore, is that the sun's orbital period, and hence G, are essentially constant over long intervals of time. Sedimentary cycles containing 11- and 22-year solar periods tend to confirm this.

The recognition of long-term rhythm in planetary events therefore could be important to cosmology, and the search for cyclicity should be a major aim of synthesis in geology and planetology.

34

Reprinted from *Nature* **267**:679–682 (1977)

Turning points in Phanerozoic history

Martin A. Whyte

Department of Geology, University of Sheffield, Sheffield, UK

During the last 450 Myr there has been concomitant variation in the Earth's rotation rate, the polarity bias of the geomagnetic field, the amount of activity at ocean ridges, sea level and climate. At turning points when trends in the variables reverse themselves, climatic instabilities have disrupted biological ecosystems and led to mass extinctions.

ONE of the most interesting cooperative efforts within the earth sciences is that between palaeontologists and geophysicists in which skeletal growth rhythms recorded in fossils are used to elucidate the Earth's rotational history. It has been established that the rotation rate has fluctuated during the Phanerozoic. This synthesis suggests that subtle long term variations in other parameters can be correlated with changes in rotation rate.

Rotational history

The modification of skeletal growth by external environmental cycles can produce within a shell a detailed record of daily, fortnightly, monthly and annual events. The potential of such structures was first fully realised by Wells[1] who made counts on Middle Devonian corals which indicated that at that time, 390 Myr, there were 400 solar days in the year. As Wells[1,2] appreciated the slowing down of the Earth's rate of rotation, which this result indicated, corroborated the geophysical prediction that the lunar tidal torque has a braking effect by which angular momentum is transferred from the Earth to the Moon. Soon afterwards Scrutton[3] who also used Middle Devonian corals, made counts indicating that there had been 30.6 d in the synodic month and Runcorn[4] showed how this pair of observations could be used to test for possible changes in the Earth's moment of inertia. Runcorn's analysis, though probably unreliable due to the statistical uncertainty inherent in the original data[5,6], suggested that any changes in the Earth's moment of inertia must have been small and subsequent workers have generally accepted tidal friction as the operative mechanism in the rotation deceleration. However, as more palaeontological data became available it became evident[7,8] that there had not been a uniform rate of decrease in spin velocity and a significant recent advance was made by Creer[5] who recognised that there have been periods of earth history when the Earth's rate of rotation has accelerated (Fig. 1). During the last 500 Myr these periods of acceleration have alternated with periods of deceleration of spin velocity and there are turning points of two types, firstly spin maxima at transitions from acceleration to deceleration and secondly spin minima at transitions from deceleration to acceleration. Revised estimates indicate that spin maxima occurred 65 Myr (end Cretaceous) and 375 Myr (middle Upper Devonian) ago and that spin minima occurred 235 Myr (end Permian) and 445 Myr (late Ordovician) ago (Fig. 1). The periods of acceleration recognised by Creer[5] cannot be easily explained by tidal friction alone and suggest that some additional mechanism or mechanisms must be operating. This additional component seems to be a fluctuating system superimposed on a general trend for deceleration of spin velocity which presumably represents the effects of tidal friction (Fig. 1).

Creer[5] also recognised that the turning points were closely associated in time with periods of instability in the palaeomagnetic field (Fig. 1). While periods of rotational acceleration are characterised by a predominance of normal polarity and periods of rotational deceleration are characterised by a predominance of reversed polarity, turning points are associated with phases of mixed polarity which show no obvious polarity bias and which may also have had high reversal frequencies. A close link between the Earth's rotation and its magnetic field has long been recognised: there seems to have been a near alignment of rotational and dipole axes throughout time and it has been suggested that the influence of rotational forces on the fluid outer core may be vital to the generation of the magnetic field[16]. On the other hand the irregular decade fluctuations in the Earth's rate of rotation have been correlated with changes in the geomagnetic field and attributed to the transfer of angular momentum between the core and the mantle[17]. The core–mantle coupling is generally interpreted as being an electromagnetic effect though viscous forces and topographic features of the boundary may also be important[18]. Though extraterrestrial sources cannot be altogether discounted, the geologically detectable rotational and magnetic fluctuations are most probably due either to changes in the Earth's moment of inertia[5,19] or to an internal redistribution of angular momentum[19] and it is a great pity that the palaeontological data are not yet sufficiently precise to allow elimination of one or other of these possibilities using an extended form of Runcorn's analysis.

In the middle Carboniferous and again during the Late Jurassic and early Cretaceous there were periods in which the palaeomagnetic field became moderately disturbed[5,20]. However, unlike the conditions at turning points, the field retained an overall polarity bias and there was no associated switch in the direction of the predominant polarity. One might predict that these events were accompanied by slight inflections in the rate of change of the Earth's rotation but the palaeontological data are at present too scanty to allow this possibility to be tested. These episodes could perhaps be interpreted as aborted turning points and it is worth noting that they are followed by the most important Phanerozoic quiet intervals namely the Permo–Carboniferous (Kiaman) Reversed Interval and the Cretaceous (Mercanton) Normal Interval. The Cenozoic magnetic field has apparently always been relatively unstable but there seems to have been within the last 10 Myr a change from a predominance of reversed polarity[5,20] (Fig.

1) and during the same period there has also been an increase in the frequency of reversals[21]. Astronomical estimates of the present rate of secular deceleration give a value lower than that which seems to have operated during the last 65 Myr (Fig. 1) and this, together with the magnetic evidence, might indicate that we are at or close to another spin minimum. Including this putative turning point and the two aborted ones it seems that turning points are separated by a characteristic time interval with an average length of 74 Myr.

Mass extinctions

Several investigations[22] of Pleistocene deposits have noted a significant correlation between magnetic reversals and extinctions of fossil species and these results are consistent with Uffen's hypothesis[23] that during reversals biological systems would be adversely affected by increased radiation doses. Simpson[24] extended this hypothesis to suggest a correlation between mass extinctions and periods of high reversal frequency and this is strongly supported by Creer's data[5] since the major mass extinctions in the late Ordovician, middle Upper Devonian, end Permian and end Cretaceous coincide with turning points (Fig. 1). Considerable doubt[25–27] has been cast on whether a magnetic reversal alone would cause a sufficient increase in radiation fluxes but it has recently been suggested that if a solar flare occurs during a reversal it might cause the ozone layer, which shields the Earth's surface from ultraviolet radiation, to become depleted[28]. Alternative theories[22] have suggested a direct relationship between the magnetic field and the well-being of organisms or an indirect relationship through climatic effects.

Correlations between climatic changes and changes in the intensity of the geomagnetic field have been demonstrated and possible linking mechanisms have been discussed[29,30]. As previously noted there is a correlation between geomagnetic field intensity and rotation rate[17] and Lambeck and Cazenave[67] have detected a correlation between climate and rotation rate. They have however calculated that the magnitude of the rotational changes is not sufficient to fully explain the associated climatic effects. It is suggested here that magnetic fluctuations cause climatic effects partly as a direct effect and partly through the associated changes in spin velocity, and that the magnetic instabilities at turning points cause climatic instabilities.

The general pattern of mass extinctions is for the communities of stable environments with their great diversity

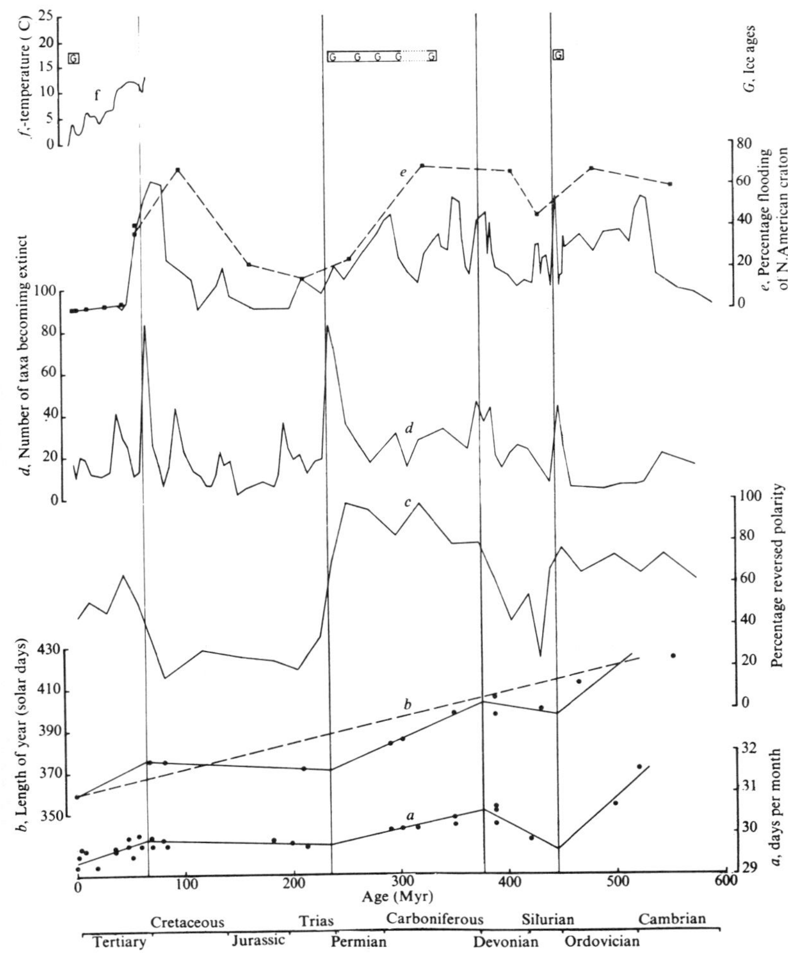

Fig. 1 Aspects of Phanerozoic history. *a*, Number of solar days per month (modified from Creer[5] using palaeontological data[7–13]). *b*, Number of solar days per year (modified from Creer[5]; using palaeontological data 1, 2, 8, 14; the dashed line is extrapolated from the present rate of deceleration[15]). *c*, Percentage of reversed polarity (based on Creer[5]). *d*, Number of higher taxa, of all biological groups, becoming extinct (based on Cutbill and Funnell[38]). *e*, Percentage flooding of the N. American craton (based on Wise[53]; the dashed curve is a period-by-period smoothed curve). *f*, Temperature of high latitude marine waters (based on Savin *et al.*[39]). *g*, Occurrence of ice ages (refs 43, 45–47).

of specialised member species to be most severely affected[31]. Thus offshore benthic communities were more affected than onshore benthic communities[32,33] and that most complex of all marine associations, the reef community, underwent at each mass extinction a dramatic collapse from which it often took millions of years to recover[34,35]. The climatic instabilities at turning points were thus probably only triggers, which by causing a few extinctions and/or a reduction in primary production, created a house of cards effect leading to the collapse of ecosystems[31].

Neither the mid-Carboniferous nor the end Jurassic events were associated with abnormally high extinction rates but Late Pleistocene and Recent extinctions, though dominantly affecting large (that is, specialised) terrestrial vertebrates, are well documented[36,37]. While man must in all probability take a large share of the blame for these extinctions it may well be that he is only exacerbating an unstable situation.

Climatic history

Apart from the possible climatic instability at turning points the periods of acceleration seem to be characterised by climatic amelioration while periods of deceleration are characterised by climatic deterioration. Thus spin minima were preceded by ice ages with marked seasonal and latitudinal climatic contrast (Fig. 1) while spin maxima were preceded by climatic optima with an equable warm climate. Isotopic evidence[39,40] (Fig. 1) shows clearly the temperature decline that has taken place, especially at high latitudes, during the Tertiary. Mesozoic isotope data[41,42] are less well controlled but indicate warm temperatures even at high latitudes and, though there were fluctuations a trend for temperature increase up to the Late Cretaceous optimum about 80 Myr ago. The Permo-Carboniferous and the Late Ordovician Ice Ages are well documented[43] and a recent synthesis[35] has suggested that there was a reduction in climatic gradients during the Silurian and Devonian. The Cretaceous optimum gave rise to the acme of reef development by scleractinian corals and the early Upper Devonian optimum was similarly the acme of the tabulate coral and stromatoporoid reef assemblage[34]. The scleractinia as a group have declined during the Tertiary[34] and the poor representation of reef corals in the Carboniferous and Permian may also be a climatic effect. As in the Permian, bryozoans and sponges are an important element of Ordovician carbonate build-up[54] formed during a phase of climatic deterioration. A pre-mass extinction diversification has been detected in the Cretaceous and in the Devonian but not in the Permian[44] which may be a consequence of its different climatic regime.

It is interesting to note that Upper Palaeozoic ice sheets seem to have developed before the mid-Carboniferous aborted event[45–47] and this may explain the long time span of the Permo–Carboniferous Ice Age.

Climate and eustasy

Though a change in the length of the day would, for instance, cause a small change in the diurnal temperature inequality[48], the overall trend evident in the rotational data suggests that the link between climate and spin velocity is not a direct one. Rotational changes may have been accompanied by other changes in the Earth's orbit, which could also have caused slight climatic effects, but the most important control on climate would seem to be sea level[49]. Hays and Pitman postulated[49] that increased climatic contrast will result from eustatic lowering of sea level and that eustatic raising of sea level will cause 'moderation and stabilisation of climate'. The global Mesozoic transgression and the subsequent Cainozoic regression are well documented[49–53] (Fig. 1). Sea level changes in the Palaeozoic are less clear and estimates based on N. American data[53] are clearly affected by tectonic events in the Appalachians. Cruder curves[53] (Fig. 1) suggest that there may have been overall transgression from the Late Ordovician to the Late Devonian followed by a regression into the end of the Permian. The Ashgill regression[54] and the widespread Ordovician–Silurian unconformity[55] represent the effects of the Late Ordovician eustatic minimum and in the Devonian a series of increasingly penetrative transgressions, which reach a maximum in the Late Devonian, have been recognised[56]. Ice ages are thus a consequence of the increased climatic contrast resulting from regression but the exact timing of the onset of glaciation and the subsequent glacial history will be affected by other factors such as distribution of land masses, atmospheric turbidity and orbital motions. The association of glaciation with aridity[57] fits well with this theory and the increasing severity of Pleistocene glacial periods[57] can be linked to the continued decrease in sea level.

Though orogenic events can cause eustatic changes in sea level[52] the dominant eustatic control seems to be the volume of oceanic ridges which is itself controlled by plate spreading rates (refs 49, 50, 58 and 59). Thus increased plate generation not only leads to an increasing volume of ocean ridges, a raising of sea level and an amelioration of climate but also correlates with an acceleration of mantle spin velocity and a predominance of normal polarity. Fast Cretaceous and slow Cenozoic spreading rates have been deduced for a number of ridge systems[49,59]. More detailed work, despite uncertainties due to imprecision of the polarity time scale, suggests that Tertiary spreading rates decreased gradually up to about 10 to 15 Myr ago, since when there has been a slight increase in spreading rate[60,61]. This may again suggest an impending turning point. The possible role of tidal and rotational forces in the plate tectonic mechanism has been discussed by several workers[62–66], and, whether the relationships are causal or not, clearly ridge activity and changes in mantle spin velocity and predominant polarity are all related to some unifying process.

A coherent picture–but gaps still remain

Though much work needs to be done to improve our knowledge of the factors considered here, the data present a coherent picture. The Cenozoic, the Late Devonian to Permian and the Ordovician were times of deceleration of mantle spin velocity, of predominantly reversed polarity, of decreasing oceanic ridge activity, of global regression and of climatic deterioration leading to ice ages. By contrast in the Silurian to early Upper Devonian and in the Mesozoic there was acceleration of mantle spin velocity, predominance of normal polarity, increasing ocean ridge activity, global transgression and climatic amelioration. Turning points between phases of different type are marked by magnetic instability, possible climatic instability and by mass extinctions. Aborted turning points occurred in the middle Carboniferous and late Jurassic. Despite the many imperfections and biased nature of the geological record a subtle but distinctive background pattern can thus be detected in it. Whatever the cause may be this pattern is not only a record of surface processes but also of deep seated events some of which may even have taken place within the Earth's core.

This synthesis has a number of important implications. First, it places important new constraints on Earth models. These must be considered especially in discussions of the origin of the geomagnetic field, the mechanisms affecting the Earth's rotation and the driving force for plate tectonics. Second, the pattern can be applied to the Precambrian as an aid to understanding events and processes in the Proterozoic (Whyte, in preparation). Finally it allows us to make some simple predictions about the future geological history of the Earth. Thus not only do we seem to be at present at or close to a turning point (spin minimum) but we can predict that it will be followed by a period lasting about 60 to 90 Myr in which there will be rotational acceleration, a dominantly normal magnetic field, increasing ridge activity, global transgression and climatic amelioration.

Received 4 February; accepted 22 April 1976.

1 Wells, J. *Nature* 197, 948–50 (1963).
2 Wells, J. in *Palaeogeophysics* (ed. Runcorn, S. K.) 3–9 (Academic, London, 1970).
3 Scrutton, C. T. *Palaeontology* **7**, 552–558 (1965).
4 Runcorn, S. K. *Nature* **204**, 823–5 (1964).
5 Creer, K. M. in *Growth Rhythms and the History of the Earth's Rotation* (eds Rosenberg, G. D. & Runcorn, S. K.) 293–318 (J. Wiley, London, 1975).
6 Scrutton, C. T. & Hipkin, R. G. *Earth-Sci. Rev.* **9**, 259–274 (1973).
7 Pannella, G., MacClintock, C. & Thompson, M. N. *Science* **162**, 792–796 (1968).
8 Pannella, G. *Astrophys. Space Sci.* **16**, 212–237 (1972).
9 Pannella, G. & MacClintock, C. *Palaeontol. Soc. Mem.* **2** 64–80 (1968).
10 Scrutton, C. T. in *Palaeogeophysics* (ed. Runcorn, S. K.) 11–16 (Academic, London, 1970).
11 Berry, W. B. N. & Barker, R. M. *Nature* **217**, 938–939 (1968).
12 Berry, W. B. N. & Barker, R. M. in *Growth Rhythms and the History of the Earth's Rotation* (eds Rosenberg, G. D. & Runcorn, S. K.) 9–25 (J. Wiley, London, 1975).
13 Johnson, G. A. L. & Nudds, J. R. in *Growth Rhythms and the History of the Earth's Rotation* (eds Rosenberg, G. D. & Runcorn, S. K.) 27–42 (J. Wiley, London, 1975).
14 McGugan, A. *Abstr. Ann. Meeting geol. Soc. Am.* **145** (1967).
15 Muller, P. M. & Stephenson, F. R. in *Growth Rhythms and the History of the Earth's Rotation* (eds Rosenberg, G. D. & Runcorn, S. K.) 459–534 (J. Wiley, London, 1975).
16 Strangway, D. W. *History of the Earth's Magnetic Field* (McGraw-Hill, New York, 1970).
17 Yukutake, T. *J. Geomag. Geoelec.* **25**, 195–212 (1973).
18 Rochester, M. G. *Trans. Am. geophys. Union* **54**, 769–781 (1973).
19 Tarling, D. H. in *Growth Rhythms and the History of the Earth's Rotation* (eds Rosenberg, G. D. & Runcorn, S. K.) 397–412 (J. Wiley, London, 1975)
20 Irving, E. & Pullaiah, G. *Earth-Sci. Rev.* **12**, 35–64 (1976).
21 Cox, A. *Science* **163**, 237–245 (1969).
22 Hays, J. D. *Bull. geol. Soc. Am.* **83**, 2433–47 (1971).
23 Uffen, R. J. *Nature* **198**, 143 (1963).
24 Simpson, J. F. *Bull. geol. Soc. Am.* **77**, 197–204 (1969).
25 Black, D. I. *Earth planet. Sci. Lett.* **3**, 225–236 (1967).
26 Waddington, C. J. *Science* **158**, 912–5 (1967).
27 Harrison, C. G. A. *Nature* **217**, 46–47 (1968).
28 Reid, G. C., Isaksen, I. S. A., Holzer, T. E. & Crutzen, P. J. *Nature* **259**, 177–179 (1976).
29 Wollin, G. *et al. Nature* **242**, 34–36 (1973).
30 Harrison, C. G. A. & Prospero, J. M. *Nature* **250** 563–6 (1974).
31 Bretsky, P. W. & Lorenz, D. M. *Bull. geol. Soc. Am.* **81**, 2449–56 (1970).
32 Bretsky, P. W. *Palaeogeog. Palaeoclim. Palaeoecol.* **6**, 45–59 (1969).
33 Walker, K. R. *J. Paleontol.* **46**, 82–93 (1972).
34 Newell, N. D. *Sci. Am.* **226**, 54–56 (1972).
35 Boucot, A. J. *Developments Palaeontol. Stratigr.* **1**, 427 (1975).
36 Martin, P. S. & Wright, H. E. *Pleistocene Extinctions: the Search for a Cause* (Yale Univ. Press, Yale, 1968).
37 Van Valen, L. *Proc. N. Am. Palaeontol. Convention* 469–485 (1969).
38 Cutbill, J. L. & Funnell, B. M. in *The Fossil Record* (eds. Harland, W. B. *et al.*) 791–820 (Geol. Soc. Lond., London, 1967).
39 Savin, S. M., Douglas, R. G. & Stehli, F. G. *Bull. geol. Soc. Am.* **86**, 1499–1510 (1975).
40 Devereux, I. *New Zealand J. Sci.* **10**, 988–1011 (1967).
41 Bowen, R. *Methods Geochem. Geophys.* **2**, 265 (1966).
42 Clayton, R. N. & Stevens, G. R. *Tuatara* **16**, 3–7 (1968).
43 Wright, A. E. & Moseley, F. *Geol. J. Spec. Issue* **6**, 320 (1975).
44 Thompson, K. S. *Nature* **261**, 578–580 (1976).
45 Harrington, H. J. *Am. Assoc. Petrol. Geol. Bull.* **46**, 1773–1814 (1962).
46 Frakes, L. A. & Crowell, J. C. *Bull. geol. Soc. Am.* **80**, 1007–42 (1969).
47 Frakes, L. A. & Crowell, J. C. *Bull. geol. Soc. Am.* **81**, 2261–86 (1976).
48 Lovenburg, M. F., Dell, C. I. & Johnson, M. J. S. *Bull. geol. Soc. Am.* **83**, 3529 (1972).
49 Hays, J. D. & Pitman, W. C. III *Nature* **246**, 18–22 (1973).
50 Hallam, A. *Am. J. Sci.* **261**, 397–423 (1963).
51 Hallam, A. *Earth-Sci. Rev.* **5**, 45–68 (1969).
52 Grasty, R. L. *Nature* **216**, 779–80 (1967).
53 Wise, D. U. *Mem. geol. Soc. Am.* **132**, 87–100 (1972).
54 Berry, W. B. N. & Boucot, A. T. *Bull. geol. Soc. Am.* **84l**, 275–284 (1973).
55 Dennison, J. M. in *The Ordovician System* (ed. Bassett, M. G.) 107–120 (University of Wales Press and Na. Mus. of Wales, Cardiff, 1976).
56 House, M. R. *Proc. Yorks. geol. Soc.* **40**, 233–287 (1975).
57 Fairbridge, R. W. in *Implications of Continental Drift to the Earth Sciences* (eds Tarling, D. H. & Runcorn, S. K.) 503–515 (Academic, London, 1973).
58 Armstrong, R. L. *Nature* **221**, 1042–43 (1969).
59 Vine, F. J. in *Implications of Continental Drift to the Earth Sciences* (eds Tarling, D. H. & Runcorn, S. K.) 831–839 (Academic Press, London, 1973).
60 Vogt, P. R. & Avery, O. E. *J. geophys. Res.* **79**, 363–389 (1974).
61 Fleming, N. C. & Roberts, D. G. *Nature* **243**, 19–22 (1973).
62 Howell, B. F. *J. geophys. Res.* **75**, 2769–72 (1970).
63 Hughes, T. *Tectonophysics* **18**, 215–30 (1973).
64 Moore, G. W. *Am. Assoc. Petrol. Geol. Bull.* **59**, 1020–1 (1975).
65 Torbett, M. V. *Tectonophysics* **30**, T17 (1976).
66 Bostrom, R. C. *The Moon* **15**, 109–119 (1976).
67 Lambeck, K. & Cazenave, A. *Geophs. J. R. astr. Soc.* **46**, 555–73. (1976).

35

Reprinted by permission of the publisher from pages 165–173 and 178–181 of *Modern Geol.* **3**:165–181 (1972)

GEOLOGICAL EVIDENCE RELATING TO THE ORIGIN AND SECULAR ROTATION OF THE SOLAR SYSTEM

GEORGE E. WILLIAMS

Department of Geology, University of Adelaide, Adelaide, South Australia 5001, *Australia*

presented by L. B. Ronca

Abstract—The Earth's glacial, sedimentary, pedological, biological and geomagnetic records provide strong evidence for secular change in the obliquity of the ecliptic. It is deduced from periodicity of the secular change that at the time of the Earth's formation approximately 4500 million years ago the spin axis lay in the orbital plane. Moreover, Palaeozoic palaeontological "clocks" provide evidence for a former more rapid spin of the Earth. These reconstructed dynamic elements of the primeval Earth lead to a new hypothesis for the origin of the solar system, which envisages the planets and satellites condensing from an attenuated, tornado-like prominence or jet of gases ejected from the primeval Sun. Secular change in the obliquity of the planets' spin axes is attributed to secular rotation, with a period of approximately 2500 million years, of the plane of the solar system, the spinning planets acting as gyroscopes and tending to maintain their attitudes in space during such rotation.

INTRODUCTION

Numerous hypotheses for the origin of the solar system now exist (see Jastrow and Cameron, 1963; Brandt and Hodge, 1964; Williams and Cremin, 1968). According to Whipple (1968, p. 243), however, "all hypotheses so far presented have failed, or remain unproved, when physical theory is properly applied." Why have astronomy and physics failed to solve this monumental problem? Can geology contribute to the solution?

Hannes Alfvén virtually issued a challenge to geologists when he wrote "To trace the origin of the solar system is archaeology, not physics. . . . During the . . . years which have elapsed since it originated . . . secular perturbations might have grown very large. If this were so no detailed theory of its origin would be possible until the present elements of the solar system were reduced to their values at a time immediately after the creation" (Alfvén, 1954, pp. 1 and 9).

Most hypotheses for the origin of the solar system assume, without deep questioning, that the primeval planets rotated in a *prograde* (direct) sense about axes roughly *perpendicular* to the general orbital plane. However, the spin axis of Uranus lies close to the orbital plane (Alexander, 1965), and Venus undergoes retrograde rotation (Whipple, 1968). At least one of two inferences must therefore be made: (1) the axes of rotation of at least some of the planets have undergone significant change in attitude with respect to the orbital plane; and (2) the attitude of the planets' spin axes at the time of formation of the solar system was not as generally assumed. Clearly, the original attitude of the planets' spin axes with respect to the orbital plane must be ascertained, as it may provide the key to understanding the origin of the solar system.

The Earth's geological record does indeed provide evidence for secular change in the obliquity of the ecliptic. From periodicity of the secular change it is deduced that at the time of the Earth's formation approximately 4500 million years (m.y.) ago the spin axis lay in the orbital plane. This original attitude of the Earth with respect to the orbital plane forms the base to a new model which accounts for the origin and observed major features of the solar system.

The palaeoclimatic model presented in this paper differs in several important ways from that of Meyerhoff (1970a, b), Meyerhoff and Teichert (1971), and Meyerhoff and Meyerhoff (1972). These authors, in attempting to explain the distribution of palaeoclimatic indicators in space and time, assume that: (1) the relative positions of the Earth's spin axis, the continents, and the ocean basins have remained essentially stable since middle Proterozoic time; (2) palaeomagnetic data are virtually worthless; and (3) tacitly, the obliquity of the ecliptic has remained constant, at least since the middle Proterozoic. The model presented here accepts that the continents have moved, by means

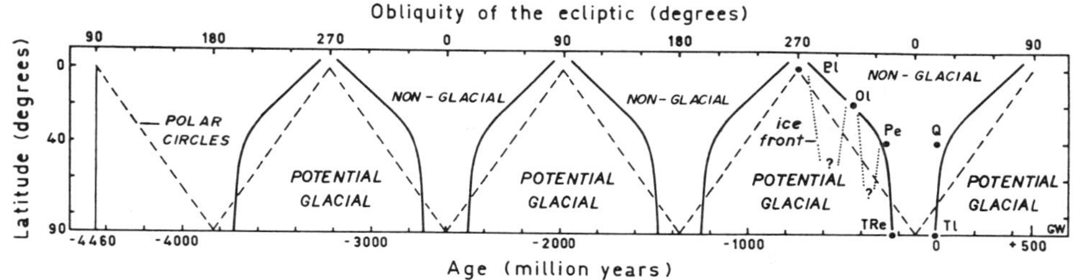

FIG. 1. Distribution of continental and marine glacial deposition in time and space; see text for full explanation. Points plotted for the late Precambrian (P1), Late Ordovician (O1), Early Permian (Pe), Early Triassic (TRe), late Tertiary (T1), and Quaternary (Q).

of continental drift and/or polar wandering, with respect to the geographic co-ordinates, and that palaeomagnetism provides a guide to palaeolatitudes of deposition of sedimentary rocks. Most earth scientists would agree that these are reasonable tenets. Further, it is argued that the third assumption of A. A. Meyerhoff and co-workers, namely the invariability of the obliquity of the ecliptic, is also invalid. The concept of continental drift is more readily compatible with the Earth's palaeoclimatic history if one accepts the concept of secular change in the obliquity of the ecliptic.

TEMPORAL AND LATITUDINAL DISTRIBUTION OF CONTINENTAL AND MARINE GLACIAL DEPOSITION

Figure 1 shows the variation in equatorward extent of continental and marine glacial deposition through geological time.

The plot for the Quaternary glaciation is placed at 40° latitude, which approximates the southernmost extent of the northern ice sheets (Frye *et al.*, 1965; Flint, 1971).

It appears that mountain glaciers existed in North and South America about 3 m.y. ago, and an ice cap in Antarctica about 10 m.y. ago (Crowell and Frakes, 1970). In the long interval between about 10 m.y. and 230 m.y. ago (start of Triassic), however, there is no record of widespread ice sheets (Crowell and Frakes, 1970). During this interval much of the world enjoyed a warm, tropical to subtropical climate.

The Tertiary, according to Fairbridge (1969a, p. 241), was "a warm-wet equatorial to subtropical world. ... Lateritic soils, product of chemical weathering under two-seasonal warm-wet conditions, extended from latitudes 50°N to 50°S." During the early Tertiary, laterite developed in Tasmania (Gill, 1961) at a palaeolatitude of about 65°S (Irving, 1964).

Throughout the Mesozoic much of Australia was within the Antarctic circle (Irving, 1964; Dietz and Holden, 1970a, b), and yet the formation of coal at a palaeolatitude of about 70°S during the Triassic (Parkin, 1969), the occurrence of deep kaolinitic weathering at a palaeolatitude of about 70°S during the Jurassic (Parkin, 1969, p. 146), and the presence of ammonites at a palaeolatitude of about 75°S during the Cretaceous (Brown *et al.*, 1968, p. 289) indicate that a temperate to warm equable climate extended well into polar regions. Furthermore, during the Mesozoic northeast Asia lay within the Arctic circle (Irving, 1964; Dietz and Holden, 1970a, b); no glacial deposits of Mesozoic age are known there, but Mesozoic coals are common and suggest a boreal climate (Holmes, 1965, p. 737).

The evidence is thus very strong that no polar ice caps existed during the Mesozoic and much of the Tertiary (about 230 to 10 m.y. ago).

Glacial deposits of Permian age include the Late Permian glaciomarine sediments of the northeastern USSR (Mikhaylov *et al.*, 1970; Ustritsky *et al.*, 1971), and the famous continental and glaciomarine deposits of Gondwanaland (Frakes and Crowell, 1969, 1970; Frakes *et al.*, 1971; Crowell and Frakes, 1971a, b). The Permian plot on Figure 1 is based on Crowell and Frakes' (1971a) palaeogeographic reconstruction of Gondwanaland in Early Permian time (about 260 m.y. ago), which depicts ice sheets extending to a palaeolatitude of about 40°.

Fairbridge (1970a, b, 1971) and Bennacef *et al.* (1971) present convincing evidence for widespread glaciation in northwest Africa during the Late Ordovician. Late Ordovician tillites also occur in the Pakhuis beds on Table Mountain, South

Africa, and may also occur in southwest Africa (Fairbridge, 1971) and Nova Scotia (Schenk, 1972). It thus appears that during the Late Ordovician (about 450 m.y. ago) ice sheets extended discontinuously from the Gondwana pole in northwest Africa (McElhinny *et al.*, 1968; Fairbridge, 1969b; McElhinny and Luck, 1970) as far as southern Africa, that is, to a palaeolatitude of about 20°.

There is unequivocal geological evidence for extensive glaciation during the late Precambrian (Mawson, 1949; Harland, 1964a, b; Harland and Rudwick, 1964; Reading and Walker, 1966; Dunn *et al.*, 1971), now recognized on every continent except Antarctica (Crittenden, 1971). The precise age at which to plot this event is difficult to determine, however, because (1) at least two world-wide glaciations occurred at this time, and (2) most isotopic dates available are for sedimentary rocks, and such dates do not necessarily reflect true ages of sedimentation. Accordingly, the median of the several ages determined for probable glacial deposits of late Precambrian age is used. The two glacial formations in Australia (Dunn *et al.*, 1971) are about 750 and 670 m.y. old respectively (Rb-Sr whole-rock isotopic dates), and thus give a median age of 710 m.y.; tillitic rocks in the USSR (Salop, 1968) range from 810 to 715 m.y. old (K-Ar dates on glauconite), with a median of 760 m.y.; and several probable glaciations took place in China from 950 to 570 m.y. ago (Saito, 1969), with a median also of 760 m.y. An overall median age of 740 m.y. is thus assumed. This is plotted at latitude 0°, as there is palaeomagnetic and geological evidence that the late Precambrian glacial deposits reached the palaeoequator (Harland, 1964a, b). A further peculiarity of the late Precambrian glacial deposits is their common association with dolomites and limestones, sediments generally associated with *warm* saline waters. Moreover, banded iron formation is associated with the late Precambrian glacial deposits and dolomites in South Australia (Mirams, 1962; Dalgarno and Johnson, 1965; Whitten, 1966) and Africa (Hatfield, 1937), and with late Precambrian dolomites and glaciogenic? conglomerates in South America (Guild, 1957).

Several early Proterozoic deposits are demonstrably glacial. The marine glacial deposits of the Transvaal Supergroup in South Africa (Visser, 1971) are associated with dolomites, dolomitic limestones, and banded iron formation (Haughton, 1969), a facies association characteristic of late Precambrian glacial deposits. The Transvaal Supergroup was deposited within the interval 2300 to 1950 m.y. ago (Haughton, 1969; Nicolaysen *et al.*, 1958; van Niekerk and Burger, 1964). The Gowganda glaciation occurred in southern Canada about 2288 ± 87 m.y. ago (Fairbairn *et al.*, 1969). The facies are typically glaciolacustrine and glaciomarine (Schenk, 1965; Lindsey, 1969, 1971); dolomites and banded iron formation are absent. The majority of other probable glaciogenic rocks of Proterozoic age are between 1700 and 2500 m.y. old (Holmes, 1965, p. 738). Moreover, I have observed isolated megaclasts in well-bedded argillites near the base of the Nullaginian System (Brown *et al.*, 1968) in Western Australia; these rocks, which are about 2250 m.y. old (Dunn *et al.*, 1966), may be partly of glacial origin.

Precambrian glacial deposits thus appear to be temporally grouped at about 2500–1700 and about 950–570 m.y. old. It seems that glacial deposition was minimal from about 1700 to 950 m.y. ago.

CAUSE OF ICE AGES

A curve sketched through the points plotted on Figure 1 shows initial symmetry about the Mesozoic-Tertiary non-glacial interval. This suggests that a rhythm exists to the waxing and waning of the ice sheets. Support for such a concept is obtained when the curve is extrapolated back in time, as the glaciations of the early Proterozoic (about 2500–1700 m.y. ago) correlate with a projected potential glacial maximum, and the interval of minimal glaciation (about 1700–950 m.y. ago) with a projected glacial minimum.

The main features of Figure 1 are: (1) the thick curve, which represents long period (1000–2000 m.y.) variation in the equatorward limit of *potential bipolar glaciation;* and (2) the dotted curve, representing moderate-period (100–200 m.y.) fluctuations of the ice sheets which are superimposed on the long-period variation.

Steiner (1967) and Meyerhoff (1971) explain fluctuations of Earth climate in terms of galactic revolution of the solar system. Since the period of revolution of the solar system about the galactic centre is approximately 200 m.y. (Clemence, 1963), such galactic motion can explain only the moderate-period fluctuations in glaciation. The long-period variation whereby the Earth, over about 600 m.y., passes in steps from a state of equatorial glaciation to that of non-glaciation, has yet to be explained.

Long-period variation of glaciation and non-glaciation may suggest a regular and drastic change

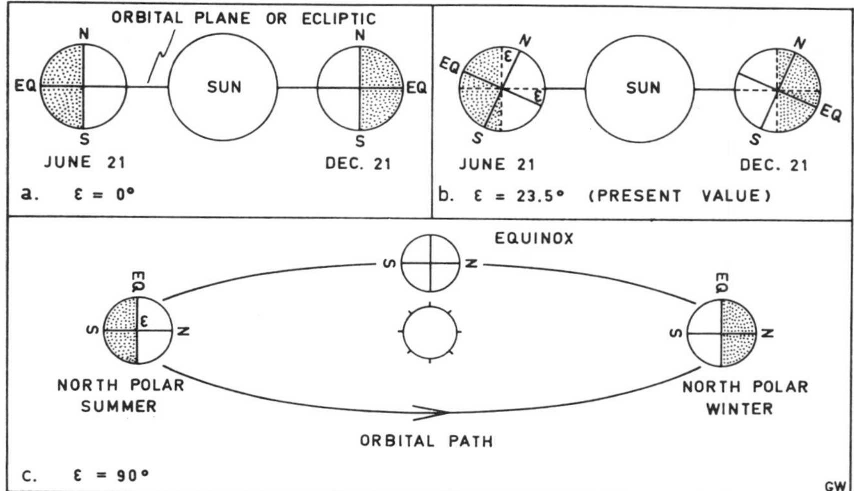

FIG. 2. Diagram illustrating change in the obliquity of the ecliptic (ε), with the Earth in midsummer and midwinter. (a) and (b) Cross-section of the orbit; stippling represents night hemisphere. Modified after Zeuner (1959, Figures 43 and 44). (c) Oblique view of the orbit; stippling represents winter hemisphere.

of external conditions such as in energy received from the Sun or Galaxy. Under such conditions, the entire Earth would be in the grip of an ice age during episodes of equatorial glaciation. This concept is incompatible with the geological and biological records (Fairbridge, 1966), particularly with the geology of Precambrian glacial deposits which are associated with dolomites and limestones.

Roberts (1971) has suggested that an "anti-greenhouse effect," due to the locking-up of atmospheric CO_2 in the carbonate rocks of late Precambrian age, caused a worldwide lowering of temperatures which triggered the rapid onset of the late Precambrian glaciation. This concept also is incompatible with the geological record: carbonate deposition was no less common during the Palaeozoic when glaciation had retreated from equatorial regions, and was far more common during the Mesozoic and early Tertiary when no ice caps existed.

One way by which the geological record and the temporal and latitudinal pattern of glaciation can be reconciled is to postulate secular change in the obliquity of the ecliptic, whereby the Earth's spin axis changes attitude with respect to the orbital plane with a period of approximately 2500 million years. Thus, when the obliquity of the ecliptic (ε) is 0° and 180° (Figure 2a) there are no polar ice caps and a warm climate without seasons extends to high latitudes. As the obliquity of the ecliptic increases (Figure 2b) so polar ice sheets form, and they and their girdles of marine glacial deposition extend equatorward. When the obliquity of the ecliptic is 90° and 270° (Figure 2c) the northern and southern hemispheres are alternating between a summer of continuous heat and a winter of extreme cold. At the equator during equinoxes the climate would be similar to that at the equator today.

The dramatic contrast of climates when the obliquity of the ecliptic was near 90° and 270° would produce some strange sedimentary bedfellows. During solstices there would be enormous evaporation at the summer pole, and evaporites might be expected there. The abundant atmospheric moisture probably would condense at middle and low latitudes within the winter hemisphere, the winter pole remaining arid. Thus, as the obliquity of the ecliptic approached 90° and 270°, continental ice sheets would be confined largely to middle and low latitudes, where tills and marine glacial deposits would accumulate; polar glaciation would be restricted mainly to mountains. When the Earth's equator was virtually perpendicular to the orbital plane, however, it is likely that the continental ice sheets would disappear due to the extreme heat of the summers, and marine glacial deposition temporarily cease. Throughout this period warm-water marine sedimentation would occur near the equator during spring and autumn, and the accumulation of marine dolomites and limestones, together with other marine sediments, might be expected there.

This model accounts for (1) the association of Precambrian glacial deposits with carbonate rocks; (2) the common two-fold subdivision of the late Precambrian glacial deposits into a lower, thicker and more widespread subunit and an upper, thinner and more localized subunit (Dunn *et al.*, 1971); and (3) the regional variation in time span of late Precambrian glaciation. As the obliquity of the ecliptic moved toward 270° in late Precambrian time, so the zones in which sea-ice melted and marine glacial deposition occurred during glacial advances approached the equator, eventually crossing into the opposite hemisphere. With the cessation of glacial deposition for about 100 m.y. when the obliquity of the ecliptic was about 270 ± 7°, however, marine deposition prevailed at the equator. As the obliquity of the ecliptic progressed beyond 270°, small glaciers reformed and marine glacial deposition recommenced. This resulted in thick, widespread transgressive and thinner, more localized regressive marine glacial facies overlapping at the equator (Figure 3). There would thus be two main glacial formations at the equator which thinned and diverged to the north and south. Glacial formations in low palaeo-latitudes would be interbedded with marine deposits including dolomites and limestones. Varves might also be expected in the interglacial sequences. It may be significant that regular laminations having a "superficial resemblance to glacial varve deposits" occur in shales and silt-stones of the interglacial sequence in South Australia (Parkin, 1969, p. 66).

The Late Ordovician glacial deposits in north-west Africa (Fairbridge, 1970a, b, 1971) also exhibit evidence of marked seasonal variation of climate, as would be expected with the proposed secular change in the obliquity of the ecliptic.

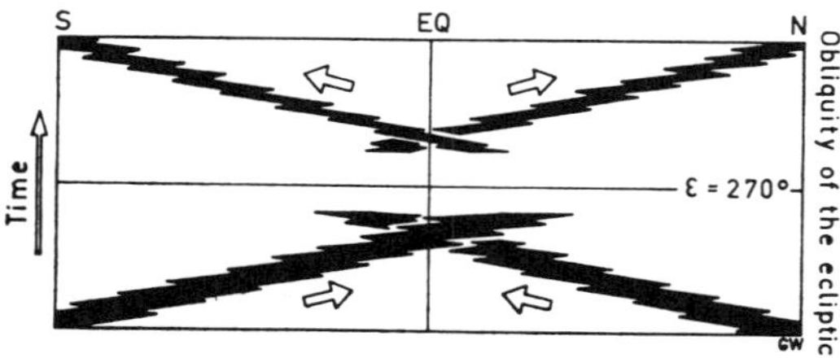

FIG. 3. Schematic section illustrating the relationship of transgressive-regressive marine glacial facies of late Precambrian age.

Features of palaeoclimatic significance are: (1) the predominance of sand throughout the entire section; (2) vast outwash sheets covering thousands of square kilometres, with sandstone units up to 20 m thick and hundreds of kilometres in extent exhibiting hydrodynamic evidence of very-high-velocity currents, and suggesting "a catastrophic decanting of meltwaters" (Fairbridge, 1970a, p. 878); (3) evidence for the grounding of ice in fossiliferous marine sediments; (4) the presence of fossils or tracks of trilobites, and traces of other marine life, systematically almost through the entire section; (5) evidence for the freezing of loose sands into temporary "bedrock" upon which tills were deposited; and (6) long parallel grooves, cut in outwash sandstones, which extend for hundreds of kilometres. Underlying Precambrian rocks are weathered and bleached to depths of 3 to 4 m and are capped by a residual hematitic crust or palaeosol.

The sedimentary features suggest severe winters when unconsolidated sands were frozen and they and marine sediments overridden by advancing ice, alternating with warm summers when rapid melting of the ice caused catastrophic floods, marine life expanded far into polar waters, and weathering of debris produced an abundance of sand. Moreover, huge blocks of ice may have calved from the melting ice sheets and slid down the outwash plains for many kilometres, producing long parallel grooves in the frozen sands beneath. The weathering of the underlying rocks may have occurred under the dramatic two-seasonal conditions of the late Precambrian.

In contrast with the glacial sequences of late Precambrian and Late Ordovician age, the Permo-Carboniferous glacial sequences of Gondwanaland (Frakes and Crowell, 1969, 1970; Frakes *et al.*, 1971; Crowell and Frakes, 1971a, b) do not exhibit evidence of drastic contrast of seasons. Although there is evidence for advances and retreats of the ice sheets and for high-velocity floods of limited extent, and marine glacial drop-stones locally rest on fossiliferous marine sediments, these features can be explained by glacial and stadial fluctuations comparable with those of the Quaternary. The climatic spectrum indicated by the Permo-Carboniferous glacial deposits is thus consistent with the postulate of secular change in the obliquity of the ecliptic, as during the Late Carboniferous and Permian the Earth's equator would have been inclined between 30° and 20° to the ecliptic (see Figure 1), values comparable with

[*Editor's Note:* The author's updated views on the cause of the common two-fold subdivison of late Precambrian glacial deposits are given in Paper 17.]

those for the Quaternary (about 21.5° to 24.5° according to Zeuner, 1959, p. 175). The Permian deserts of the northern hemisphere may have expanded equatorward during episodes of glacial advance, as did their Quaternary counterparts (Fairbridge, 1964, 1970c; Damuth and Fairbridge, 1970). Australia's deglaciation toward the end of the Permian cannot be linked with the breakup of Gondwanaland, as Australia remained in high latitudes close to Antarctica until the early Tertiary (Dietz and Holden, 1970a, b).

It appears that ice caps rapidly developed during the late Cenozoic when the obliquity of the ecliptic exceeded about 16°. Stadial advances of ice sheets far beyond the polar circles may have been triggered by perturbations of the Earth's motions as postulated by Milankovitch Theory, modified by circulation patterns in the atmosphere and hydrosphere.

Finally, it is pointed out that the dashed line shown in Figure 1, representing the latitude of palaeo-polar circles, may be slightly displaced to the right with respect to its true position. This is because key points for the late Precambrian and the Early Cretaceous were selected as *midpoints* of respective climatic intervals, and therefore take no account of lag effects due to the high albedo of ice sheets and to the Earth's thermal inertia. Any error, however, is probably no more than a few tens of millions of years.

SEDIMENTARY AND PEDOLOGICAL RECORDS

An immediate sedimentological application of the hypothesis of secular change in the obliquity of the ecliptic is the explanation it provides for the accumulation of Precambrian banded iron formation (BIF), the rhythmic and laterally persistent microbanding of which has previously defied interpretation using uniformitarian principles. A global episode of deposition of BIF, where dated with reasonable accuracy, falls in the general range of 1800 to 2200 m.y. old, and commonly around 2000 to 2100 m.y. (Cloud and Licari, 1972, p. 140). According to the present hypothesis, these BIF were deposited when the Earth's spin axis lay in the orbital plane ($\varepsilon \simeq 90°$, see Figure 1), that is, when the northern and southern palaeohemispheres alternately baked and froze (Figure 2c). So also were the banded sedimentary ironstones associated with early Proterozoic glacial deposits of the Transvaal ($\varepsilon \simeq 90°$), and with late Precambrian glacial deposits ($\varepsilon \simeq 270°$).

The stupendous BIF of the Hamersley Group, Western Australia, is close to 2000 m.y. old (Trendall and Blockley, 1970). Features of the Hamersley Group BIF which cannot be explained in terms of modern sedimentary processes include (1) the abundant iron and its source, and (2) macro-, meso- and microbanding, mainly of silica and iron oxide, displaying incredible regularity and lateral persistence, some microbands about 0.5 mm thick having been correlated over a distance of 300 km. The present hypothesis provides plausible explanations for these features.

Much iron and silica was probably derived from the intense weathering and oxidation of associated basic rocks during the summers, thence being carried basinward in solution and colloidal suspension by surface waters and groundwater; solubility of iron may have been significantly increased by the high temperatures of basin waters. Freezing of basin waters during the winters may have caused widespread precipitation of the iron and silica. Trendall and Blockley (1970), on the other hand, consider that the microbands are annual varves deposited by seasonal evaporation in an arid climate. Whether winter or summer precipitation, the model still holds and accounts for the lateral persistence of the microbands. Although the final answers, particularly those geochemical (see Hem, 1972), must await a full re-evaluation of data, I consider that the concept of secular change in the obliquity of the ecliptic may provide the key to the BIF puzzle. Proterozoic BIF should be re-examined in this light.

The Torridon Group of northwest Scotland (Selley, 1965; Stewart, 1969; Williams, 1969) is another sedimentary sequence which must be re-examined in the light of secular change in the obliquity of the ecliptic. These rocks were deposited at a palaeolatitude of about 30° (Irving, 1956) approximately 751 ± 24 m.y. ago (Moorbath, 1969) during the "interglacial" when, according to the present model, the obliquity of the ecliptic was about 270°. The marked seasonal changes of climate that would have prevailed at that time account for the rapid weathering and extensive pedimentation of the crystalline source rocks (Williams, 1968, 1969). Flash floods derived from annual rapid thaw of snows in the highland source area better account for the deposition of the extensive piedmont alluvial fans of the Applecross Formation (Williams, 1969). The scree breccias near the base of the sequence suggest frost shattering; associated "highly fissile" grey shales (Selley, 1965) may be annual varves. Palaeoclimatic interpretation of the entire Group is thrown back into the melting pot, as is that of many other continental red bed sequences.

Palaeotidal deposits also should be re-examined. For example, tidal ranges would have been greater and tidal currents stronger than those of today when the obliquity of the ecliptic was about 0° and 180°, because of the combined pull of the Moon and Sun directly in the Earth's equatorial plane. Thus, ordinary spring tides at such times would be comparable with equinoctial spring tides today. The widespread marine transgressions which commenced in the Early Cretaceous may have been assisted by a vigorous tidal environment. For instance, the Lower Albian Folkestone Beds of southeastern England (Allen and Narayan, 1964) are probable tidal deposits presenting many remarkable features, in particular relatively coarse-grained cross-stratified sets up to and exceeding 5 m in thickness, which are consistent with vigorous tidal conditions. Tidal currents without modern analogue may have been generated when the obliquity of the ecliptic was near 90° and 270°.

Short-period (25,000 year) changes of attitude in space of the Earth's spin axis (precession and nutation; see Munk and MacDonald, 1960) probably would have been pronounced when the obliquity of the ecliptic was near 90° and 270°, due to the greatly increased pull of the Moon and Sun tangentially on the Earth's equatorial bulge. Such perturbations may be revealed by the climatic oscillations, with periods of up to several millennia, indicated by sedimentary rhythms in the BIF of the Hamersley Group (Trendall and Blockley, 1970). Other sedimentary cycles deposited near 2000 and 750 m.y. ago, such as the unique coarsening-upward cycles in fluvial deposits of the Torridon Group (Williams, 1969), may also reflect pronounced perturbations of the Earth's spin axis.

Palaeopedological evidence for secular change in the obliquity of the ecliptic is provided by the temporal distribution of bauxites and laterites, which are generally regarded as the products of chemical weathering under *two-seasonal* warm-wet conditions. Most bauxite deposits are of Late Cretaceous to middle Tertiary age, and are known also from the Devonian, Permo-Carboniferous and Triassic (Bateman, 1958, pp. 218–220; Overseas Geological Surveys, 1962; Grubb, 1970). Apparently no true bauxite developed during the Jurassic, despite the widespread peneplanation and deep weathering that occurred during this period (Parkin, 1969; King, 1967; Ollier, 1969, p. 263). Moreover, most laterites are of Tertiary age, many having developed on Mesozoic peneplains (Maignien, 1966). This temporal distribution of bauxites and laterites, and the deep weathering of the Jurassic and Early Cretaceous, accords with the concept of a warm climate with virtually no seasonal changes during the Jurassic and Early Cretaceous, and the return to a two-seasonal climate in the Late Cretaceous.

BIOLOGICAL RECORD

The dramatic changes of climate and seasonal contrast resulting from secular change in the obliquity of the ecliptic would have a profound influence on the evolution of life.

The sudden proliferation of animal life in Early Cambrian time, "a biological event of profound significance in the history of life" (Harland and Rudwick, 1964, p. 36), may be plausibly explained by the present model. With the amelioration of Earth climate and the severity of seasons in the Early Cambrian, marine faunas would have migrated outward from their late Precambrian equatorial haven. This migration and concomitant proliferation of life marks the start of the present *Life Megacycle*. Neritic faunas may have developed external shells to protect them from the solar radiation of the long, hot summers. Moreover, the radiation may have triggered dramatic changes in the evolutionary pattern by its acceleration of organic mutation rates.

As the life cycles of many modern species are wholly dependent on the seasons, major change in present seasonal contrast would be catastrophic for many forms of life. By analogy, the faunal and floral extinctions at the end of the Permian and Cretaceous, and the rise of new forms in Triassic and Late Cretaceous–early Tertiary times, are readily explained in terms of changing contrast of seasons. Secular change in the obliquity of the ecliptic may therefore explain the divisibility of the Phanerozoic into Palaeozoic, Mesozoic and Cenozoic Eras. Moderate-period (80–90 m.y.) waves of extinctions which appear to be superimposed on these long-period faunal and floral variations may be controlled by periodicity of the Sun's orbit about the galactic centre (see Hatfield and Camp, 1970; Meyerhoff, 1971).

The rapid evolution and abrupt extinction of the dinosaurs, for example, accords with change in contrast of seasons. According to the present hypothesis there would have been minimal seasonal contrast during most of the Mesozoic, and the Earth's climate would have been uniformly warm over wide areas. Such a warm, uniform Mesozoic climate, so often stressed by stratigraphers and palaeontologists (Arkell, 1935, 1969; Colbert, 1962), would have provided optimum conditions for the evolution and physical growth of cold-blooded reptiles. It is significant that, according to Figure 1, the obliquity of the ecliptic was 0° about 120 m.y. ago (Late Jurassic–Early Cretaceous), the time when the sauropods, the greatest of all dinosaurs, attained their maximum physical dimensions (Colbert, 1962). It is postulated that the dinosaurs were unable to adapt to the seasonal climate which commenced toward the end of the Cretaceous, and rapidly became extinct. In contrast, warm-blooded mammals readily adapted to the changing climate.

Ma (1933, 1934) was first to study annual growth increments in fossil and recent corals, concluding that seasonal changes of temperature occurred during the Silurian, Devonian and Carboniferous. Since Ma's pioneering work, annual growth increments have been recognized in Devonian and Carboniferous corals (Wells, 1963; Scrutton, 1964), Silurian and Devonian corals and brachiopods (Mazzullo, 1971), and Late Cretaceous bivalves (Berry and Barker, 1968). Despite the prolific shelly fauna of the Mesozoic, Jurassic and Early Cretaceous data are not included in these studies. It is noteworthy in this regard that Arkell (1935, p. 102), in a study of British Jurassic corals as palaeoclimatic indicators, "cut longitudinal sections of some dendroid corals (*Thecosmilia annularis*) from the Rag [Late Jurassic] of Wiltshire and Dorset (Ringstead Coral Bed), but without finding any banding of the kind discovered by Ma in, among others, Carboniferous corals from the Avon gorge."

The above palaeontological studies are in accord with the present model, which requires strong seasonal contrast during the Palaeozoic and minimal seasonal contrast over most of the Earth during the Jurassic and Early Cretaceous. However, short-period (40,000 year) perturbations of the obliquity of the ecliptic similar to those of the Pleistocene (Zeuner, 1959) may have maintained moderate seasonal conditions at high palaeolatitudes. Such perturbations may account for the occurrence of growth rings in fossil wood, dissepiment rings in certain fossil corals, and ledging or ringing of the growth lines of certain pelecypods of Jurassic age "in what are now temperate and Arctic regions" (Arkell, 1956, p. 617). The study of floral and faunal annual growth structures, combined with accurate determination of species, age and palaeolatitude of growth, may provide the best palaeontological test of the hypothesis of secular change in the obliquity of the ecliptic.

Extension of the curve in Figure 1 into the future predicts that the present Life Megacycle will end within the next 500 m.y. During this period most terrestrial life will gradually migrate equatorward in pace with the shift of equable environments.

GEOMAGNETIC RECORD

In view of the influence of solar flares and the solar wind on the strength and figure of the Earth's magnetic field (Ellison, 1968), it is possible that secular change in the obliquity of the ecliptic, whereby the attitude of the geomagnetic axis probably changes with respect to such solar radiation, would be reflected in some way by the geomagnetic record.

As discussed above, the present model of long-period (1000–2000 m.y.) variation in attitude of the Earth's spin axis with respect to the orbital plane, combined with moderate-period (100–200

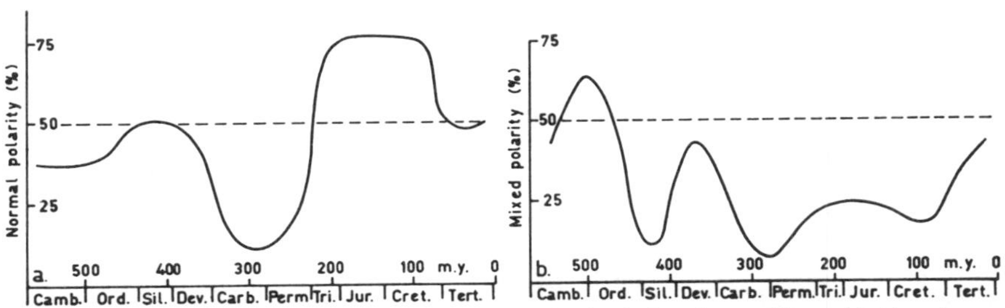

FIG. 4. (a) Percentage of normal polarity, and (b) percentage of mixed polarity, observed in worldwide palaeomagnetic investigations for the Phanerozoic. After McElhinny (1971, Figures 1 and 2).

m.y.) galactic revolution of the solar system (Steiner, 1967; Meyerhoff, 1971), may account for the "inclined sawtooth" distribution pattern of glaciation during the Phanerozoic (Figure 1). This pattern has broad similarities with those obtained by McElhinny (1971) for percentage normal polarity and percentage mixed polarity observed in worldwide palaeomagnetic investigations for the Phanerozoic (Figure 4). The temporal patterns for geomagnetic reversals may therefore reflect (1) long-period change of attitude of the Earth's geomagnetic axis, relative to solar radiation, resulting from secular change in the obliquity of the ecliptic; and (2) moderate-period vibrations of the solar system with respect to the galactic magnetic field and the band of high energy gamma radiation coincident with the galactic plane (see Clark *et al.*, 1968; Shen and Berkey, 1968).

[*Editor's Note:* In pages 173 to 178 Williams presents an hypothesis for the origin of the solar system which incorporates features (ejection of matter by an exploding star) in accord with current consensus on this subject (see G. J. Wasserburg, *Geotimes*, June 1978, pp. 12–13).]

REFERENCES

Abetti G. 1962. *The Sun.* Faber and Faber, London.

Alexander A. F. O'D. 1965. *The Planet Uranus.* Faber and Faber, London.

Alfvén H. 1954. *On the Origin of the Solar System.* Clarendon Press, Oxford.

Allen J. R. L. and Narayan J. 1964. Cross-stratified units, some with silt bands, in the Folkestone Beds (Lower Greensand) of southeast England. *Geol. Mijnbouw* **43,** 451–461.

Arkell W. J. 1935. On the nature, origin, and climatic significance of the coral reefs in the vicinity of Oxford. *Quart. J. Geol. Soc. London* **91,** 77–110.

Arkell W. J. 1956. *Jurassic Geology of the World.* Oliver and Boyd, Edinburgh.

Arkell W. J. 1969. Jurassic System. *Encyclopaedia Britannica* **13,** 145–149.

Arp H. 1971. Observational paradoxes in extragalactic astronomy. *Science* **174,** 1189–1200.

Arrhenius G. and Alfvén H. 1971. Fractionation and condensation in space. *Earth Planet. Sci. Letters* **10,** 253–267.

Bateman A. M. 1958. *Economic Mineral Deposits* (2nd edn.). Wiley, New York.

Bennacef A., Beuf S., Biju-Duval B., de Charpal O., Gariel O. and Rognon P. 1971. Example of cratonic sedimentation: Lower Paleozoic of Algerian Sahara. *Bull. Amer. Assoc. Petroleum Geologists* **55,** 2225–2245.

Berry W. and Barker R. 1968. Fossil bivalve shells indicate longer month and year in Cretaceous than present. *Nature* **217,** 938–939.

Birch, F. 1965. Speculations on the Earth's thermal history. *Bull. Geol. Soc. Amer.* **76,** 133–154.

Black L. P., Gale N. H., Moorbath S., Pankhurst R. J. and McGregor V. R. 1971. Isotopic dating of very early Precambrian amphibolite facies gneisses from the Godthaab district, west Greenland. *Earth Planet. Sci. Letters* **12,** 245–259.

Brandt J. C. and Hodge P. W. 1964. *Solar System Astrophysics.* McGraw-Hill, New York.

Bray R. J. and Loughhead R. E. 1964. *Sunspots.* Chapman and Hall, London.

Brown D. A., Campbell K. S. W. and Crook K. A. W. 1968. *The Geological Evolution of Australia and New Zealand.* Pergamon, Oxford.

Brown W. K. 1971. A solar system formation model based on supernova shell fragmentation. *Icarus* **15,** 120–134.

Bruzek A. 1969. Flare associated optical phenomena. In *Solar Flares and Space Research,* C. de Jager and Z. Švestka, Editors. North-Holland, Amsterdam, pp. 61–77.

Clark G. W., Garmire G. P. and Kraushaar W. L. 1968. Observation of high-energy cosmic gamma rays. *Astrophys. J.* **153,** L203–L207.

Clarke G. R. 1968. Mollusk shell: daily growth lines. *Science* **161,** 800–802.

Clemence G. M. 1963. Astronomical reference systems. In *Basic Astronomical Data,* K. A. Strand, Editor. Univ. Chicago Press, Chicago, pp. 1–10.

Cloud P. and Licari G. R. 1972. Ultrastructure and geologic relations of some two-aeon old Nostocacean algae from northeastern Minnesota. *Amer. J. Sci* **272,** 138–149.

Colbert E. H. 1962. *Dinosaurs. Their Discovery and their World.* Hutchinson, London.

Crittenden M. D. 1971. Tillites as potential time lines in the Precambrian. *Geol. Soc. Amer. Abs. with Prog., Ann. Meetings* **3**(7), 534.

Crowell J. C. and Frakes L. A. 1970. Phanerozoic glaciation and the causes of ice ages. *Amer. J. Sci.* **268,** 193–224.

Crowell J. C. and Frakes L. A. 1971a. Late Palaeozoic glaciation of Australia. *J. Geol. Soc. Aust.* **17,** 115–155.

Crowell J. C. and Frakes L. A. 1971b. Late Paleozoic glaciation: Part IV, Australia. *Bull. Geol. Soc. Amer.* **82,** 2515–2540.

Dalgarno C. R. and Johnson J. E. 1965. The Holowilena Ironstone, a Sturtian glacigene unit. *Geol. Surv. South Aust. Quart. Geol. Notes* **13,** 2–4.

Damuth J. E. and Fairbridge R. W. 1970. Equatorial Atlantic deep-sea arkosic sands and ice-age aridity in tropical South America. *Bull. Geol. Soc. Amer.* **81**, 189–206.

Dearnley R. 1966. Orogenic fold belts and a hypothesis of Earth evolution. *Phys. Chem. of the Earth* **7**, 1–114.

Dicke R. H. 1964. The Sun's rotation and relativity. *Nature* **202**, 432–435.

Dietz R. S. and Holden J. C. 1970a. Reconstruction of Pangaea: breakup and dispersion of continents, Permian to present. *J. Geophys. Res.* **75**, 4939–4956.

Dietz R. S. and Holden J. C. 1970b. The breakup of Pangaea. *Sci. Amer.* **223**(4), 30–41.

Dietz R. S. and Sproll W. P. 1966. Equal areas of Gondwana and Laurasia (ancient supercontinents). *Nature* **212**, 1196–1198.

Dingle H. D. and Mohler O. C. 1969. Sun. *Encyclopaedia Britannica* **21**, 414–418.

Dunn P. R., Plumb K. A. and Roberts H. G. 1966. A proposal for time-stratigraphic subdivision of the Australian Precambrian. *J. Geol. Soc. Aust.* **13**, 593–608.

Dunn P. R., Thomson B. P. and Rankama K. 1971. Late Precambrian glaciation in Australia as a stratigraphic boundary. *Nature* **231**, 498–502.

Dyson F. J. 1971. Energy in the Universe. *Sci. Amer.* **224**(3), 51–59.

Egyed L. 1957. A new dynamic conception of the internal constitution of the Earth. *Geol. Rundschau* **46**, 101–121.

Ellison M. A. 1968. *The Sun and its Influence* (3rd edn.). Routledge and Kegan Paul, London.

Fairbairn H. W., Hurley P. M., Card K. D. and Knight C. J. 1969. Correlation of radiometric ages of Nipissing diabase and Huronian metasediments with Proterozoic orogenic events in Ontario. *Canad. J. Earth Sci.* **6**, 489–497.

Fairbridge R. W. 1964. Eiszeitklima in Nordafrica. *Geol. Rundschau* **54**, 399–414.

Fairbridge R. W. 1966. Endospheres and interzonal coupling. *Ann. New York Acad. Sci.* **140**, 133–148.

Fairbridge R. W. 1969a. On the absolute fall of sea-level during the Quaternary: a discussion. *Palaeogeog., Palaeoclimatol., Palaeoecol.* **6**, 241–242.

Fairbridge R. W. 1969b. Early Paleozoic South Pole in northwest Africa. *Bull. Geol. Soc. Amer.* **80**, 113–114.

Fairbridge R. W. 1970a. South Pole reaches the Sahara. *Science* **168**, 878–881.

Fairbridge R. W. 1970b. An ice-age in the Sahara. *Geotimes* **15**(6), 18–20.

Fairbridge R. W. 1970c. World paleoclimatology of the Quaternary. *Rev. Géog. Phys. Geól. Dynam.* **12**, 97–104.

Fairbridge R. W. 1971. Upper Ordovician glaciation in northwest Africa? Reply. *Bull. Geol. Soc. Amer.* **82**, 269–274.

Flint R. F. 1971. *Glacial and Quaternary Geology.* Wiley, New York.

Frakes L. A. and Crowell J. C. 1969. Late Paleozoic glaciation: I, South America. *Bull. Geol. Soc. Amer.* **80**, 1007–1042.

Frakes L. A. and Crowell J. C. 1970. Late Paleozoic glaciation: II, Africa exclusive of the Karroo Basin. *Bull. Geol. Soc. Amer.* **81**, 2261–2286.

Frakes L. A., Matthews J. L. and Crowell J. C. 1971. Late Paleozoic glaciation: Part III, Antarctica. *Bull. Geol. Soc. Amer.* **82**, 1581–1604.

Frye J. C., Willman H. B. and Black R. F. 1965. Outline of glacial geology of Illinois and Wisconsin. In *The Quaternary of the United States*, H. E. Wright and D. G. Frey, Editors. Princeton Univ. Press, Princeton, pp. 43–61.

Gill E. D. 1961. The climates of Gondwanaland in Kainozoic times. In *Descriptive Palaeoclimatology*, A. E. M. Nairn, Editor. Interscience, London, pp. 332–353.

Grubb P. L. C. 1970. Mineralogy, geochemistry, and genesis of the bauxite deposits on the Gove and Mitchell Plateaux, northern Australia. *Mineral. Deposita* **5**, 248–272.

Guild P. W. 1957. Geology and mineral resources of the Congonhas District, Minas Gerais, Brazil. *U.S. Geol. Surv. Prof. Paper* **290.**

Harland W. B. 1964a. Critical evidence for a great Infra-Cambrian glaciation. *Geol. Rundschau* **54**, 45–61.

Harland W. B. 1964b. Evidence of late Precambrian glaciation and its significance. In *Problems in Palaeoclimatology*, A. E. M. Nairn, Editor. Interscience, New York, pp. 119–149.

Harland W. B. and Rudwick M. J. S. 1964. The great Infra-Cambrian ice age. *Sci. Amer.* **211**(2), 28–36.

Hatfield C. B. and Camp M. J. 1970. Mass extinctions correlated with periodic galactic events. *Bull. Geol. Soc. Amer.* **81**, 911–914.

Hatfield W. C. 1937. The geology of the Solwezi District, Northern Rhodesia. *Quart. J. Geol. Soc. London* **93**, 127–155.

Haughton S. H. 1969. *Geological History of South Africa.* Geol. Soc. South Africa, Cape Town.

Hays J. F. 1972. Radioactive heat sources in the lunar interior. *Phys. Earth Planet. Interiors* **5**, 77–84.

Hem J. D. 1972. Chemical factors that influence the availability of iron and manganese in aqueous systems. *Bull. Geol. Soc. Amer.* **83**, 443–450.

Hinners N. W. 1971. The new Moon: a view. *Rev. Geophys. Space Phys.* **9**, 447–522.

Holmes A. 1965. *Principles of Physical Geology.* Nelson, London.

Hoyle F. 1965. *Galaxies, Nuclei, and Quasars.* Heinemann, London.

Hurley P. M. and Rand J. R. 1969. Pre-drift continental nuclei. *Science* **164**, 1229–1242.

Irving E. 1956. Palaeomagnetic and palaeoclimatological aspects of polar wandering. *Rev. Geofisica Pura Applicata* **33**, 23–41.

Irving E. 1964. *Paleomagnetism.* Wiley, New York.

Jastrow R. and Cameron A. G. W. (Editors) 1963. *Origin of the Solar System.* Academic Press, New York.

Jeffreys H. 1970. *The Earth, its Origin, History and Physical Constitution* (5th edn.). Cambridge Univ. Press, Cambridge.

King L. C. 1967. *The Morphology of the Earth* (2nd edn.). Oliver and Boyd, Edinburgh.

Kuenen P. H. 1950. *Marine Geology.* Wiley, New York.

Kuhi L. V. 1966. T Tauri mass ejection. In *Stellar Evolution*, R. F. Stein and A. G. W. Cameron, Editors. Plenum Press, New York, pp. 373–376.

Larimer J. W. and Anders E. 1970. Chemical fractionations in meteorites—II. Abundance patterns and their interpretation. *Geochim. Cosmochim. Acta* **31**, 1239–1270.

Lindsey D. A. 1969. Glacial sedimentology of the Precambrian Gowganda Formation, Ontario, Canada. *Bull. Geol. Soc. Amer.* **80**, 1685–1702.

Lindsey D. A. 1971. Glacial marine sediments in the Precambrian Gowganda Formation at Whitefish Falls, Ontario (Canada). *Palaeogeog., Palaeoclimatol., Palaeoecol.* **9**, 7–25.

Ma T. Y. H. 1933. On the seasonal change of growth in some Palaeozoic corals. *Proc. Imp. Acad. Tokyo* **9**, 407–409.

Ma T. Y. H. 1934. On the seasonal change of growth in a reef coral, *Favia speciosa* (Dana), and the water temperature of the Japanese seas during the latest geological times. *Proc. Imp. Acad. Tokyo* **10**, 353–355.

MacDonald G. J. F. 1966. Origin of the Moon: dynamical considerations. In *The Earth-Moon System*, B. G. Marsden and A. G. W. Cameron, Editors. Plenum Press, New York, pp. 165–209.

Maignien R. 1966. *Review of Research on Laterites.* UNESCO Natural Resources Research IV, Paris.

Mawson D. 1949. The late Precambrian ice age and glacial record of the Bibliando Dome. *J. Proc. Roy. Soc. New South Wales* **82**, 150–174.

Mazzullo S. J. 1971. Length of the year during the Silurian and Devonian Periods: new data. *Bull. Geol. Soc. Amer.* **82**, 1085–1086.

McElhinny M. W. 1971. Geomagnetic reversals during the Phanerozoic. *Science* **172**, 157–159.

McElhinny M. W., Briden J. C., Jones D. L. and Brock A. 1968. Geological and geophysical implications of paleomagnetic results from Africa. *Rev. Geophys.* **6**, 201–238.

McElhinny M. W. and Luck G. R. 1970. Paleomagnetism and Gondwanaland. *Science* **168**, 830–832.

Meyerhoff A. A. 1970a. Continental drift: implications of paleomagnetic studies, meteorology, physical oceanography, and climatology. *J. Geol.* **78**, 1–51.

Meyerhoff A. A. 1970b. Continental drift, II: high latitude evaporite deposits and geologic history of Arctic and North Atlantic Oceans. *J. Geol.* **78**, 406–444.

Meyerhoff A. A. 1971. Galactic motions, world climate, and mass biotal extinctions: possible interrelations (abs.). *Bull. Canad. Petroleum Geol.* **19**, 340.

Meyerhoff A. A. and Meyerhoff H. A. 1972. "The new global tectonics": major inconsistencies. *Bull. Amer. Assoc. Petroleum Geologists* **56**, 269–336.

Meyerhoff A. A. and Teichert C. 1971. Continental drift, III: late Paleozoic glacial centers, and Devonian-Eocene coal distribution. *J. Geol.* **79**, 285–321.

Mikhaylov Y. A., Ustritskiy V. I., Chernyak G. Y. and Yavshits G. P. 1970. Upper Permian glaciomarine sediments of the northeastern USSR. *Doklady Akad. Nauk SSSR* **190**, 100–102.

Mirams R. C. 1962. The geology of the Manunda military sheet. *Geol. Surv. South Aust. Rep. Invest.* **19.**

Moorbath S. 1969. Evidence for the age of deposition of the Torridonian sediments of north-west Scotland. *Scott. J. Geol.* **5**, 154–170.

Munk W. H. and MacDonald G. J. F. 1960. *The Rotation of the Earth.* Cambridge Univ. Press, Cambridge.

Nicolaysen L. O., de Villiers J. W. L., Burger A. J. and Strelow F. W. E. 1958. New measurements relating to the absolute age of the Transvaal System and the Bushveld Igneous Complex. *Trans. Geol. Soc. South Africa* **61**, 137–163.

Ollier C. D. 1969. *Weathering.* Oliver and Boyd, Edinburgh.

Oversby V. M. and Ringwood A. E. 1971. Time of formation of the Earth's core. *Nature* **234**, 463–465.

Overseas Geological Surveys Mineral Resources Division 1962. *Bauxite, Alumina and Aluminium.* H.M. Stationery Office, London.

Parkin L. W. 1969. *Handbook of South Australian Geology.* Geol. Surv. South Aust., Adelaide.

Payne-Gaposchkin C. 1961. *Stars in the Making.* Eyre and Spottiswoode, London.

Reading H. G. and Walker R. G. 1966. Sedimentation of Eocambrian tillites and associated sediments in Finnmark, northern Norway. *Palaeogeog., Palaeoclimatol., Palaeoecol.* **2**, 177–212.

Roberts J. D. 1971. Late Precambrian glaciation: an anti-greenhouse effect. *Nature* **234**, 216–217.

Rossner L. F. 1972. Blast waves in rotating media. *Nature* **235**, 68–69.

Runcorn S. K. 1966. Corals as paleontological clocks. *Sci. Amer.* **215(4)**, 26–33.

Runcorn S. K. 1970. Palaeontological measurements of the changes in the rotation rates of Earth and Moon and of the rate of retreat of the Moon from the Earth. In *Palaeogeophysics*, S. K. Runcorn, Editor. Academic Press, London, pp. 17–23.

Russell R. C. H. and MacMillan D. H. 1952. *Waves and Tides.* Hutchinson, London.

Saito R. 1969. Glacier problems of late Pre-Cambrian eon. *Kumamoto J. Sci.*, Ser. B, Sect. 1, **8**, 7–44.

Salop L. I. 1968. Pre-Cambrian of the U.S.S.R. *Internat. Geol. Congr., Rep. Twenty-third Session, Czechoslovakia, 1968, Proc. Sect. 4*, 61–73.

Sanders R. H., Wrixon G. T. and Penzias A. A. 1972. Further evidence of explosive events in the galactic nucleus. *Astron. Astrophys.* **16**, 322–326.

Schatzman E. 1962. A theory of the role of magnetic activity during star formation. *Ann. d'Astrophysique* **25**, 18–29.

Schenk P. E. 1965. Depositional environment of the Gowganda Formation (Precambrian) at the south end of Lake Timagami, Ontario. *J. Sediment. Petrol.* **35**, 309–318.

Schenk P. E. 1972. Possible Late Ordovician glaciation of Nova Scotia. *Canad. J. Earth Sci.* **9**, 95–107.

Scrutton C. T. 1964. Periodicity in Devonian coral growth. *Palaeontology* **7**, 552–558.

Selley R. C. 1965. Diagnostic characters of fluviatile sediments of the Torridonian Formation (Precambrian) of northwest Scotland. *J. Sediment. Petrol.* **35**, 366–380.

Shen C. S. and Berkey G. 1968. Cosmic gamma rays from inverse Compton scattering. *Astrophys. J.* **151**, 895–900.

Steiner J. 1967. The sequence of geological events and the dynamics of the Milky Way Galaxy. *J. Geol. Soc. Aust.* **14**, 99–131.

Stewart A. D. 1969. Torridonian rocks of Scotland reviewed. *Mem. Amer. Assoc. Petroleum Geologists* **12**, 595–608.

Trendall A. F. and Blockley J. G. 1970. The iron formations of the Precambrian Hamersley Group, Western Australia. *Bull. Geol. Surv. Western Aust.* **119.**

Urey H. C. 1952. *The Planets, their Origin and Development.* Yale Univ. Press, New Haven.

Urey H. C. 1954. The origin of the Earth. In *Nuclear Geology*, H. Faul, Editor. Wiley, New York, pp. 355–371.

Ustritsky V. I., Andrianov V. N., Arkhipov Y. V., Ganelin V. G., Korostelyov V. I. and Chernyak G. E. 1971. Permian formation of northeast U.S.S.R. (abs.). *Bull. Canad. Petroleum Geol.* **19**, 368–369.

van Niekerk C. B. and Burger A. J. 1964. The age of the Ventersdorp System. *Ann. Geol. Surv. South Africa* **3**, 75–86.

Visser J. N. J. 1971. The deposition of the Griquatown Glacial Member in the Transvaal Supergroup. *Trans. Geol. Soc. South Africa* **74**, 187–199.

Wasserburg G. J., Schramm D. N. and Huneke J. C. 1969. Nuclear chronologies for the galaxy. *Astrophys. J.* **157**, L91–L96.

Wells J. W. 1963. Coral growth and geochronometry. *Nature* **197**, 948–950.

Whipple F. L. 1968. *Earth, Moon and Planets*. Harvard Univ. Press, Cambridge, and Penguin Books.

Whitten G. F. 1966. Suggested correlation of iron ore deposits within South Australia. *Geol. Surv. South Aust. Quart. Geol. Notes* **18**, 7–11.

Williams G. E. 1968. Torridonian weathering, and its bearing on Torridonian palaeoclimate and source. *Scott. J. Geol.* **4**, 164–184.

Williams G. E. 1969. Characteristics and origin of a Precambrian pediment. *J. Geol.* **77**, 183–207.

Williams I. P. and Cremin A. W. 1968. A survey of theories relating to the origin of the solar system. *Quart. J. Roy. Astron. Soc.* **9**, 40–62.

Zeuner F. E. 1959. *The Pleistocene Period*. Hutchinson, London.

36

Reprinted from pages 7–9 and 20–22 of *Royal Soc. London Philos. Trans.* ser. A, **285**:7–22 (1977)

OUTLINE OF A LUNAR CHRONOLOGY

By G. J. Wasserburg, D. A. Papanastassiou, F. Tera and J. C. Huneke

The Lunatic Asylum of the Charles Arms Laboratory, Division of Geological and Planetary Sciences,†
California Institute of Technology, Pasadena, California 91125, *U.S.A.*

Introduction

We present here an outline of lunar chronology and evolution based on analyses of the isotopic parent-daughter systems ^{87}Rb-^{87}Sr, U-Th-^{206}Pb-^{207}Pb-^{208}Pb and ^{40}K-^{40}Ar. An overview of the chronology will first be given, followed by an outline of the observational basis. A more complete discussion of ^{40}K-^{40}Ar results and their interpretation is presented in the paper by G. Turner in this volume. While the body of data on lunar materials is limited, the chronology for lunar evolution appears to be rather well defined. The samples which have been investigated represent mare basalts [returned by the Apollo missions (11, 12, 15 and 17) and by the Soviet Luna 16 mission] and terra rocks, which include non-mare basalts, anorthosites, troctolites and norites but are predominantly comprised of complex breccias [returned by Apollo 12, 14, 16 and 17 and Luna 20]. The mare basalts are associated with the late stage lava flows which covered the mare basins. These flood basalts have been broken up by impact processes but for the most part are associated with the local areas and have not been subject to major transport or metamorphism by impact. The highland rocks predate the mare lava flows but are not clearly associated with a particular magmatic or impact process. They may have been excavated from considerable depths and transported over wide distances. Impact metamorphism is certainly one of the critical stages in their development.

The techniques which permit the precise dating of lunar rocks are the result of a rapid development over the past nine years in mass spectrometric techniques, in refined chemical separation methods and mineral separation procedures, and in mineral identification and analysis. These methods have been applied most extensively to lunar problems, but in the future they will undoubtedly be utilized in a new generation of studies of meteorites and of terrestrial samples.

General outline

An overall summary of lunar chronology is given in the cartoon in figure 1. Mare volcanism is a clear manifestation of lunar igneous activity and is directly dated by analyses of basalt samples. The period over which mare basalts were extruded extends from 3.9 to 3.1 Ga. Clasts of mare basalts in the more ancient terra breccias are extremely rare. The rate of basalt flooding appears to have decreased with time between 3.9 and 3.1 Ga and mare volcanism seems to have terminated at about 3.0 Ga. The many suggestions of 'young' mare volcanism

† Contribution no. 2705.

predicted by several workers based on photogeology have not been susbtantiated. In particular, the Apollo 17, site which was predicted before the mission to yield very young rocks, has instead yielded many basalt samples with ages of 3.8 Ga. The cartoon shown in figure 1, which was originally drafted before Apollo 17, showed a question mark over the small young volcano; the question mark has been deleted. The ages of lunar samples demonstrate that there has been no major lunar volcanism in the last 3.0 Ga. However, this does not necessarily mean that all lunar volcanic phenomena have ceased. This matter is not yet clear and it is important to keep an open mind to the possible continuation of lunar igneous processes into recent times.

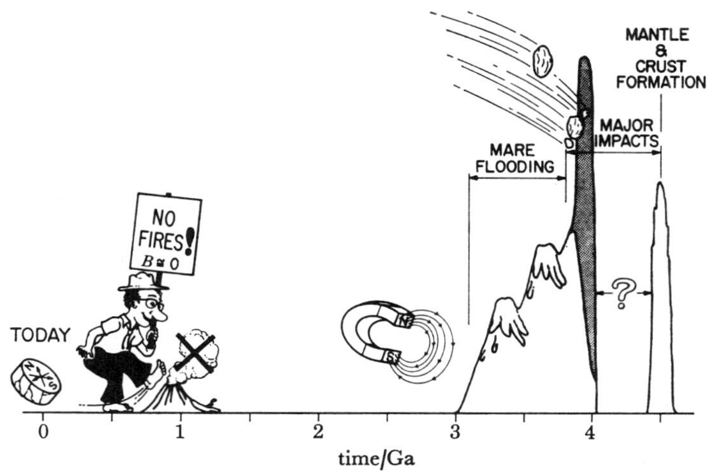

FIGURE 1. Cartoon showing the chronology of major lunar events as presently known or surmised.

Observations of 'transient' phenomena by several workers have suggested that some gaseous emanations from the lunar interior are occurring today. As we shall argue later, a special mechanism is required which permits the moon to be partially molten over its earlier evolution and then to become cool or rigid enough so as to suppress further volcanism. Rigidity of the lunar crust to support mascons is required while some mare volcanism was still going on.

A distinct period of major lunar igneous activity occurred approximately 4.5 Ga ago. This period is close to the time of formation of the moon and is associated with major lunar differentiation. The evidence for this igneous activity is both inferential and direct. Until recently there had been very little direct evidence of truly ancient lunar magmatic rocks. These ancient rocks occur as fragments within younger breccias or as isolated samples and are not obviously related to broad scale igneous morphologic features. Indeed, our ability to recognize primary lunar morphologic features other than impact basins and late mare floods has been very limited. The inferential data on early lunar processes is strong and is obtained from model ages of total rocks and soils. Demonstration of the existence of major early lunar differentiation and examples of the characteristic rock types produced during this process are of great importance since there are no observations of such processes on the earth and only fragmentary data from diverse meteorites. The duration of this early epoch of planetary differentiation is uncertain but appears to be about 0.2 Ga.

In the interval between about 4.4 and 3.8 Ga there is no *strong* evidence either direct or indirect for igneous activity on the moon. This lacuna in lunar history is indicated by a query mark in figure 1. The validity of assigning a long interval ($\sim$ 0.6 Ga) of quiescence following early lunar differentiation remains a fundamental question to be tested by careful observations. Some samples of a highland breccia from the Apollo 16 site yield ^{40}Ar-^{39}Ar apparent ages of up to 4.3 Ga (Schaeffer & Husain 1973). Comparable ^{40}Ar-^{39}Ar ages are observed in samples from highland breccias from the Apollo 17 site. It has not yet been possible to confirm these results by other methods. In some cases the nature of the events dated by the ^{40}Ar-^{39}Ar method is unclear. Because of the steep thermal gradient in the Moon, rocks that are originally at depth may undergo continuous ^{40}Ar loss and the ^{40}K-^{40}Ar system may be measuring the time at which samples were excavated from depth and brought to the surface, where the temperatures are much lower (Podosek *et al.* 1973). In this model the ^{40}Ar-^{39}Ar ages are interpreted as determining the time when fragments were excavated by large impacts during 4.6–4.0 Ga and which escaped recrystallization during the period 4.0–3.8 Ga.

There is abundant evidence of widespread shock metamorphism at a time of *about* 3.9 Ga, apparently in an interval from 4.0 to 3.8 Ga. This episode is reflected in partial to complete isotopic equilibration for all the isotopic systems so far studied. The rock types range from basaltic clasts in breccias to partially and locally recrystallized breccia matrices and to isolated fragments of plutonic rocks. Evidence for metamorphism in this period has so far been found in samples from all sites except Apollo 11 and Luna 16. This metamorphic episode represents a period of major lunar bombardment which has left a distinct imprint on materials covering the whole earth-facing side of the moon. It was first identified by our studies and was called the *terminal lunar cataclysm* (Tera, Papanastassiou & Wasserburg 1974). The question as to whether this was a distinct episode over a very narrow time interval or a series of events over $\sim$ 0.2 Ga is not yet resolved. The terminal lunar cataclysm could represent the Imbrium impact, with a much larger ejecta blanket than predicted or, more plausibly, the formation of several major lunar basins in a highly restricted time interval. The discovery of major impacts at a time of 0.6 Ga after the initial accretion of planetary bodies (with a time constant of 0.01 to possibly 0.10 Ga) is itself of considerable importance. This phenomenon was certainly not predicted and has attracted several workers to propose storage mechanisms for the source of bombarding objects with long time constants (cf. Wetherill 1975). The implications of this late stage lunar bombardment for the cratering histories of all the inner planets have now been recognized by various workers, although the clear-cut applicability of the lunar bombardment time scale to Mercury, Mars and Venus is not self-evident.

The observation of a fossil lunar magnetic field is of major significance (cf. Strangway, Larson & Pearce 1970; Runcorn *et al.* 1970). As indicated in figure 1, the magnetic field was apparently present during the period of lunar volcanism between 3.9 and 3.0 Ga and is absent today. The existence of a field before 3.9 Ga is uncertain, since all of the rocks which are more ancient have been subjected to metamorphism during the terminal lunar cataclysm. The source of the ancient lunar magnetic field is a subject of intense interest, and a possible mechanism for its origin has been proposed (Runcorn & Urey 1973).

[*Editor's Note:* In pages 10 to 20 Wasserburg et al. discuss principally the Rb-Sr and Pb-U-Th dating and systematics of lunar material.]

Lunar chronology evolution and structure

Isotopic analyses of returned lunar samples have permitted the establishment of a rather comprehensive time scale for the Moon. The Earth and the Moon are the only planets for which extensive information now exists on their history.

The true importance of the lunar time scale is in the inferred relationships between age determinations on particular rocks and the significant events in lunar history. Insofar as the connection between the ages, the rock types and major events in lunar history may be properly ascertained, it should be possible to construct a true model of lunar evolution and structure.

We will attempt here to outline a history of lunar evolution based on the arguments outlined in this paper. Such an attempt must be considered preliminary since far more extensive studies will be required before a true understanding is possible.

An oversimplified view

The protoplanetary bodies from which the Moon formed were separated out of the solar nebula with major chemical fractionation between 5 and 10 Ma after the earliest known condensates as determined from initial Sr values. The actual time of aggregation of the Moon is not precisely known, but the Moon existed as a planetary body at 4.45 Ga, based on mutually consistent Rb-Sr and U-Pb data. This is remarkably close to the ^{207}Pb-^{206}Pb age of the Earth and suggests that the Moon and the Earth were formed or differentiated at the same time. If the Moon formed at 4.55 Ga, the radiogenic Pb produced before 4.45 Ga was either lost by volatilization or was buried in a lunar core with other siderophile elements. In the neighbourhood of 4.45 Ga, the Moon underwent major differentiation to produce a crust enriched in K, Rb, U, Th and associated elements and a mantle which was correspondingly depleted. Fractionation of the rare earths took place during this process and the Eu-depleted and Eu-enriched magma reservoirs were concurrently generated. The distinctive magma reservoirs from which later mare basalts derived were formed about this time. At least a small percentage of the mass of the Moon must have been molten in order to provide a sufficient separation of the fractionated chemical elements. Magmatic processes produced not only a low melting fraction but were extensive enough to produce large masses of cumulate rocks characteristic of basic to ultrabasic magmas. The recent discovery by Lee *et al.* (1976) of abundant ^{26}Al in the early solar nebula at the time of condensation and aggregation can provide the basic heat source for early and rapid melting of planets and small planetesimals.

Major lunar differentiation was completed in 0.1 to 0.2 Ga, and the outer ~ 100 km of the Moon became quite rigid. The rate of volcanism after the initial lunar differentiation was low, and the nature of the volcanic rocks produced were distinct from the later mare basalts. The bombardment rate between 4.4 and 4.0 Ga was very high so that the old lunar crust was highly comminuted and the major primary features were destroyed. An epoch of massive bombardment occurred at about 3.95 Ga. This was the terminal lunar cataclysm which produced sufficient debris of metamorphosed materials as to dominate the age pattern over wide areas of the Moon. After 3.8 Ga the bombardment rate suddenly became quite low, thereby preserving the features of younger lava flows. The large impacts at ~ 3.95 Ga generated zones of weakness that penetrated to depths of 200 km or more and triggered the mare volcanism which flooded the impact basins and large depressions. The basaltic rocks represent partial melts derived from magma reservoirs formed during early lunar differentiation. The basalt flows were in general thin sheets, and thick piles only accumulated in deep basins. The mare basins were repeatedly flooded with basalts over a period of a few hundred million years. Many of the basalts penetrated highly fractured ancient radioactive crustal rocks which were thoroughly assimilated. The basalt flooding peaked immediately after the terminal lunar cataclysm and subsequently decreased as the zones of weakness were healed. At about 3.0 Ga the major fracture zones were closed, and the outer mantle and crust became a rigid barrier.

The deep interior of the Moon today remains at elevated temperatures just at the liquidus, but no tectonic forces exist to permit further volcanism.

The time notation used in this paper is the Ga, which is equal to 1 AE or 10^9 years.

In addition to the data obtained at this laboratory, we have utilized a large amount of data from the literature. These include the Pb-U data of P. D. Nunes and M. Tatsumoto, Sr-Rb data of L. E. Nyquist and of C. J. Allegre on Juvinas. We have profited much from models of lunar differentiation proposed by J. A. Wood and colleagues, by S. R. Taylor and P. Jakes, and by D. Walker and colleagues. All of us have profited from the many ideas and contributions of P. W. Gast.

This work was supported by a grant from the National Aeronautics and Space Administration Grant no. NGL-05-002-188.

REFERENCES (Wasserburg *et al.*)

Albee, A. L. & Gancarz, A. J. 1976 *Proceedings of the Soviet-American Conference on Cosmochemistry of the Moon and Planets*, NASA, Washington (in the press).

Allegre, C. J., Birck, J. L., Fourcade, S. & Semet, M. P. 1975 *Science, N.Y.* **187**, 436–438.

Jessberger, E. K., Huneke, J. C. & Wasserburg, G. J. 1974 *Nature, Lond.* **248**, 199–201.

Lee, T., Papanastassiou, D. & Wasserburg, G. J. 1976 *Geophys. Res. Lett.* **3**, 109–112.

Lunatic Asylum 1970*a* *Science, N.Y.* **170**, 463–466.

Lunatic Asylum 1970*b* *Earth Planet. Sci. Lett.* **9**, 137–163.

Nunes, P. D., Tatsumoto, M. & Unruh, D. M. 1974 *Geochim. cosmochim. Acta*, suppl 5, **2**, 1487–1514.

Nunes, P. D., Tatsumoto, M., Knight, R. J., Unruh, D. M. & Doe, B. R. 1973 *Geochim. cosmochim. Acta*, suppl. 4, **2**, 1797–1822.
Nyquist, L. E., Bansal, B. M. & Wiesmann, H. 1975 *Geochim. cosmochim. Acta*, suppl. 6, **2**, 1445–1465.
Papanastassiou, D. A. & Wasserburg, G. J. 1972*a* *Earth Planet. Sci. Lett.* **17**, 52–63.
Papanastassiou, D. A. & Wasserburg, G. J. 1972*b* *Earth Planet. Sci. Lett.* **16**, 289–298.
Papanastassiou, D. A. & Wasserburg, G. J. 1975 *Geochim. cosmochim. Acta*, suppl. 6, **2**, 1467–1489.
Papanastassiou, D. A. & Wasserburg, G. J. 1976 *Geochim. cosmochim. Acta* (in the press).
Podosek, F. A., Huneke, J. C., Gancarz, A. J. & Wasserburg, G. J. 1973 *Geochim. cosmochim. Acta* **37**, 887–904.
Runcorn, S. K. & Urey, H. C. 1973 *Science, N.Y.* **180**, 636–638.
Runcorn, S. K., Collinson, D. W., O'Reilly, W., Stephenson, A., Greenwood, N. N. & Battey, M. H. 1970 *Science, N.Y.* **167**, 697–700.
Schaeffer, A. O. & Husain, L. 1973 *Geochim. cosmochim. Acta*, suppl. 4, **2**, 1847–1863.
Strangway, D. W., Larson, E. E. & Pearce, G. W. 1970 *Science, N.Y.* **167**, 691–693.
Tatsumoto, M., Nunes, P. D., Knight, R. J., Hedge, C. E. & Unruh, D. M. 1973 *Trans. Am. Geophys. Union* **54**, 614–615.
Taylor, S. R. & Jakes, P. 1974 *Geochim. cosmochim. Acta*, suppl. 5, **2**, 1287–1305.
Tera, F., Papanastassiou, D. A. & Wasserburg, G. J. 1974 *Earth Planet. Sci. Lett.* **22**, 1–21.
Tera, F. & Wasserburg, G. J. 1975 *Lunar Sci.* **6** (*abstracts*), 807–809.
Tera, F. & Wasserburg, G. J. 1972 *Earth Planet. Sci. Lett.* **14**, 281–304.
Turner, G. 1971 *Earth Planet. Sci. Lett.* **11**, 169–191.
Walker, D., Grove, T. L., Longhi, J., Stolper, E. M. & Hays, J. F. 1973 *Earth Planet. Sci. Lett.* **20**, 325–336.
Wasserburg, G. J. & Papanastassiou, D. A. 1976 *Nature, Lond.* **259**, 159–160.
Wetherill, G. W. 1956 *Trans. Am. Geophys. Union*, **37**, 320–324.
Wetherill, G. W. 1975 *Geochim. cosmochim. Acta*, suppl. 6, **2**, 1539–1561.
Wood, J. A., Dickey, Jr, J. S., Marvin, U. V. & Powell, B. N. 1970 *Geochim. cosmochim. Acta*, suppl. 1, **1**, 965–988.

37

Reprinted from *Nature* **257**:776–778 (1975)

GALACTIC DUST LANES AND LUNAR SOIL

J. F. Lindsay and L. J. Srnka

IT has been proposed by McCrea[1], Shapley[2] and Hoyle and Lyttleton[3] that passages of the Solar System through interstellar clouds have appreciable effects on the Earth. McCrea argues that the recurrence of ice epochs[4] every ~ 250 Myr coincides with the passage of the Solar System through a galactic spiral arm approximately every 10^8 yr. We report here studies on the character and grain-size distribution of texturally-mature lunar soils which support these views. The evidence shows the flux of micrometeoroids ($\lesssim 10^{-6}$ g) at the lunar surface has remained in quasi-equilibrium near the present-day value over the past 2×10^9 yr, but that significant increases have occurred. Three near-cyclical enhancements are superimposed on the drill core stratigraphy, with separations of ~ 10^8 yr. The magnitudes, durations, and periodicity of the flux increases suggest their origin may be the passage of the Solar System through dust lanes in the galactic spiral arms.

It has been noted[5] that a segment of the cumulative grain-size distribution of "texturally-mature"[6] lunar soils closely parallels the present meteoroid flux distribution in the soil grain-size range 44–177 µm, at the maximum concentration of impact-derived constructional glass particles, or "agglutinates"[7,8]. We believe that this segment of the soil distribution provides information on the meteoroid flux distribution that produced it.

Most of the kinetic energy in the present meteoroid flux[9] is in particles of masses $< 10^{-6}$ g which strike the lunar soil at a velocity of ~ 20 km s^{-1} and generate enough heat to fuse a volume of soil. Some melt is lost as spray during crater excavation, but morphological information in the form of ring- (Fig. 1) or bowl-shaped agglutinates[10] suggests that only one large agglutinate particle is formed by any one micrometeoroid impact. Thus the parallelism between grain size and meteoroid flux is understandable: one simply mirrors the other. Experimental data on the ratio of mass of the melt to mass of the meteoroid at 20 km s^{-1} is lacking, but theory[9] suggests that an agglutinate represents ~ 5 times the mass of the impacting micrometeoroid at 20 km s^{-1}. Using mean agglutinate diameters and a particle density of 3.1 g cm^{-3} the parallel segment of the soil curve converts to a micrometeoroid mass range of 0.08×10^{-6}–2.0×10^{-6} g.

A meteoroid distribution is $N = \alpha m^{\beta}$, where N is the total number of particles larger than mass m, α is the intercept and β is the slope of a plot of log flux against log mass for masses larger than ~ 10^{-7} g (refs 9 and 11). We convert the bulk grain-size distribution for the 45–177-µm range from weight % retained in sieves to particle number equivalents assuming spherical particles. The numbers of particles in five size intervals are normalised to the total number, and we construct a cumulative particle number frequency distribution. Using equivalent meteoroid masses and cumulative particle numbers in log–log form, a linear least squares fit gives β. But α (which gives total flux) cannot be found without knowing the meteoroid exposure time of the soil. We derived β for 61 samples—12 surface soils and 49 from the Apollo 15 deep drill core. In all cases the linear fit explains > 97% of the variance, thus demonstrating a strong linear dependence between log N and log m.

Soils from the lunar surface should give β estimates near the modern flux value derived from other methods, thus offering an independent test of the model. The present meteoroid flux in our mass range has $\beta = -1.213$ (ref. 9). Estimates of β for the surface samples range from -1.266 to -1.142 with a mean of -1.213 ± 0.039. A one-tailed t test indicates no significant difference between the modern β value and our estimates from the surface soils ($t_{0.05} = 2.201 > 0.043$, d.f. = 11). That is, the model works.

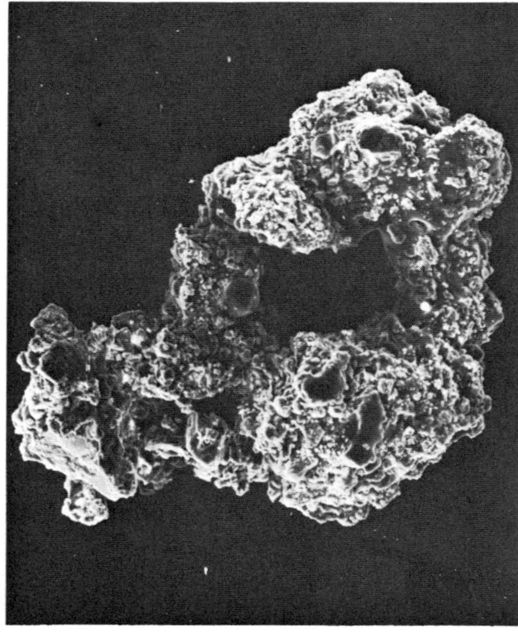

Fig. 1 A ring-shaped agglutinate from the lunar soil. The agglutinate is approximately 180 µm in diameter.

Large meteoroids produce ejecta blankets of sufficient thickness and extent to survive as layers. This creates three interpretation problems: (1) inverted stratigraphy may appear in the record[12]; (2) old agglutinates may survive excavation from an earlier layer; and (3) the soil accumulation rate is nonlinear[8,13].

First, inverted stratigraphic sequences of small thickness will not destroy regular trends in the data, but will increase deviations about the trend. The range in β values is not affected. Second Lindsay[5] investigated agglutinate survival under reworking and found most pre-existing agglutinates are destroyed in excavation by layer-forming events (that is, crushed to sizes $\lesssim$ 15 μm, below our range). Third, soil accumulation rates are nonlinear[8,13], but modelling the growth requires assumptions of the meteoroid flux history, so it is best to model linear accumulation. The age of the substrate[14] and total soil thickness at the Apollo 15 site[15] give an average accumulation rate of 1.35×10^{-7} cm yr^{-1}. Nonlinear soil models[13] fortunately give similar rates for this time period. Since our sampling interval is 0.5 cm our time resolution is $\sim 4\times10^{6}$ yr.

The core samples cover a depth of 120–240 cm in known stratigraphy (Fig. 2). Estimates of β for these samples range from -1.199 to -1.381 with a mean of -1.248 ± 0.036. The data do not show a monotonically decreasing flux distribution but suggest a quasi-steady-state background level near the modern flux value, with enhancements superimposed. Below 180 cm the deviations are a series of disordered 'spikes'; above 180 cm three regular deviations each extend over 15–20 cm of the record. Each cycle crosses several stratigraphic units, indicating that the β variations have extra-lunar, time-dependent sources, rather than being determined by independent depositional processes operating on individual stratigraphic units.

Micrometeoroid experiments on the Pioneer 8, 9, 10 and 11 spacecraft[18,19] show the asteroid belt is not a small particle source. Unless the particle populations and dynamics in the Solar System were quite different 1–2 eons ago, we suggest that the flux enhancements are not redistributions of matter by asteroidal collisions. We propose the enhancements record passages of the Solar System through dense interstellar clouds, plus active cometary episodes. These two sources may be manifestations of the same event: namely the passage of the Solar System through compression lanes in the arms of the Galaxy[1] and the concomitant generation of comets and their subsequent mass loss to the Solar System[20].

Lyttleton[20,22,25] suggests that comets are "new" members of the Solar System formed by the interaction of the Sun with interstellar clouds. The Sun encounters dense clouds ($\sim 10^{-19}$ g cm^{-3}) in "compression lanes" of interstellar matter in the galactic spiral arms[20]. Mass loss from long-period comets so formed could enhance the small particle flux for 10^{7}–10^{8} yr (ref. 17).

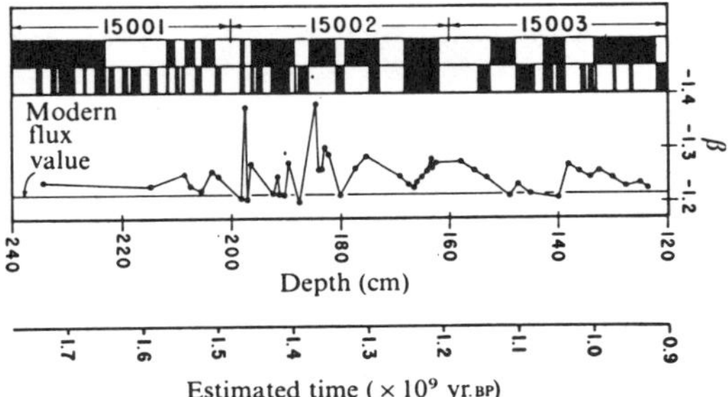

Fig. 2 Estimates of β as a function of depth in the lunar soil. Estimates of β seldom exceed the modern flux value and are not correlated with the soil stratigraphy. The top half of the stratigraphic column shows major units only—the bottom half shows all units identified.

Hartung and Störzer[16] suggest the modern small particle flux is increasing due to activity by Comet Encke. A micrometeoroid detector on the Heos-2 spacecraft[21] shows that Comet Kohoutek deposits small particles into the inner Solar System. Lyttleton[20] suggests that particles $\lesssim 10^{-6}$ g are lost from comets due to the solar wind. Since comets have masses of 10^{17}–10^{20} g and lose $\gtrsim 10^{-4}$ of this in each solar encounter[20,22], one comet could increase the particulate material density within the orbit of Jupiter[20] by $\sim 3\times10^{-24}$ g cm^{-3} if the debris is in the ecliptic. For matter collected by the Moon at 20 km s^{-1}, the surface flux would increase by $\sim 10^{-10}$ g cm^{-2} yr^{-1} above the present level of 1.2×10^{-9}–4×10^{-9} cm^{-2} yr^{-1} in small particles[9,11,23,24]. Further flux enhancements are possible for cometary periods which are short compared with the Solar System debris retention time[17]. Such enhancements could explain brief flux excursions. The 'peak' at $\sim$ 200 cm corresponds to a small particle flux increase of $\sim$ 15 times the present value.

The speed[26] of the Sun through compression lanes is 5–24 km s^{-1}, so cloud densities as above give enhancements 5×10^{-6} g cm^{-2} yr^{-1}. Although interstellar clouds have variable constitution, "typical clouds" contain $\simeq 1\%$ grains by mass[27,28]. For an encounter speed of 20 km s^{-1}, these particles striking the Moon would increase the small particle flux by $\sim 10^{-8}$ g cm^{-2} yr^{-1}. So such encounters are numerically capable of producing the flux changes deduced from lunar soil parameters.

From 120 to 180 cm in the core tube the β variations are smooth, cyclical deviations from the present-day value. Although it is difficult to correlate depth with age exactly, the three cycles have periodicities of 1×10^{8}–2×10^{8} yr and comparable durations. On this basis, the cycles cover an age 0.9×10^{9}–1.5×10^{9} yr (all Precambrian), so unfortunately our data cannot be compared with known glacial epochs. McCrea[1] points out that the Solar System crosses a spiral arm every 10^{8} year or so, taking $\sim 10^{6}$ yr to cross the compression lane and spending a total of $\sim 10^{7}$ yr in the arm. This agrees well with our estimates of β cycles in the soil, except for their longer durations ($\sim 10^{8}$ yr) which may be due to mass loss to the inner Solar System by long-period comets born in such encounters.

The Lunar Science Institute is operated by the Universities Space Research Association under contract with NASA.

1 McCrea, W. H., *Nature*, 255, 607–609 (1975).
2 Shapley, H., *J. Geol.*, 29, 502–504 (1921).
3 Hoyle, F., and Lyttleton, R. A., *Proc. Camb. Phil. Soc.*, 35, 405–415 (1939).
4 Seyfert, C. K., and Sirkin, L. A., *Earth History and Plate Tectonics*, 455 (Harper and Row, New York, 1973).
5 Lindsay, J. F., *Bull. Geol. Sóc. Am.* (in the press).
6 Lindsay, J. F., *Proc. Fifth Lunar Sci. Conf., Geochim. Cosmochim. Acta, Suppl. 5*, 1, 861–878 (1974).
7 Lindsay, J. F., *J. Sedim. Petrol.*, 41, 780–797 (1971).
8 Lindsay, J. F., *J. Sedim. Petrol.*, 42, 876–888 (1972).
9 Gault, D. E., Hörz, F., and Hartung, J. B., *Proc. Third Lunar Sci. Conf., Geochim. Cosmochim. Acta, Suppl. 3*, 3, 2713–2734 (1972).
10 McKay, D. S., Heiken, G. H., Taylor, R. M., Clanton, U. S., and Morrison, D. A., *Proc. Third Lunar Sci. Conf., Geochim. Cosmochim. Acta, Suppl. 3*, 1, 983–994 (1972).
11 Dohnany, J. S., *Icarus*, 17, 1–48 (1972).
12 Peace, W., Gose, W., and Lindsay, J., *Proc. Sixth Lunar Sci. Conf., Geochim. Cosmochim. Acta, Suppl. 6* (in the press).
13 Quaide, W., and Oberbeck, V., *The Moon*, 13 (in the press).
14 Papanastassiou, D. A., and Wasserburg, G. J., *Earth planet. Sci. Lett.*, 17, 324–337 (1973).
15 Nakamura, Y., Dorman, J., Duennebier, F., and Latham, G., *The Moon*, 13 (in the press).
16 Hartung, J. B., and Störzer, D., *Proc. Fifth Lunar Sci. Conf., Geochim. Cosmochim. Acta, Suppl. 5*, 3, 2527–2541 (1974).
17 Biermann, L., *The Zodical Light and the Interplanetary Medium*, 279–286 (NASA SP-150, Washington DC, 1967).
18 Rhee, J. W., Berg, O. E., and Richardson, F. F., *Geophys. Res. Lett.*, 1, 345–346 (1974).
19 Humes, D. H., Alvarez, J. M., Kinard, W. H., and O'Neal, R. L., *Science*, 188, 473–474 (1975).
20 Lyttleton, R. A., *Astrophys. Space Sci.*, 34, 491–510 (1975).
21 Hoffman, H. J., Fechtig, H., Green, E., and Kissel, J. (unpublished).
22 Lyttleton, R. A., *Astrophys. Space Sci.*, 31, 385–401 (1974).
23 Gault, D. E., Hörz, F., Brownlee, D. E., and Hartung, J. B., *Proc. Fifth Lunar Sci. Conf., Geochim. Cosmochim. Acta, Suppl. 5*, 3, 2365–2386 (1974).
24 Laul, J. C., Morgan, J. W., Ganapathy, R., and Anders, E., *Proc. Second Lunar Sci. Conf., Geochim. Cosmochim. Acta, Suppl. 2*, 2, 1139–1158 (1971).
25 Oort, J. H., *Bull. Astr. Inst. Neth.*, 11, 91 (1950).
26 Allen, C. W., *Astrophysical Quantities* (third ed.), (Athlone Press, London, 1973).
27 Spitzer, L., in *Nebulae and Interstellar Matter*, (edit. by Kuiper, G., and Middlehurst, B.), 1–58 (Univ. Chicago Press, Chicago, 1968).
28 Greenberg, J. M., in *Intersteller Gas Dynamics*, 306–315 (D. Reidel, 1970).

38

Reprinted from *Icarus* **22**:301–311 (1974)

Martian and Terrestrial Paleoclimatology: Relevance of Solar Variability

WILLIAM K. HARTMANN

Planetary Science Institute, 252 W. Ina Road, Suite D, Tucson, Arizona 85704

I. Nature of the Problem

Mariner 9 has impressed a curious observational fact on scientists: Mars, a planet noted for extreme aridity, has arroyo-like features which have not been adequately explained other than by flowing water (Milton, 1973; Sagan, Toon and Gierasch, 1973; Carr, 1974; Schumm, 1974). Braiding and other deposition patterns observed at resolutions of a few hundred meters give strong evidence for water flow. While some arroyos appear to spring from thermokarst topography, suggesting local melting of permafrost and thus *internal* sources of heat and water, systems of minor sinuous arroyos exist, particularly in equatorial regions. Some major arroyos have secondary and tertiary tributary systems. These latter features suggest a widespread, probably atmospheric, mechanism to deliver liquid water to the surface. Further, Hartmann (1973a) gives evidence for a much higher past Martian erosion rate (this may or may not be only the last cycle of an episodic Martian erosion history).

At recent conferences the consequences of the "rivers" have been little discussed. This is perhaps because of their radical implications and not simply because of lack of tractable investigation, since geomorphological analyses, sinuosity indices, crater counts in river floors and crater counts in background regions are all underway in various institutions.

In this paper, the nature of the problem is assumed to lie in four assertions:

1. The Martian arroyos exist and imply that Mars was wetter and warmer in the past.
2. Simultaneously with this discovery, a deficiency in solar neutrinos has been found, indirectly implying that the sun was more luminous in the past than now.
3. A large body of geological literature indicates that the earth's climate is "normally" warmer than in the present.
4. A small increase in solar luminosity in the past or future would convert Mars

to a more favorable climate for liquid water, by releasing more CO_2 and increasing advective heat transport to the poles (Sagan, Toon, and Gierasch, 1973).

Assertions (1) and (4) will not be further discussed, as they are dealt with in the papers already mentioned. Some discussion of recent debate about assertions (2) and (3) will be given, and in particular, a review will be given of the terrestrial geological evidence for climatic changes driven by solar changes, because this portion of the problem has not been advanced in the recent Martian literature. A working hypothesis of solar variation is developed by adopting terrestrial and Martian observations, and the paper will discuss whether this hypothesis can be firmly rejected on the basis of present data.

II. Historical Factors in the Question of Solar Variability

The problem of solar variations has been bound up in a special way with the geological problem of plate tectonics or "continental drift." As is well-known (see Wyllie, 1971), the first half of this century saw a raging debate on this problem. One of the empirical facts was a set of geological data showing such curiosities as ancient widespread glaciation in Africa and northern South America and ancient palms, figs, and corals in locations such as Alaska, Greenland, and Spitzbergen Island. Of the two schools, "drifters" asserted that these data showed the continents had moved in latitude, while some "fixists" asserted that they showed that solar luminosity had varied in such a way as to cause world-wide climate changes. The compelling evidence in the mid-1960's for continental and sea-floor rifting and plate motion led to a widespread assumption that the paleoclimatic data had been adequately explained. However, several analysts have indicated that the plate tectonic theory in itself is inadequate to explain all paleoclimatic data, as will be discussed. This paragraph is intended simply to warn readers that the discovery of plate tectonics, though irrelevant to solar variability, has for historical reasons had the sociological effect of "suppressing" discussion of the latter during the last ten years.

III. Review of Öpik's Paleoclimatic Theory

One of the analysts of terrestrial paleoclimatic data in pre-Mariner years deserves special attention here because of his direct concern with planetary problems. This is E. J. Öpik, whose work over many decades has shown remarkable prescience and intuition in the light of recent results. Öpik (1958) reviewed paleoclimatic data and concluded:

> "It seems that variations in the amount of heat received by the earth as a whole can only account for the major variations in past climates . . ."
>
> "For a long time there were doubts about the globe-wide character of the cold periods. Scientists were more inclined to consider them of hemispherical extent, taking place alternately in the northern and southern hemispheres and caused by minor changes in the eccentricity of the orbit and obliquity of the axes of the earth (also an early and current attempt to explain the Martian features—WKH). (Öpik) showed that this cause is inadequate. Only a simultaneous decrease of solar heat all over the globe can lead to such results, in which case the temperature will decrease everywhere."
>
> "All doubts in this respect have been removed by the determination of 'fossil' temperatures of the surface of tropical oceans, based on Urey's method of oxygen isotopes. . . ."

Reviewing such data and geologic reports, Öpik (1958) presented a table of geologic observations which implied recurrent, dated cold waves every 250 m.y., including the present time and traced by geologic data back to 1000 m.y. ago. He asserted at least three older undated ice ages occurred. Theoretically, Öpik calculated that a 9% higher rate of solar radiation would account for an increased global average temperature of about 8°C, accounting for the observations. A drop of 5% in present solar luminosity, according to this calculation, would drop the temperature by about 6°K, causing the very recent ice ages when glaciers advanced from the poles to low latitudes. (Budyko, 1970, refers to two additional models that predict complete glaciation of the earth with a solar luminosity only 2 to 5% below the *present* value.) In the same 1958 paper, Öpik hypothesized that these changes result from disturbances in radiation transfer rate in the sun's core, causing temporary expansions of the sun, lowering surface temperature, and decreasing luminosity. The durations of low-luminosity would equal the Helmholtz–Kelvin contraction time for the sun to re-adjust its structure. This is about 6 m.y., taken by Öpik to be consistent with the duration of the "cool periods" noted in earth history. Öpik estimated the time *between* "cool periods" to be "several hundred million years," consistent with the geologic record. As shown below, these ideas are quite consistent with recent observations.

Öpik (1965) updated this model, producing exactly the same basic results, and additionally calling for a "flickering" of the sun due to convective processes in the outer sun, not amenable to a detailed, predictive theory. The "flickers" were advanced to account for fine structure in recent glacial paleoclimatology, and could cause, for example, "a rapid rise in solar luminosity, followed by a gradual decline. "

In a recent note Öpik (1973) reiterated his belief in the recurrent variations of terrestrial climate driven by solar coolings separated by intervals theoretically estimated at 600 m.y. and observationally estimated at 250 m.y. He commented that the recent neutrino work "agrees completely with Öpik's conclusions . . ."

IV. Terrestrial Evidence of Recurrent Climatic Changes

The geological literature abounds with evidence that the earth was once warmer than at present, just as Mars appears to have been. This evidence has been underemphasized in recent literature on the Mars problem. Furthermore, it causes some concern among traditional geologists. For example, Kummel (1970) states in his detailed text on historical geology:

> "The doctrine of uniformitarianism has been such a strong guiding principle in the reconstruction of the earth's history that it may come as a surprise to state that the world today is in a very abnormal condition. . . . The world's climate at the moment . . . is quite exceptional, for the 'normal' climate of the geologic past appears to have varied from moderate to warm."

Precisely the same could be said of Mars.

Figure 1 presents some examples of the terrestrial data in question.

The top record is after Öpik's, Kummel's (1970, p. 542), and Dunbar's (1963) figures and tables, based on paleontological and geological data. This record is qualitative, in that no absolute temperature measures are made apart from "warmer" or "colder" than present.

The essential question about the paleontological and geological data is whether continental drift can remove all the discrepancies. For example, much evidence for a Permian cold period about 260 m.y. ago comes from glacial evidence in Africa, South America, and India, at sites presently located near latitudes −20°, −10° and +10°. However, utilizing a widely accepted plate tectonic reconstruction (Dietz and Holden, 1970) we find these areas in the Permian were near −65°, −35°, and −70°, respectively, so that the glaciation does not appear so anomalous. Thus, uncritical listing of fossil evidence for low-latitude glaciation or high-latitude tropical plants is not sufficient to establish world-wide climatic variations.

The writer has thus gone through a list or more than two dozen reported fossil flora and fauna (Kummel, 1970; Dunbar, 1963) and reconstructed the sites' original latitudes from Dietz and Holden (1970). About 25 of the sites involve fossil corals.

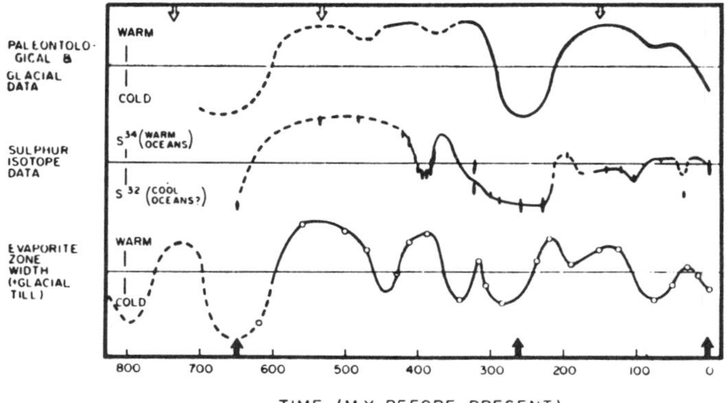

FIG. 1. Example of three types of published paleoclimatic terrestrial data, all suggesting major climatic fluctuations with several 10^8 yr timescales. Black arrows (bottom) show suggested major cold periods; open arrows (top) show suggested major warm periods. Other less marked maxima and minima may reflect nonglobal climate variations, depending on technique used. See text for discussion and sources.

Corals now have a global distribution symmetric with the equator, primarily limited to latitudes $\pm 30°$, dependent on water temperature and coastal conditions (originally shown by Darwin, 1897). About 70% of these coral cases are brought by continental drift to original positions within the 30° latitude range, and about 80% are brought within a 35° range.

In other words, most of the problem is solved by continental drift, but some marked anomalies remain. For example, Cretaceous corals are reported in Argentina at a site which should have been about 49° south of the equator, and Triassic corals are reported in Alaska when it should have been about 60° north of the equator. Cretaceous figs and palms in Greenland come from a site estimated to have been near 42°N. Tertiary figs and magnolias are reported in Alaska at sites reconstructed to lie near 70°N. All these anomalies suggest that some other factor besides continental drift has affected paleoclimatology, and has caused the earth to be globally warmer in the past than at present, as widely reported in geological literature.

The middle graph in Fig. 1 is quantitative and serves as an example of the isotopic data. It is a measure of sulphur isotope ratios based on a world wide inventory of sulphate sediments by Holser and Kaplan (1966). Here, many sources of fine structure could be present, such as changes in sediment delivery rates of ancient rivers emptying into the ocean, bacteriological activity, and mixing of bottom sediments by currents. In spite of all these local sources of fine structure which might wipe out long term trends, Holser and Kaplan find,

> "For sulphates of oceanic origin, variation with geological age is so large that it nearly masks all other factors."

The bottom record in Fig. 1 is that of Meyerhoff (1970) and Meyerhoff and Teichert (1971) who made a comprehensive inventory of evaporites, desert eolian deposits, and glacial tillite beds from around the world, and produced maps of them for different geologic periods. Latitudinal belts were found, and as Meyerhoff notes,

> "When the evaporite beds have the greatest latitudinal spread (up to 125°) the earth must have been very warm. . . . The great fluctuation of the widths of these belts illustrates the . . . time-related pattern: specifically, the average world temperature has changed considerably during earth history, and these changes are epi-

> sodic; ultimately a periodicity may be established."

Meyerhoff and Teichert (1971) confirm this and stress that the widths of these climatic belts are not constant:

> ". . . average earth temperature (or climate zonation)—not latitude—is a very critical element in determining whether or not glaciation will take place."

The graph produced by this last study is semiquantitative; the original ordinate is the width of the evaporite zone in degrees. This can be called *semi*quantitative, since the measured width depends to some extent on the observer's judgement of the data from the given geologic period. Sources of noise or fine structure in this graph include statistical limitations on available field data, and local climatic factors on the ancient continents, such as coastal ranges.

The tie between climatic data and continental drift data is illustrated, incidentally, by Meyerhoff's assertion that his climatic bands are so well defined with respect to the present equatorial regions that the continents must have not moved in latitude nearly as much as in the widely accepted models, such as that of Dietz and Holden (1970). To the extent that Meyerhoff's conclusion is correct, the data in the graphs become even stronger in arguing for world-wide climate shifts.

In conclusion, it would appear that just as on Mars, geological evidence on earth indicates that the world-wide average climate varied in the past. A warm period occurred (centered) about 150 m.y. ago and a still warmer period occurred about 500 m.y. ago. Cold periods occurred now, about 260 and about 650 m.y. ago. A great variety of data, including historical records, sunspot correlations, Öpik's "flickering" theory, planetary dynamical studies, and the very recent (30 000 yr BP) glacial advances and retreats (e.g. Gribbin, 1973), also suggest that periodic or non-periodic variations of climate occur on all timescales as implied by Öpik's proposed "flickering," and that the further back in time we go, the greater the excursions of climate. In any case, noticeable major excursions toward unusually warm or unusually cold conditions occur at intervals on the order of a few hundred million years, and these may not be precisely periodic. In addition to the above data, the evidence for several ("at least three"—Öpik, 1958) preCambrian epochs of widespread glaciation should be noted, although the dating and spacing of these is very uncertain (R. Siever, private communication).

V. Astrophysical Evidence

Any attempt to disprove solar-induced Martian climatic change by establishing the stability of the earth's paleoclimate is hardly successful, as the above section shows. Rather, the terrestrial evidence is parallel to the Martian evidence in suggesting a cold period now and noticeably warmer periods earlier. Hence the question becomes, can we rule out solar variations as a cause of these phenomena on the basis of astrophysical evidence? Note that we are here discussing episodic variations; Sagan and Mullen (1972) have already reviewed astrophysical evidence that a long-term increase of solar luminosity by 40% occurred over geologic time. We are discussing smaller superimposed oscillations in luminosity, not necessarily sinusoidal.

The lack of solar neutrinos, as compared to current theoretical models, and the interpretation that the sun was more luminous in the past, are well-known (Dilke and Gough, 1973; Cameron, 1972) and the applications to the problem of Mars and earth paleoclimates has been more widely discussed than the terrestrial geological evidence itself (Cameron, 1973; Fowler, 1973; Sagan and Young, 1973). This material will not be reviewed here except to consider whether any facets of it are strong enough to disprove the idea that solar variations caused the Martian climate changes.

Direct evidence for solar variation would be very shaky if the neutrino deficiency could be explained without the sun's luminosity being affected. No such models have been entirely successful (Fowler, 1973; Cameron, 1973), although Demarque *et al.* (1973) assert that a Dicke-type fast-rotating solar core would explain the neutrino observations. This model has a drawback of predicting much lower early solar luminosities than other theories, for example $<0.2L_{\odot}$ during the first b.y., and about $0.85L_{\odot}$ in the Cambrian. This prediction, which would imply early freezing of sea water (Budyko, 1970; Sagan and Mullen, 1972), has not been confirmed by available geologic data.

The most popular model is similar to Öpik's original suggestion, and calls for episodic mixing which changes the energy production of the sun. A new equilibrium configuration is then reached in a time governed by Helmholtz–Kelvin contraction. Cameron (1973) has given perhaps the most complete review of the theories and derives a sudden-mixing model calling for a 6 m.y. drop in luminosity to 0.7 its normal value. This model is said to be an improvement over other sudden-mixing models, but all sudden-mixing models have the virtue of getting minimum luminosities that readily account for the neutrino deficiency. However, such drastic changes in solar luminosity appear to be inconsistent with the terrestrial paleoclimatic data. Öpik (1958) estimated that a reduction of solar luminosity to only 0.84 the normal value would cause the coldest ice ages; Budyko (1970) and Öpik invoke reductions of only a few percent to account for the recent ice ages.

If, as in this paper, we try to account for the paleoclimatic data by solar theory, we would prefer a model with a longer and less pronounced cool period. Along these lines, Cameron notes evidence by Fenyves *et al.* (1973), that the earth's ocean temperatures are anomalously low and have been low, not for 6 m.y., but for 50 m.y. Cameron points out that slow mixing in the sun can be invoked arbitrarily to extend the cool periods. But current versions of such slow-mixing models make it more difficult or impossible to account for the low observed neutrino flux, since the neutrino minimum becomes less deep.

If solar-type stars vary in brightness, a broadening of the main sequence for G stars could result. Such a broadening has been sought and found by Sagan and Young (1973) in the careful photometry by Johnson (1952) of the Praesepe star cluster. This cluster is said by Sagan and Young to be the optimum cluster for this test. However, Cameron (1973) earlier pointed out that according to the best models of the variability process, the sun changes size, surface temperature, and luminosity in such a way as to move primarily *along* the main sequence, thus producing very little apparent broadening. Cameron (private communication, 1973) suggests that this detracts from the applicability of the Sagan–Young results. The strongest case against the Sagan–Young conclusion is that (1) Johnson had to estimate and discard other sources of dispersion, particularly the known influence of stars with unresolved companions that affect photometry, and (2) Johnson's result was expressed merely as an *upper limit* on intrinsic, or "cosmic" dispersion of 0.03 magnitude, equivalent to excursions of individual stars of about 0.07 in luminosity), whereas Sagan and Young, re-analyzing Johnson's data, propose *actual* excursions of 0.15 in luminosity. These arguments are not strong enough to dispose of the Sagan–Young conclusion, because the fact remains that dispersion in the Praesepe main sequence is equivalent to 0.07 to 0.15 excursions in luminosity and is unaccounted for by any known sources of error. Further, as Sagan and Young point out in reversing Cameron's argument, if we empirically detect excursions of 0.07 to 0.15, the true excursions of an individual star are still considerably larger because of the large theoretical component of evolution along the main sequence.

Smith (1973) has responded to the Sagan-Young conclusion by confirming the dispersion with newer data, but pointing out that broadening of the main sequence can occur due to varying rotation rates among stars. Rotation rates affect luminosity. In the case of Praesepe, Smith concludes that a normal spread of uniform rotations (as in solid rotators) among stars would account for less than half the observed luminosity spread. However, nonuniform rotations (fast-rotating cores) of the type hypothesized by Dicke could be invoked to account for all the observed spread. We conclude here that, in view of the controversial nature of Dicke's model, the luminosity spread in Praesepe is still unaccounted for, and may indicate luminosity variations.

Smith (1973) makes the further interesting point that the scatter of luminosities in Praesepe is inconsistent with the sudden-mixing models that call for 6 m.y. dim periods and much longer bright periods. Instead, the scatter is more uniform and would be consistent with more sinusoidal variations in luminosity among individual stars.

In summary, the present case against solar luminosity variations is certainly no stronger than the case for it. Neutrino-related sudden-mixing models do not provide the time durations that are consistent with terrestrial or Praesepe evidence, and they invoke purely ad hoc choices of mixing rates inside the sun. Further modified models are called for. Given the present state of flux in solar models, it seems most prudent simply to say that present-day solar theory and observations suggest intermittent cold periods, including the present cold periods observed on earth and Mars. Indeed, Dilke and Gough (1973) have already proposed a solar periodicity of 250 m.y. on grounds of identification with terrestrial climate cycles, and Öpik (1973) has claimed a theoretical estimate of solar periodicity of about 600 m.y., based on his own model of the sun.

VI. Reconciling Interplanetary Paleoclimatic Timescales

The remaining avenue for disproving the idea that solar variations cause simultaneous planetary climate variations is to show that the time-scales of observed or suspected variations are incompatible. Here, the hypothesis faces greatest vulnerability because a superficial reading of the most popular *published* interpretations of the *hitherto independent* lines of evidence reveals some inconsistencies:

1. The sun is said to be cool now and stay cool for about 6 m.y. (sudden-mixing model) or longer at unknown intervals.

2. The earth appears to be cool now and to have been cool for about 50 m.y., to have been warmer roughly 150 m.y. ago, and to have had cool episodes roughly 260 and 650 m.y. ago.

3. Mars is cool now. The Martian timescale is extremely uncertain, although Hartmann (1973a) derived a "best" estimate of the end of the last high-erosion interval on Mars of roughly 600 m.y. (factor three uncertainty). Using the same best estimate of timescale, Hartmann (1973b) has suggested that some Martian river flows have occurred in the last 300 m.y.

At first glance, one is tempted to argue that the timescales of cool intervals on the sun are much shorter than permissible to account for terrestrial cold intervals, and that cool intervals on Mars are too long to be correlated with the sun or earth. However, it is just as instructive to consider the state of uncertainty of all these results, and to see whether a synthesis can be achieved by considering them *not independently*. For example, the 6 m.y. cool period of the sun is only the most currently popular result of the simplest model of mixing: a sudden mixing. It is not completely satisfying. Cameron suggests that slower mixing and longer cool periods might occur, Öpik argues for rather unpredictable "flickering" superimposed on this due to not-easily-tractable convection processes, and the Praesepe data remind us that empirical evidence may be at hand for luminosity excursions not wholly accounted for by any of the theoretical models.

As for the Martian timescales, the ages derived by Hartmann depend on an assumed cratering rate and are the shortest of a range of ages permitted by rough cratering rate estimates based on statistical dynamic theory (Wetherill, 1973, private communication). The oldest Martian ages permitted on this view are about five times those quoted and occur if comets are the primary source of Martian cratering. Chapman (1974, in press) has concluded that only one erosive episode occurred on Mars, probably 1 to 2 b.y. ago. Should these old ages be proven correct, the synchronism of earth and Mars climatic changes could probably be ruled out and solar variations as the prime cause be disproven.

However, a qualifier is in order on the use of statistical dynamical meteoritic theories to estimate Martian cratering rates and ages. We know that a substantial percentage of our own meteorites result from a small number of specific parent-body fragmentation events within the last 10^9 years. Thus, we are at the mercy of small-number statstics; it is always conceivable that the Martian cratering rate is many times different from that predicted by statistical models. This could be the case if a major fragmentation in the last few 10^8 yr produced a family of Mars-crossing asteroid fragments that do not cross the earth's orbit. In such a case, the cratering rate in the last few 10^8 yr could be anomalously high on Mars (a "spike" with some 10^8 yr decay time), and features be younger than suspected. It is unlikely that features could be many times older than suspected, due to such a mechanism, because the high cratering rate "spikes" due to collisions decay back to a normal background rate of meteorite impacts; the rate should never drop many times below normal.

Alternative hypotheses on the cause of Martian climatic variations face an even greater difficulty with timescale than the solar variation hypothesis. The most prominent example is Ward's (1973) analysis of obliquity variations in the history of Mars. He concludes that obliquity is now near its median value but has varied widely in the past, causing greater possible polar heating and strongly affecting the Martian climate. Ward, as well as Sagan, Toon, and Gierasch (1973) have considered this a possible explanation of the Martian climate variations, in which case solar variations would be irrelevant. The problem is that obliquity variations occur on a 1.2 m.y. period, implying a timescale for Martian river history of 10^6 yr. This conflicts with the stratigraphic timescales proposed by investigators of Martian geology. Median river ages appear to compare to ages of the major volcanic systems of the Tharsis area; *major* channels do not post-date Tharsis. But as noted above, the youngest currently-acceptable ages for these features are several 10^8 yr, over 100 times as ancient as obliquity variations alone would suggest. To put it another way, finite numbers of distinct craters overlie features that should be only some 10^6 yr old on the obliquity hypothesis, and it is difficult to imagine such craters forming in such short time periods (Fig. 2).

VII. Conclusions: A Working Hypothesis

To form a working hypothesis that solar variations are the dominant, but not the only, cause of climatic variations on the earth and Mars, we need only bring the three timescales into agreement. Terrestrial data is most reliable for this purpose. Such a working hypothesis thus states:

1. Superimposed on any secular change in solar luminosity are variations occurring on a time-scale of several 10^8 yr and having an amplitude of 7 to 35% in total bolometric luminosity.

2. The sun's luminosity has been reduced to 0.65 to 0.93 of its normal values, the range suggested from solar, ice age, and Praesepe data, for the last 6–50 m.y.

3. Öpik's slow "flickering" probably occurs on various timescales, but it is not possible to distinguish it in the planetary records from other variable effects, such as obliquity changes, changes in ocean current patterns, continental drift, etc. *Vestigial* Martian river activity might be associated with some of Ward's recent obliquity cycles.

4. About 150 m.y. ago, the sun was nearer its normal or peak luminosity, and Mars and earth were warmer than today.

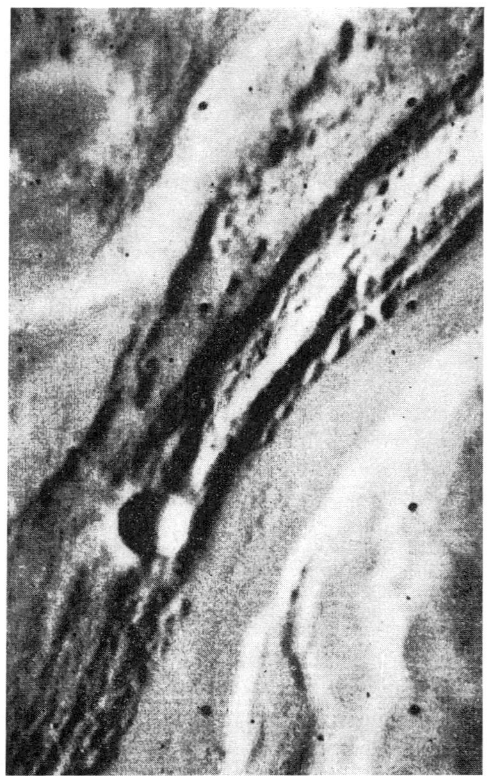

Fig. 2. Example of a crater (diameter about 2.5 km) overlapping deposits in a Martian channel system. This is an argument for considerable age of the system, Vallis Mangala, near 151°, −7°. The crater appears *not* to postdate the most recent, central channel, and is thus an example of evidence for long time intervals between episodes of fluvial activity.

Some fluvial activity may have occurred on Mars at this time.

5. About 260 m.y. ago, the sun, earth, and Mars were again cooler than normal, a result suggested by all three records in Fig. 1.

6. About 500 m.y. ago, for an interval of several 10^8 m.y., the sun was near its peak luminosity. Higher luminosities may have been reached in this period than at any subsequent time, according to the sulphur isotope and evaporite data in Fig. 1. The ancient high rates of Martian erosion detected by Hartmann (1973a), Soderblom *et al.* (1973), and Chapman (1974), may have occurred at, or prior to, this warm period.

7. Earlier cold and warm cycles occurred, including a probable cold period around 660 m.y. ago suggested by all there records in Fig. 1.

8. The Martian arroyos, or river-like channels, represent episodic vestigial fluvial activity (Fig. 2) that occurred after even higher rates of erosion ceased. This statement is independent of the solar data, and rests only on the Martian crater-count and morphological evidence that the channels saw activity more recently than the end of the high-erosion period (Hartmann, 1973a; Chapman, 1974; Soderblom, 1974).

As already emphasized by Fowler (1973), Cameron (1973), Sagan, Toon and Gierasch (1973), and Hartmann (1973b), this working hypothesis has several extraordinary consequences and is worthy of further efforts towards disproof or confirmation. Among the consequences are:

1. Planetary climate changes are primarily controlled by solar changes. These planetary changes are not the trivial grey-body changes that might be expected, but are catastrophic "flips" from one equilibrium state to another through advective instabilities (Sagan, Toon, and Gierasch, 1973). These reversible effects deserve further theoretical attention to determine their magnitudes for planetary atmospheres composed of various constituents, representing different planets and different evolutionary times.

2. Speciation and biological evolution on earth may be driven by these solar-induced climatic changes. Cameron (1973) assuming the 6-m.y. time-scale, suggests that the recent decline in temperature may have driven the natural selection of intelligence in man, and Hartmann (1973b), assuming a longer timescale more compatible with Martian evidence, suggests that it caused the transition from dominance of large cold-blooded reptiles to warm-blooded mammals. Further analysis of paleontological evidence is warranted.

3. The geologic tenet, "uniformitarianism," may have to be modified due to planetological progress. The present is not necessarily the key to the past if solar

variations cause the main climate variations on planets.

4. If planets commonly "flip" from states with liquid volatiles to arid and/or frozen conditions, as Mars appears to have done, the evolution of substantial, H_2O-based intelligent life-forms on planets around most stars may be hindered, even though speciation of more primitive life-forms could be encouraged. Periodic 10^8-yr disappearances of liquid water would not favor evolution of large, intelligent beasts, a process which took more than 500 m.y. on earth. This could be a factor in our current loneliness in the universe.

Note added in proof: Additional data are in accord with this paper's hypothesis that the sun may be driving simultaneous climatic changes on all the planets. In a paper being published simultaneously (*J. Geophys. Res.*, in press), I have described Martian arroyo crater counts that imply ages of the order 10^8 yr (not the 10^6 yr that would be implied by climate variations driven only by Ward's obliquity changes). The hypothesis that speciation is governed by major climatic change is supported by Van Valen and Sloan (1972, Abstr. 24th Internat. Geol. Cong.), who report Montana fossil sequences that show dinosaur extinction due to faunal changes caused by cooling winters in an "environmental change...probably world-wide" about 60 m.y. ago. Newell (1972, *Sci. Amer.*, June, p. 54) gives a nontechnical example of the appeal to plate tectonics alone to explain "episodes of sweeping mass extinctions (that) simultaneously affected such disparate organisms as ammonites at sea and dinosaurs ashore," suggesting "simultaneous world-wide evolutions." Such data are more readily explainable if solar variations on a time-scale of some 10^8 yr produce climatic instabilities. Recent papers grappling with the neutrino problem continue to suggest room for improvement in traditional solar models.

Acknowledgments

The writer thanks various colleagues, especially A. Cameron, C. Chapman, G. Hartmann, F. Herbert, C. Sagan, L. Soderblom, and an anonymous referee for discussions. The writer also also thanks E. Anders for encouragement. This work was supported by private funds and through the Planetary Science Institute of Science Applications, Inc.

References

Budyko, M. I. (1970). Comments on "A global climatic model based on the energy balance of the earth-atmosphere system." *J. Appl. Met.* **9**, 310.

Camerson, A. G. W. (1973). Major variations in solar luminosity? *Rev. Geophys. Space Phys.* **11**, 505.

Carr, M. H. (1974). The role of lava erosion in the formation of lunar and Martian channels. *Icarus* **22**, 1–23.

Chapman, C. R. (1974). Cratering on Mars, I. Cratering and obliteration history. *Icarus* **22**, 272–291.

Darwin, C. R. (1897). "The Structure and Distribution of Coral Reefs." D. Appleton, New York.

Demarque, P., Mengel, J., and Sweigert, A. (1973). Solar rotation and the neutrino flux. *Nature Phys. Sci.* **246**, 33.

Dilke, F., and Gough, D. (1972). The solar spoon. *Nature* **240**, 262.

Fenyves, E., Jani, Y., Bozoki, K., and Lee, C. (1973). The solar neutrino puzzle and the temperature history of the earth. *In* "Proc. 6th Texas Symp. on Rel. Astrophys.", N.Y. Acad. Sci., in press.

Fowler, W. A. (1973). "What SNU? The Case of the Missing Solar Neutrinos." Cal. Tech. preprint OAP-339.

Gribbin, J. (1973). Planetary alignments, solar activity, and climatic change. *Nature* **246**, 453.

Hartmann, W. K. (1973a). Martian cratering. IV: Mariner 9 initial analysis of cratering chronology. *JGR* **78**, 4096.

Hartmann, W. K. (1973b). Martian surface history. Paper presented at International Mars Colloquium, Pasadena, Nov. 30, 1973.

Holser, M., and Kaplan, I. (1966). Isotope geochemistry of sedimentary sulfates. *Chemical Geology* **1**, 93.

Johnson, H. L. (1952). Praesepe: magnitudes and colors. *Astrophys. J.* **116**, 640.

Kummel, B. (1970). "History of the Earth." W. Freeman, San Francisco.

Meyerhoff, A. A. (1970). Continental drift: Implications of paleomagnetic studies, meteorology, physical oceanography, and climatology. *J. Geology* **1**, 78.

Meyerhoff, A., and Teichert, C. (1971). Continental drift. III: Late Paleozoic glacial centers, and Devonian–Eocene coal distribution. *J. Geol.* **79**, 285.

Milton, D. (1973). Water and processes of degradation in the Martian landscape. *JGR* **78**, 4037.

Murray, B., Ward, W., and Yeung, S. (1973). Periodic isolation variations on Mars. *Science* **180**, 638.

Öpik, E. J. (1958). Solar variability and paleoclimatic changes. *Irish A.J.* **5**, 97.

Öpik, E. J. (1965). Climatic change in cosmic perspective. *Icarus* **4**, 289.

Öpik, E. J. (1973). Terrestrial climate and solar evolution. *Irish A.J.* **10**, 295.

Sagan, C., and Mullen, G. (1972). Earth and Mars: evolution of atmospheres and surface temperatures. *Science* **177**, 52.

Sagan, C., Toon, O., and Gierasch, P. (1973). Climatic change on Mars. *Science* **181**, 1045.

Sagan, C., and Young, A. (1973). Solar neutrinos, Martian rivers, and Praesepe. *Nature* **243**, 459.

Schumm, S. A. (1974). Martian "channels." *Icarus* **22**, 371–384.

Soderblom, L. A., Condit, C. D., West, R. A., Herman, B. M., and Kreidler, T. J. (1974). Martian planetwide crater distributions: Implications for geologic history and surface processes. *Icarus* **22**, 239–263.

Smith, R. C. (1973). Praesepe and the solar neutrino problem. *Nature* **244**, 560.

Ward, William R. (1973). Large scale variations in the obliquity of Mars. *Science* **181**, 260.

Wyllie, P. J. (1971). "The Dynamic Earth." Wiley, New York.

39

Reprinted from *Jour. Geophys. Research* **80**:2508–2514 (1975)

Surface History of Mercury: Implications for Terrestrial Planets

BRUCE C. MURRAY,[1] ROBERT G. STROM,[2] NEWELL J. TRASK,[3] AND DONALD E. GAULT[4]

INTRODUCTION

Our goal in this paper is to portray a plausible history of Mercury as suggested by the Mariner 10 data and to call attention to the implications for the history of the other terrestrial planets. We view this first 'working hypothesis' for the history of Mercury not as a definitive statement but as an initial framework to stimulate and focus subsequent analyses and interpretation.

A convenient way to discuss the history of Mercury as suggested by the Mariner 10 television pictures is to postulate five periods delineated by successive variations in the modification of the surface by external and internal processes: (1) accretion and differentiation, (2) terminal heavy bombardment, (3) formation of the Caloris basin, (4) flooding of that basin and other areas, and (5) light cratering accumulated on the smooth plains.

The sequence of events recorded on Mercury's surface closely resembles that of the moon, although there are interesting differences in detail. Whether the absolute time periods represented by those sequences also are closely similar is one principal question in the interpretation of the Mariner 10 photography. The other major issue is whether the smooth plains of Mercury, like those of the moon, are primarily of volcanic origin. Our working hypothesis is that Mercury is indeed moonlike in both respects, absolute history and extensive volcanism, and we will outline the basis for our judgments. However, very divergent interpretations, including major differences in the early history of the two objects and even nonvolcanic origin of the plains, cannot be ruled out entirely on the basis of the analyses carried out so far.

ACCRETION AND DIFFERENTIATION

Strom et al. [1975*a*] have argued the case for at least some volcanic extrusions on Mercury of morphology similar to the lunar maria and of thousands of kilometers in horizontal dimension, emphasizing stratigraphic, geographic, volumetric, and albedo relationships. Figure 1 presents additional data supporting a volcanic rather than impact origin of the plains east of the Caloris basin. By analogy with the earth [*Bowen*, 1928] and the moon [*Ringwood and Essene*, 1970], large-scale volcanism on Mercury would imply the existence of a silicate mantle probably of at least several hundred kilometers in depth [*Murray et al.*, 1974, note 20]. Yet the models of *Reynolds and Summers* [1969] indicate that a silicate outer shell of only 500- or 600-km thickness requires all the silicate available in a planet with a mean density of 5.45. Hence on the basis of the television data alone it can be inferred that the planet is most likely composed of a large iron core enclosed by a relatively thin silicate layer from which the smooth plains originated.

The existence of an iron core on Mercury also has been inferred from the magnetic field observations [*Ness et al.*, 1974, 1975]. Ness et al. argue that either an active dynamo or a fossil magnetic field must be involved. We would emphasize here that regardless of the origin of the magnetic field, a substantial iron core is indicated by Mercury's interaction with the solar wind. Thus the conclusion that Mercury is a differentiated planet is suggested by two independent observations.

Granted that Mercury very probably is composed of a large iron core and a silicate outer shell, when did that chemical segregation take place?

Regardless of how Mercury accumulated as a planet, chemical differentiation must have been complete well before any of the existing impact craters formed, since melting or any subsequent crustal plasticity would have modified the lunarlike appearance that they still retain. In addition, if any atmosphere was generated during or after accretion, it too must have disappeared entirely before any of the present topographic features formed. A very small amount of atmosphere will markedly modify the appearance of secondary craters and ejecta blankets surrounding a primary impact crater. For example, Mars, with an average surface pressure at present of only about 5 mbar, very rarely exhibits the superficial features characteristic of fresh impact craters on the moon and Mercury. Thus we conclude that profound chemical

[1] Division of Geological Sciences, California Institute of Technology, Pasadena, California 91125.

[2] University of Arizona, Tucson, Arizona 85721.

[3] U.S. Geological Survey, Reston, Virginia 22070.

[4] NASA/Ames Research Center, Mountain View, California 94040.

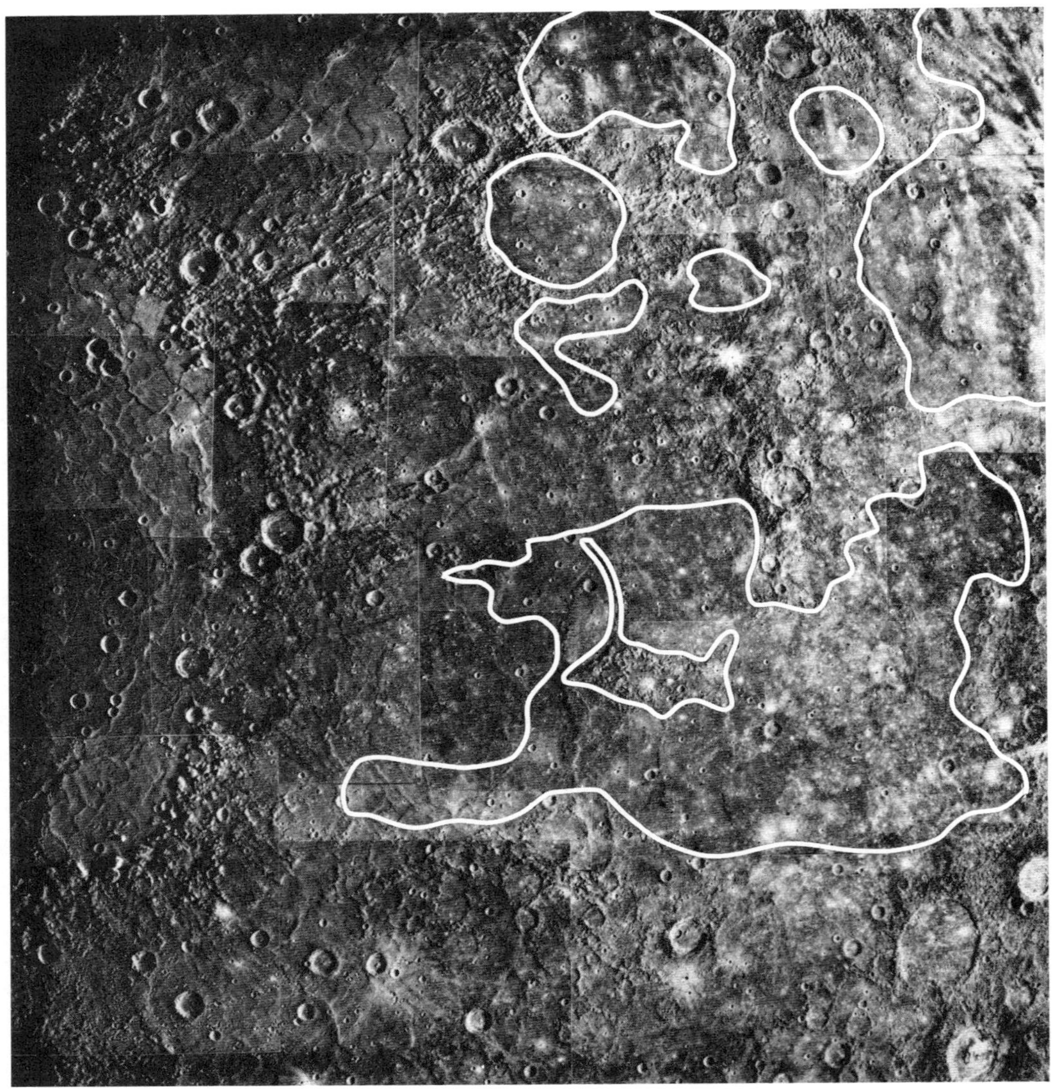

Fig. 1. Volcanic plains east of Caloris basin. In this photomosaic of the Caloris basin and adjacent areas, north is at the top. The area of Mercury shown is about 2000 km in dimension. On the basis of the distribution of large (20–50 km) fresh impact craters the smooth plains east of Caloris basin outlined in white are inferred by us to be younger than both the mountainous rim and the surrounding ejecta deposits, as well as the older plains material which fills Caloris basin. These relationships appear inconsistent with the possibility that the smooth plains in question could have originated as fluidized ejecta from the Caloris impact event. A later volcanic origin is indicated instead.

segregation into a large iron core and relatively thin silicate mantle took place sufficiently before the formation of the large craters, so that any lingering atmosphere had largely dissipated and a rigid lithosphere had come into existence.

Terminal Heavy Bombardment

A period of heavy bombardment is represented by the large craters hundreds of kilometers in diameter grading into large basins, including the Caloris basin itself. As is discussed by *Trask and Guest* [1975], large craters appear in many areas to be emplaced into an older host material, although in a few regions the surface appears more nearly saturated with large craters (>50 km). In most of the older terrains, macroscopically smooth intercrater areas are discerned which must in part predate the formation of those large craters (see Figure 2, upper left).

On the moon, by contrast, the architecture of the present surface is dominated by a heavy bombardment that ended about 4×10^9 years ago and by subsequent lava filling. Older surfaces saturated with large craters as well as small areas similar to the extensive intercrater plains of Mercury also can be recognized. Terminal bombardment relationships may be more evident on Mercury than on the moon because the greater surface gravity on Mercury restricts the areal coverage of the debris from the impact formation of the basin [*Gault et al.*, 1975]. Also there may have been fewer large impacts in comparison to the greater surface area, although that point is not clear. In any case, the existence of large intercrater areas on Mercury implies that some earlier process obliterated nearly all topographic remnants of the saturation bombardment that necessarily constituted the final stages of planetary accumulation (accretion). Otherwise, a surface saturated by large crater forms would still survive (see Figure 2, lower left).

Just what crater obliteration process is recorded by that old smooth surface? Conceivably, it could be a relic from melting associated with accretion itself or a record of erosion from an

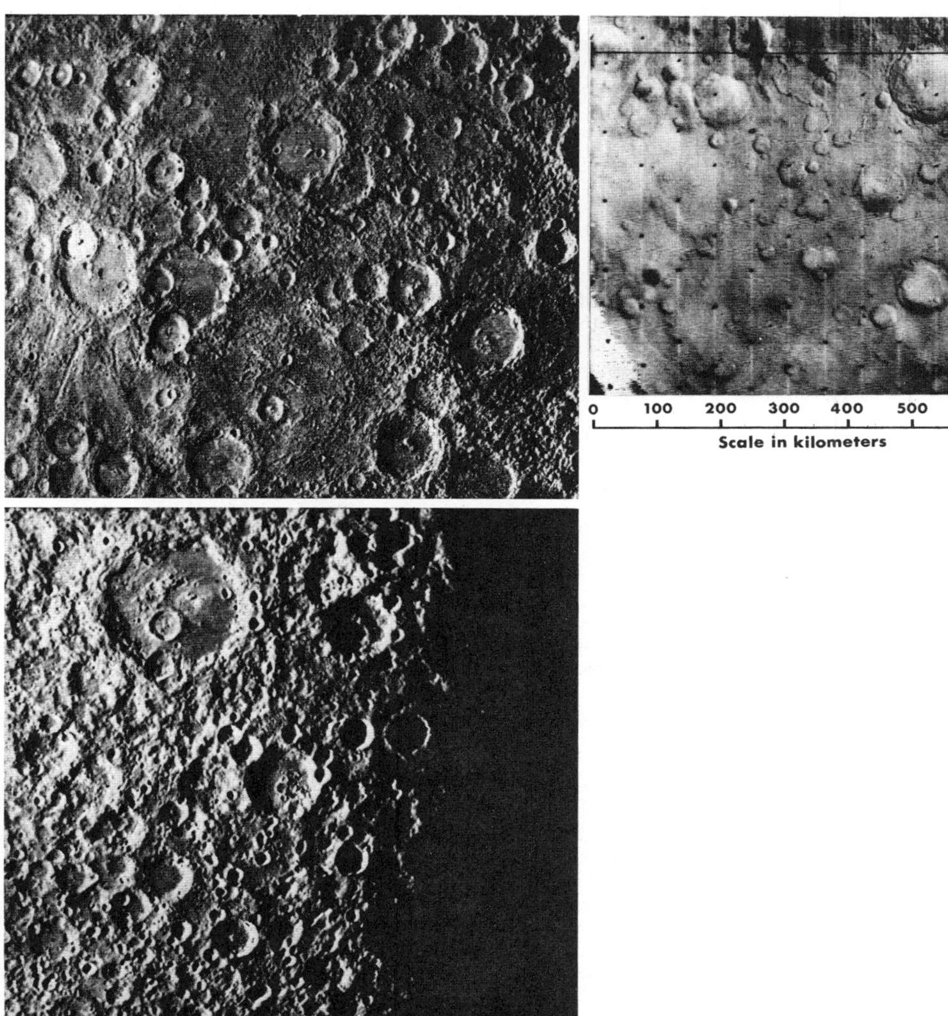

Fig. 2. Cratered terrains on Mercury, the moon, and Mars. (Upper left) Mariner 10 photograph of the bright-rayed crater Kuiper and surrounding cratered terrain on Mercury centered at 13°S, 25°W. The solar elevation angle varies from about 5° to 27° (Flight Data Subsystem 27,256). (Lower left) Mariner 10 photograph region in the northern uplands of the moon centered at 60°N, 130°E. The solar elevation angle varies from 0° (i.e., the terminator) to about 25° (Flight Data Subsystem 2669). (Upper right) Mariner 9 photograph of cratered terrain in the Deuteronilus region on Mars centered at 30°N, 340°W. The solar elevation angle varies from about 17° to 33° (DAS number 7,290,743). All three images have been projected at the surface scale indicated. Both Mars and Mercury exhibit extensive intercrater plains, deficient in 10- to 50-km craters in comparison with the moon. The Mercurian craters, however, resemble in distribution of morphology those of the moon rather than the highly degraded population present on Mars. From these relationships we infer a period of extensive crater obliteration on Mercury generally preceding the terminal heavy bombardment responsible for the present population of large craters on Mercury.

associated atmospheric event. However, those processes should have operated globally, not leaving even slight remnants of saturated surfaces. For this reason, and on the basis of local morphology, we feel that it is somewhat more likely that the intercrater plains on Mercury constitute a very early volcanic surface. Petrologic data from the moon summarized by *Albee* [1975] suggest to some that there were early volcanic episodes on the moon now represented by occasional fragments in the breccias created by later bombardment. Another possible obliteration process is fluidized flow generated by large impacts, as was suggested recently for the origin of some smooth plains on the moon previously believed to be volcanic (D. E. Wilhelms, personal communication, 1975). Obliteration of early crater forms over some fraction of Mercury's surface by hypothetical catastrophic basin formation cannot entirely be ruled out at present. However, we consider the distribution and morphology of the Mercurian intercrater plains to be inconsistent with such an origin. In addition, the surface gravity of Mercury requires this cataclysmic process to be more localized on Mercury than on the moon.

It is useful at this point to consider Figure 2. The most heavily cratered areas of Mars also exhibit intercrater plains, i.e., also record a period of crater obliteration, as is shown in Figure 2, upper right. But in contrast to Mercury, most large Martian craters are characteristically modified, probably as a result of atmospheric erosion and deposition during and subsequent to the terminal heavy bombardment. The Mercurian craters (Figure 2, upper left), on the other hand, reflect the full distribution of morphologies seen on the moon (Figure 2, lower left); i.e., they record merely the normal degradation associated with the impact process itself.

This crucial question of the distribution of morphologies of the Mercurian and lunar craters warrants, and surely will receive, a major quantitative effort (which is now made much more tractable by recent success in creating useful stereo pairs from originally isolated Mariner 10 frames). In the meantime, careful examination of many individual Mariner 10 photographs has failed to reveal to us any consistent difference in the distribution of lunar and Mercurian crater morphologies other than the difference identified by *Gault et al.* [1975] as arising from the difference in surface gravity. As a consequence, we infer that the craters formed during the final stages of accretion had been nearly completely erased before the present population of large lunarlike craters was formed. Otherwise, there would still be present an obvious, even dominant, population of degraded crater forms, as there is on Mars, reflecting the persistence of the crater obliteration process. This means that if the bombardment presently recorded on Mercury were simply the end of a continuous bombardment process persisting since accretion itself, the process of crater obliteration must have (1) operated at a rate comparable to that of crater formation and (2) terminated abruptly. It seems difficult to require hypothetical melting, early volcanism, or atmospheric erosion to cease so abruptly. Conversely, the impacting flux recorded on Mercury may not have been continuous since accretion; instead there may have been an extensive period of crater obliteration by volcanic or other processes followed by a distinct episode of late bombardment. Most important, an episodic early history of Mercury is required either way. We find an episodic bombardment history more plausible at this point than the specially constrained episodic obliteration history which would be required if the bombardment had been continuously decreasing from accretion. Accordingly, we interpret the intercrater plains as recording one or more periods of crater obliteration separating the main accumulation of the planet from a heavy terminal bombardment. Obviously, future studies can be expected to challenge this interpretation and to search for an alternate explanation based, perhaps, on an episodic crater removal process such as catastrophic impact ejecta.

Another aspect of the heavily cratered terrains on Mercury suggests significant differences from the moon (and also from Mars). The intercrater areas especially exhibit compressional features, probably thrust or high-angle reverse faults [*Strom et al.*, 1975*a*]; crustal shortening must have been taking place during the terminal heavy bombardment but evidently not much after that. Strom et al. suggest a net crustal shortening during this period corresponding to a few tenths of a percent reduction in the radius of Mercury. In contrast, significant crustal shortening on a global scale has not been recognized on the moon or Mars. Since the core mantle boundary is apparently quite shallow on Mercury and the proportional core volume is so large in comparison with the core volumes of the moon and Mars, it is plausible that these unique compressional features correspond to an early period of minor core shrinkage. Strom et al. note that a phase change from liquid to solid iron involving less than 6% of a large core, or even just the consequence of a higher coefficient of thermal expansion of solid metal compared to solid silicate, may be sufficient to produce the necessary radius decrease.

Thus the oldest terrains on Mercury probably represent remnants of an early period of crater obliteration followed by a distinct episode of terminal heavy bombardment. Crustal shortening was taking place at that time, perhaps corresponding to core shrinkage.

FORMATION OF CALORIS BASIN

The impact which created the Caloris basin was a major event in the history of Mercury. It modified the landscape over much of the illuminated hemisphere observed by Mariner 10 and certainly affected the other hemisphere as well. In association with basin excavation a ring of mountains resembling the lunar Apennine mountains and hummocky terrain similar to that seen in the Orientale basin on the moon formed. Orientale is a particularly useful lunar ringed basin with which to compare because it was not flooded by extensive lava filling afterward, like most other lunar basins. Strom et al. show that many characteristic features of a lunar ringed basin also can be recognized around Caloris. Accordingly, all these features are ascribed to a single time interval associated with the Caloris basin impact.

It is also possible to correlate speculatively the peculiar hilly and lineated terrain recognized on the incoming hemisphere of the Mercury 1 photography with the Caloris basin event. That terrain is apparently restricted to a location approximately antipodal to Caloris: none was observed in the south polar region in the Mercury 2 photography [*Strom et al.*, 1975*b*] nor on the outgoing hemisphere as viewed in the first encounter. Furthermore, local sequence relationships indicate that the terrain must have developed near the end of heavy bombardment [*Strom et al.*, 1975*a*], i.e., approximately at the time of the Caloris impact. *Schultz and Gault* [1975] have speculated that the lineated and hilly terrain is the result of the focusing of the seismic energy from the Caloris impact at the antipodal point and therefore the terrain formed at the time of the Caloris impact. Smaller areas of similar terrain have been recognized antipodal to the Imbrium and to the Orientale basins on the moon and attributed to focusing of ejecta rather than seismic energy [*Moore et al.*, 1974]; this mechanism is inhibited in the case of Mercury by the higher surface gravity. The less obvious nature of the lunar occurrences is consistent with the relationship discussed by *Gault et al.* [1975] that the material between ancient basins is more visible on Mercury because the basin ejecta was restricted to narrower rings.

BASIN FLOODING

The smooth plains of much of the outgoing hemisphere and parts of the incoming hemisphere of Mercury represent a secondary filling of earlier basins, generally at a time when the impact flux had greatly decreased. There is plausible evidence [*Trask and Guest*, 1975; *Strom et al.*, 1975*a*] that filling was not simultaneous and that much of the fill must be of volcanic origin, rather than from ejecta sheets, impact melt, or mass wasting processes (also see Figure 1). Nevertheless, the total time period represented by these plains compared to the total time interval recorded on the surface of Mercury may be relatively brief. No difference as large as 1:1.5 in crater density of the various Mercurian plains in the equatorial regions has been recognized so far [*Murray et al.*, 1974].

In addition to these planet-wide filling episodes, probably relatively close in time but nevertheless discrete, there was a subsequent structural modification of the Caloris basin itself. Deformation continued after the end of filling, resulting in open fractures now several kilometers across which cut across earlier ridges to comprise the present surface. This pattern of tensional deformation is localized to Caloris basin: whether it was a subsequent uplift or subsidence, it can be ascribed plausibly to adjustment to the principal Caloris event [*Strom et al.*, 1975*a*].

POST-FILLING PERIOD

The last and presumably longest episode recorded on the surface of Mercury is represented by the light cratering present over the entire planet and especially apparent on the smooth plains. Bright-rayed craters are very common and tend to be bluish in color in comparison to the underlying material [*Hapke et al.*, 1975]. They resemble in morphology and in size-frequency spectrum the post-mare cratering of the moon; the Mercurian crater-frequency relationships measured so far all plot between those of the Apollo 12 and Apollo 14 sites [*Murray et al.*, 1974]. Thus since the formation of the plains, Mercury has accumulated a light cratering similar to that of the moon.

Equally significant is the absence of substantial surface modification arising from tectonic, volcanic, or atmospheric processes. No tangible atmosphere accumulated sufficiently to transport fine particles, nor has there been any observable deformation of the silicate outer shell of Mercury arising from motions of any hypothetical fluid core. The heat machine apparently was turned off after formation of the plains, and that was the end of it, although future exploration of the 50% of the Mercurian surface not visible to Mariner 10 could reveal more recent episodes of localized internal activity.

Thus to us the most plausible sequence of events recorded on the surface of Mercury is (1) major chemical differentiation either during or very shortly after accretion, (2) after decline of the original bombardment, a period of crater obliteration, perhaps through extensive volcanism, followed by an episode of terminal heavy bombardment, global crustal shortening taking place at this time, (3) formation of the Caloris basin near the end of terminal bombardment, (4) volcanic flooding of large areas of the planet to form most of the smooth plains, and (5) light cratering subsequent to the filling episodes recording an accumulated flux comparable to that recorded on the lunar maria.

Other interpretations of Mercury's history are not ruled out by the Mariner 10 data, although we consider them less plausible. For example, the crater obliteration process might have been catastrophic, such as might accompany early large basin formation. In that case the terminal heavy bombardment need not have been episodic. It is also still possible, in the absence of truly definitive morphological evidence, that none of the smooth plains are of volcanic origin, although we think it quite unlikely.

COMPARISON WITH SURFACE HISTORIES OF VENUS, THE MOON, AND MARS

Studies of returned lunar samples have demonstrated that heavy bombardment occurred on the moon as late as 500 m.y. after accretion [*Tera et al.*, 1974]. However, there is not a consensus as to whether such bombardment was episodic [*Stewart*, 1975] or continuous and whether it was an event restricted to the moon or was solar system wide [*Hartmann*, 1975; *Wetherill*, 1975]. The Mercury results seem relevant to both questions. The terminal heavy bombardment recorded on Mercury probably could not have occurred appreciably more recently than the emplacement of the oldest lunar mare surfaces (~3.8 b.y. ago); it is difficult to imagine a population of large objects (i.e., capable of producing the Caloris basin) impacting Mercury long after accretion which would not also have impacted the moon. Most important, episodic terminal bombardment evidently occurred on Mercury. Thus we find it plausible to correlate the terminal bombardments on both the moon and Mercury as resulting from a distinct episode that affected at least the inner solar system about 4 b.y. ago.

There still remains the seemingly contrived possibility that the Mercurian surface records an earlier episodic bombardment and the moon records a later heavy bombardment which somehow did not affect Mercury. But Mars also exhibits a heavily cratered surface (in equatorial and southern hemispheric areas) which strikingly resembles the Mercurian cratered terrain; i.e., it is only partially saturated with large craters (see Figure 2). Very large ringed basins also have been recognized [*Wilhelms*, 1973]. The simplest and, to us, the most plausible explanation is that all three surfaces record the same episode of solar system bombardment, presumably from objects originating in large aphelia orbits. It is not possible, however, on the basis of the picture data alone (i.e., ignoring dynamical considerations), to exclude entirely the possibility that the old cratered surfaces instead record late bombardment of accretionary objects, objects in heliocentric orbits similar to those of the three planets.

If a terminal heavy bombardment affected both Mercury and the moon, then Venus must have been bombarded as well. Recent radar images of portions of the equatorial regions of Venus [*Rumsey et al.*, 1974] show large circular craterlike topographic forms in some areas. If these features prove upon higher-resolution examination to be of probable impact origin, then they very probably have survived at least 4×10^9 years, since any more recent episode of bombardment should have affected Mercury and the moon as well. Such a conclusion is most significant, since it would mean that Venus, despite its similarity to earth in size and mass, has not manifested a global tectonic and volcanic style sufficient to obliterate all such ancient topography. Similarly, severe constraints on the erosional capacity of the extremely dense hot Venus atmosphere would be implied.

Extensive areas on the moon and apparently on Mercury were flooded by volcanism after the end of late heavy bombardment, and internal activity on both planets evidently ceased after that volcanic episode. (Similar volcanic flooding also occurred on Mars following the end of heavy bombardment [*Soderblom et al.*, 1974; *Wilhelms*, 1973] but continued longer, along with subsequent shield volcanism and extensional deformation.) The timing and mechanism of volcanic flooding on the moon has been related to specific models of lunar thermal history and partial melting. (For a review of these ideas, see *Albee* [1975].) In view of the profoundly different structure of the Mercurian interior as compared to that of the moon, it seems most fortuitous for a closely similar history of surface igneous activity, and cessation of activity, to have developed on both Mercury and the moon. It can be speculated that somehow the subsequent volcanic flooding on both planets (and perhaps Mars as well) is a delayed but direct consequence of the heavy bombardment itself. We can offer no plausible mechanism, however, to explain such a delayed effect. Alternatively, in the unlikely (to us) event that the smooth plains of Mercury are determined confidently to be entirely of impact, rather than volcanic, origin, then the difficulties raised by the apparent similarities in igneous history of the moon and Mercury would be avoided.

It is quite significant that the light cratering on the flooded plains of Mercury is similar to that on the maria of the moon in diameter-frequency relationships; similar flux histories over the last 3–4 b.y. are implied if the Mercurian surfaces are of comparable age. Relatively uniform impact flux histories throughout the inner solar system for the last 3–4 b.y. were inferred recently by *Wetherill* [1974], who concluded that the impacting objects probably originated in large aphelia orbits (like comets) rather than directly from the asteroid belt. Similar flux histories for Mars and the moon were independently hypothesized by *Soderblom et al.* [1974] on the basis of the similarity of diameter-frequency versus areal coverage relationships for those two objects. In contrast, many earlier analyses had hypothesized that the apparent impact flux on Mars was 6–20 times greater than that which had affected the moon [*Hartmann*, 1973, 1966; *Anders and Arnold*, 1965]. As a result, the age of the large shield volcano Olympus Mons, for instance, was initially placed at 40 m.y. with an uncertainty of a factor of 2 or 3 [*McCauley et al.*, 1972]. The same diameter-frequency data now would be inferred to indicate by the Soderblom-Wetherill hypothesis an average age of the surface materials closer to 500 m.y. Similarly, the oldest cratered plains on Mars, which strongly resemble the lunar maria in morphology, albedo, and degree of cratering, would now be deemed to be of comparable age as well, i.e., 3–4 b.y.

It is still possible, of course, that the Mercurian plains are older than the lunar maria, just enough older to have accumulated virtually all of their present crater populations sufficiently close to the peak of the terminal heavy bombardment episode to mimic the flux accumulated on the lunar maria over 3½ b.y. Only under such coincidental conditions can the lunar and Martian plains crater populations be ascribed to a source of objects exhibiting strong depletion between the orbits of Mars and Mercury, and the Martian surfaces consequently be assigned ages much younger than comparably cratered lunar surfaces.

IMPLICATIONS FOR EARTH

The Mercury pictures support the notion that a late heavy bombardment, dated from lunar samples at about 4 b.y., was an episode which characterized the entire inner solar system. The earth therefore very probably underwent a comparable bombardment as well. Such a conjecture is not new [*Safronov*, 1972]. However, now the bombardment event 4 aeons ago must be regarded as being most probable and included in any reconstruction of the earth's history.

What would be the effects of major basin-forming impacts on earth? Imbrium and Caloris were probably excavated initially to a depth of 100–200 km; permanent modification of the host material certainly transpired to some appreciably greater depth. Such events on earth 4 b.y. ago must have created physical and chemical heterogeneities in the upper mantle. The oldest confidently dated rocks are 3.7–3.9 b.y. old [*Goldich and Hedge*, 1974; *Moorbath et al.*, 1973]. Whether the absence of earlier samples reflects a planet-wide metamorphism, as it does on the moon, or is merely a matter of chance survival is an open question. In any case, the Mercury picture data, combined with lunar sample analyses, serve to emphasize that planetary history and geological history are almost convergent nearly 4 b.y. ago. Continuing study of ancient terrestrial rocks, as well as eventual sample return from other terrestrial planets, may help unite the two subjects firmly.

The existence of a large iron core, perhaps one even sustaining a currently active dynamo process [*Ness et al.*, 1975], also carries implications for the origin and history of earth's core. The formation of earth's core is sometimes linked to a concurrent exhalation of volatiles to produce much of earth's primary atmosphere during planet-wide differentiation following homogeneous accretion [*Ringwood*, 1966]. The present terrestrial atmosphere is then ascribed to a long-term degassing. Such a process is not excluded for Mercury by the Mariner 10 data but is burdened by the need both to develop a thick cool lithosphere and to remove entirely any secondary atmosphere on Mercury before the end of heavy bombardment.

Atmospheric escape is governed by solar irradiation and by surface gravity; Mercury's nearness to the sun, in comparison to Mars, enhances certain atmospheric escape processes, for example. However, Mercury and Mars have similar surface gravities, so that differential solar irradiation would have to be afforded extraordinary effectiveness in governing the very different atmospheric histories of the two planets if Mercury ever evolved even minor amounts of volatiles to its surface. Conversely, if Mercury accumulated heterogeneously, i.e., with very early differentiation, then volatiles may never have accumulated on the surface of Mercury in significant amounts (See, for example, *Anderson* [1973] and *Kaula* [1975].) Generally, we find heterogeneous accumulation to be more probable for Mercury and, by analogy, for the earth and the other terrestrial planets.

Regardless of how Mercury formed, very probably it has been a differentiated planet for well over 4 b.y., the inferred age of the large impact craters still preserved on its surface. Yet there is no evidence after the end of heavy bombardment of any modification of the silicate outer layers by 'hot spots' or fluid convection within that core. In contrast, hot spots and convection in the terrestrial mantle have been attributed speculatively to an origin associated with the core-mantle boundary. Of course, Mercury's core may have cooled to a stable solid configuration by the end of heavy bombardment; indeed the lobate scarps may record the end of that process. If so, Mercury's present magnetic field could not easily be attributed to an active dynamo field, the explanation preferred by *Ness et al.* [1975]. Likewise, there may be some difficulties in finding a plausible thermal and electrical model of Mercury to permit a 'fossil' magnetic field to survive 4 b.y. Such difficulties in finding any plausible interpretation of the Mercury magnetic field perhaps may be interpreted as providing new clues to otherwise ad hoc circumstances hypothecated within the earth's core to account for earth's magnetic field.

Mariner 10's voyage of exploration to the unknown planet Mercury has enlarged our knowledge not just of that planet but of the family of terrestrial planets, including our own.

Acknowledgments. An earlier version of this manuscript was thoroughly and constructively reviewed by Clark Chapman and William Hartmann of the Planetary Science Institute, Tucson, Arizona; by William Kaula of the University of California at Los Angeles; and by Don Wilhelms of the U.S. Geological Survey, Menlo Park, California. As a consequence the paper has been substantially rewritten to present our interpretations more clearly and to call attention more conspicuously to alternative possibilities. We feel that the resulting paper has been substantially improved through response to the reviewers' efforts and wish to express our appreciation for their efforts. Contribution 2579 of the Division of Geological and Planetary Sciences, California Institute of Technology, Pasadena, California 91125.

REFERENCES

Albee, A. L., A review of lunar sample studies and their application to studies of the terrestrial planets, *Rev. Geophys. Space Phys.*, in press, 1975.

Anders, E., and J. Arnold, Age of craters on Mars, *Science, 149,* 1494, 1965.

Anderson, D. L., The moon as a high temperature condensate, *Moon, 8,* 33–57, 1973.

Bowen, N. L., *The Evolution of the Igneous Rocks,* Princeton University Press, Princeton, N. J., 1928.

Gault, D. E., J. E. Guest, J. B. Murray, D. Dzurisin, and M. C. Malin, Some comparisons of impact craters on Mercury and the moon, *J. Geophys. Res., 80,* this issue, 1975.

Goldich, S. S., and C. E. Hedge, 3800-Myr granite gneiss in southwestern Minnesota, *Nature, 252,* 467, 1974.

Hapke, B., G. E. Danielson, Jr., K. Klaasen, and L. Wilson, Photometric observations of Mercury from Mariner 10, *J. Geophys. Res., 80,* this issue, 1975.

Hartmann, W. K., Martian cratering, *Icarus, 5,* 565, 1966.

Hartmann, W. K., Martian cratering, 4, Mariner 9 initial analysis of cratering chronology, *J. Geophys. Res., 78,* 4096, 1973.

Hartmann, W. K., Lunar cataclysm: A misconception?, *Icarus,* in press, 1975.

Kaula, W. H., The seven ages of a planet, submitted to *Icarus,* 1975.

McCauley, J. F., M. H. Carr, J. A. Cutts, W. K. Hartmann, H. Masursky, D. J. Milton, R. P. Sharp, and D. E. Wilhelms, Preliminary Mariner 9 report on the geology of Mars, *Icarus, 17,* 289, 1972.

Moorbath, S., R. K. O'Nions, and R. J. Parkhurst, Early Archaean age for the Isua iron formation, West Greenland, *Nature, 245,* 138, 1973.

Moore, H. J., C. A. Hodges, and D. H. Scott, Multi-ring basins (illustrated by Orientale and associated features), Proceedings of the 5th Lunar Science Conference, *Geochim. Cosmochim. Acta, Suppl. 5, 1,* 71–100, 1974.

Murray, B. C., M. J. S. Belton, G. E. Danielson, M. E. Davies, D. E. Gault, B. Hapke, B. O'Leary, R. G. Strom, V. Suomi, and N. Trask, Mercury's surface: Preliminary description and interpretation from Mariner 10 pictures, *Science, 185,* 169, 1974.

Ness, N. F., K. W. Behannon, R. P. Lepping, Y. C. Whang, and K. H. Schatten, Magnetic field observations near Mercury: Preliminary results from Mariner 10, *Science, 185,* 151, 1974.

Ness, N. F., K. W. Behannon, R. P. Lepping, and Y. C. Whang, Magnetic field of Mercury confirmed, submitted to *Nature,* 1975.

Reynolds, R. T., and A. L. Summers, Calculations on the composition of the terrestrial planets, *J. Geophys. Res., 74,* 2494, 1969.

Ringwood, A. E., Chemical evolution of the terrestrial planets, *Geochim. Cosmochim. Acta, 30,* 41, 1966.

Ringwood, A. E., and E. Essene, Petrogenesis of Apollo 11 basalts: Internal constitution and origin of the moon, Proceedings Apollo 11 Lunar Science Conference, *Geochim. Cosmochim. Acta, Suppl. 1, 1,* 769, 1970.

Rumsey, H. C., G. A. Morris, R. R. Green, and R. M. Goldstein, A radar brightness and altitude image of a portion of Venus, *Icarus, 23,* 1, 1974.

Safronov, V. S., The initial state of the earth and certain features of its evolution, *Izv. Acad. Sci. USSR Phys. Solid Earth,* no. 7, 444, 1972.

Schultz, P. H., and D. E. Gault, Seismic effects from major basin formation on the moon and Mercury, *NASA Tech. Memo. TM X-62,* 388, 1974. (Also *Moon,* in press, 1975.)

Soderblom, L. A., C. D. Condit, R. A. West, B. M. Herman, and T. J. Kreidler, Martian planetwide crater distribution: Implications for geologic history and surface processes, *Icarus, 22,* 239, 1974.

Stewart, D. B., Apollonian metamorphic rocks—The products of prolonged subsolidus equilibration, in *Lunar Science VI,* Lunar Science Institute, Houston, Tex., 1975.

Strom, R. G., N. J. Trask, and J. E. Guest, Tectonism and volcanism on Mercury, *J. Geophys. Res., 80,* this issue, 1975*a*.

Strom, R. G., B. C. Murray, M. J. S. Belton, G. E. Danielson, M. E. Davies, D. E. Gault, B. Hapke, B. O'Leary, N. Trask, J. E. Guest, J. Anderson, and K. Klaasen, Preliminary imaging results from the second Mercury encounter, *J. Geophys. Res., 80,* this issue, 1975*b*.

Tera, F., D. A. Papanastassiou, and G. J. Wasserburg, Isotopic evidence for a terminal lunar cataclysm, *Earth Planet. Sci. Lett., 22,* 1, 1974.

Trask, N. J., and J. E. Guest, Preliminary geologic/terrain map of Mercury, *J. Geophys. Res., 80,* this issue, 1975.

Wetherill, G. W., Problems associated with estimation of the relative impact rate on Mars and moon, *Moon, 9,* 227, 1974.

Wetherill, G. W., Pre-mare cratering and early solar system history, in *Proceedings of the Soviet-American Conference on Cosmochemistry of Moon and Planets,* NASA, Greenbelt, Md., 1975.

Wilhelms, D. E., Comparison of Martian and lunar multiringed circular basins, *J. Geophys. Res., 78,* 4084, 1973.

(Received February 14, 1975;
revised March 3, 1975;
accepted March 5, 1975.)

40

Reprinted from *Nature* **214**:1317–1318 (1967)

GEOLOGICAL EVIDENCE FOR A PULSATING GRAVITATION

F. Machado

DIRAC[1] suggested in 1937 that the so-called gravitation "constant" f could have been decreasing with time. A decrease of f has important implications—namely, an expansion of the Earth's volume and a decrease of solar radiation[2]. Expansion will produce regression of the oceans and fracturing of the crust (with apparent migration of continents), and decreasing radiation will make the climate become cooler.

This is the sort of evolution which certainly occurred during some intervals of the Earth's history, and consequently Egyed, Jordan and other authors[3] considered that the geological record confirmed the hypothesis of an expanding (and cooling) Earth.

Expansion, however, is no explanation for thickening of the crust in orogenic belts (as emphasized by Jeffreys[4]), or for the production of oceanic transgressions. Both these phenomena definitely occurred in the past history of the Earth, and fairly warm climates appear to have alternated with the cold ones (even when allowance is made for polar wandering). It is therefore worth reviewing what the geological evidence really suggests for the variation of f.

In Table 1, some of the principal geological features of the past 400×10^6 yr are schematically indicated (compare Brinkmann[5]). The events which correspond to the intervals (I) and (II) are reasonably repeated, in the same order, during the intervals (III) and (IV), which is quite remarkable. Ocean transgressions and regressions, together with climate variation, suggest a pulsating gravitation "constant" with maxima at Carboniferous and Cretaceous, and minima at Triassic and at present time.

The tectonic phenomena are more complicated. Formation of oceans may have been a discontinuous process, but orogeny is not so easy to understand. It appears that crustal shortening (with corresponding increase of thickness) occurred during the geosyncline phase and produced a root of less dense rocks; folding, however, seems to have been a consequence of the following upheaval, as believed by many writers (see, for example, Beloussov[6]); this latter phase occurred already during the intervals of expansion.

Table 1. GEOLOGICAL EVOLUTION (DEVONIAN TO PRESENT)

Time interval	Main tectonic events	Variation of sea level	Variation of climate
(I) Devonian to Carboniferous	Beginning of Variscan orogeny (geosyncline phase)	Mainly transgressive (rise of sea level)	Generally getting warmer (?)
(II) Carboniferous to Triassic	Upheaval and folding of Variscan chains; probable division of Gondwanaland and of Laurasia	Mainly regressive (lowering of sea level)	Generally getting cooler (Permian glaciations)
(III) Triassic to Cretaceous	Beginning of Alpine orogeny (geosyncline phase)	Mainly transgressive	Generally getting warmer (spreading of corals)
(IV) Cretaceous to present	Upheaval and folding of Alpine chains; formation of present mid-oceanic ridges	Mainly regressive	Generally getting cooler (Pleistocene glaciations)

Notwithstanding the possible difficulties of interpretation, the geological record suggests a pulsating variation of f rather than a monotonic decrease.

In order to obtain quantitative values, it can be tentatively assumed that the variation of the Earth's radius R follows the sine curve (compare Fig. 1)

$$\delta R/R = A \sin [2\pi(t+50)/200] \qquad (1)$$

where t is the time in millions of years. The period is assumed to be 200×10^6 yr. Using, as a measure of the last expanding phase, the width indicated by Heezen[7] for the actual mid-oceanic ridges, the approximate value $A = 0{\cdot}03$ (with a mean radius of 6,200 km) could be obtained.

The relative variation of the gravitation "constant" is

$$\delta f/f = -k^{-1}\delta R/R \qquad (2)$$

k being a constant given by the expression (valid for small $\delta f/f$)

$$k = \frac{1}{R^3}\int_0^R \frac{pr^2}{\varkappa}\,dr \qquad (3)$$

where p and $\varkappa$ are, respectively, the pressure and the bulk modulus at the level of radius r. Bullen's values[8] for p and

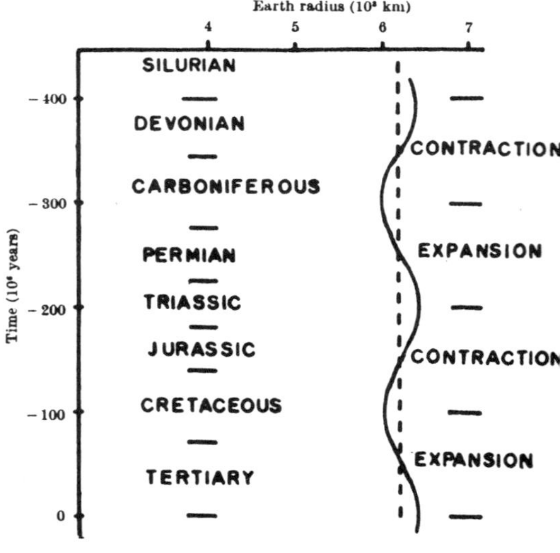

Fig. 1. Probable variation of the Earth's radius since the Devonian.

$\varkappa$ can be substituted into equation (3), giving $k = 0{\cdot}05$. Noting that R now appears to be at a maximum and therefore f at a minimum, equation (2) shows that the gravitation "constant" must have varied between the present value and more than twice as much.

If, according to the estimate of Teller[2], the absolute temperature on the Earth were proportional to $f^{2 \cdot 25}$, that large value of f would have produced a very large increase of temperature, which is certainly incompatible with the moderately warm climates prevailing during Carboniferous and Cretaceous times. Either that estimate is wrong or the variation of the Earth's volume (and climate) is not due to the variation of f.

Pulsations in the geological history are not a new idea, having been pointed out a long time ago by Bucher[9]. According to the model here proposed, shortening of the crust in orogenic belts was achieved when compression was available, but upheaval of the mountains, occurring during the following phase of expansion, would have been responsible for most of the folding.

Opening of fractures in the mid-oceanic ridges is certainly caused by tension; subsequent intrusion of dykes, as suggested by Walker[10], makes the process irreversible, the fractures being blocked by the solidified lava. This then provides a mechanism for a discontinuous continental "drift", which probably started in Carboniferous time but was interrupted during the Mesozoic phase of contraction (when the horizontal compressive displacements were shifted to the geosynclines). That pause in continental "drift" is also suggested by Creer's palaeomagnetic data[11].

The present state (as well as the Triassic one) corresponds fairly well to maximum expansion, and this explains why most measurements of ancient Earth's radii have shown values smaller than at present, a fact which apparently suggested expansion (without any contraction).

Finally, it must be emphasized that equation (1) is a simple crude approximation, and that the events indicated in Table 1 correspond only to the general trends of geological evolution. Minor variations have occurred which may be caused by other controlling agents (erosion, sedimentation, isostatic adjustment), or to a small variation of R, with much shorter period, superimposed on the curve of Fig. 1.

This investigation was supported by Instituto de Alta Cultura, in Lisbon.

[1] Dirac, P. A. M., *Nature*, **139**, 323 (1937).
[2] Teller, E., *Phys. Rev.*, **73**, 801 (1948).
[3] Egyed, L., *Nature*, **178**, 534 (1956). Jordan, P., *Die Expansion der Erde* (Vieweg, Braunschweig, 1966). Holmes, A., *Principles of Physical Geology*, second ed., chapt. 27 (Nelson, London, 1965).
[4] Jeffreys, H., *The Earth*, fourth ed., chapter 11 (Univ. Press, Cambridge, 1959).
[5] Brinkmann, R., *Abriss der Geologie*, ninth ed., **2** (Enke, Stuttgart, 1966).
[6] Beloussov, V. V., *Basic Principles of Geotectonics* (McGraw-Hill, New York, 1962).
[7] Heezen, B. C., *Sci. Amer.*, **203**, 99 (1960).
[8] Bullen, K. E., *An Introduction to the Theory of Seismology*, third ed., 233 and 235 (Univ. Press, Cambridge, 1963).
[9] Bucher, W. H., *The Deformation of the Earth's Crust* (Univ. Press, Princeton, 1933).
[10] Walker, G. P. L., *Phil. Trans. Roy. Soc.*, A, **258**, 199 (1965).
[11] Creer, K. M., *Phil. Trans. Roy. Soc.*, A, **258**, 27 (1965).

REFERENCES

Ager, D. V. 1973. *The Nature of the Stratigraphical Record.* London: Macmillan, 114 pp.

Ager, D. V. 1976. The nature of the fossil record. *Geol. Assoc. London Proc.* **87**:131–159.

Allard, H. A. 1948. Length of day in the climates of past geological eras and its possible effects upon changes in plant life. In *Vernalization and Photoperiodism,* A. E. Murneek and R. O. Whyte, eds. Waltham, Mass.: Chronica Botanica Co., pp. 101–119.

Allègre, C. J. 1972. Géochronologie isotopique et développement de la croute terrestre au cours des temps géologiques. In *Structure et Dynamique de la Lithosphere,* C. J. Allègre and M. Mattauer, eds. Paris: Hermann, pp. 374–433.

Allen, C. W. 1973. *Astrophysical Quantities,* 3rd ed. Univ. of London: Athlone Press, 310 pp.

Anders, E. 1964. Origin, age, and composition of meteorites. *Space Sci. Rev.* **3**:583–714.

Anhaeusser, C. R., and A. Button. 1976. A review of southern African stratiform ore deposits—their position in time and space. In *Handbook of Strata-Bound and Stratiform Ore Deposits. Vol. 5: Regional Studies,* K. H. Wolf, ed. Amsterdam: Elsevier, pp. 257–319.

Aparin, V. P., and V. S. Vedenkov. 1975. Periodic variations in the rate of migration of paleomagnetic poles in the Phanerozoic. *Akad. Nauk SSSR Doklady Earth Sci. Sec.* **222**:49–50.

Aparin, V. P., and V. S. Vedenkov. 1977. Curves of the rate of displacement of paleomagnetic poles. *Geomagnetism and Aeronomy* **16**:476–477.

Armstrong, R. L. 1971. Glacial erosion and the variable isotopic composition of strontium in sea water. *Nature Phys. Sci.* **230**:132–133.

Armstrong, R. L., and W. G. McDowall. 1975. Proposed refinement of the Phanerozoic time scale. *Geodynam. Highlights* **2**:33–34.

Arvidson, R. E., A. B. Binder, and K. L. Jones. 1978. The surface of Mars. *Sci. American* **238**(3):76–89.

Auman, J. R., and W. H. McCrea. 1976. Solar neutrinos and galactic contamination of the Sun. *Nature* **262**:560–561.

Austin, P. M., and G. E. Williams. 1978. Tectonic development of late Precambrian to Mesozoic Australia through plate motions possibly influenced by the Earth's rotation. *Geol. Soc. Australia Jour.* **25**:1–21.

Austin, S. A. 1978. Uniformitarianism—doctrine which needs rethinking (abs.). *Am. Assoc. Petroleum Geologists Bull.* **62**:492.

Awramik, S. M. 1971. Precambrian columnar stromatolite diversity: reflection of metazoan appearance. *Science* **174**:825–827.

Awramik, S. M., and E. S. Barghoorn. 1977. The Gunflint microbiota. *Precambrian Research* **5**:121–142.

Awramik, S. M., L. Margulis, and E. S. Barghoorn. 1976. Evolutionary processes in the formation of stromatolites. In *Stromatolites,* M. R. Walter, ed. Amsterdam: Elsevier, pp. 149–162.

Baadsgaard, H., and P. A. Mueller. 1973. K-Ar and Rb-Sr ages of intrusive Precambrian mafic rocks, southern Beartooth Mountains, Montana and Wyoming. *Geol. Soc. America Bull.* **81**:3635–3644.

Baldwin, R. B. 1974. Was there a "Terminal Lunar Cataclysm" $3.9–4.0 \times 10^9$ years ago? *Icarus* **23**:157–166.

Balukhovskiy, N. F. 1966. *Geological Cycles.* Kiev: Naukova Dumka, 168 pp. (in Russian).

Baragar, W. R. A. 1977. Volcanism of the stable crust. *Geol. Assoc. Canada Spec. Paper 16,* pp. 377–405.

Barrell, J. 1917. Rhythms and the measurement of geologic time. *Geol. Soc. America Bull.* **28**:745–904.

Beaudet, G., G. Fontaine, A. Sirois, and M. Tassoul. 1977. The solar neutrino problem: limitations of energy transport by mechanical means. *Astron. Astrophys.***54**:213–218.

Beloussov, V. V. 1940. Rhythm of oscillatory movements of the Earth's crust. *Akad. Nauk SSSR Izv., Ser. Geog. Geophys.* **4**:533–558 (in Russian).

Beloussov, V. V. 1962. *Basic Problems in Geotectonics.* New York: McGraw-Hill, 809 pp.

Berggren, W. A., and C. D. Hollister. 1977. Plate tectonics and paleocirculation—commotion in the ocean. *Tectonophysics* **38**:11–48.

Berggren, W. A., D. P. McKenzie, J. G. Sclater, and J. E. van Hinte. 1975. World-wide correlation of Mesozoic magnetic anomalies and its implications: discussion. *Geol. Soc. America Bull.* **86**:267–269.

Berry, E. W. 1929. Shall we return to cataclysmal geology? *Am. Jour. Sci.,* ser. 5, **17**:1–12.

Berry, W. B. N., and R. M. Barker. 1975. Growth increments in fossil and modern bivalves. In *Growth Rhythms and the History of the Earth's Rotation,* G. D. Rosenberg and S. K. Runcorn, eds. London: Wiley, pp. 9–24.

Bgatov, V. I., and V. P. Kazarinov. 1970. Sedimentary series as principal stages in cyclical development of sedimentation. *Internat. Geology Rev.* **12**:298–308.

Bingquan, Z., and Z. Xinhua. 1976. The frequency spectra of isotopic ages and the Phanerozoic time scale. *Geochimica* **6**:83–94 (in Chinese, with English summary).

Blackwelder, E. 1914. A summary of the orogenic epochs in the geologic history of North America. *Jour. Geology* **22**:633–654.

Bogard, D. D., L. Husain, and R. J. Wright. 1976. ^{40}Ar-^{39}Ar dating of collisional events in chondrite parent bodies. *Jour. Geophys. Research* **81**:5664–5678.

Bond, H. E. 1974. Possible galactic causes for geologic events: comment. *Geology* **2**:279–280.

Borchardt, G., and J. P. Platt. 1978. Catastrophe theory: application to the Permian mass extinction. Comments. *Geology* **6**:453–454.

Born, A. 1936. Periodicity in epeirogenic movements of the Earth's crust. *Internat. Geol. Congr., 16th, Rep.* **1**:169–189.

Briden, J. C. 1966. Variation of intensity of the palaeomagnetic field through geological time. *Nature* **212**:246–247.

Briden, J. C., and I. G. Gass. 1974. Plate movement and continental magmatism. *Nature* **248**:650–653.

Brookfield, M. 1971. Periodicity of orogeny. *Earth and Planetary Sci. Letters* **12**:419–424.

Brooks, C., S. R. Hart, A. Hofmann, and D. E. James. 1976. Rb-Sr mantle isochrons from oceanic regions. *Earth and Planetary Sci. Letters* **32**:51–61.

Brooks, J., M. D. Muir, and G. Shaw. 1973. Chemistry and morphology of Precambrian microorganisms. *Nature* **244**:215–217.

Bubnoff, S. von. 1941. Geotektonik und Rhythmus der Erdgeschichte. *Scientia* **69**:147–153.

Bubnoff, S. von. 1948a. Der Rhythmus der Erde. *Universitas* **3**:961–967.

Bubnoff, S. von. 1948b. Rhythmen, Zyklen und Zeitrechnung in der Geologie. *Geol. Rundschau* **35**:6–22.

Bubnoff, S. von. 1949. Osteuropa und die zyklische Gliederung de Erdgeschichte. *Geol. Rundschau* **37**:60–71.

Buchardt, B. 1978. Oxygen isotope palaeotemperatures from the Tertiary period in the North Sea area. *Nature* **275**:121–123.

Bucher, W. H. 1933. *The Deformation of the Earth's Crust*. Princeton, N.J.: Princeton Univ. Press, 518 pp.

Bucher, W. H. 1939. Deformation of the Earth's crust. *Geol. Soc. America Bull.* **50**:421–432.

Budanov, V. I., A. M. Meskhi, V. N. Volkov, and S. P. Kirillov. 1962. On the epochs of granitoid magmatism of the Pamirs and Darvas. *Akad. Nauk SSSR Doklady* **136**:68–70 (AGI translation).

Burke, K. 1977. Aulacogens and continental breakup. *Annual Rev. Earth and Planetary Sci.* **5**:371–396.

Burke, K., J. F. Dewey, and W. S. F. Kidd. 1976. Precambrian palaeomagnetic results compatible with contemporary operation of the Wilson Cycle. *Tectonophysics* **33**:287–299.

Burke, K., J. F. Dewey, and W. S. F. Kidd. 1977. World distribution of sutures—the sites of former oceans. *Tectonophysics* **40**:69–99.

Burns, J. A., W. Ward, and O. Toon. 1977. Mars before Tharsis: much larger obliquity in the past? *Am. Astron. Soc. Bull.* **9**:449.

Burwash, R. A. 1969. Comparative Precambrian geochronology of the North American, European and Siberian Shields. *Canadian Jour. Earth Sci.* **6**:357–365.

Busse, F. H. 1978. A model of time-periodic mantle flow. *Royal Astron. Soc. Geophys. Jour.* **52**:1–12.

Button, A. 1973. Algal stromatolites of the early Proterozoic Wolkberg Group, Transvaal Sequence. *Jour. Sed. Petrology* **43**:160–167.

Cameron, A. G. W. 1973. Major variations in solar luminosity? *Rev. Geophys. Space Phys.* **11**:505–510.

Carmichael, C. M. 1967. An outline of the intensity of the paleomagnetic field of the Earth. *Earth and Planetary Sci. Letters* **3**:351–354.

Carmichael, C. M. 1970. The intensity of the Earth's paleomagnetic field from 2.5 × 10^9 years ago to the present. In *Palaeogeophysics,* S. K. Runcorn, ed. London: Academic Press, pp. 73–77.

Chadwick, P. 1977. The perception and interpretation of continuity and discontinuity. *Catastrophist Geol.* **2**(1):35–48.

Chamberlin, R. T. 1914. Diastrophism and the formative processes. VII. Periodicity of Paleozoic orogenic movements. *Jour. Geology* **22**: 315–345.

Chamberlin, T. C. 1898. The ulterior basis of time divisions and the classification of geologic history. *Jour. Geology* **6**:449–462.

Chamberlin, T. C. 1909. Diastrophism as the ultimate basis of correlation. *Jour. Geology* **17**:685–693.

Chapman, C. R. 1976. Chronology of terrestrial planet evolution: the evidence from Mercury. *Icarus* **28**:523–536.

Chilingar, G. V. 1956. Relationship between Ca/Mg ratio and geologic age. *Am. Assoc. Petroleum Geologists Bull.* **40**:2256–2266.

Chillingworth, D., and P. Furness. 1975. Reversals of the Earth's magnetic field. In *Dynamical Systems,* Lecture Notes in Mathematics, vol. 468, A. Manning, ed. Berlin: Springer-Verlag, pp. 91–98.

Choubert, B. 1967. Réflexions sur la finalité des mesures géochronologiques (périodicité des événements du Précambrien et dérive des continents). *Soc. Géol. France Bull.* **9**:809–819.

Cloud, P. 1972. A working model of the primitive Earth. *Am. Jour. Sci.* **272**:537–548.

Cloud, P. 1974. Evolution of ecosystems. *Am. Scientist* **62**:54–66.

Coleman, A. P. 1926. *Ice Ages, Recent and Ancient.* London: Macmillan, 296 pp.

Coney, P. J. 1971. Cordilleran tectonic transitions and motion of the North American plate. *Nature* **233**:462–465.

Cooper, J. A., J. R. Richards, and F. D. Stacey. 1967. Possible new evidence bearing on the lunar capture hypothesis. *Nature* **215**:1256.

Cox, A., and R. R. Doell. 1960. Review of paleomagnetism. *Geol. Soc. America Bull.* **71**:645–768.

Crain, I. K. 1971. Possible direct causal relation between geomagnetic reversals and biological extinctions. *Geol. Soc. America Bull.* **82**:2603–2606.

Crain, I. K., and P. L. Crain. 1970. New stochastic model for geomagnetic reversals. *Nature* **228**:39–41.

Crain, I. K., P. L. Crain, and M. G. Plaut. 1969. Long period Fourier spectrum of geomagnetic reversals. *Nature* **223**:283.

Creer, K. M. 1975. On a tentative correlation between changes in the geomagnetic polarity bias and reversal frequency and the Earth's rotation through Phanerozoic time. In *Growth Rhythms and the History of the Earth's Rotation,* G. D. Rosenberg and S. K. Runcorn, eds. London: Wiley, pp. 293–317.

Crowell, J. C. 1978. Gondwanan glaciation, cyclothems, continental positioning, and climate change. *Am. Jour. Sci.* **278**:1345–1372.

Crowell, J. C., and L. A. Frakes. 1970. Phanerozoic glaciation and the causes of ice ages. *Am. Jour. Sci.* **268**:193–224.

Crozaz, G., G. Poupeau, R. M. Walker, E. Zinner, and D. A. Morrison. 1977. The record of solar and galactic radiations in the ancient lunar regolith and their implications for the early history of the Sun and Moon. *Royal Soc. London Philos. Trans.* ser. A, **285**:587–592.

Cubitt, J. M., and B. Shaw. 1976. The geological implications of steady-state mechanisms in catastrophe theory. *Math. Geology* **8**:657–662.

Cutbill, J. L., and B. M. Funnell. 1967. Numerical analysis of *The Fossil Record.* In *The Fossil Record,* W. B. Harland et al., eds. Geol. Soc. London, pp. 791–820.

Damon, P. E. 1968. Potassium-argon dating of igneous and metamorphic rocks with applications to the Basin Ranges of Arizona and Sonora. In *Radiometric Dating for Geologists,* E. I. Hamilton and R. M. Farquhar, eds. London: Interscience, pp. 1–71.

Damon, P. E. 1971. The relationship between late Cenozoic volcanism and tectonism and orogenic-epeirogenic periodicity. In *The Late Cenozoic Glacial Ages,* K. K. Turekian, ed. New Haven, Conn.: Yale Univ. Press, pp. 15–35.

Damon, P. E., and R. L. Mauger. 1966. Epeirogeny-orogeny viewed from the Basin and Range Province. *Am. Inst. Min. Metall. Petrol. Engrs., Soc. Min. Engrs. AIME Trans.* **235**:99–112.

Dana, J. D. 1856. On American geological history. *Am. Jour. Sci.* ser. 2, **22**:305–349.

Dana, J. D. 1870. *Manual of Geology.* New York: Ivison, Blakeman, Taylor, 800 pp.

Davies, T. A., O. E. Weser, B. P. Luyendyk, and R. B. Kidd. 1975. Unconformities in the sediments of the Indian Ocean. *Nature* **253**:15–19.

Dawson, Sir J. William. 1868. *Acadian Geology.* London: Macmillan, 694 pp.

Dearnley, R. 1966. Orogenic fold-belts and a hypothesis of Earth evolution. *Phys. and Chem. Earth* **7**:1–114.

Dennison, B., and V. N. Mansfield. 1976. Glaciations and dense interstellar clouds. *Nature* **261**:32–34.

de Vaucouleurs, G., and W. D. Pence. 1978. An outsider's view of the Galaxy: photometric parameters, scale lengths, and absolute magnitudes of the spheroidal and disk components of our Galaxy. *Astron. Jour.* **83**:1163–1173.

Devereux, I. 1967. Oxygen isotope paleotemperature measurements on New Zealand Tertiary fossils. *New Zealand Jour. Sci.* **10**:988–1011.

Dewey, J. F., and K. Burke. 1974. Hot spots and continental break-up: implications for collisional orogeny. *Geology* **2**:57–60.

Dewey, J. F., and H. Spall. 1975. Pre-Mesozoic plate tectonics: how far back in Earth history can the Wilson Cycle be extended? *Geology* **3**:422–424.

Dietz, R. S. 1977. Plate tectonics: a revolution in geology and geophysics. *Tectonophysics* **38**:1–6.

Donaldson, J. A., J. C. McGlynn, E. Irving, and J. K. Park. 1973. Drift of the Canadian Shield. In *Implications of Continental Drift to the Earth Sciences,* vol. 1. D. H. Tarling and S. K. Runcorn, eds. London: Academic Press, pp. 3–17.

Donn, W. L., and D. M. Shaw. 1977. Model of climate evolution based on continental drift and polar wandering. *Geol. Soc. America Bull.* **88**:390–396.

Donovan, D. T., and E. J. W. Jones. 1979. Causes of world-wide changes in sea level. *Geol. Soc. London Jour.* **136**:187–192.

Dontsova, E. I., A. A. Migdisov, and A. B. Ronov. 1973. On the causes of variation of oxygen isotopic composition in the carbonate strata of the sedimentary column. *Geochem. Internat. 1972*, pp. 885–891.

Duff, P. McL. D., A. Hallam, and E. K. Walton. 1967. *Cyclic Sedimentation.* Dev. in Sedimentol. 10, Amsterdam; Elsevier, 280 pp.

Dunbar, C. O., and J. Rodgers. 1957. *Principles of Stratigraphy.* New York: Wiley, 356 pp.

Edwards, A. R. 1973. Calcareous nannofossils from the southwest Pacific, Deep Sea Drilling Project, Leg 21. *Deep Sea Drilling Proj. Init. Rept.* **21**:641–691.

Egyed, L. 1956a. Determination of changes in the dimensions of the Earth from palaeogeographical data. *Nature* **178**:534.

Egyed, L. 1956b. The change of the Earth's dimensions determined from paleogeographical data. *Geofis. Pura Applicata* **33**:42–48.

Egyed, L. 1957. A new dynamic conception of the internal constitution of the Earth. *Geol. Rundschau* **46**:101–121.

Eichler, J. 1976. Origin of the Precambrian banded iron-formations. In *Handbook of Strata-Bound and Stratiform Ore Deposits. Vol. 7: Au, U, Fe, Mn, Hg, Sb, W, and P Deposits*, K. H. Wolf, ed. Amsterdam: Elsevier, pp. 157–201.

Elie de Beaumont, M. L. 1829, 1830. Recherches sur quelques-unes des Révolutions de la surface du globe. *Annales Sci. Nat.* **18**:5–25, 284–416; **19**:5–99.

Elliot, D. H. 1975a. Tectonics of Antarctica: a review. *Am. Jour. Sci.* **275–A**: 45–106.

Elliot, D. H. 1975b. Gondwana basins of Antarctica: *Gondwana Symp., 3d, Proc.* Canberra, pp. 493–536.

Emiliani, C. 1966. Isotopic paleotemperatures. *Science* **154**:851–857.

Engel, A. E. J., S. P. Itson, C. G. Engel, D. M. Stickney, and E. J. Cray. 1974. Crustal evolution and global tectonics: a petrogenic view. *Geol. Soc. America Bull.* **85**:843–858.

Evernden, J. F., and R. W. Kistler. 1970. Chronology of emplacement of Mesozoic batholithic complexes in California and western Nevada. *U.S. Geol. Survey Prof. Paper 623*, 42 pp.

Ewing, J., and M. Ewing. 1967. Sediment distribution on the mid-ocean ridges with respect to spreading of the sea floor. *Science* **156**:1590–1592.

Ewing, J., M. Ewing, T. Aitken, and W. J. Ludwig. 1968. North Pacific sediment layers measured by seismic profiling. *Am. Geophys. Union Geophys. Mon.* **12**:147–173.

Fairbridge, R. W. 1958. What is a consanguineous association? *Jour. Geol.* **66**:319–324.

Fairbridge, R. W. 1960. The changing level of the sea. *Sci. American* **202**(5): 70–79.

Fairbridge, R. W. 1961. Eustatic changes in sea level. *Phys. and Chem. Earth* **4**:99–185.

Fairbridge, R. W. 1967. Carbonate rocks and paleoclimatology in the biogeochemical history of the planet. In *Carbonate Rocks. Origin, Occurrence and Classification*, G. V. Chilingar, H. J. Bissell, and R. W. Fairbridge, eds. Dev. in Sedimentol. 9A. Amsterdam: Elsevier, pp. 399–432.

Fairbridge, R. W. 1972. Planetary spin-rate and evolving cores. *New York Acad. Sci. Annals* **187**:88–107.

Fairbridge, R. W. 1978. Exo- and endogenetic geomagnetic modulation of climates on decadal to galactic scale (abs.). *Eos* **59**:269.

Faul, H., T. W. Stern, H. H. Thomas, and P. L. D. Elmore. 1963. Ages of intrusion and metamorphism in the northern Appalachians. *Am. Jour. Sci.* **261**:1–19.

Firsov, L. V. 1977. Galactic periodicity in the evolution of the terrestrial organic world. In *Basic Theoretical Questions of Cyclic Sedimentation,* A. A. Trofumik, M. F. Murchink, and U. N. Karozobin, eds. Moscow: Nauka, pp. 104–116 (in Russian).

Fitch, F. J., S. C. Forster, and J. A. Miller. 1974. Geological time scale. *Rept. Prog. Phys.* **37**:1433–1496.

Fitch, F. J., and J. A. Miller. 1965. Major cycles in the history of the Earth. *Nature* **206**:1023–1027.

Fitch, F. J., J. A. Miller, and J. G. Mitchell. 1969. A new approach to radioisotopic dating in orogenic belts. In *Time and Place in Orogeny,* P. E. Kent, G. E. Satterthwaite, and A. M. Spencer, eds. *Geol. Soc. London Spec. Publ.* **3**:157–195.

Flemming, N. C. and D. G. Roberts. 1973. Tectono-eustatic changes in sea level and seafloor spreading. *Nature* **243**:19–22.

Flessa, K. W., and J. Imbrie. 1973. Evolutionary pulsations: evidence from Phanerozoic diversity patterns. In *Implications of Continental Drift to the Earth Sciences,* vol. 2, D. H. Tarling and S. K. Runcorn, eds. London: Academic Press, pp. 247–285.

Forbes, W. T. M. 1931. The great glacial cycle. *Science* **74**:294–295.

Frakes, L. A. 1979. *Climates Throughout Geologic Time.* Amsterdam: Elsevier, 310 pp.

Francis, S., L. Margulis, and E. S. Barghoorn. 1978. On the experimental silicification of microorganisms II. On the time of appearance of eukaryotic organisms in the fossil record. *Precambrian Research* **6**:65–100.

Freedman, M. S., et al. 1976. Solar neutrinos: proposal for a new test. *Science* **193**:1117–1119.

French, D. K., and J. L. Osborne. 1976. An interpretation of the galactic continuum radiation using a new model of galactic spiral structure. *Royal Astron. Soc. Monthly Notices* **177**:569–581.

Frerichs, W. E., and P. N. Shive. 1971. Tectonic implications of variations in sea floor spreading rates. *Earth and Planetary Sci. Letters* **12**:406–410.

Fujimoto, M., and Y. Sofue. 1977. The Large and Small Magellanic Clouds in a binary state, the bending of the galactic disk and the Magellanic Stream. *Astron. Astrophys.* **61**:199–215.

Garrels, R. M., and F. T. Mackenzie. 1969. Sedimentary rock types: relative proportions as a function of geological time. *Science* **163**:570–571.

Garrels, R. M., and F. T. Mackenzie. 1971. *Evolution of Sedimentary Rocks.* New York: Norton, 397 pp.

Garrels, R. M., F. T. Mackenzie, and R. Siever. 1972. Sedimentary cycling in relation to the history of the continents and oceans. In *The Nature of the Solid Earth*, E. C. Robertson, ed. New York: McGraw-Hill, pp. 93–121.

Garrels, R. M., and E. A. Perry. 1974. Cycling of carbon, sulfur, and oxygen through geological time. In *The Sea*, vol. 5, E. D. Goldberg, ed. New York: Wiley, pp. 303–336.

Gastil, R. G. 1960. The distribution of mineral dates in time and space. *Am. Jour. Sci.* **258**:1–35.

Gates, T. M., and P. M. Hurley. 1973. Evaluation of Rb-Sr dating methods applied to the Matachewan, Abitibi, Mackenzie, and Sudbury dike swarms in Canada. *Canadian Jour. Earth Sci.* **10**:900–919.

Geiss, J., P. Eberhardt, N. Grögler, S. Guggisberg, P. Maurer, and A. Stettler. 1977. Absolute time scale of lunar mare formation and filling. *Royal Soc. London Philos. Trans.* ser. A, **285**:151–158.

Gilluly, J. 1949. Distribution of mountain building in geologic time. *Geol. Soc. America Bull.* **60**:561–590.

Gilluly, J. 1950. Distribution of mountain building in geologic time. (Discussion and reply with H. Stille.) *Geol. Rundschau* **38**:89–111.

Gilluly, J. 1967. Chronology of tectonic movements in the western United States. *Am. Jour. Sci.* **265**:306–331.

Gilluly, J. 1973. Steady plate motion and episodic orogeny and magmatism. *Geol. Soc. America Bull.* **84**:499–514.

Goguel, J. 1962. *Tectonics.* San Francisco: Freeman, 384 pp.

Goldich, S. S. 1968. Geochronology in the Lake Superior region. *Canadian Jour. Earth Sci.* **5**:715–724.

Goldich, S. S., A. O. Nier, H. Baadsgaard, J. H. Hoffman, and H. W. Krueger. 1961. *The Precambrian Geology and Geochronology of Minnesota.* Minneapolis: Univ. Minnesota Press, 193 pp.

Gould, S. J. 1965. Is uniformitarianism necessary? *Am Jour. Sci.* **263**:223–228.

Grabau, A. W. 1936a. Oscillation or pulsation. *Internat. Geol. Congr., 16th, Rept.* **1**:539–553.

Grabau, A. W. 1936b. *Palaeozoic Formations in the Light of the Pulsation Theory*, vol. 1. Peking: University Press, 680 pp.

Grabau, A. W. 1940. *The Rhythm of the Ages.* Peking: Vetch, 561 pp.

Grasty, R. L. 1967. Orogeny, a cause of world-wide regression of the seas. *Nature* **216**:779–780.

Gregor, C. B. 1968. The rate of denudation in post-Algonkian time. *Koninkl. Nederlandse Akad. Wetensch. Proc. B* **71**:22–30.

Gregor, C. B. 1970. Denudation of the continents. *Nature* **228**:273–275.

Gribbin, J. 1978. Astronomical influences. Long-term effects. In *Climatic Change*, J. Gribbin, ed. Cambridge: Cambridge Univ. Press, pp. 133–138.

Griggs, D. 1939. A theory of mountain-building. *Am Jour. Sci.* **237**:611–650.

Haarmann, E. 1930. *Die Oszillationstheorie.* Stuttgart: Enke, 260 pp.

Hailwood, E. A. 1977. Configuration of the geomagnetic field in early Tertiary times. *Geol. Soc. London Jour.* **133**:23–36.

Hall, B. A. 1969. Pre-Middle Ordovician unconformity in northern New England and Quebec. *Am. Assoc. Petroleum Geologists Mem.* **12**: 467–476.

Hallam, A. 1977. Secular changes in marine inundation of USSR and North America through the Phanerozoic. *Nature* **269**:769–772.

Hallam, A. 1978. Seismic stratigraphy and global changes. *Nature* **272**:400.

Haller, J. 1971. *Geology of the East Greenland Caledonides.* London: Interscience, 413 pp.

Harland, W. B., and K. N. Herod. 1975. Glaciations through time. In *Ice Ages: Ancient and Modern,* A. E. Wright and F. Moseley, eds. Liverpool: Seel House Press, pp. 189–216.

Hartmann, W. K. 1974. Martian and terrestrial paleoclimatology: relevance of solar variability. *Icarus* **22**:301–311.

Hartmann, W. K. 1975. Lunar "cataclysm": a misconception? *Icarus* **24**: 181–187.

Hartmann, W. K. 1977. Relative crater production rates on planets. *Icarus* **31**:260–276.

Hartmann, W. K. 1978. Martian cratering V: Toward an empirical Martian chronology, and its implications. *Geophys. Research Letters* **5**:450–452.

Hatcher, R. D. 1974. North American Paleozoic foldbelts and deformational histories: a plate tectonics anomaly? *Am. Jour. Sci.* **274**:135–147.

Hatfield, C. B., and M. J. Camp. 1970. Mass extinctions correlated with periodic galactic events. *Geol. Soc. America Bull.* **81**:911–914.

Hays, J. D. 1971. Faunal extinctions and reversals of the Earth's magnetic field. *Geol. Soc. America Bull.* **82**:2433–2447.

Hébert, E. 1857. Les mers anciennes et leurs rivages dans le bassin de Paris, ou classification des terrains par les oscillations du sol. *1e Partie, Terrain Jurassique*, Paris, 88 pp.

Hein, J. R., D. W. Scholl, and J. Miller. 1978. Episodes of Aleutian Ridge explosive volcanism. *Science* **199**:137–141.

Henley, S. 1976. Catastrophe theory models in geology. *Math. Geology* **8**:649–655.

Herz, N. 1977. Timing of spreading in the South Atlantic: information from Brazilian alkalic rocks. *Geol. Soc. America Bull.*, **88**:101–112.

Hoefs, J. 1973. *Stable Isotope Geochemistry.* Berlin: Springer-Verlag, 140 pp.

Hoffman, P. F., M. St-Onge, D. M. Carmichael, and I. de Bie. 1978. Geology of the Coronation Geosyncline (Aphebian), Hepburn Lake Sheet (86J), Bear Province, District of Mackenzie. *Current Res., Part A, Geol. Survey Canada Paper 78-1A* pp. 147–151.

Holmes, A. 1925. Radioactivity and the Earth's thermal history. Part V: the control of geological history by radioactivity. *Geol. Mag.* **62**:529–544.

Holmes, A. 1926. Contributions to the theory of magmatic cycles. *Geol. Mag.* **63**:306–329.

Holmes, A. 1937. *The Age of the Earth,* rev. ed. London: Nelson, 263 pp.

Holmes, A. 1951. The sequence of the Pre-Cambrian orogenic belts in south and central Africa. *Internat. Geol. Congr., 18th, Rept.* part XIV, 254–269.

Holmes, A. 1959. A revised geological time-scale. *Edinburgh Geol. Soc. Trans.* **17**:183–216.

Holmes, A. 1965. *Principles of Physical Geology.* London: Nelson, 1288 pp.

Holser, W. T. 1977. Catastrophic chemical events in the history of the ocean. *Nature* **267**:403–408.

Holser, W. T., and I. R. Kaplan. 1966. Isotope geochemistry of sedimentary sulfates. *Chem. Geology* **1**:93–135.

Hoyle, F. 1972. The history of the Earth. *Royal Astron. Soc. Quart. Jour.* **13**:328–345.

Hudson, J. D. 1964. Sedimentation rates in relation to the Phanerozoic time-scale. *Geol. Soc. London Quart. Jour.* **120S**:37–42.

Huntington, E. 1914. Crustal deformation as the cause of climatic changes. In *The Climatic Factor,* E. Huntington, ed. Carnegie Inst. Washington, Publ. No. 192, pp. 255–262.

Hurley, P. M., and J. R. Rand. 1969. Pre-drift continental nuclei. *Science* **164**:1229–1242.

Hutton, J. 1788. Theory of the Earth. *Royal Soc. Edinburgh Trans.* **1**(2): 209–304.

Irving, E. 1964. *Paleomagnetism.* New York: Wiley, 399 pp.

Irving, E., and J. C. McGlynn. 1976. Proterozoic magnetostratigraphy and tectonic evolution of Laurentia. *Royal Soc. London Philos. Trans.* ser. A, **280**:433–468.

Irving, E., and J. K. Park. 1972. Hairpins and superintervals. *Canadian Jour. Earth Sci.* **9**:1318–1324.

Irving, E., and G. Pullaiah. 1976. Reversals of the geomagnetic field, magnetostratigraphy, and relative magnitude of paleosecular variation in the Phanerozoic. *Earth-Sci. Rev.* **12**:35–64.

Jackson, T. A. 1973. 'Humic' matter in the bitumen of ancient sediments: variations through geologic time. *Geology* **1**:163–166.

Jackson, T. A., and C. B. Moore. 1976. Secular variations in kerogen structure and carbon, nitrogen and phosphorus concentrations in pre-Phanerozoic and Phanerozoic sedimentary rocks. *Chem. Geology* **18**:107–136.

Jahn, B.-M., P. V. Chen, and T. P. Yen. 1976. Rb-Sr ages of granitic rocks in southeastern China and their tectonic significance. *Geol. Soc. America Bull.* **86**:763–776.

James, H. L. 1972. Subdivision of Precambrian: an interim scheme to be used by U.S. Geological Survey. *Am. Assoc. Petroleum Geologists Bull.* **56**:1128–1133.

Janle, P. 1977. History of the terrestrial planets. *Geol. Rundschau* **66**: 277–288.

Jeffery, P. M., W. Compston, D. Greenhalgh, and J. de Laeter. 1955. On the carbon-13 abundance of limestones and coals. *Geochim. et Cosmochim. Acta* **7**:255–286.

Jeffreys, H. 1926. On Professor Joly's theory of Earth history. *Philos. Mag.* ser. 7, **1**:923–931.

Jeffreys, H. 1928. Prof. Joly and the Earth's thermal history. *Philos. Mag.* ser. 7, **5**:208–214.

Joly, J. 1923. The surface movements of the Earth's crust. *Nature* **111**: 603–606.

Joly, J. 1925. *The Surface-History of the Earth.* Oxford: Clarendon Press, 192 pp.

Joly, J. 1928. The theory of thermal cycles. *Gerlands Beitr. Geophysik* **19**. (Reprinted, with some additions and corrections, in J. Joly, *The Surface-History of the Earth,* 2d ed., Oxford: Clarendon, 1930, pp. 185–206.)

Jones, G. M. 1977. Thermal interaction of the core and the mantle and long-term behaviour of the geomagnetic field. *Jour. Geophys. Research* **82**:1703–1709.

Kaula, W. M. 1975. The seven ages of a planet. *Icarus* **26**:1–15.

Kawano, Y., and Y. Ueda. 1967. Periods of the igneous activities of the granitic rocks in Japan by K-A dating method. *Tectonophysics* **4**: 523–530.

Kaz'min, V. G. 1975. Epochs of rift formation and some problems of the origin of rift structures. *Soviet Geol. Geophys.* **16**(9):1–11.

Keith, M. L., and J. N. Weber. 1964. Carbon and oxygen isotopic composition of selected limestones and fossils. *Geochim. et Cosmochim. Acta* **28**:1787–1816.

Kennett, J. P., et al. 1972. Australian-Antarctic continental drift, palaeocirculation changes and Oligocene deep-sea erosion. *Nature Phys. Sci.* **239**:51–55.

Kennett, J. P., et al. 1975. Cenozoic paleoceanography in the southwest Pacific Ocean, Antarctic glaciation, and the development of the circum-Antarctic current. *Deep Sea Drilling Proj. Init. Rept.* **29**: 1155–1169.

Kennett, J. P., A. R. McBirney, and R. C. Thunell. 1977. Episodes of Cenozoic volcanism in the circum-Pacific region. *Jour. Volcan. Geotherm. Research* **2**:145–163.

Khain, V. Ye. 1939. Oscillatory rhythm of the Earth's crust. *Moskov. Obshch. Ispytateley Prirody Byull., Otdel Geol.* **17**(1):56–82 (in Russian).

Khain, V. Ye. 1963. Lomonosov and modern geology. *Internat. Geology Rev.* **5**:706–715.

Khain, V. Ye. 1964. Fundamental stages in the evolution of the crust, recent continental regions. *Internat. Geology Rev.* **6**:439–449.

Khramov, A. N. 1977. Paleomagnetism and problems of geodynamics. *Izv. Phys. Solid Earth* **13**:791–801.

King, P. B. 1976. Precambrian geology of the United States; an explanatory text to accompany the geologic map of the United States. *U.S. Geol. Survey Prof. Paper 902,* 85 pp.

Kistler, R. W., J. F. Evernden, and H. R. Shaw. 1971. Sierra Nevada plutonic cycle: Part 1, origin of composite granitic batholiths. *Geol. Soc. America Bull.* **82**:853–868.

Knauth, L. P., and S. Epstein. 1976. Hydrogen and oxygen isotopic ratios in nodular and bedded cherts. *Geochim. et Cosmochim. Acta* **40**: 1095–1108.

Knauth, L. P., and D. R. Lowe. 1978. Oxygen isotope geochemistry of cherts from the Onverwacht Group (3.4 billion years), Transvaal, South Africa, with implications for secular variations in the isotopic composition of cherts. *Earth and Planetary Sci. Letters* **41**:209–222.

Kölbel, H. 1971. Zur absoluten Dauer erdgeschichtlicher Ereignisse. *Deutsch. Gesell. Geol. Wiss. Ber. A, Geol. Paläont.* **16**:221–229.

Kröner, A. 1977. Non-synchoneity of late Precambrian glaciations in Africa. *Jour. Geology* **85**:289–300.

Kropotkin, P. N. 1970. The possible role of cosmic factors in geotectonics. *Geotectonics 1970,* pp. 80–88.

Kropotkin, P. N., and Yu. A. Trapeznikov. 1965. Variation of the angular velocity of the Earth's rotation, the oscillation of the pole and the rate of drift of the geomagnetic field and their possible association with geotectonic processes. *Internat. Geology Rev.* **7**:1949–1962.

Krylov, N. A., and A. K. Mal'tseva. 1977. Tectonic control of the cyclicity of sedimentation in young platforms. *Geotectonics* **10**:339–349.

Kuenen, P. H. 1939. Quantitative estimations relating to eustatic movements. *Geol. en Mijnbouw* **1**:194–201.

Kuenen, P. H. 1941. Major geological cycles. *Nederlands Akad. Wetensch. Proc.* **44**:333–338.

Landis, C. A., and D. S. Coombs. 1967. Metamorphic belts and orogenesis in southern New Zealand. *Tectonophysics* **4**:501–518.

Langseth, M. G., X. Le Pichon, and M. Ewing. 1966. Crustal structure of the mid-ocean ridges. 5. Heat flow through the Atlantic Ocean floor and convection currents. *Jour. Geophys. Research* **71**:5321–5355.

Lanphere, M. A., and B. L. Reed. 1973. Timing of Mesozoic and Cenozoic plutonic events in circum-Pacific North America. *Geol. Soc. America Bull.* **84**:3773–3782.

Lantzy, R. J., M. F. Dacey, and F. T. Mackenzie. 1977. Catastrophe theory: application to the Permian mass extinction. *Geology* **5**:724–728.

Lantzy, R. J., M. F. Dacey, and F. T. Mackenzie. 1978. Catastrophe theory: application to the Permian mass extinction. Reply to comments by G. Borchardt and J. P. Platt. *Geology* **6**:454.

Larson, R. L., and W. C. Pitman, III. 1972. World-wide correlation of Mesozoic magnetic anomalies, and its implications. *Geol. Soc. America Bull.* **83**:3645–3622.

Larson, R. L., and W. C. Pitman, III. 1975. World-wide correlation of Mesozoic magnetic anomalies and its implications. Reply to discussion by W. A. Berggren et al. *Geol. Soc. America Bull.* **86**:270–272.

Lattimore, R. K., P. A. Rona, and O. E. DeWald. 1974. Magnetic anomaly sequence in the central North Atlantic. *Jour. Geophys. Research* **79**:1207–1209.

Lavrov, A. A. 1962. Some consequences of the motion of the Earth with respect to the Galaxy. In *Geogr. Symp. 15: Astrogeology.* Moscow and Leningrad: Geogr. Soc. USSR, pp. 162–167 (in Russian).

Le Conte, J. 1895. Critical periods in the history of the Earth. *Univ. California Dept. Geol. Bull.*, **1**:313–336.

Leonov, Yu. G. 1976. Global tectonic impulses in areas of Devonian mountain building. *IGCP Global Correlation of Epochs of Tectogenesis*, Meet. of Working Group, IGCP Proj. 107, Moscow, Rept., 8 pp.

Le Pichon, X. 1968. Sea-floor spreading and continental drift. *Jour. Geophys. Research* **73**:3661–3695.

Lindsay, J. F., and L. J. Srnka. 1975. Galactic dust lanes and lunar soil. *Nature* **257**:776–778.

Logvinenko, N. W. 1976. *Periodic Processes in Geology.* Leningrad: Nedra, 264 pp. (in Russian).

Lull, R. S. 1918. The pulse of life. In *The Evolution of the Earth and its Inhabitants,* R. S. Lull, ed. New Haven, Conn.: Yale Univ. Press, pp. 109–146.

Lull, R. S. 1929. The pulse of life. In *Organic Evolution.* New York: Macmillan, pp. 693–698.

Lungershausen, G. F. 1957. Periodic changes in climate and the major glaciations of the globe (some problems of historical palaeogeography and absolute chronology). *Sovetskaya Geologiya* **59**:88–115 (in Russian).

Machado, F. 1967. Geological evidence for a pulsating gravitation. *Nature* **214**:1317–1318.

Macintyre, R. M. 1971. Apparent periodicity of carbonatite emplacement in Canada. *Nature* **230**:79–81.

Macintyre, R. M. 1977. Anorogenic magmatism, plate motion and Atlantic evolution. *Geol. Soc. London Jour.* **133**:375–384.

Malinovskiy, U. M. 1977. Dependence of the periodicity of sediment formation on the position of the solar system in the Galaxy. In *Basic Theoretical Questions of Cyclic Sedimentation,* A. A. Trofumik, M. F. Murchink, and U. N. Karozobin, eds. Moscow: Nauka, pp. 117–123 (in Russian).

Martin, H. 1965. *The Precambrian Geology of South West Africa and Namaqualand.* Cape Town: Precambrian Res. Unit, Univ. Cape Town, 159 pp.

Masarovich, A. N. 1940. On the rhythm in the history of the Earth (English summary). *Soc. Naturalists, Moscow, Bull.* new ser., **48**:41–47.

Masursky, H. 1978. Terrestrial evolution—insights from the Moon and planets (abs.). *Am. Assoc. Petroleum Geologists Bull.* **62**:540.

Masursky, H., J. M. Boyce, A. L. Dial, G. G. Schaber, and M. E. Strobell, 1977. Classification and time of formation of Martian channels based on Viking data. *Jour. Geophys. Research,* **82**:4016–4038.

Mathewson, D. S., and J. R. Healey. 1964. Continuum radio emission from the Magellanic Clouds. In *The Galaxy and the Magellanic Clouds,* IAU Symp. 20, F. J. Kerr and A. W. Rodgers, eds. Canberra: Aust. Acad. Sci., pp. 245–255.

Mats, V. D. 1972. Precambrian weathered crust of the Siberian and Russian Platforms. *Akad. Nauk SSSR Doklady Earth Sci. Sec.* **200**:64–67.

McAlester, A. L. 1970. Animal extinctions, oxygen consumption, and atmospheric history. *Jour. Paleontology* **44**:405–409.

McAlester, A. L. 1973. Phanerozoic biotic crisis. In *The Permian and Triassic Systems and Their Mutual Boundary,* Canadian Soc. Petrol. Geol. Mem. 2, A. Logan and L. V. Hills, eds., pp. 11–15.

McClay, K. R., and D. G. Carlile. 1978. Mid-Proterozoic sulphate evaporites at Mount Isa mine, Queensland, Australia. *Nature* **274**:240–241.

McCrea, W. H. 1975. Ice ages and the Galaxy. *Nature* **255**:607–609.

McCulloch, M. T., and G. J. Wasserburg. 1978. Sm-Nd and Rb-Sr chronology of continental crust formation. *Science* **200**:1003–1011.

McElhinny, M. W. 1971. Geomagnetic reversals during the Phanerozoic. *Science* **172**:157–159.

McElhinny, M. W. 1973. *Palaeomagnetism and Plate Tectonics.* Cambridge: Cambridge Univ. Press, 358 pp.

McElhinny, M. W., and M. O. McWilliams. 1977. Precambrian geodynamics—a palaeomagnetic view. *Tectonophysics* **40**:137–159.

McWilliams, M. O. 1978. Paleolatitude distribution of late Precambrian glacial deposits (abs.). *Eos* **59**:1061

Meyerhoff, A. A. 1970. Continental drift, II: High-latitude evaporite deposits and geologic history of Arctic and North Atlantic Oceans. *Jour. Geology* **78**:406–444.

Meyerhoff, A. A. 1973. Mass biotal extinctions, world climate changes, and galactic motions: possible interrelations. In *The Permian and Triassic Systems and Their Mutual Boundary,* Canadian Soc. Petrol. Geol. Mem. 2, A. Logan and L. V. Hills, eds. pp. 745–758.

Moorbath, S. 1967. Recent advances in the application and interpretation of radiometric age data. *Earth-Sci. Rev.* **3**:111–133.

Moorbath, S. 1975. Evolution of Precambrian crust from strontium isotopic evidence. *Nature* **254**:395–398.

Moorbath, S. 1976. Age and isotopic constraints for the evolution of Archaean crust. In *The Early History of the Earth,* B. F. Windley, ed. London: Wiley, pp. 351–360.

Moorbath, S. 1977a. Ages, isotopes and evolution of Precambrian continental crust. *Chem. Geology* **20**:151–187.

Moorbath, S. 1977b. The oldest rocks and the growth of continents. *Sci. American* **236**(3):92–104.

Moore, T. C., and G. R. Heath. 1977. Survival of deep-sea sedimentary sections. *Earth and Planetary Sci. Letters* **37**:71–80.

Moore, T. C., Tj. H. van Andel, C. Sancetta, and N. Pisias. 1978. Cenozoic hiatuses in pelagic sediments. *Micropaleontology* **24**:113–138.

Morris, W. A., P. W. Schmidt, and J. L. Roy. 1979. A graphical approach to polar paths: paleomagnetic cycles and global tectonics. *Physics Earth and Planetary Interiors* **19**:85–99.

Muller, J. E. 1961. Notes on age determinations made on Cordilleran rocks. *Canada Geol. Survey Paper 61–17,* pp. 98–107.

Murray, B. C., R. G. Strom, N. J. Trask, and D. E. Gault. 1975. Surface history of Mercury: implications for terrestrial planets. *Jour. Geophys. Research* **80**:2508–2514.

Murray, B. C., W. R. Ward, and S. C. Yeung. 1973. Periodic insolation variations on Mars. *Science* **180**:638–640.

Nalivkin, D. V., and N. V. Tupitsyn, eds. 1963. *Problems in Planetary Geology—Some General Causes of Orogeny and Distribution Patterns for Economic Mineral Deposits.* Moscow: Gosgeoltekhizdat, 344 pp. (in Russian).

Neruchev, S. G. 1976. Effects of radioactive epochs on the evolution of the biosphere. *Soviet Geol. Geophys.* **17**(2):1–9.

Neukum, G., and D. U. Wise. 1976. Mars: a standard crater curve and possible new time scale. *Science* **194**:1381–1387.

Newell, N. D. 1962. Paleontological gaps and geochronology. *Jour. Paleontology* **36**:592–610.

Newell, N. D. 1967. Revolutions in the history of life. *Geol. Soc. America Spec. Paper 89,* pp. 63–91.

Newell, N. D. 1977. Extinction (biology). *Encyclopedia of Science and Technology,* 4th ed. New York: McGraw-Hill, pp. 166–167.

Nielsen, H. 1965. Schwefelisotope im marinen kreislauf und das δ ^{34}S der früheren meere. *Geol. Rundschau* **55**:160–172.

Nielsen, H. 1972. Sulphur isotopes and the formation of evaporite deposits. In *Geology of Saline Deposits,* G. Richter-Bernburg, ed. Paris: UNESCO, pp. 91–102.

Nunes, P. D., M. Tatsumoto, R. J. Knight, D. M. Unruh, and B. R. Doe. 1973. U-Th-Pb systematics of some Apollo 16 lunar samples. Proc. Fourth Lunar Sci. Conf., *Geochim. et Cosmochim. Acta, Suppl.* **4**:1797–1822.

Nunes, P. D., M. Tatsumoto, and D. M. Unruh. 1974. U-Th-Pb systematics of some Apollo 17 lunar samples and implications for a lunar basin excavation chronology. Proc. Fifth Lunar Sci. Conf., *Geochim. et Cosmochim. Acta, Suppl.* **5**:1487–1514.

Odesskiy, I. A., and A. I. Aynemer. 1969. Harmonic analysis of sedimentary rocks to determine the periodicity of sedimentation. *Geotectonics 1969,* pp. 405–410.

Oehler, D. Z., J. W. Schopf, and K. A. Kvenvolden. 1972. Carbon isotopic studies of organic matter in Precambrian rocks. *Science* **175**:1246–1248.

Oehler, D. Z. and J. W. Smith. 1977. Isotopic composition of reduced and oxidized carbon in early Archaean rocks from Isua, Greenland. *Precambrian Research* **5**:221–228.

Öpik, E. J. 1976. Solar structure, variability, and the ice ages. *Irish Astron. Jour.* **12**:253–276.

Öpik, E. J. 1977. The morphological method in research with astrophysical, climatological and meteorological applications. *Irish Astron. Jour.* **13**:95–108.

Orlova, A. V. 1963. Climatic change as an indicator of variable terrestrial rotation. In *Problems in Planetary Geology,* D. V. Nalivkin and N. V. Tupitsyn, eds. Moscow: Gosgeoltekhizdat, pp. 50–121 (in Russian).

Oversby, V. M. and P. W. Gast. 1970. Isotopic composition of lead from oceanic islands. *Jour. Geophys. Research* **75**:2097–2114.

Pannella, G. 1972. Paleontological evidence on the Earth's rotational history since early Precambrian. *Astrophys. Space Sci.* **16**:212–237.

Pannella, G. 1975. Palaeontological clocks and the history of the Earth's rotation. In *Growth Rhythms and the History of the Earth's Rotation,* G. D. Rosenberg and S. K. Runcorn, eds. London: Wiley, pp. 253–284.

Pannella, G., C. MacClintock, and M. N. Thompson. 1968. Paleontological evidence of variations in length of synodic month since Late Cambrian. *Science* **162**:792–796.

Papanastassiou, D. A., R. S. Rajan, J. C. Huneke, and G. J. Wasserburg. 1974. Rb-Sr ages and lunar analogs in a basaltic achondrite; implications for early solar system chronologies. *The Moon* **11**:255–257.

Papike, J. J., F. N. Hodges, A. E. Bence, M. Cameron, and J. M. Rhodes. 1976. Mare basalts: crystal chemistry, mineralogy, and petrology. *Rev. Geophys. Space Phys.* **14**:475–540.

Park, J. K., E. Irving, and J. A. Donaldson. 1973. Paleomagnetism of the Precambrian Dubawnt Group. *Geol. Soc. America Bull.* **84**:859–870.

Payton, C. E., ed. 1977. *Seismic Stratigraphy—Applications to Hydrocarbon Exploration.* Am. Assoc. Petroleum Geologists Mem. **26**, 516 pp.

Perry, E. C., S. N. Ahmad, and T. M. Swulius. 1978. The oxygen isotope composition of 3,800 m.y. old metamorphosed chert and iron formation from Isukasia, west Greenland. *Jour. Geology* **86**:223–239.

Peterman, Z. E., C. E. Hedge, and H. A. Tourtelot. 1970. Isotopic composition of strontium in sea water throughout Phanerozoic time. *Geochim. et Cosmochim. Acta* **34**:105–120.

Peyve, A. V. 1961. Tectonics and magmatic phenomena. *Akad. Nauk SSSR Izv. Ser. Geol.* **1961**(3):28–39 (in English).

Pimm, A. C. 1974. Sedimentology and history of the northeastern Indian Ocean from Late Cretaceous to Recent. *Deep Sea Drilling Proj. Init. Rept.* **22**:717–803.

Piper, D. Z., and L. A. Codispoti. 1975. Marine phosphorite deposits and the nitrogen cycle. *Science* **188**:15–18.

Piper, J. D. A. 1974. Proterozoic crustal distribution, mobile belts and apparent polar movements. *Nature* **251**:381–384.

Piper, J. D. A. 1976. Palaeomagnetic evidence for a Proterozoic supercontinent. *Royal Soc. London Philos. Trans.* ser. A, **280**:469–490.

Pitman, W. C., and M. Talwani. 1972. Sea-floor spreading in the North Atlantic. *Geol. Soc. America Bull.* **83**:619–646.

Podosek, F. A., and J. C. Huneke. 1973. Argon 40-argon 39 chronology of four calcium-rich achondrites. *Geochim. et Cosmochim. Acta* **37**: 667–684.

Quilty, P. G. 1977. Cenozoic sedimentation cycles in Western Australia. *Geology* **5**:336–340.

Råheim, A., and W. Compston. 1977. Correlations between metamorphic events and Rb-Sr ages in metasediments and eclogite from western Tasmania. *Lithos* **10**:271–289.

Rast, N., and T. P. Crimes. 1969. Caledonian orogenic episodes in the British Isles and northwestern France and their tectonic and chronological interpretation. *Tectonophysics* **7**:277–307.

Rees, C. E. 1970. The sulphur isotope balance of the ocean: an improved model. *Earth and Planetary Sci. Letters* **7**:366–370.

Rice, A., and R. W. Fairbridge. 1975. Thermal runaway in the mantle and neotectonics. *Tectonophysics* **29**:59–72.

Rodgers, J. 1967. Chronology of tectonic movements in the Appalachian region of eastern North America. *Am. Jour. Sci.* **265**:408–427.

Rona, P. A. 1973a. Worldwide unconformities in marine sediments related to eustatic changes of sea level. *Nature Phys. Sci.* **244**:25–26.

Rona, P. A. 1973b. Relations between rates of sediment accumulation on continental shelves, sea-floor spreading, and eustacy inferred from the central North Atlantic. *Geol. Soc. America Bull.* **84**:2851–2872.

Ronov, A. B. 1959. On the post-Precambrian geochemical history of the atmosphere and hydrosphere. *Geochemistry 1959,* pp. 493–506.

Ronov, A. B. 1968. Probable changes in the composition of sea water during the course of geological time. *Sedimentology* **10**:25–43.

Ronov, A. B. 1972. Composition evolution in rocks and geochemical processes in the sedimentary shell of the Earth. *Geochem. Internat. 1972,* pp. 85–94.

Ronov, A. B., and A. A. Migdisov. 1971. Geochemical history of the crystalline basement and the sedimentary cover of the Russian and North American platforms. *Sedimentology* **16**:137–185.

Ronov, A. B., A. A. Migdisov, and N. V. Barskaya. 1969. Tectonic cycles and regularities in the development of sedimentary rocks and paleographic environments of sedimentation of the Russian Platform (an approach to a quantitative study). *Sedimentology* **13**:179–212.

Ronov, A. B., A. A. Migdisov, N. T. Voskresenskaya, and G. A. Korzina. 1970. Geochemistry of lithium in the sedimentary cycle. *Geochem. Internat. 1970,* pp. 75–102.

Ronov, A. B., M. S. Mikhaylovskaya, and I. I. Solodkova. 1965. The evolution of the chemical and mineral compositions of sandy rocks. *Geochem. Internat.* **2**:318–345.

Rosenberg, G. D., and S. K. Runcorn, eds. 1975. *Growth Rhythms and the History of the Earth's Rotation*. London: Wiley, 559 pp.

Roubault, M., et al. 1967. A comparative table of recently published geological time-scales for the Phanerozoic time—explanatory notice. *Norsk Geol. Tidsskr.* **47**:375–380.

Roy, J. L. 1977. Problems in determining paleointensities from very old rocks. *Physics Earth and Planetary Interiors* **13**:319–324.

Roy, J. L., and P. L. Lapointe. 1976. The paleomagnetism of Huronian red beds and Nipissing diabase; post-Huronian igneous events and apparent polar path for the interval -2300 to -1500 Ma for Laurentia. *Canadian Jour. Earth Sci.* **13**:749–773.

Rubinshteyn, M. M. 1967. Orogenic phases and the periodicity of folding in the light of absolute age measurements. *Geotectonics 1967*, pp. 80-85.

Runcorn, S. K. 1962a. Towards a theory of continental drift. *Nature* **193**: 311–314.

Runcorn, S. K. 1962b. Convection currents in the Earth's mantle. *Nature* **195**:1248–1249.

Runcorn, S. K. 1962c. Palaeomagnetic evidence for continental drift and its geophysical cause. In *Continental Drift*, S. K. Runcorn, ed. New York: Academic Press, pp. 1–40.

Runcorn, S. K. 1966. Changes in the convection pattern in the Earth's mantle and continental drift: evidence for a cold origin of the Earth. *Royal Soc. London Philos. Trans.* ser. A, **258**:228–251.

Rutland, R. W. R. 1973a. Tectonic evolution of the continental crust of Australia. In *Implications of Continental Drift to the Earth Sciences*, vol. 2, D. H. Tarling and S. K. Runcorn, eds. London: Academic Press, pp. 1011–1033.

Rutland, R. W. R. 1973b. On the interpretation of Cordilleran orogenic belts. *Am. Jour. Sci.* **273**:811–849.

Rutten, L. M. R. 1949. Frequency and periodicity of orogenetic movements. *Geol. Soc. America Bull.* **60**:1755–1770.

Sabine, P. A., and J. V. Watson, 1965. Isotopic age-determinations of rocks from the British Isles, 1955–1964. 1. Introduction. *Geol. Soc. London Quart. Jour.* **121**:478–487.

Sagan, C., O. B. Toon, and P. J. Gierasch, 1973. Climatic change on Mars. *Science* **181**:1045–1049.

Sakai, H. 1972. Oxygen isotopic ratios of some evaporites from Precambrian to recent ages. *Earth and Planetary Sci. Letters* **15**:201–205.

Salop, L. I. 1964. Pre-Cambrian geochronology and some features of the early stage of the geological history of the Earth. *Internat. Geol. Congr., 22nd, Proc.* **10**:131–149.

Salop, L. I. 1968. Pre-Cambrian of the U.S.S.R. *Internat. Geol. Congr., 23rd,* **4**:61–73.

Salop, L. J. 1972. A unified stratigraphic scale of the Precambrian. *Internat. Geol. Congr., 24th,* **sec. 1**:253–259.

Salop, L. J. 1977. *Precambrian of the Northern Hemisphere*. Amsterdam: Elsevier, 378 pp.

Sarkar, S. N. 1972. Present status of Precambrian geochronology of peninsular India. *Internat. Geol. Congr., 24th,* **sec. 1**:260–272.

Savin, S. M. 1977. The history of the Earth's surface temperature during the past 100 million years. *Annual Rev. Earth and Planetary Sci.* **5**: 319–355.

Schindewolf, O. H. 1950. *Der Zeitfactor in Geologie und Paläontologie.* Stuttgart: Schwietzerbart, 114 pp.

Schindewolf, O. H. 1962. Neokatastrophismus? *Deutsch. Geol. Gesell. Zeitschr.* **114**:430–445. (English translation in *Catastrophist Geol.* **2(2)**: 9–21, 1977.)

Schneider, E. D., and P. R. Vogt. 1968. Discontinuities in the history of sea-floor spreading. *Nature* **217**:1212–1222.

Schopf, J. W. 1977. Biostratigraphic usefulness of stromatolitic Precambrian microbiotas: a preliminary analysis. *Precambrian Research* **5**: 143–173.

Schopf, J. W., and D. Z. Oehler. 1976. How old are the eukaryotes? *Science* **193**:47–49.

Schopf, T. J. M. 1974. Permo-Triassic extinctions: relation to sea-floor spreading. *Jour. Geology* **82**:129–143.

Schuchert, C. 1910. Paleogeography of North America. *Geol. Soc. America Bull.* **20**:427–606.

Schuchert, C. 1914. Climates of geologic time. In *The Climatic Factor,* Carnegie Inst., Washington, Publ. No. 192, E. Huntington, ed., pp. 265–298.

Schuchert, C. 1916. Correlation and chronology in geology on the basis of paleogeography. *Geol. Soc. America Bull.* **27**:491–514.

Schuchert, C. 1932. The periodicity of oceanic spreading, mountain-making, and paleogeography. In *Physics of the Earth—V. Oceanography.* Natl. Research Council, Washington, Bull. **85**:537–557.

Schuchert, C. 1955. *Atlas of Paleogeographic Maps of North America.* New York: Wiley, 177 pp.

Schwab, F. L. 1978. Secular trends in the composition of sedimentary rock assemblages—Archean through Phanerozoic time. *Geology* **6**:532–536.

Sclater, J. G., S. Hellinger, and C. Tapscott. 1977. The paleobathymetry of the Atlantic Ocean from the Jurassic to the present. *Jour. Geology* **85**:509–552.

Semenenko, N. P. 1964. Correlation of the history of Pre-Cambrian by the data of absolute geochronology. *Internat Geol. Congr., 22nd, Proc.* **10**:150–160.

Semenenko, N. P. 1972. Precambrian geochronology and problems. *Internat. Geology Rev.* **14**:947–953.

Semenenko, N. P., A. P. Scherbak, A. P. Vinogradov, A. I. Tougarinov, G. D. Eliseeva, F. I. Cotlovskay, and S. G. Demidenko. 1968. Geochronology of the Ukrainian Precambrian. *Canadian Jour. Earth Sci.* **5**:661–671.

Semenenko, N. P., N. P. Shcherbak, and E. N. Bartnitsky. 1972. Geochronology, stratigraphy and tectonic structure of the Ukrainian Shield. *Internat. Geol. Congr., 24th,* **sec. 1**:363–370.

Shapley, H. 1921. Note on a possible factor in changes of geological climate. *Jour. Geology* **29**:502–504.

Shapley, H. 1949. Galactic rotation and cosmic seasons. *Sky and Telescope* **9**:36–37.

Shaw, H. R., R. W. Kistler, and J. F. Evernden. 1971. Sierra Nevada plutonic cycle: Part II, tidal energy and a hypothesis for orogenic-epeirogenic periodicities. *Geol. Soc. America Bull.* **82**:869–896.

Shepard, F. P. 1923. To question the theory of periodic diastrophism. *Geol. Mag.* **31**:599–613.

Siedner, G., and J. G. Mitchell. 1976. Episodic Mesozoic volcanism in Namibia and Brazil: a K-Ar isochron study bearing on the opening of the South Atlantic. *Earth and Planetary Sci. Letters* **30**:292–302.

Sigal, J. 1974. Comments on Leg 25 sites in relation to the Cretaceous and Paleogene stratigraphy in the eastern and southeastern Africa coast and Madagascar regional setting. *Deep Sea Drilling Proj. Init. Rept.* **25**:687–723.

Simonds, C. H., P. H. Schultz, and S. C. Solomon. 1978. Comparison of Mercury and the Moon: a conference. *Eos* **59**:43–48.

Simpson, G. G. 1953. *The Major Features of Evolution*. New York: Columbia Univ. Press, 434 pp.

Simpson, J. F. 1966. Evolutionary pulsations and geomagnetic polarity. *Geol. Soc. America Bull.* **77**:197–204.

Sloss, L. L. 1963. Sequences in the cratonic interior of North America. *Geol. Soc. America Bull.* **74**:93–114.

Sloss, L. L. 1964. Tectonic cycles of the North American craton. In *Symposium on Cyclic Sedimentation*, D. F. Merriam, ed. *Kansas Geol. Survey Bull.* **169**:449–460.

Sloss, L. L. 1972. Synchrony of Phanerozoic sedimentary-tectonic events of the North American craton and the Russian Platform. *Internat. Geol. Congr., 24th*, **sec. 6**:24–32.

Sloss, L. L. 1973. Tectonic and eustatic (?) factors in late Precambrian-Phanerozoic global sea-level changes (abs.). *Geol. Soc. America Abs. with Programs* **5**:813–814.

Sloss, L. L. 1976. Areas and volumes of cratonic sediments, western North America and eastern Europe. *Geology* **4**:272–276.

Sloss, L. L., and R. C. Speed. 1974. Relationships of cratonic and continental-margin tectonic episodes. In *Tectonics and Sedimentation*, W. R. Dickinson, ed., SEPM Spec. Publ. **22**, pp. 98–119.

Smith, A. G., and J. C. Briden. 1977. *Mesozoic and Cenozoic Paleocontinental Maps*. Cambridge: Cambridge Univ. Press, 63 pp.

Smith, A. G., J. C. Briden, and G. E. Drewry. 1973. Phanerozoic world maps. *Spec. Papers Paleontology* **12**:1–42.

Smith, P. J. 1967a. Intensity of the Earth's magnetic field in the geological past. *Nature* **216**:989–990.

Smith, P. J. 1967b. The intensity of the ancient geomagnetic field: a review and analysis. *Royal Astron. Soc. Geophys. Jour.* **12**:321–362.

Smith, P. J. 1970. The intensity of the ancient geomagnetic field: a summary of conclusions. In *Palaeogeophysics*, S. K. Runcorn, ed. London: Academic Press, pp. 79–90.

Smith, P. J. 1977. The return of whole-mantle convection. *Nature* **268**: 687–688.

Snider, L. C. 1932. *Earth History*. New York: Century, 683 pp.

Soares, P. C., P. M. B. Landim, and V. J. Fulfaro. 1978. Tectonic cycles and sedimentary sequences in the Brazilian intracratonic basins. *Geol. Soc. America Bull.* **89**:181–191.

Stashkov, H. M. 1977. Application of periodicity of tectonic processes in the construction of an absolute geochronological time scale. In *Basic Theoretical Questions of Cyclic Sedimentation*, A. A. Trofumik, M. F. Murchink, and U. N. Karozobin, eds. Moscow: Nauka, pp. 137–148 (in Russian).

Steiner, J. 1967. The sequence of geological events and the dynamics of the Milky Way Galaxy. *Geol. Soc. Australia Jour.* **14**:99–131.

Steiner, J. 1973. Possible galactic causes for synchronous sedimentation sequences of the North American and eastern European cratons. *Geology* **1**:89–92.

Steiner, J. 1978. Lead isotope events of the Canadian Shield, *ad hoc* solar galactic orbits and glaciations. *Precambrian Research* **6**:269–274.

Steiner, J. 1979. Regularities of the revised Phanerozoic time scale and the Precambrian time scale. *Geol. Rundschau* **68**:825–831.

Steiner, J., and E. Grillmair. 1973. Possible galactic causes for periodic and episodic glaciations. *Geol. Soc. America Bull.* **84**:1003–1018.

Stewart, A. D. 1970. Palaeogravity. In *Palaeogeophysics*, S. K. Runcorn, ed. London: Academic Press, pp. 413–434.

Stewart, A. D. 1977. Quantitative limits to palaeogravity. *Geol. Soc. London Jour.* **133**:281–291.

Stewart, A. D. 1978. Limits to palaeogravity since the late Precambrian. *Nature* **271**:153–155.

Stille, H. 1924. *Grundfragen der vergleichenden Tektonik*. Berlin: Gebruder Borntraeger, 443 pp.

Stille, H. 1936. The present tectonic state of the Earth. *Am. Assoc. Petroleum Geologists Bull.* **20**:849–880.

Stille, H. 1950. Distribution of mountain building in geologic time. (Discussion and reply with J. Gilluly.) *Geol. Rundschau* **38**:89–111.

Stockwell, C. H. 1961. Structural provinces, orogenies, and time classification of rocks of the Canadian Precambrian Shield. *Canada Geol. Survey Paper 61-17*, pp. 108–118.

Stockwell, C. H. 1963a. Second report on structural provinces, orogenies, and time-classification of rocks of the Canadian Precambrian Shield. *Canada Geol. Survey Paper 62-17(2)*, pp. 123–133.

Stockwell, C. H. 1963b. Third report on structural provinces, orogenies, and time-classification of rocks of the Canadian Precambrian Shield. *Canada Geol. Survey Paper 63-17(2)*, pp. 125–131.

Stockwell, C. H. 1964. Fourth report on structural provinces, orogenies, and time-classification of rocks of the Canadian Precambrian Shield. *Canada Geol. Survey Paper 64-17(2)*, pp. 1–21.

Stockwell, C. H. 1968. Geochronology of stratified rocks of the Canadian Shield. *Canadian Jour. Earth Sci.* **5**:693–698.

Stockwell, C. H. 1973. Revised Precambrian time scale for the Canadian Shield. *Canada Geol. Survey Paper 72-52*, 4 pp.

Stockwell, C. H., et al. 1970. Geology of the Canadian Shield. In *Geology and Economic Minerals of Canada*, Canada Geol. Survey Econ. Geology Rept. 1, pp. 43–150.

Strakhov, N. M. 1949. Periodicity and irreversibility of evolution of sedimentation in the Earth's history. *Akad. Nauk. SSSR Izv. Ser. Geol.* **6**:70–111 (in Russian).

Strakhov, N. M. 1967, 1969. *Principles of Lithogenesis*. New York: Consultants Bureau, vol. 1, 245 pp.; vol. 2, 609 pp.

Strakhov, N. M. 1971. *Development of Lithogenetical Ideas in Russia and USSR*. Moscow: Nauka (in Russian).

Stueber, A. M., R. A. Heimlich, and M. Ikramuddin. 1976. Rb-Sr ages of Precambrian mafic dikes, Bighorn Mountains, Wyoming. *Geol. Soc. America Bull.* **87**:909–914.

Suess, E. 1904, 1906, 1908, 1909, 1924. *The Face of the Earth*, vols. 1–5. Oxford: Clarendon Press (English translation by H. B. C. Sollas).

Sutton, J. 1963. Long-term cycles in the evolution of the continents. *Nature* **198**:731–735.

Sutton, J. 1967. The extension of the geological record into the Pre-Cambrian. *Geol. Assoc. London Proc.* **78**:493–534.

Sutton, J. 1970. Migration of high temperature zones in the crust. In *Palaeogeophysics*, S. K. Runcorn, ed. London: Academic Press, pp. 365–375.

Sutton, J. 1973. Some changes in continental structure since early Precambrian time. In *Implications of Continental Drift to the Earth Sciences*, vol. 2. D. H. Tarling and S. K. Runcorn, eds. London: Academic Press, pp. 1071–1081.

Sutton, J. 1976a. Tectonic relationships in the Archean. In *The Early History of the Earth*, B. F. Windley, ed. London: Wiley, pp. 99–104.

Sutton, J. 1976b. Introductory remarks. In *A Discussion on Global Tectonics in Proterozoic Times*, J. Sutton, R. M. Shackleton, and J. C. Briden, organizers. *Royal Soc. London Philos. Trans.* ser. A, **280**: 399–403.

Sutton, J. 1977. Some consequences of horizontal displacements in the Precambrian. *Tectonophysics* **40**:161–181.

Tamrazyan, G. P. 1954. Geological revolutions and the cosmic life of the Earth. *Akad. Nauk Azerbaydzhan. SSR Doklady* **10**:433–438 (in Russian).

Tamrazyan, G. P. 1959. Periodic changes in climate and some problems of paleogeography. *Sovetskaya Geologiya* **7**:143–149 (in Russian).

Tamrazyan, G. P. 1967. The global historical and geological regularities of the Earth's development as a reflection of its cosmic origin (as a sequence of interaction in the course of galactic movement of the solar system). *Ostrava Vysoké Škola Baňske Sbornik Věd. Praci. Řada Horn.—Geol.* **13**:5–24.

Tappan, H. 1976. Possible eucaryotic algae (Bangiophycidae) among early Proterozoic microfossils. *Geol. Soc. America Bull.* **87**:633–639.

Tarling, D. H. 1978. The geological-geophysical framework of ice ages. In *Climatic Change*, J. Gribbin, ed. Cambridge: Cambridge Univ. Press, pp. 3–24.

Tera, F., and G. J. Wasserburg. 1974. U-Th-Pb systematics on lunar rocks and inferences about lunar evolution and the age of the Moon. Proc. Fifth Lunar Sci. Conf., Geochim. et Cosmochim. Acta, Suppl. **5**:1571–1599.

Thierstein, H. R., and W. H. Berger. 1978. Injection events in ocean history. *Nature* **276**:461–466.

Thode, H. G., and J. Monster. 1965. Sulfur-isotope geochemistry of petroleum, evaporites, and ancient seas. *Am. Assoc. Petroleum Geologists Mem.* **4**:367–377.

Thom, R. 1975. *Structural Stability and Morphogenesis*. Reading, Mass.:

Benjamin, 348 pp.

Thom, R. 1978. Plate tectonics and catastrophe theory. *Catastrophist Geol.* **3**(1):30–48.

Tilton, G. R., and G. L. Davis. 1959. Geochronology. In *Researches in Geochemistry,* P. H. Abelson, ed. New York: Wiley, pp. 190–216.

Trendall, A. F. 1972. Revolution in Earth history. *Geol. Soc. Australia Jour.* **19**:287–311.

Trofumik, A. A., and U. N. Karozobin, eds. 1976. *Geocyclicity.* Novosibirsk: Acad. Sci. USSR, 123 pp. (in Russian).

Trofumik, A. A., M. F. Murchink, and U. N. Karozobin, eds. 1977. *Basic Theoretical Questions of Cyclic Sedimentation.* Moscow: Nauka, 263 pp. (in Russian).

Turcotte, D. L., and K. Burke. 1978. Global sea-level changes and the thermal structure of the Earth. *Earth and Planetary Sci. Letters* **41**: 341–346.

Turner, G. 1977. The early chronology of the Moon: evidence for the early collisional history of the solar system. *Royal Soc. London Philos. Trans.* ser. A, **285**:97–103.

Ulrich, R. K. 1975. Solar neutrinos and variations in the solar luminosity. *Science* **190**:619–624.

Ulrych, T. J. 1969. Lead isotopes, lunar capture and mantle evolution. *Nature* **224**:766–768.

Ulrych, T. J. 1972. Maximum entropy power spectrum of long period geomagnetic reversals. *Nature* **235**:218–219.

Umbgrove, J. H. F. 1939a. The relation between magmatic cycles and orogenic epochs. *Geol. Mag.* **76**:444–450.

Umbgrove, J. H. F. 1939b. On rhythms in the history of the Earth. *Geol. Mag.* **76**:116–129.

Umbgrove, J. H. F. 1942, 1947. *The Pulse of the Earth.* The Hague: Nijhoff, 1st ed., 179 pp.; 2d ed., 358 pp.

Umbgrove, J. H. F. 1945. Periodical events in the North Sea Basin. *Geol. Mag.* **82**:237–244.

Vail, P. R., R. M. Mitchum, and S. Thompson. 1977. Seismic stratigraphy and global changes of sea level, Part 4: Global cycles of relative changes of sea level. *Am. Assoc. Petroleum Geologists Mem.* **26**:83–97.

Valentine, J. W. 1973. *Evolutionary Paleoecology of the Marine Biosphere.* Englewood Cliffs, N. J.: Prentice-Hall, 511 pp.

Valentine, J. W. 1977. General patterns of metazoan evolution. In *Patterns of Evolution as Illustrated by the Fossil Record,* A. Hallam, ed. Amsterdam: Elsevier, pp. 27–57.

Valentine, J. W., and E. M. Moores. 1970. Plate-tectonic regulation of faunal diversity and sea-level: a model. *Nature* **228**:657–659.

Valentine, J. W., and E. M. Moores. 1972. Global tectonics and the fossil record. *Jour. Geology* **80**:167–184.

Valeton, I. 1972. *Bauxites.* Amsterdam: Elsevier, 226 pp.

van Andel, Tj. H., J. Thiede, J. G. Sclater, and W. W. Hay. 1977. Depositional history of the South Atlantic Ocean during the last 125 million years. *Jour. Geology* **85**:651–698.

van Andel, Tj. H., G. R. Heath, and T. C. Moore. 1975. Cenozoic history and paleoceanography of the central equatorial Pacific Ocean. *Geol. Soc. America Mem. 143,* 134 pp.

Van Houten, F. B. 1969. Molasse facies: records of worldwide crustal stresses. *Science* **166**:1506–1508.

Veizer, J. 1973. Sedimentation in geologic history: recycling vs. evolution or recycling with evolution. *Contr. Mineralogy and Petrology* **38**:261–278.

Veizer, J. 1976. Evolution of ores of sedimentary affiliation through geologic history; relations to the general tendencies in evolution of the crust, hydrosphere, atmosphere and biosphere. In *Handbook of Strata-Bound and Stratiform Ore Deposits. Vol. 3: Supergene and Surficial Ore Deposits; Textures and Fabrics*, K. H. Wolf, ed. Amsterdam: Elsevier, pp. 1–41.

Veizer, J. 1977. Diagenesis of pre-Quaternary carbonates as indicated by tracer studies. *Jour. Sed. Petrology* **47**:565–581.

Veizer, J. 1978. Secular variations in the composition of sedimentary carbonate rocks, II. Fe, Mn, Ca, Mg, Si and minor constituents. *Precambrian Research* **6**:381–413.

Veizer, J., and W. Compston. 1974. $^{87}Sr/^{86}Sr$ composition of seawater during the Phanerozoic. *Geochim. et Cosmochim. Acta* **38**:1461–1484.

Veizer, J., and W. Compston. 1976. $^{87}Sr/^{86}Sr$ in Precambrian carbonates as an index of crustal evolution. *Geochim. et Cosmochim. Acta* **40**: 905–914.

Veizer, J., and D. E. Garrett. 1978. Secular variations in the composition of sedimentary carbonate rocks, I. Alkali metals. *Precambrian Research* **6**:367–380.

Veizer, J., and J. Hoefs. 1976. The nature of O^{18}/O^{16} and C^{13}/C^{12} secular trends in sedimentary carbonate rocks. *Geochim. et Cosmochim. Acta* **40**:1387–1395.

Vine, F. J. 1973. Continental fragmentation and ocean floor evolution during the past 200 m.y. In *Implications of Continental Drift to the Earth Sciences*, vol. 2, D. H. Tarling and S. K. Runcorn, eds. London: Academic Press, pp. 831–839.

Vine, F. J., and H. H. Hess. 1970. Sea-floor spreading. In *The Sea*, vol. 4, pt. II, A. E. Maxwell, ed. New York: Wiley, pp. 587–622.

Vinogradov, A. P., and A. B. Ronov. 1956a. Composition of the sedimentary rocks of the Russian Platform in relation to the history of its tectonic movements. *Geochemistry* **6**:533–559.

Vinogradov, A. P., and A. B. Ronov, 1956b. Evolution of the chemical composition of clays of the Russian Platform. *Geochemistry 1956(2)*, pp. 123–139.

Vinogradov, A. P., and A. I. Tugarinov. 1961. The geologic age of Pre-Cambrian rocks of the Ukrainian and Baltic shields. *New York Acad. Sci. Annals* **91**:500–513.

Vinti, J. P. 1974. Classical solution of the two-body problem if the gravitational constant diminishes inversely with the age of the Universe. *Royal Astron. Soc. Monthly Notices* **169**:417–427.

Vogt, P. R. 1975. Changes in geomagnetic reversal frequency at times of tectonic change: evidence for coupling between core and upper mantle processes. *Earth and Planetary Sci. Letters* **25**:313–321.

Vogt, P. R., O. E. Avery, E. D. Schneider, C. N. Anderson, and D. R. Bracey. 1969. Discontinuities in sea-floor spreading. *Tectonphysics* **8**:285–317.

Vogt, P. R., G. L. Johnson, T. L. Holcombe, J. G. Gilg, and O. E. Avery. 1971. Episodes of sea-floor spreading recorded by the North Atlantic basement. *Tectonophysics* **12**:211–234.

Voitkevich, G. V. 1958. A unified time-scale for the Precambrian. *Priroda 1958(5)*, pp. 77–79 (in Russian).

Walker, R. N., M. D. Muir, W. L. Diver, N. Williams, and N. Wilkins. 1977. Evidence of major sulphate evaporite deposits in the Proterozoic McArthur Group, Northern Territory, Australia. *Nature* **265**:526–529.

Walzer, U. 1974. A convection mechanism for explaining episodicity of magmatism and orogeny. *Pure and Appl. Geophysics* **112**:106–117.

Ward, W. R. 1973. Large-scale variations in the obliquity of Mars. *Science* **181**:260–262.

Ward, W. R. 1979. Present obliquity oscillations of Mars: fourth-order accuracy in orbital e and *I. Jour Geophys. Research* **84**:237–241.

Ward, W. R., J. A. Burns, and O. B. Toon. 1979. Past obliquity oscillations of Mars: the role of the Tharsis uplift. *Jour. Geophys. Research* **84**: 243–259.

Wasserburg, G. J., D. A. Papanastassiou, F. Tera, and J. C. Huneke. 1977. Outline of a lunar chronology. *Royal Soc. London Philos. Trans.* ser. A, **285**:7–22.

Weaver, C. E. 1967. Potassium, illite and the ocean. *Geochim. et Cosmochim. Acta* **31**:2181–2196.

Weber, J. N. 1964. Palaeoclimatic significance of δ-oxygen-18-time trends observed by oxygen isotopic analysis of freshwater limestones. *Nature* **203**:969–970.

Weber, J. N. 1967. Possible changes in the isotopic composition of the oceanic and atmospheric carbon reservoir over geologic time. *Geochim. et Cosmochim. Acta* **31**:2343–2351.

Wegener, A. 1924. *The Origin of Continents and Oceans*. London: Methuen, 212 pp. (English translation of 3rd German ed. by J. G. A. Skerl.)

Wetherill, G. W. 1975. Late heavy bombardment of the moon and terrestrial planets. Proc. Sixth Lunar Sci. Conf., *Geochim. et Cosmochim. Acta, Suppl.* **6**:1539–1561.

Whyte, M. A. 1977. Turning points in Phanerozoic history. *Nature* **267**: 679–682.

Willard, B. 1950. Paleozoic continental phases of sedimentation in the northern Appalachians. *Internat. Geol. Congr., 18th Rept.*, pt. IV, sec. C, pp. 29–39.

Williams, G. E. 1972. Geological evidence relating to the origin and secular rotation of the solar system. *Modern Geol.* **3**:165–181.

Williams, G. E. 1973. Geotectonic cycles, lunar evolution, and the dynamics of the Earth-Moon system. *Modern Geol.* **4**:159–183.

Williams, G. E. 1975a. Possible relation between periodic glaciation and the flexure of the Galaxy. *Earth and Planetary Sci. Letters* **26**:361–369.

Williams, G. E. 1975b. Late Precambrian glacial climate and the Earth's obliquity. *Geol. Mag.* **112**:441–465.

Williams, G. E., and P. M. Austin. 1973. Global tectonics and the Earth's rotation. *Modern Geol.* **4**:185–199.

Wilson, J. F., M. J. Bickle, C. J. Hawkesworth, A. Martin, E. G. Nisbet, and

J. L. Orpen. 1978. Granite-greenstone terrains of the Rhodesian Archaean craton. *Nature* **271**:23–27.

Wilson, J. T. 1966. Did the Atlantic close and then re-open? *Nature* **211**: 676–681.

Wilson, J. T. 1968. Static or mobile Earth: the current scientific revolution. *Am. Philos. Soc. Proc.* **112**:309–320.

Wilson, J. T. 1970. Continental drift, transcurrent and transform faulting. In *The Sea,* vol. 4, pt. II, A. E. Maxwell, ed. New York: Wiley, pp. 623–644.

Wilson, J. T. 1974. The life cycle of ocean basins: stages of growth and stages of decline. In *Physics and Geology,* 2d ed., J. A. Jacobs, R. D. Russell, and J. T. Wilson. New York: McGraw-Hill, pp. 397–470.

Wilson, R. L., and M. W. McElhinny. 1974. Investigation of the large scale palaeomagnetic field over the past 25 million years. Eastward shift of the Icelandic spreading ridge. *Royal Astron. Soc. Geophys. Jour.* **39**:570–586.

Wise, D.U. 1974. Continental margins, freeboard and the volumes of continents and oceans through time. In *The Geology of Continental Margins,* C. A. Burk and C. L. Drake, eds. Berlin: Springer-Verlag, pp. 45–58.

Wolfe, J. A. 1969. Paleogene floras from the Gulf of Alaska region. *U.S. Geol. Survey Open-file Rept. 374,* 114 pp.

Wolfe, J. A. 1977. Paleogene floras from the Gulf of Alaska region. *U.S. Geol. Survey Prof. Paper 997,* 108 pp.

Wolfe, J. A. 1978. A paleobotanical interpretation of Tertiary climates in the Northern Hemisphere. *Am. Sci.* **66**:694–703.

Yanshin, A. L. 1974. The so-called global transgressions and regressions. Internat. Geol. Rev. **16**:617–646.

York, D., and R. M. Farquhar. 1972. *The Earth's Age and Geochronology.* Oxford: Pergamon, 178 pp.

Young, G. M. 1976a. Iron-formation and glaciogenic rocks of the Rapitan Group, Northwest Territories, Canada. *Precambrian Research* **3**: 137–158.

Young, G. M. 1976b. Late Precambrian mixtites: glacial and/or nonglacial? Discussion. *Am. Jour. Sci.* **276**:366–370.

Zahler, R. S., and H. J. Sussmann. 1977. Claims and accomplishments of applied catastrophe theory. *Nature* **269**:759–763.

Zeller, E. J. 1964. Cycles and psychology. In Symposium on Cyclic Sedimentation, D. F. Merriam, ed. *Kansas Geol. Survey Bull.* **169**:631–636.

ADDITIONAL READINGS

Balukhovskiy, N. F. 1963. Geologic cycles. *Priroda 1963(2),* pp. 54–59 (in Russian).

Bubnoff, S. von. 1963. *Fundamentals of Geology.* Edinburgh: Oliver and Boyd, 287 pp.

Gussow, W. C. 1963. Metastasy. In *Polar Wandering and Continental Drift,* A. C. Munyan, ed. SEPM Spec. Publ. 10, pp. 146–169.

Hargraves, R. B. 1978. Punctuated evolution of tectonic style. *Nature* **276**:459–461.

Holmes, A. 1933. The thermal history of the Earth. *Washington Acad. Sci. Jour.* **23**:169–195.

Huang, T. K., and J. Chun-fa. 1962. A preliminary investigation of the evolution of the Earth's crust from the point of view of polycyclic tectonic movements. *Sci. Sinica* **11**:1377–1442.

Machado, F. 1975. Pulsation of tectonic phenomena and tectonophysical mechanisms. *Geol. Rundschau* **64**:74–84.

Nevesskiy, Ye. N. 1961. Rhythm of marine transgressions. *Okeanologiya* **1**:63–77 (in Russian).

Newell, N. D. 1956. Catastrophism and the fossil record. *Evolution* **10**: 97–101.

Obrutchev, V. A. 1940. The pulsation hypothesis of geotectonics. *Akad. Nauk SSSR Izv. Ser. Geol. 1940(1),* pp. 12–30 (in Russian; English summary pp. 29–30).

Ozard, J. M., W. F. Slawson, and R. D. Russell. 1973. An integrated model for lead isotopic evolution for samples from the Canadian Shield. *Canadian Jour. Earth Sci.* **10**:529–537.

Pessagno, E. A. 1972. Pulsations, interpulsations, and sea-floor spreading. *Geol. Soc. America Mem.* **132**:67–73.

Pollack, J. B. 1979. Climatic change on the terrestrial planets. *Icarus* **37**: 479–553.

Saito, R. 1968. The regularity of the occurrence of late Pre-Cambrian orogenic cycles. *Japan Jour. Geology and Geography* **39**:149–166.

Schuchert, C. 1917. Hébert's views of 1857 regarding the periodic submergence of Europe. *Am. Jour. Sci.* 4th ser., **43**:35–41.

Stille, H. 1955. Recent deformations of the Earth's crust in the light of those of earlier epochs. *Geol. Soc. America Spec. Paper 62,* pp. 171–191.

Tamrazyan, G. P. 1964. Cyclicity—a reflection of the Earth's development. *Priroda 1964(1),* pp. 107–110 (in Russian).

Theobald, N. 1969. Rhythmes et cycles en géologie. *Annales Sci. Univ. Besançon Geol.,* 3e ser., fasc. 6, pp. 9–16.

Umbgrove, J. H. F. 1940. Periodicity in terrestrial processes. *Am. Jour. Sci.* **238**:573–576.

Umbgrove, J. H. F. 1950. Rhythm and synchronism of tectonic movements. *Am. Jour. Sci.* **248**:521–526.

Van Bemmelen, R. W. 1966. On mega-undations: a new model for the Earth's evolution. *Tectonophysics* **3**:83–127.

Vogt, P. R. 1979. Global magmatic episodes: New evidence and implications for the steady-state mid-oceanic ridge. *Geology* **7**:93–98.

AUTHOR CITATION INDEX

SUBJECT INDEX

About the Editor

GEORGE E. WILLIAMS is senior research geologist in the Exploration Department of The Broken Hill Pty. Company Limited's Melbourne Research Laboratories, Clayton, Victoria. Prior to taking up this position he held a Queen Elizabeth II Post-Doctoral Fellowship at the University of Adelaide, and also undertook research and mineral exploration in north Africa, Scotland, Australia and the Middle East.

Dr. Williams received the B. Sc. with highest honors from the University of Melbourne in 1961 and the M.Sc. from that university in 1963. He then held an 1851 Exhibition Scholarship at the University of Reading, England, where he received the Ph. D. in 1966. He has published in a number of fields in the earth sciences, including sedimentology, paleoclimatology, isotopic dating, structural geology, geotectonics and planetology. Dr. Williams is a member of the editorial board of the Geological Society of Australia and is a Fellow of the Geological Society of America.